To Lassi Kivioja

Hope that you will find it interesting to read.

Edward M.
Sept. 1976

Observations and Least Squares

The **IEP** Series in Civil Engineering

Series Editor
Russell C. Brinker
New Mexico State University

Brinker and Wolf	*ELEMENTARY SURVEYING*, 6th edition
Clark, Viessman, and Hammer	*WATER SUPPLY AND POLLUTION CONTROL*, 3rd edition
Ghali and Neville	*STRUCTURAL ANALYSIS: A UNIFIED CLASSICAL AND MATRIX APPROACH*
Hill	*FUNDAMENTALS OF STRUCTURAL DESIGN: STEEL, CONCRETE, AND TIMBER*
Jumikis	*FOUNDATION ENGINEERING*
McCormac	*STRUCTURAL ANALYIS*, 3rd edition
McCormac	*STRUCTURAL STEEL DESIGN*, 2nd edition
Meyer	*ROUTE SURVEYING AND DESIGN*, 4th edition
Moffitt	*PHOTOGRAMMETRY*, 2nd edition
Moffitt and Bouchard	*SURVEYING*, 6th edition
Salmon and Johnson	*STEEL STRUCTURES: DESIGN AND BEHAVIOR*
Spangler and Handy	*SOIL ENGINEERING*, 3rd edition
Ural	*FINITE ELEMENT METHOD: Basic Concepts and Applications*
Viessman, Harbaugh, Knapp, and Lewis	*INTRODUCTION TO HYDROLOGY*, 2nd edition
Wang	*COMPUTER METHODS IN ADVANCED STRUCTURAL ANALYSIS*
Wang	*MATRIX METHODS OF STRUCTURAL ANALYSIS*, 2nd edition
Wang and Salmon	*REINFORCED CONCRETE DESIGN*, 2nd edition
Winfrey	*ECONOMIC ANALYSIS FOR HIGHWAYS*

OBSERVATIONS AND LEAST SQUARES

Edward M. Mikhail

Purdue University

with contributions by

F. Ackermann

Stuttgart University

IEP—A Dun-Donnelley Publisher

New York

Published simultaneously in Canada by Fitzhenry & Whiteside, Ltd., Toronto.

Library of Congress Cataloging in Publication Data

Mikhail, Edward M.
Observations and least squares.

Bibliography: p.
Includes index.
1. Least squares. 2. Errors, Theory of. I. Ackermann, Friedrich E., 1929- joint author. II. Title.
QA275.M62 1976 519.5 76-10277
ISBN 0-7002-2481-5

Tables D1, D2, D3, D4(a), D4(b), and D4(c) adapted with permission of Addison-Wesley, from *Handbook of Statistical Tables* by Donald B. Owen. © 1962 by Addison-Wesley, Reading Mass. courtesy of the U.S. Energy Research and Development Adm. Adapted and reproduced with permission of John Wiley & Sons, Inc. from *Random Data: Analysis and Measurement* by Julius S. Bendat and Allan G. Piersol, © 1971 by John Wiley & Sons, Inc.

IEP—A Dun-Donnelley Publisher
666 Fifth Avenue
New York, New York 10019

Manufactured in the United States of America

Contents

Preface

This book is an outgrowth of lecture material that evolved over many years of teaching the subject of least squares at the undergraduate and the graduate levels by the primary and contributing authors. The presentation is such that the beginner may acquire a thorough understanding of the basic concepts. There is also sufficient material to challenge the advanced student. The book is therefore suitable for a first course and for one or two advanced courses on observations and least squares. In addition, the numerous examples worked out in detail together with complete derivations make it equally useful for self-study or as a reference.

The book is divided into three parts: Part I deals with fundamental concepts, Part II covers in detail the common techniques of least squares adjustment, and Part III discusses advanced topics. The appendices following the main text provide necessary supplementary material.

Chapter 1, which introduces the concepts of observation and the mathematical model, should be read first. Chapter 2 includes a review of statistical concepts that are pertinent to the main theme of the book. Since it is not intended to substitute for statistical literature, it contains very few theoretical derivations. As a review, however, it is sufficiently complete, covering both one- and multidimensional cases. The beginner need not read this chapter in its entirety before proceeding. Chapters 3 and 4 are concerned with "errors" and their propagation. "Error," a commonly used term, is related to the more appropriate term "observation" in the context of statistical notions. For propagation the reader needs familiarity with matrix manipulation. Prior review of Appendix A is necessary before reading Chapter 4.

An introductory undergraduate course on the subject might include Chapter 1, parts of Chapter 2, Chapters 3, 4 in Part I; Chapter 5, 7 and parts of Chapters 8 and 11 in Part II. If the first course is dual-level undergraduate-graduate, Chapter 6 should be added to the list. Thus the student would be exposed to basic statistics, propagation, and two or three techniques of least squares adjustment using condition equations. The term "condition equations" has been generalized in this book to mean all equations involving observations.

A second course might include Chapter 2; a review of Chapters 4 and 6, Chapters 9, 10 and 11 in Part II; and parts of Chapters 12 and 13 in Part III. At the end of such a course the student would have learned techniques of adjustment with constraints (equations involving only parameters); unified adjustment where all variables in the model are treated as observations; adjustment in steps; and sequential data processing.

For more advanced study, the latter parts of Chapters 12 and 13, as well as 14 in Part III provide the reader with further material. In particular, Chapter 14 introduces the concepts of interpolation (prediction), filtering, and collocation. After reading this chapter, the student will be better prepared to further his knowledge in these advanced subjects by studying the technical literature.

The development of the topics is sytematic in that each technique is progressively shown to be an extension of a preceding simpler technique. Essential to the following of the development is familiarity with matrix algebra, a review of which is given in Appendix A. Although generalized matrix algebra may be used instead, it is not considered necessary to the topics covered in this book.

The presentation is complete in itself. However, a bibliographical list is given at the end of the book to provide a source of supplementary reading. The reader should consult this list when seeking additional information on specific subjects.

The terms used are usually defined when introduced. However, some common statistical and adjustment terms may have been employed without specific definition. Readers unfamiliar with such common terms may consult a dictionary of statistical terms or the *International Society for Photogrammetry Multi-Lingual Dictionary*, 1961.

There are many colleagues whose encouragement and assistance made the completion of this book possible and are gratefully acknowledged: Mr. D. C. Brown reviewed a major part of the manuscript, made valuable recommendations, and allowed the use of some paragraphs and figures and adaptation of a section on recursive partitioning; Dr. E. Stark programmed most of the numerical examples; Dr. G. W. Marks edited Chapter 2 and computed some of its numerical examples; Professor K. Kraus contributed examples and illustrations to Chapter 14 which was reviewed by Dr. F. Leberl; Professor P.-L. Baetslé assisted with the material in Appendix C; Drs. B. Makarovič and P. Stefanovič read parts of the manuscript; Dr. D. L. Gifford had many discussions with the primary author and made significant recommendations; Professor K. Linkwitz allowed the use of Example 9.4; Professor H. Moritz permitted the use of the example in Chapter 14; and Mr. R. S. Johnson who assisted in reviewing the galley proofs. The primary author also wishes to acknowledge his past and present graduate students who taught him more than he is willing to admit and who continuously encouraged him to get this book published; and to thank his wife LaVerne without whose faithful support this book would not have become a reality.

Observations and Least Squares

PART I

1

Concepts of Observation and the Model

1.1. INTRODUCTION

Adjustment is meaningful only in those cases in which the data available exceed the minimum necessary for a unique determination (redundant data). Since the data are usually obtained from observations (or measurements), which are subject to probabilistic fluctuations (classically known as errors), redundant data are usually inconsistent in the sense that each sufficient subset yields different results from another subset. To obtain a unique solution, an additional criterion (such as that of least squares) is applied.

In order that we may understand the details of adjustment techniques, a number of basic concepts and their interrelations are introduced in this chapter.

1.2. OBSERVATIONS (MEASUREMENTS)

Both the terms *measurement* and *observation* will be used alternatively hereafter as having the same meaning. The term "observation" (or measurement) is often used in practice to refer to both the operation or process itself, as well as the actual outcome of such operation. With regard to adjustment the outcomes, and in particular *numerical* outcomes, will be designated as the "observations." Such numerical observational data are fundamental to science and engineering because they supply the instrument for analysis and manipulation.

Casual consideration of the concept of measurement is not problematic, since we can easily accept such seemingly simple notions as measuring distances, angles, temperature, or speed. However, when we take a closer look at the operation called "measurement," matters do not turn out to be so easy. As a matter of fact, even the measuring of a distance with a tape is a relatively complex operation, the analysis of which leads to the following fundamental properties of measurement:

1. To measure always means to perform a physical operation; and the process of measurement often consists of several more elementary operations, such as preparations, either instrument setup or calibration (or both), pointing, matching, and comparing.
2. The numerical reading or result obtained from the process is considered to represent the measurement. Thus the figure obtained as the measurement carries with it the circumstances and relevant historical data of how it originated.
3. Measurements are almost always performed with the help of instruments, no matter how simple, or with the aid of physical events. (An exception would be simple counting of a certain event.)
4. Measurements refer to standards, which are established by convention and which are rather arbitrary. Thus, in essence, to measure is to compare with a standard. Consequently, one is concerned with units and dimensions because the results of measurement always have a dimension of some sort.
5. Deeper consideration of measurements reveals that they refer to rather theoretical concepts, such as the geometrical abstractions used for distances and angles that have no real, direct equivalent in physical nature. We choose such concepts in order to describe certain elements of nature, such as location, area, or extension, that may be of interest to us.
6. Although performing measurements is an operation or a process, the results obtained from such processes take their meaning of a measurement only by association with the theoretical concepts to which they refer and on which they are based.

The theoretical abstractions to which measurements refer are called the *model.* In science and engineering, which are the subjects considered in this book, the model is almost always mathematical. Although it may appear that such a model concept is rather theoretical, it is in fact of basic importance to the subject of adjustment. A thorough understanding of the concept of the model considerably facilitates the task of adjustment.

1.3. THE MATHEMATICAL MODEL

Since most of the activities treated in this book pertain to quantitative problems, consideration will always be given to the *mathematical model.* It is defined here as a theoretical system or an abstract concept by which one describes a physical situation or a set of events. Such a description is not necessarily meant to be complete or exhaustive, but to relate only to those aspects or properties that are under consideration. Since a model serves a particular purpose, its setup can vary widely from one point of view to another. Thus the same physical system may be described by more than one model. The model then replaces the physical situation for the purpose of assessing it.

The mathematical model is often thought of as being composed of two parts: the functional model and the stochastic model. The *functional model* will in general describe the deterministic properties of the physical situation or event under consideration. On the other hand, the *stochastic model* designates and describes the nondeterministic or stochastic (probabilistic) properties of the variables involved, particularly those representing the observations. Both functional and stochastic models must be considered together at all times as there may be several possible combinations, each representing a possible mathematical model.

1.3.1. The Functional Model Whenever measurements are planned, some functional model is usually chosen to represent either a physical or a fictitious system with which the measurements are associated. In fact measurements are usually made in order to assess values for some or all of the parameters of the functional model. In photogrammetry, geodesy, and surveying we generally deal with geometric time-independent models and occasionally with dynamic models, such as the following:

1. Geometric model in surveying: A plane triangle, in Euclidian space, characterized by three angles, three corner points, three sides, and perhaps also orientation with respect to a coordinate system.
2. Geometric model in photogrammetry: Aerial photographs considered to be perspective images of (geometrical) terrain points.
3. Dynamic model in geodesy: Gravity field of the earth.
4. Dynamic models in photogrammetry: Orbital photography and time-dependent scanning photography.

Because the geometric models used are rather simple and easy to visualize, the elements of the model and the physical elements to which they refer are often not clearly distinguished. It must be recognized, however, that there are in nature no such objects as points, angles, distances, or coordinates. These are only elements of the functional model that are used to describe corresponding features of natural objects or their relationships of location.

Functional models are very often not stated explicitly, as they may refer to conventional models which are mostly defined by implication. If, for instance, a surveyor says that he measured a distance, then he refers to two objects that are abstracted and considered as two geometrical points. Even then he may not refer to the distance in its direct geometrical meaning, but rather to its projection onto a plane or even on an ellipsoid. Similarly, an angle CAB from point C to B about another point A is usually not the angle in the plane ABC, but its projection in another plane such as the horizontal (XY) plane.

It is perhaps the role of education in a given field to expound different workable models for different tasks. It is part of the technical skill of a scientist or engineer to know in which cases to operate with certain models and in which cases to construct new ones.

In general the functional model should conform to the physical reality with sufficient accuracy for the intended purpose.

1.3.2. Relating Observations to the Model A functional model is a completely fictitious construction used to describe physical events by an intelligible system suitable for analysis. It is linked to physical realities by measurements or observations that are themselves physical operations. In simpler situations, measurements refer directly to at least some elements of the functional model. However, it is not necessary, and often not practical, that all elements of the model be observable. In more involved situations, measurements do not actually relate directly to the elements of the model in question. For example, electronic distance measurement is actually the measurement of time, or time differences, not of distances. Here, a lot more theory becomes involved that must be incorporated into the model when the measurements are linked to the model. In fact, additional variables are necessarily included and new functional relationships added, thus expanding the model concept beyond the apparently simple task of measuring a distance. Consequently, in order to associate the results of the measurements with the elements of the model, the model must be extended.

The augmentation of the functional model due to measurements and their properties is of basic importance. The evaluation of observations depends on how and by what instruments and methods they have been acquired. The "measured" length of a line depends to a great extent on the measuring process, on whether the calibration of the instrument is considered to be known, on which reductions are to be applied, and so on. The same applies to the case of measuring angular directions. For instance, we must consider the zero direction and determine its place in the model, whether it is to be taken as known or unknown.

It may be a relatively long process, therefore, until the outcome of a measurement can be associated with an element of the model. Very often the direct readings from the measuring operation have to be "reduced" or preprocessed (which implements a correction model) before they can be considered appropriate.

In linking the observations to the model many of their features are often discarded for the sake of simplicity. For example, the special circumstances under which the actual measurements are acquired are seldom taken into account. In the light of the extension of the functional model, the modeling of this part is simplified. However, whenever a change is made in the functional model, the stochastic model must be modified accordingly.

1.3.3. The Stochastic Model—Statistical Properties of Measurements As the practitioner knows from experience, measurements are always subject to some unaccountable influences. They may be subject to physical influences that cannot be completely controlled, resulting in a certain variability of outcomes when observations are repeated. The variability of the results of measurements may or may not be attributed to some distinct physical causes. Such statistical variations, either due to neglecting physical or time-dependent influences or due to more inherent qualities of physical processes, are a basic property of measurements.

In the past these variations were said to be due to observational errors. At present, however, we accept variability or randomness of the results of measurements as a principal property of the observations and refer to statistical concepts in order to account for it.

From a practical standpoint it is rather difficult to assess the statistical properties of the observations. One way is to obtain repeated observations and derive the required properties, but this is usually demanding. Another way, which is often used, is to assume the statistical properties on the basis of a general reference to similar observations that were performed under similar circumstances in the past. Therefore when measurements are made, all relevant physical and environmental circumstances should be recorded during the measuring period in order to be able to judge the results properly. Actually, we often accept in practice rather coarse approximations for the statistical properties of the observations. For example, in geodesy the observations are usually considered (statistically) independent and often of equal precision (weight). The same applies in photogrammetry to image coordinates, although it is rather well known that film shrinkage causes (physical) correlation between image points. The assumption of independent image coordinates is made mainly because of its simplicity and because of the practical difficulty experienced so far in determining the correlation.

The totality of the assumptions on the statistical properties of the variables involved is called the *stochastic model*. It includes all model variables and designates those that are considered fixed (that is, constant during the adjustment, either known or a priori determined) and those that are considered free (that is, the parameters to be determined in the adjustment).

The classical theory of least squares adjustment did not explicitly specify the concept of stochastic model. Instead, the terms *observational errors* or *error properties of observations* were used. To the present practitioner, however, the term "measurement" or "observation" is not ambiguous and is used with very general meaning. In this book the term "observation" will refer to any quantity that is considered as a stochastic (random) variable and for which an estimate is available a priori. Such an estimate may be derived from direct or previous measurements or otherwise.

Examples of Variable Designations

1. The refraction coefficient in trigonometric height observation may be treated as a constant, as a free parameter to be determined from the adjustment, or as a (stochastic) variable (to which a "standard deviation" is associated). In the latter case it is considered as an observation and treated accordingly.
2. In fitting polynomials to a set of data one or more coefficients can be treated as observations (with "expectation," or mean, zero and a given standard deviation). This practice can often alleviate the possibility of an indeterminate solution.

As has been alluded to above, in the theory of observation and adjustment we deal with random variables whose treatment requires knowledge of statistical concepts. Therefore the remainder of Part I, particularly Chapter 2, is devoted to those aspects of statistics that are useful in adjustment. The treatment of such subjects will by necessity be concise and the reader is advised to refer to statistical textbooks for more detailed study and deeper insight (see the Bibliography at the end of the book).

Chapter 3 of Part I deals with an elaboration on the properties of observations. It attempts to relate classical theory of errors and modern theory of observations. Chapter 4 covers the principle and techniques of propagation, from one set of random variables to another.

2

Review of Statistical Concepts

2.1. INTRODUCTION

This chapter summarizes briefly a number of statistical concepts that are needed for the understanding of the theory of observations as will be treated in this book. These concepts are also necessary for the introductory foundation of the method of least squares as well as its assessment and practical application. The concepts to be introduced will be extracted from the two related mathematical fields of probability and statistics. In simple terms probability involves the laws of chance regarding the outcome of experiments. Its theories deal with averages of mass phenomena that may occur sequentially or simultaneously. They attempt to describe and predict those averages that would have been realized if the experiments were repeated many times. Statistics, on the other hand, deals with applying the laws of probability either to obtain estimates or to make inferences based on a given set of observational data. It is therefore concerned with methods of collecting, analyzing, and interpreting data. Both of these disciplines are used to assist the experimenter in making proper judgments in the face of uncertainty. They yield criteria on the basis of which decisions may be made. However, they never actually lead directly to the decisions to be made.

2.2. PROBABILITY, DISTRIBUTIONS, AND DENSITY FUNCTIONS

2.2.1. Probability Classical definitions relate the concept of probability of an event to the frequency of its occurrence when repeating an experiment; this repetition may be real or hypothetical. This leads to defining probability as the limit of the frequency of occurrence when the number of repetitions approaches infinity. If, for instance, the relative frequency of throwing a 6 with a die tends toward $\frac{1}{6}$ (with number of throws n approaching ∞), then the probability of throwing a 6 with this die is considered to be $\frac{1}{6}$ or 0.166.

In modern statistics this concept of probability as expected frequency is no longer in common use. Instead, probability is now considered as an independent basic concept associated with statistical events, and its properties are being founded axiomatically. In order to appreciate this concept, we first introduce the idea of the random variable.

2.2.2. Random Variable Probabilities are associated with statistical events, whether real or hypothetical, and an event is the outcome of a statistical experiment (such as throwing dice, measuring angles, or counting defective units of a product). If a statistical event has several possible outcomes, we associate with that event a *stochastic or random variable*, $\tilde{x}$. A random variable may be defined as a variable that takes on several possible values to each of which is associated a probability.

In probability we normally seek the behavior of a system on the basis of a known mathematical model with specified parameters. The totality of elements to be studied and about which information regarding its behavior is sought is called the *population*. In the theoretical sense the population is assumed to contain an infinite number of observations, whereas in practice it may have a very large, but finite, number of observations. The population includes *all* possible values of the random variables under consideration. Said another way, the population is the total of all possible outcomes of the statistical event associated with the random variable. Because of the very large size of the population, it is either impossible or entirely impractical to study each and every one of its elements to evaluate its characteristics. Therefore we select only a certain number of observations from the population, which we call a *sample*, and study it. From the results of sample study, we may draw inferences and make statements regarding the population from which the sample was taken.

The method of drawing a sample, as well as its size, will influence the conclusions that may be made. Obviously, the larger the sample size, to a certain extent, the more confident we will be regarding the results. As far as the method of drawing the sample is concerned, we must be careful that it does not follow a regular pattern. When this does occur, we run the risk of having the elements of the sample exhibit systematic effects, and the extension of the sample results to the population may not be entirely valid. To alleviate such a difficulty it must be ascertained that the sample is drawn *at random*, which means that each element of the population has an equal chance of being chosen into the sample. This is equivalent to saying that the selection of each element of the sample is *independent* of each other selection.

The total set of possible values of a random variable, $\tilde{x}$, together with their probabilities, constitute what is termed a *probability distribution* associated with the random variable. Consequently, a probability distribution describes the various probabilities related to the possible values of a random variable. These distributions are specified as two types: *cumulative distribution functions and probability density functions.*

2.2.3. The Cumulative Distribution Function The cumulative distribution function $F(x)$ for the random variable $\tilde{x}$ is defined by the relationship

$$P(\tilde{x} \leq x) = F(x) \tag{2.1}$$

which says: The probability that the random variable $\tilde{x}$ takes values less than or equal to x (x being a running variable) defines $F(x)$ or the cumulative distribution function. Because probabilities are, by axiomatic definition, limited to values between 0 and $+1$ (that is, $0 \leq P \leq +1$), the distribution function $F(x)$ satisfies the following marginal conditions:

$$\lim_{x \to -\infty} F(x) = 0 \qquad \lim_{x \to \infty} F(x) = +1 \tag{2.2}$$

Although the basic definition in equation (2.1) holds for discrete as well as continuous functions, we shall, in the context of this book, be mainly concerned with continuous distribution functions.

2.2.4. The Probability Density Function The probability density function $f(x)$ is formulated in a manner analogous to the concept of density in physics. It can be defined as the probability for an interval Δx. Assuming continuous differentiable functions, the relationship between $F(x)$ and $f(x)$ is

$$F(x) = \int_{-\infty}^{x} f(r)\, dr \tag{2.3}$$

and

$$f(x) = \frac{\partial F(x)}{\partial x} \tag{2.4}$$

From these two relationships we can easily deduce the following statements:

1. The probability that a random variable $\tilde{x}$ takes values in the interval between x_1 and x_2, where $x_2 > x_1$ is given by

$$P(x_1 < \tilde{x} < x_2) = F(x_2) - F(x_1) = \int_{x1}^{x2} f(r)\, dr \tag{2.5}$$

which is depicted by the shaded area in Figure 2.1.

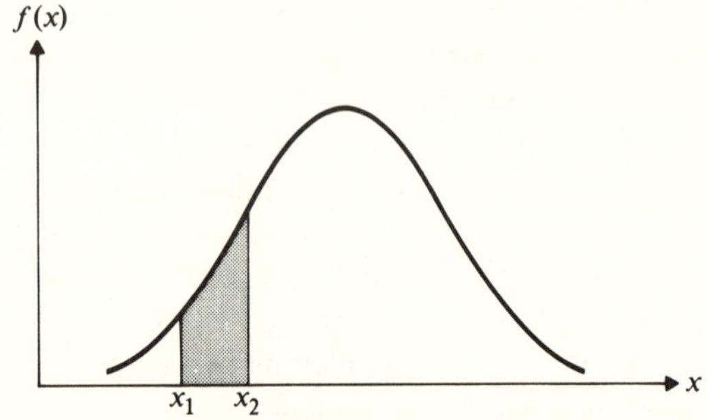

Figure 2.1

2. The probability that a random variable $\tilde{x}$ takes values smaller than x_1 is given by

$$P(-\infty < \tilde{x} < x_1) = F(x_1) - F(-\infty) = F(x_1)$$
$$= \int_{-\infty}^{x_1} f(r)\, dr \tag{2.6}$$

Note that $F(-\infty) = 0$ according to equation (2.2).

3. Similar to (2) we can write for the probability that $\tilde{x}$ takes on values larger than x_1.

$$P(x_1 < \tilde{x} < \infty) = F(\infty) - F(x_1) = 1 - F(x_1)$$
$$= \int_{x_1}^{\infty} f(r)\, dr \tag{2.7}$$

Note again that $F(\infty) = 1$ according to equation (2.2).

From the preceding development two conditions stand out to characterize $f(x)$ as a probability density function:

1. $f(x) \geq 0$ for all values of x.
2. $\int_{-\infty}^{\infty} f(x)\, dx = 1$, since infinite values of x have been implied and since the total value of the probability, being in this case the entire area under the density curve, must equal 1.

The two criteria (1) and (2) given above must be strictly satisfied in order that a function may be considered as a probability density function.

Before addressing cases involving more than one random variable (variate), it should be mentioned that the probability density function expresses the whole *population.* It is usually represented by a number of variables which are called *parameters* in statistical terminology. Knowing the parameters totally specifies the density function. By contrast, the *sample* values drawn for a population are used to compute quantities that are known as *statistics*. Each statistic usually represents an estimate of the corresponding population parameter.

2.3. MULTIDIMENSIONAL DISTRIBUTIONS, MARGINAL AND CONDITIONAL DISTRIBUTIONS, INDEPENDENCE

The concepts introduced in Section 2.2 involved only one random variable. In many practical situations we often encounter cases with more than one random variable. For example, the case of two random variables $\tilde{x}$ and $\tilde{y}$ has a two-dimensional cumulative distribution function $F(x, y)$. The joint probability of $\tilde{x}$ taking values smaller than x and of $\tilde{y}$ taking values smaller than y (x and y are both running variables) is given by

$$F(x, y) = P(\tilde{x} < x, \tilde{y} < y) \tag{2.8}$$

In a manner similar to the one-dimensional case, if $f(x, y)$ refers to the *two-dimensional probability density function*, and assuming continuous differentiable functions, then

$$f(x, y) = \frac{\partial^2 F(x, y)}{\partial x \, \partial y} \tag{2.9}$$

$$F(x, y) = \int_{-\infty}^{x} \int_{-\infty}^{y} f(u, v) \, du \, dv \tag{2.10}$$

As before, since $0 \leq P \leq 1$ is the allowed region of values of probability, therefore

$$\lim_{\substack{x \to -\infty \\ y \to -\infty}} F(x, y) = 0 \qquad \lim_{\substack{x \to +\infty \\ y \to +\infty}} F(x, y) = +1 \tag{2.11}$$

Finally, the probability that the two random variables $\tilde{x}$ and $\tilde{y}$ take values between x_1 and x_2, and y_1 and y_2, respectively, is given by

$$P(x_1 < \tilde{x} < x_2; y_1 < \tilde{y} < y_2) = \int_{x_1}^{x_2} \int_{y_1}^{y_2} f(u, v) \, du \, dv \tag{2.12}$$

Geometric representation for two-dimensional joint probability distribution can be effected in a way similar to the one-dimensional case (Figure 2.2). A probability within the interval $x_1 \rightarrow x_2$ and $y_1 \rightarrow y_2$ is equal to the volume that is bounded by the surface $f(x, y)$, the $x - y$ plane and the four vertical planes, $x = x_1$, $x = x_2$, $y = y_1$, and $y = y_2$.

For the n-dimensional case, the n random variables are collected in a *random* vector $\tilde{\mathbf{x}}$, or

$$\tilde{\mathbf{x}} = [\tilde{x}_1 \tilde{x}_2 \cdots \tilde{x}_n]^t \tag{2.13}$$

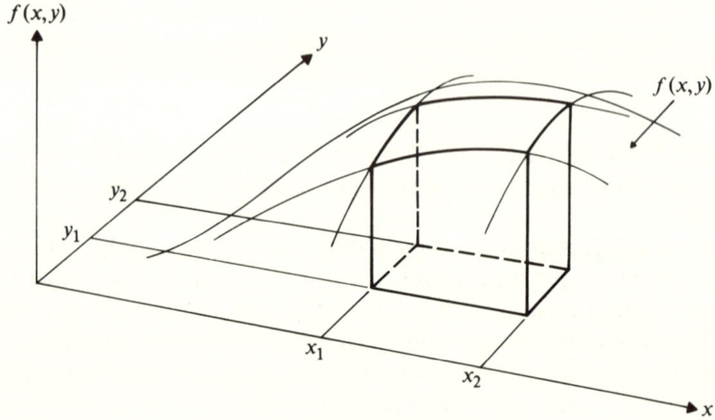

Figure 2.2

For that random vector, an n-dimensional joint density distribution function $f(\mathbf{x})$ is

$$f(\mathbf{x}) = f(x_1, x_2, \ldots, x_n) \tag{2.14}$$

and a corresponding cumulative distribution function $F(\mathbf{x})$, is

$$F(\mathbf{x}) = F(x_1, x_2, \ldots, x_n) \tag{2.15}$$

The probability that the random vector $\tilde{\mathbf{x}}$ takes values smaller than $\mathbf{x}$ is given by

$$\begin{aligned} F(\mathbf{x}) &= P(\tilde{\mathbf{x}} < \mathbf{x}) \\ &= P(\tilde{x}_1 < x_1, \tilde{x}_2 < x_2, \ldots, \tilde{x}_n < x_n) = \int_{-\infty}^{x_1} \int_{-\infty}^{x_2} \\ &\quad \cdots \int_{-\infty}^{x_n} f(u_1, u_2, \ldots, u_n)\, du_1\, du_2 \cdots du_n \end{aligned} \tag{2.16}$$

Although these relations show a relatively straightforward extension of the one-dimensional case, the two-dimensional and multidimensional cases are sufficiently different to warrant a few new concepts. These concepts include *marginal distribution, conditional distribution, independence (and correlation)*, all of which are only relevant when speaking of situations other than one-dimensional cases.

2.3.1. Marginal Distributions A marginal distribution is obtained from an n-dimensional distribution ($n \geq 2$) by disregarding the distribution of one or more of the components of the random vector $\tilde{\mathbf{x}}$. For example, the two-dimensional joint distribution for $\tilde{x}$ and $\tilde{y}$ may be reduced to the one-dimensional marginal distribution for $\tilde{x}$ by disregarding its relation, if any, with $\tilde{y}$. This is accomplished technically by selecting for $\tilde{y}$ the upper limit value of $y = +\infty$ in the cumulative distribution function of equation (2.8) or (2.10). Consequently, the *marginal cumulative distribution function* of the random variable $\tilde{x}$ is given as the joint probability of $\tilde{x}$ taking values smaller than x_1 and $\tilde{y}$ taking values less than $+\infty$, or

$$F_m(x) = F(x, \infty) = P(\tilde{x} < x, \tilde{y} < \infty) = \int_{-\infty}^{x} \int_{-\infty}^{\infty} f(u, v)\, du\, dv \quad (2.17)$$

The *joint density distribution function* $f(x, y)$ is reduced to the marginal form

$$f_m(x) = \int_{-\infty}^{\infty} f(u, v)\, dv \quad (2.18)$$

which leads to the marginal cumulative distribution function for $\tilde{x}$ as

$$F_m(x) = \int_{-\infty}^{x} f_m(x)\, dx = P(\tilde{x} < x) \quad (2.19)$$

Extending to the n-dimensional case, we can readily see that marginal distributions exist for all possible combinations of the random variables in groups of 1, 2, ..., and $(n - 1)$.

2.3.2. Independence Let $F(x, y)$ denote the joint two-dimensional cumulative distribution function of the two random variables $\tilde{x}$ and $\tilde{y}$, and $F(x)$ and $F(y)$ denote the marginal cumulative distribution functions for $\tilde{x}$ and $\tilde{y}$, respectively. The two random variables $\tilde{x}$ and $\tilde{y}$ are said to be *independent* if

$$F(x, y) = F(x) \cdot F(y) \quad (2.20)$$

Equation (2.20) also leads to a relation for density functions, provided they are continuous. Thus for two *independent* random variables

$$f(x, y) = f(x) \cdot f(y) \quad (2.21)$$

In fact, the relation in equation (2.21) can be directly extended to the case of n-dimensional density functions, or

$$f(x_1, x_2, \ldots, x_n) = f(x_1) \cdot f(x_2) \cdots f(x_n) \quad (2.22)$$

The concept of *independence* of two or more random variables is a consequence of the concept of independence of statistical events. It may be recalled that two events are said to be statistically independent if the probability of the joint event is equal to the product of the probabilities of the separate events, that is,

$$P(ab) = P(a) \cdot P(b) \quad (2.23)$$

We should distinguish between statistical or stochastical independence and functional independence of variables. Two variables may be functionally independent but may not necessarily be stochastically independent.

2.3.3. Conditional Distribution The concept of conditional distribution is defined in a manner similar to that of conditional probability. For example, $P(b \mid a)$, which reads probability of b given a, is the probability of the statistical event b on condition that the statistical event a has happened. Analogously, the *conditional* distribution of $(\tilde{y} \mid \tilde{x})$ (this reads the distribution of $\tilde{y}$ given $\tilde{x}$) is the distribution of $\tilde{y}$ for a given value of $\tilde{x}$ from a two-dimensional distribution of the random variables $\tilde{x}$ and $\tilde{y}$.

The conditional density function of $\tilde{y}$ given $\tilde{x}$ is

$$f(y \mid x) = \frac{f(x, y)}{f_m(x)} \tag{2.24}$$

The two concepts of independence and conditional distribution are directly related to each other. If two random variables $\tilde{x}$ and $\tilde{y}$ are independent, then the conditional distribution of $\tilde{y}$ given $\tilde{x}$ is the same for any value of $\tilde{x}$ and vice versa. In terms of the relationships already given

$$f(y \mid x) = \frac{f(x, y)}{f_m(x)} = \frac{f_m(x) \cdot f_m(y)}{f_m(x)} = f_m(y) = f(y) \tag{2.25}$$

Thus the values of $\tilde{x}$ have no influence on the conditional distributions of $\tilde{y}$ that are then all equal and identical to the marginal distribution of $\tilde{y}$.

The concepts of marginal and conditional distributions and of correlation are illustrated for the two-dimensional case in Section 2.7.1 on sampling by a numerical example on the stereogram (bidirectional histogram). Therefore the reader may consult that section at this point in order to gain a better understanding of these otherwise apparently abstract topics.

2.4. EXPECTATIONS, MOMENTS, AND CORRELATION

The distribution and density functions of random variables are characterized (and determined) by a number of parameters which are useful for the understanding of their behavior. The first of these expresses the intuitive concept of the mean and is referred to by one of several terms, such as expectation, expected value, mean, or average.

2.4.1. Expectation The expectation $E(\tilde{x})$ of a random variable $\tilde{x}$, if it exists, is defined as the average value μ_x of the variable over all possible values. It is computed by taking the sum of all possible values x_i of $\tilde{x}$ each multiplied by its corresponding probability $P(x_i)$. Thus

$$E(\tilde{x}) = \mu_x = \sum_{i=1}^{n} x_i \cdot P(x_i) \tag{2.26}$$

For a random variable $\tilde{x}$ with a continuous density function $f(x)$, the expectation is given by

$$E(\tilde{x}) = \mu_x = \int_{-\infty}^{\infty} xf(x)\, dx \tag{2.27}$$

where $f(x)$ is the density function of $\tilde{x}$. The relation in equation (2.27) may be extended to a more general form when we consider the expectation of a function $g(\tilde{x})$ of a random variable $\tilde{x}$ whose density function is $f(x)$. In this case

$$E(g(\tilde{x})) = \int_{-\infty}^{\infty} g(x)f(x)\,dx \tag{2.28}$$

The expression of equation (2.26) can be used to show how the name of mean of $\tilde{x}$ comes about. If there are n possible values x_i of $\tilde{x}$, each having equal probability $P(x_i) = 1/n(=\text{constant})$, then the computed expectation μ_x from equation (2.26) becomes identical to the arithmetic mean of the n values x_i. If the probabilities $P(x_i)$ are not equal, it can be easily shown that equation (2.26) would provide the weighted mean of the n values, taking the weights as proportional to the probabilities of occurrence of each value x_i.

Equation (2.28) may be extended to cases of two or more dimensions. If $\tilde{\mathbf{x}} = [\tilde{x}_1\, \tilde{x}_2]^t$ is a two-dimensional random vector with the corresponding joint density function $f(x_1, x_2)$, which is assumed to be continuous, then for any function $g(\tilde{x}_1, \tilde{x}_2)$ the expectation is defined as

$$E(g(\tilde{x}_1, \tilde{x}_2)) = \int_{-\infty}^{\infty}\int_{-\infty}^{\infty} g(x_1, x_2)\cdot f(x_1, x_2)\,dx_1\,dx_2 \tag{2.29}$$

Similarly, for the general case of n dimensions

$$E(g(\tilde{x}_1, \tilde{x}_2, \ldots, \tilde{x}_n)) = \int_{-\infty}^{\infty}\int_{-\infty}^{\infty}\cdots\int_{-\infty}^{\infty} g(x_1, x_2, \ldots, x_n) \cdot f(x_1, x_2, \ldots, x_n)\,dx_1\,dx_2\cdots dx_n \tag{2.30}$$

When working with expectations, several simple rules can be obtained as a consequence of equations (2.26) through (2.30). These rules are given here without proofs.

$$E(E(\tilde{x})) = E(\tilde{x}) \tag{2.31}$$

$$E(\tilde{x} + \tilde{y}) = E(\tilde{x}) + E(\tilde{y}) \tag{2.32}$$

$$E(c) = c \tag{2.33}$$

where c is constant,

$$E(c\tilde{x}) = cE(\tilde{x}) \tag{2.34}$$

where c is constant.

$$E(\tilde{x}\cdot\tilde{y}) = E(\tilde{x})\cdot E(\tilde{y}) \tag{2.35}$$

if and only if $\tilde{x}$ and $\tilde{y}$ are *independent* random variables.

$$E(\tilde{x}^2) \neq (E(\tilde{x}))^2 \tag{2.36}$$

in general.

Example 2.1. If $\tilde{x}$, $\tilde{y}$, and $\tilde{z}$ are three independent random variables and $\tilde{w} = 3\tilde{x} + 5\tilde{y}\tilde{z} - 2$, what is the expectation of $\tilde{w}$ if we are given the expectations μ_x, μ_y, and μ_z.

Solution: Applying the relations above, we can then write

$$E(\tilde{w}) = \mu_w = 3\mu_x + 5\mu_y\mu_z - 2$$

Before leaving this section it is perhaps worth pointing out that the combined equations (2.31) through (2.34) express the so-called laws of propagation of the mean as will be explained later (see Section 4.3.2).

2.4.2. Variance Let $g(\tilde{x})$ be defined as

$$g(\tilde{x}) = (\tilde{x} - E(\tilde{x}))^2 = (\tilde{x} - \mu_x)^2 \tag{2.37}$$

The expectation of this special function $g(\tilde{x})$ is called the *variance* of the random variable $\tilde{x}$, and is given by

$$\text{var}\,(\tilde{x}) = \sigma_x^2 = E(g(\tilde{x})) = E[(\tilde{x} - E(\tilde{x}))^2] = \int_{-\infty}^{\infty} (x - \mu_x)^2 \cdot f(x)\,dx \tag{2.38}$$

in which $f(x)$ is the density function of $\tilde{x}$, assumed to be continuous. The square root of σ_x^2, denoted by σ_x, is usually given the name *standard deviation.* The variance (or the standard deviation) is a measure of dispersion.

In view of the few rules of expectation given previously, an alternative form for the variance is possible, as follows:

$$\begin{aligned} \text{var}\,(\tilde{x}) = \sigma_x^2 &= E[(\tilde{x} - E(\tilde{x}))^2] = E[(\tilde{x} - \mu_x)^2] \\ &= E[\tilde{x}^2 - 2\tilde{x}\mu_x + \mu_x^2] \end{aligned}$$

$$\text{var}\,(\tilde{x}) = E(\tilde{x}^2) - 2\mu_x E(\tilde{x}) + E(\mu_x^2)$$

or

$$\sigma_x^2 = E(\tilde{x}^2) - \mu_x^2 = E(\tilde{x}^2) - (E(\tilde{x}))^2 \tag{2.38a}$$

2.4.3. Covariance and Correlation In a manner similar to defining the variance of one random variable, the covariance may be introduced when considering two random variables. Let the two random variables $\tilde{x}$ and $\tilde{y}$ have the joint density function $f(x, y)$, and define the function

$$h(\tilde{x}, \tilde{y}) = [(\tilde{x} - E(\tilde{x}))(\tilde{y} - E(\tilde{y}))] = [(\tilde{x} - \mu_x)(\tilde{y} - \mu_y)] \tag{2.39}$$

The covariance between $\tilde{x}$ and $\tilde{y}$ is defined as the expectation of the function $h(\tilde{x}, \tilde{y})$, or

$$\begin{aligned} \text{cov}\,(\tilde{x}, \tilde{y}) = \sigma_{xy} &= E[h(\tilde{x}, \tilde{y})] \\ &= E[(\tilde{x} - \mu_x)(\tilde{y} - \mu_y)] \end{aligned}$$

or

$$\sigma_{xy} = \int_{-\infty}^{\infty} \int_{-\infty}^{\infty} (x - \mu_x)(y - \mu_y) f(x, y)\, dx\, dy \tag{2.40}$$

provided that the distributions are continuous. Although the variance of a random variable expresses the variation of its distribution, the covariance describes the mutual variation of two random variables. It reflects their interrelationship or *mutual correlation* for which a correlation coefficient ρ is constructed as

$$\rho_{xy} = \frac{\sigma_{xy}}{\sigma_x \sigma_y} = E\left\{\frac{(\tilde{x} - E(\tilde{x}))}{\sigma_x} \cdot \frac{(\tilde{y} - E(\tilde{y}))}{\sigma_y}\right\} \tag{2.41}$$

where σ_x and σ_y designate the standard deviations of the marginal distributions of $\tilde{x}$ and $\tilde{y}$, respectively. The correlation between two random variables describes some interdependence between them. It should not, however, be confused with stochastic dependence or independence that is defined by the concept of conditional distributions [see equation (2.25)]. Correlation and statistical dependence are not the same, although both concepts are often used synonymously. It can be shown that the covariance σ_{xy} is always zero when the random variables $\tilde{x}$ and $\tilde{y}$ are statistically independent. But the reverse is not true in general. Zero covariance does not necessarily imply statistical independence. Nevertheless for multivariate *normal* probability distribution, zero covariance (lack of correlation) is a sufficient condition for statistical independence (see Hamilton, p. 31; in the bibliography at the back of the book).

2.4.4. Moments The concepts of mean (expectation), variance, and covariance are special cases of a more general concept termed *statistical moments.* The expected value of the general function $g(\tilde{x}) = (\tilde{x} - c)^k$, where c is a constant that may or may not be zero, is called the *statistical moment of the kth order* of the random variable $\tilde{x}$ (or, in short, the kth moment of $\tilde{x}$) about c. Thus

$$\bar{m}_k = E((\tilde{x} - c)^k) \tag{2.42}$$

Using the two definitions of expectation given by equations (2.26) and (2.27), we get for the case of $c = 0$

$$\bar{m}_k = E(\tilde{x}^k) = \sum_{i=1}^{n} x_i^k P(x_i) \tag{2.43}$$

for a discrete random variable, and

$$\bar{m}_k = E(\tilde{x}^k) = \int_{-\infty}^{\infty} x^k f(x)\, dx \tag{2.44}$$

for the continuous case. The class of moments given by equation (2.42) includes a group called *central moments* which is of particular importance in practice. For this group the constant c takes the value

$$c = \mu_x = E(\tilde{x}) \tag{2.45}$$

hence central moments are expectations with respect to the mean, or

$$m_k = E[(\tilde{x} - E(\tilde{x}))^k] \tag{2.46}$$

If we now refer back to the definition of the variance in equation (2.38), we find that it is a special case of equation (2.46) in which $k = 2$. This is to say that the variance $\sigma_{\tilde{x}}^2$ of the *random variable* $\tilde{x}$ *is its second central moment*, or $\sigma_{\tilde{x}}^2 = m_2$.

The different moments of a random variable $\tilde{x}$ reflect the properties of its density function. When the density function is symmetric with respect to the first moment $\bar{m}_1 = E(\tilde{x}) = \mu_x$, all central moments of odd order vanish; that is, $m_k = 0$ for k odd. Conversely, when such odd-order central moments are not zero, their values reflect the degree of asymmetry or skewness of the density function.

As with the other concepts, moments may be extended to multidimensional random variables. Suppose that $(\tilde{x}, \tilde{y})$ is a two-dimensional random vector. The expected value of the general function $g(\tilde{x}^l, \tilde{y}^n)$ is called the moment of $(l + n)$ order. The corresponding central moment of the $(l + n)$ order is given by

$$\begin{aligned} m_{ln} &= E[(\tilde{x} - E(\tilde{x}))^l(\tilde{y} - E(\tilde{y}))^n] \\ &= \int_{-\infty}^{\infty}\int_{-\infty}^{\infty} (x - E(\tilde{x}))^l(y - E(\tilde{y}))^n f(x, y)\, dx\, dy \end{aligned} \tag{2.47}$$

for continuous distributions. For such a two-dimensional random vector there are three central moments of order two. Let $\bar{m}_{10}$ and $\bar{m}_{01}$ denote the first moments (that is, expectations) of the variables $\tilde{x}$ and $\tilde{y}$, respectively. Then the following are the three central moments of second order

$$\begin{aligned} m_{20} &= E\{(\tilde{x} - \bar{m}_{10})^2\} = E((\tilde{x} - \mu_x)^2) = \sigma_{\tilde{x}}^2 = \text{variance of } \tilde{x} \\ m_{02} &= E\{(\tilde{y} - \bar{m}_{01})^2\} = E((\tilde{y} - \mu_y)^2) = \sigma_{\tilde{y}}^2 = \text{variance of } \tilde{y} \\ m_{11} &= E\{(\tilde{x} - \bar{m}_{10})(\tilde{y} - \bar{m}_{01})\} = E((\tilde{x} - \mu_x)(\tilde{y} - \mu_y)) = \sigma_{xy} \\ &= \text{covariance of } \tilde{x} \text{ and } \tilde{y} \end{aligned} \tag{2.48}$$

The central moments of order two, which according to equations (2.38) and (2.40) are identical to the variance and covariance, are of special importance for the statistical assessment of observations and for least squares adjustment.

In the general case of an n-dimensional random vector $\tilde{\mathbf{x}}$, second-order central moments can be set up in all combinations between the elements of the vector. The totality of these moments can be arranged in a matrix array, which is termed *the matrix of second central moments, the variance–covariance matrix*, or simply the *covariance matrix*. Its elements may be written symbolically as

$$M_{xx} = \begin{bmatrix} m_{x1x1} & m_{x1x2} & \cdots & m_{x1xn} \\ m_{x2x1} & m_{x2x2} & \cdots & m_{x2xn} \\ \vdots & & & \\ m_{xnx1} & m_{xnx2} & \cdots & m_{xnxn} \end{bmatrix} = \begin{bmatrix} \sigma_{x1}^2 & \sigma_{x1x2} & \cdots & \sigma_{x1xn} \\ \sigma_{x2x1} & \sigma_{x2}^2 & \cdots & \sigma_{x2xn} \\ \vdots & & & \\ \sigma_{xnx1} & \sigma_{xnx2} & \cdots & \sigma_{xn}^2 \end{bmatrix} \tag{2.49}$$

which is a square symmetric matrix, because

$$\begin{aligned} m_{x_ix_j} &= E\{(\tilde{x}_i - E(\tilde{x}_i))(\tilde{x}_j - E(\tilde{x}_j))\} \\ &= E\{(\tilde{x}_j - E(\tilde{x}_j))(\tilde{x}_i - E(\tilde{x}_i))\} = m_{x_jx_i} \end{aligned} \tag{2.50}$$

It should be mentioned, however, that $\mathbf{M}_{xx}$ is square and symmetric only because it relates to the one n-dimensional random vector $\tilde{\mathbf{x}}$. If, on the other hand, we construct the cross-covariance matrix $\mathbf{M}_{xy}$ for the n-dimensional vector $\tilde{\mathbf{x}}$ and m-dimensional vector $\tilde{\mathbf{y}}$, it will be an $n \times m$ rectangular matrix. All its elements will be covariances; that is, there will be no variances included. The symbol $\mathbf{M}$ is used to denote the covariance matrix, to be consistent with the symbol m used for the individual elements. Later it will be convenient to use $\mathbf{\Sigma}$ instead, in keeping with the lower case σ being used for variances (σ^2) and covariances.

Example 2.2. (a) Derive the relationship between the central moments m_1, m_2, m_3, and the moments $\bar{m}_i$ (about zero).

Solution: By definition

$$m_1 = E(\tilde{x} - E(\tilde{x})) = E(\tilde{x}) - E(E(\tilde{x})) = E(\tilde{x}) - E(\tilde{x}) = 0$$

$$\begin{aligned} m_2 &= E[(\tilde{x} - E(\tilde{x}))^2] = E[(\tilde{x} - \bar{m}_1)^2] \\ &= E(\tilde{x}^2) - 2\bar{m}_1 E(\tilde{x}) + \bar{m}_1^2 \\ &= \bar{m}_2 - 2\bar{m}_1^2 + \bar{m}_1^2 \\ &= \bar{m}_2 - \bar{m}_1^2 \end{aligned}$$

This result is identical to that derived in equation (2.38a) (see Section 2.4.2).

$$
\begin{aligned}
m_3 &= E[(\tilde{x} - E(\tilde{x}))^3] = E[(\tilde{x} - \bar{m}_1)^3] \\
&= E(\tilde{x}^3) + 3\bar{m}_1^2 E(\tilde{x}) - 3\bar{m}_1 E(\tilde{x}^2) - \bar{m}_1^3 \\
&= \bar{m}_3 + 3\bar{m}_1^3 - 3\bar{m}_1 \bar{m}_2 - \bar{m}_1^3 \\
&= \bar{m}_3 - 3\bar{m}_1 \bar{m}_2 + 2\bar{m}_1^3
\end{aligned}
$$

(b) Derive the relationship between the central moment m_{11} and the moments $\bar{m}_{01}$, $\bar{m}_{10}$, and $\bar{m}_{11}$ (about zero) for a case of two random variables.

Solution: Again, by definition

$$
\begin{aligned}
m_{11} &= E[(\tilde{x} - E(\tilde{x}))(\tilde{y} - E(\tilde{y}))] \\
&= E[\tilde{x}\tilde{y} - \tilde{x}\bar{m}_{01} - \tilde{y}\bar{m}_{10} + \bar{m}_{01}\bar{m}_{10}] \\
&= E(\tilde{x}\tilde{y}) - E(\tilde{x})\bar{m}_{01} - E(\tilde{y})\bar{m}_{10} + \bar{m}_{01}\bar{m}_{10} \\
&= \bar{m}_{11} - \bar{m}_{10}\bar{m}_{01} - \bar{m}_{01}\bar{m}_{10} + \bar{m}_{01}\bar{m}_{10} \\
&= \bar{m}_{11} - \bar{m}_{01}\bar{m}_{10}
\end{aligned}
$$

2.5. SOME OFTEN USED DISTRIBUTIONS

In connection with the theory of errors of observation and least squares adjustment there are a few (one-dimensional) distributions that are often used. Only continuous distributions are discussed, particularly those used for statistical testing.

2.5.1. The Gaussian or Normal Distribution The one-dimensional normal distribution is the most frequently used distribution in statistical theory and application. Its density function is given by

$$
f(x) = \frac{1}{\sigma_x\sqrt{2\pi}} \exp\left\{-\frac{(x - \mu_x)^2}{2\sigma_x^2}\right\} \tag{2.51}
$$

In equation (2.51) there are two parameters that specify the distribution (Figure 2.3):

$$\mu_x = \text{expectation or mean of } \tilde{x}$$

and

$$\sigma_x = \text{standard deviation of } \tilde{x} = +\sqrt{\text{var}(\tilde{x})} = \sqrt{\sigma_x^2}$$

The cumulative normal distribution function of the standardized random variable,

$$
z = \frac{x - \mu_x}{\sigma_x} \tag{2.52}
$$

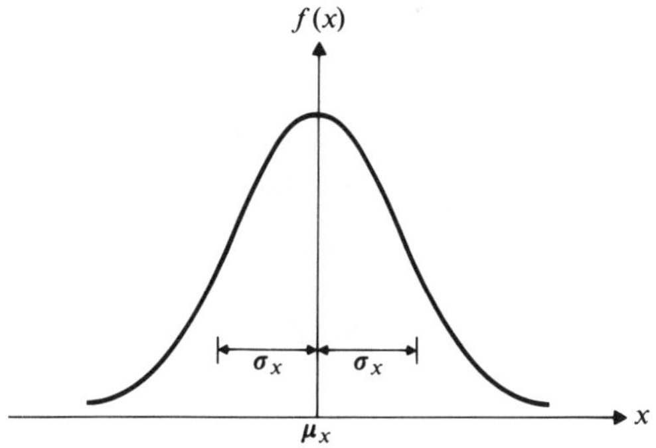

Figure 2.3. Normal Distribution Density Function

(with zero mean and unit standard deviation) is given by

$$F(z) = \frac{1}{\sqrt{2\pi}} \int_{-\infty}^{z} \exp\left(-\frac{u^2}{2}\right) du \tag{2.53}$$

Appendix Table D.1 gives the values of $[1 - F(z)]$. A few remarks may point out the essential features of the normal distribution.

1. The normal density function $f(x)$ is symmetric about the mean μ_x. Therefore all odd central moments are zero. Also the median† and the mode,‡ which are two parameters of location sometimes used in practice, are equal to the mean μ_x.
2. The maximum density value for the standardized variable is 0.399.
3. The density function approaches zero asymptotically as x goes to $\pm\infty$.
4. The density function has two points of inflection at $x = \mu_x \pm \sigma_x$.
5. The probability for $\tilde{x}$ taking values within x_1 and x_2 is given by the area between the x axis, the density function curve, and the boundaries of the interval $x = x_1$, $x = x_2$. In particular the probabilities for deviation from the mean within some multiples of σ are as follows:

$$\begin{aligned} P[-\sigma_x < x - \mu_x < +\sigma_x] &= 0.6827 \\ P[-2\sigma_x < x - \mu_x < +2\sigma_x] &= 0.9545 \\ P[-3\sigma_x < x - \mu_x < +3\sigma_x] &= 0.9973 \end{aligned} \tag{2.54}$$

† The median x_m is defined for the continuous distribution such that

$$\int_{-\infty}^{x_m} f(x)\, dx = \tfrac{1}{2}$$

Thus x_m is the abscissa through which a vertical line divides the area under the density distribution into two equal parts.

‡ The mode, another measure of location or central tendency, is for continuous distributions the value of the random variable at which the probability density function has a relative maximum.

6. The abscissae associated with intervals covering probabilities of 0.90, 0.95, and 0.99 are

$$P[-1.645\sigma_x < x - \mu_x < +1.645\sigma_x] = 0.90$$
$$P[-1.960\sigma_x < x - \mu_x < +1.960\sigma_x] = 0.95$$
$$P[-2.576\sigma_x < x - \mu_x < +2.576\sigma_x] = 0.99 \tag{2.55}$$

7. The probability that x takes on values on either side of μ_x (that is, either larger or smaller than μ_x) is equal to 0.5.

The theoretical and practical importance of the normal distribution is due to the "central limit theorem" which states that the sum $\sum_{i=1}^{n} \tilde{x}_i$ of n independent variables $\tilde{x}_1, \ldots, \tilde{x}_n$ will be asymptotically normally distributed as $n \to \infty$. In practical applications, normal distributions are encountered very often. In particular, random variables that represent measurements in photogrammetry, geodesy, or surveying are often nearly normally distributed.

2.5.2. The *t* (Student) Distribution The t distribution is used in connection with sampling (testing hypotheses using sample statistics instead of population parameters).

Let $\tilde{x}_1, \tilde{x}_2, \ldots, \tilde{x}_n$ be n independent stochastic (random) variables of identical normal distribution with mean μ and standard deviation σ. Then the random variable $\tilde{t}$ is defined by

$$t = \frac{\bar{x} - \mu}{s}\sqrt{n} \tag{2.56}$$

where

$$\bar{x} = \frac{1}{n}\sum_{k=1}^{n} x_k \qquad \text{and} \qquad s^2 = \frac{1}{n-1}\sum_{k=1}^{n}(x_k - \bar{x})^2$$

The statistics $\bar{x}$ and s refer to sample mean and sample standard deviation, respectively (see Equation (2.89) and (2.92)). The density function of t is

$$f(t) = \frac{1}{\sqrt{(n-1)\pi}}\,\frac{\Gamma(n/2)}{\Gamma[(n-1)/2]}\,\frac{1}{\left(1 + \dfrac{t^2}{n-1}\right)^{n/2}} \tag{2.57}$$

Γ = gamma function

$$\Gamma(n) = \int_0^\infty e^{-t}t^{n-1}\,dt$$

$$\Gamma(n+1) = n \cdot \Gamma(n) \qquad \text{and} \qquad \Gamma(1) = 1 \tag{2.58}$$

For whole, positive numbers n,

$$\Gamma(n) = (n-1)! \qquad \text{or} \qquad \Gamma(n+1) = n!$$

Appendix D Table D.2 lists the values of $[1 - F(t)]$. The t distribution, as defined in equation (2.57) is said to have $(n-1)$ degrees of freedom. The density function is symmetrical with respect to the mean. With increasing degrees of freedom the t distribution approaches the normal distribution with mean 0 and variance 1. For $n \geq 30$ the t distribution can be replaced by the normal distribution.

2.5.3. The χ^2 Distribution Let $\tilde{x}_1, \tilde{x}_2, \ldots, \tilde{x}_n$ be n independent stochastic (random) variables, each of which is normally distributed with mean 0 and variance 1. Then the sum of their squares yields a random variable for which the symbol χ^2 is customarily used

$$\chi_n^2 = \tilde{x}_1^2 + \tilde{x}_2^2 + \cdots + \tilde{x}_n^2 \tag{2.59}$$

This random variable has a χ^2 distribution (chi square) of n "degrees of freedom."

The density function of the random variable χ^2 is

$$f(\chi_n^2) = f(x) = c_n \cdot x^{(n-2)/2} \cdot e^{-x/2} \qquad \text{for } x > 0 \tag{2.60}$$

and

$$f(x) = 0 \qquad \text{for } x \leq 0$$

where

$$c_n = \frac{1}{2^{n/2} \cdot \Gamma\left(\frac{n}{2}\right)}$$

$$n = \text{degrees of freedom}$$

The χ^2 distribution has particularly simple central moments as follows:

Mean:

$$E(\chi_n^2) = n \tag{2.61}$$

Variance:

$$\text{var}\,(\chi_n^2) = \sigma^2(\chi_n^2) = 2n \tag{2.62}$$

In Appendix Table D.3 the values of $[1 - F(\chi^2)]$ are given for several degrees of freedom. The graphical representations in Figure 2.4 show that χ^2 density functions are markedly skew for small degrees of freedom. Beyond $n \sim 10$, however, they approach more and more a normal distribution. The asymptotic convergence for $n \to \infty$ approaches normal density function with mean n and variance equal to $2n$.

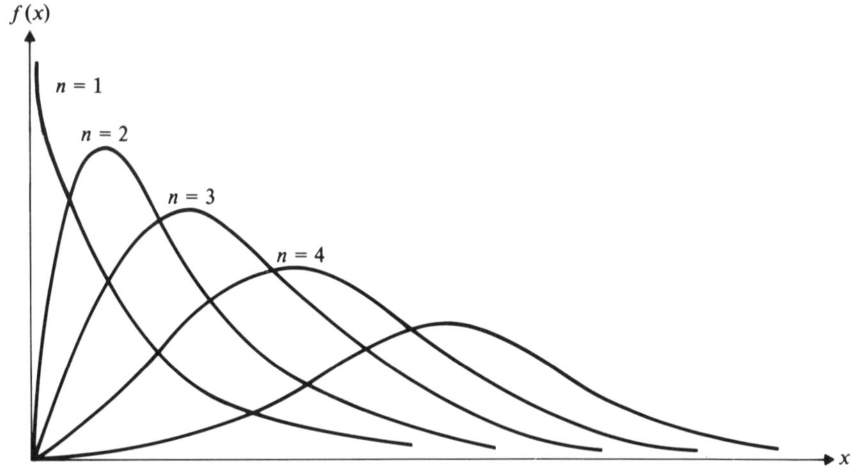

Figure 2.4

2.5.4. The *F* (Fisher) Distribution Let $\tilde{x}_1, \tilde{x}_2, \ldots, \tilde{x}_m$ and $\tilde{y}_1, \tilde{y}_2, \ldots, \tilde{y}_n$ be two sets of independent, normally distributed, normalized random variables, each with mean 0 and variance 1. Consider the sum of the squares of each set,

$$\tilde{\chi}_m^2 = \tilde{x}_1^2 + \tilde{x}_2^2 + \cdots + \tilde{x}_m^2 = \sum_{i=1}^{m} \tilde{x}_i^2 \tag{2.63}$$

$$\tilde{\chi}_n^2 = \tilde{y}_1^2 + \tilde{y}_2^2 + \cdots + \tilde{y}_n^2 = \sum_{j=1}^{n} \tilde{y}_j^2 \tag{2.64}$$

The random variable F,

$$F_{m,n} = \frac{\tilde{\chi}_m^2/m}{\tilde{\chi}_n^2/n} \tag{2.65}$$

is said to have an F distribution of m and n degrees of freedom. The density function $f_{m,n}(u)$ and the cumulative distribution function $F_{m,n}(u)$ are, for $u > 0$, given by

$$f_{m,n}(u) = \frac{\Gamma[(m+n)/2]}{\Gamma(m/2)\cdot\Gamma(n/2)} \cdot \left(\frac{m}{n}\right)^{m/2} \frac{u^{(m-2)/2}}{[1+(m/n)u]^{(m+n)/2}} \tag{2.66}$$

$$F_{m,n}(u) = \int_0^u f_{m,n}(t)\,dt \tag{2.67}$$

The value of $[1 - F]$ (where F is a cumulative function) for different degrees of freedom are given in Appendix Tables D.4(a) through D.4(c). The values of the expectation (mean) and the variance of an F-distributed random variable, with m, n degrees of freedom are

$$\text{mean} = \frac{n}{n-2} \qquad \text{for } n > 2$$

$$\text{variance} = \frac{2n^2(m+n-2)}{m(n-2)^2(n-4)} \qquad \text{for } n > 4 \tag{2.68}$$

The F distribution with m, n degrees of freedom approaches asymptotically for $n \to \infty$ the χ^2 distribution with m degrees of freedom.

The practical application of the F distribution in least squares adjustment is concerned with the comparison of variances such as those obtained from two adjustments. In some particular cases the comparison may be between a variance obtained from the adjustment and an a priori given reference variance. This case refers to the $F_{m,\infty} = \chi_m^2$ distribution.

2.6. MULTINORMAL DISTRIBUTION

Although least squares theory of adjustment does not require a specified distribution, most of the statistical testing following the adjustment is concerned with multinormal distribution of random vectors with density function

$$f(x_1, \ldots, x_n) = f(\mathbf{x}) = \left[\frac{1}{(2\pi)^{n/2}\sqrt{|\boldsymbol{\Sigma}|}}\right] \times \exp\left[-\frac{1}{2}(\mathbf{x}-\boldsymbol{\mu}_x)'\boldsymbol{\Sigma}^{-1}(\mathbf{x}-\boldsymbol{\mu}_x)\right] \tag{2.69}$$

with mean vector $\boldsymbol{\mu}_x$ and covariance matrix $\boldsymbol{\Sigma}$. From equation (2.69) the expanded form for the two-dimensional normal distribution of a random vector $(\tilde{x}, \tilde{y})$ is defined by the density function,

$$f(x, y) = \frac{1}{2\pi\sqrt{\sigma_1^2\sigma_2^2 - \sigma_{12}^2}} \exp\left\{-\frac{\sigma_1^2\sigma_2^2}{2(\sigma_1^2\sigma_2^2 - \sigma_{12}^2)} \left[\frac{(x-\mu_1)^2}{\sigma_1^2} - 2\sigma_{12}\frac{(x-\mu_1)(y-\mu_2)}{\sigma_1^2\sigma_2^2} + \frac{(y-\mu_2)^2}{\sigma_2^2}\right]\right\} \tag{2.70}$$

where μ_1, μ_2 are the means, σ_1^2, σ_2^2 are the variances for $\tilde{x}$ and $\tilde{y}$, respectively, and σ_{12} is the covariance.

The graphical representation of the two-dimensional normal density function in Figure 2.5 demonstrates that its intersection with any vertical plane $y = ax + b$ will be a one-dimensional Gauss function, that is, a one-dimensional normal density function. Any plane parallel to the x, y plane and cutting the bell-shaped surface intersects it in an elliptical curve.

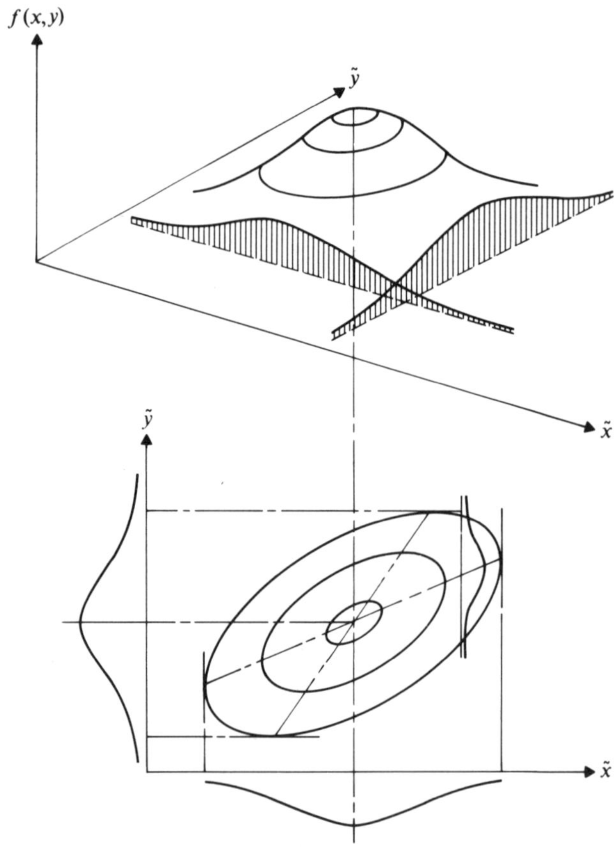

Figure 2.5

The conditional density function of y given x can be considered as that of a one-dimensional normal distribution

$$f(y\,|\,x) = \frac{1}{\bar{\sigma}\sqrt{2\pi}} \exp\left\{-\frac{(y-\mu(x))^2}{2\bar{\sigma}^2}\right\} \tag{2.71}$$

the mean of which is a function of x,

$$\mu(x) = \mu_2 + \frac{\rho\sigma_2}{\sigma_1}(x-\mu_1) = \mu_2 + \frac{\sigma_{12}}{\sigma_1^2}(x-\mu_1) \tag{2.72}$$

The variance in equation (2.71) can be seen to be

$$\bar{\sigma}^2 = \sigma_2^2(1-\rho^2) = \frac{\sigma_1^2\sigma_2^2 - \sigma_{12}^2}{\sigma_1^2} \tag{2.73}$$

Thus the mean (or expectation) of the conditional distribution of y for given values of x is a linear function of x. This function $\mu(x)$ is called the *regression line* of y on x. This provides for an interpretation of correlation and of regression between two random variables. In case of correlation the regression line is the functional dependence of the expectation of the conditional distribution of y on x and similarly for the other regression line.

2.6.1. Ellipses of Constant Probability Referring to equation (2.70) of the two-dimensional normal distribution, the properties of the density function $f(x, y)$ can also be geometrically interpreted by intersecting the surface $f(x, y)$ by horizontal planes $f(x, y)$ equal to constant. The resulting curves of intersection lines form a family of ellipses (because $\sigma_1 > 0$, $\sigma_2 > 0$, and $-1 < \rho < 1$), the equation of which is

$$h(x, y) = \frac{1}{(1-\rho^2)}\left[\frac{(x-\mu_1)^2}{\sigma_1^2} - 2\rho\frac{(x-\mu_1)(y-\mu_2)}{\sigma_1\sigma_2} + \frac{(y-\mu_2)^2}{\sigma_2^2}\right] = k^2 \quad (2.74)$$

The common center of this family of ellipses is defined by the expectations μ_1 and μ_2 (of the marginal distributions of $\tilde{x}$ and $\tilde{y}$) as coordinates. For simplicity in the analysis to follow, we shift the origin of the coordinate system to the point of common ellipses' center (μ_1, μ_2).

The equation $h(x, y) = k^2$ in equation (2.74), for a specific value of k, is an ellipse bounded by a rectangle having the dimensions $2k\sigma_1$ in the direction of x, and $2k\sigma_2$ in the direction of y, as shown in Figure 2.6. Such an ellipse is

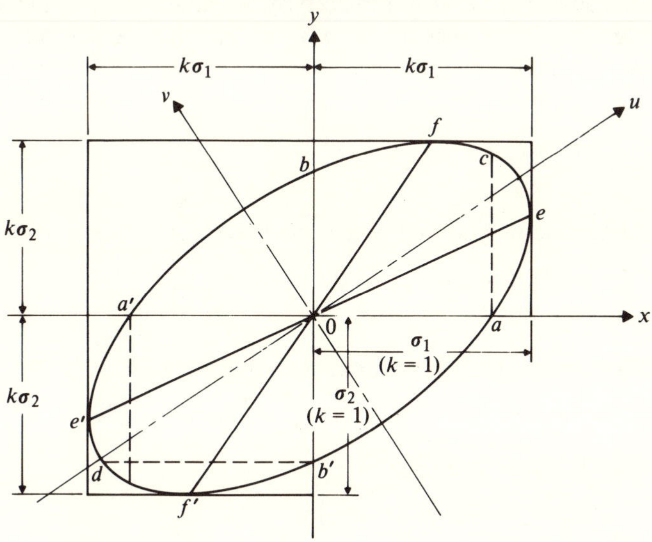

Figure 2.6. The Standard Ellipse. Relations for $k = 1$: $x_a = -x_{a'} = \sigma_1\sqrt{1-\rho^2} = \sigma(x|y)$; $y_b = -y_{b'} = \sigma_2\sqrt{1-\rho^2} = \sigma(y|x)$; $y_c = 2\rho\sigma_2\sqrt{1-\rho^2}$; $x_d = 2\rho\sigma_1\sqrt{1-\rho^2}$; $x_e = -x_{e'} = \sigma_1 y_e$ $= -y_{e'} = \rho\sigma_2$; $x_f = x_{f'} = \rho\sigma_1 y_f = -y_{f'} = \sigma_2$

known as an *ellipse of constant probability*; the value of such probability depends on the selected value of k.

In the shifted coordinate system considering the ellipse with $k = 1$ as a representative ellipse, its equation becomes

$$\left(\frac{x}{\sigma_1}\right)^2 - 2\rho\left(\frac{x}{\sigma_1}\right)\left(\frac{y}{\sigma_2}\right) + \left(\frac{y}{\sigma_2}\right)^2 = (1 - \rho^2) \tag{2.75}$$

This particular ellipse is named the *standard ellipse* and its shape is determined by σ_1, σ_2, and ρ (see Figure 2.6).

The points of intersection with the x axis are a and a' with $x = \pm\sigma_1\sqrt{1 - \rho^2}$, and with the y axis they are b and b' with $y = \pm\sigma_2\sqrt{1 - \rho^2}$. These values can be easily obtained from equation (2.75) by setting $y = 0$ and $x = 0$, respectively. Consequently, they represent the square roots of the corresponding conditional second moments $m(x \mid y = 0)$ and $m(y \mid x = 0)$. The vertical tangents to the ellipse are at points e and e' with coordinates $(\sigma_1, \rho\sigma_2)$ and $(-\sigma_1, -\rho\sigma_2)$, whereas the horizontal tangents are at points f and f' with corresponding coordinates $(\rho\sigma_1, \sigma_2)$ and $(-\rho\sigma_1, -\sigma_2)$, respectively.

From equation (2.75) and well-established relations for the geometry of an ellipse, the semimajor and semiminor axes may be computed from

$$a^2 = \tfrac{1}{2}(\sigma_1^2 + \sigma_2^2) + \sqrt{\tfrac{1}{4}(\sigma_1^2 - \sigma_2^2)^2 + \sigma_{12}^2} \tag{2.76}$$

$$b^2 = \tfrac{1}{2}(\sigma_1^2 + \sigma_2^2) - \sqrt{\tfrac{1}{4}(\sigma_1^2 - \sigma_2^2)^2 + \sigma_{12}^2} \tag{2.77}$$

where ρ in equation (2.75) was replaced by its value $\sigma_{12}/\sigma_1\sigma_2$ in terms of the covariance σ_{12}. The values of a and b obtained from equations (2.76) and (2.77) can be readily shown to be the square roots of the eigenvalues of the covariance matrix,

$$\boldsymbol{\Sigma} = \begin{bmatrix} \sigma_1^2 & \sigma_{12} \\ \sigma_{12} & \sigma_2^2 \end{bmatrix}$$

The characteristic polynomial of $\boldsymbol{\Sigma}$ is evaluated as

$$(-\lambda)^2 + \mathrm{tr}(\boldsymbol{\Sigma})(-\lambda) + |\boldsymbol{\Sigma}| = 0$$

where

$$\mathrm{tr}(\boldsymbol{\Sigma}) = \text{trace of the covariance matrix} = \sigma_1^2 + \sigma_2^2$$

$$|\boldsymbol{\Sigma}| = \text{determinant of the covariance matrix} = \sigma_1^2\sigma_2^2 - \sigma_{12}^2$$

Thus the polynomial becomes

$$\lambda^2 - (\sigma_1^2 + \sigma_2^2)\lambda + (\sigma_1^2\sigma_2^2 - \sigma_{12}^2) = 0 \tag{2.78}$$

The roots of equation (2.78) are those given by equations (2.76) and (2.77), which verifies the relationship between the eigenvalues of $\boldsymbol{\Sigma}$ and the semimajor and semiminor axes of the standard ellipse.

The angle γ between the semimajor axis of the ellipse and the x axis is obtained from the following relationship:

$$\tan 2\gamma = \frac{2\sigma_{12}}{\sigma_1^2 - \sigma_2^2} \tag{2.79}$$

The proper quadrant for 2γ is determined such that $\sin 2\gamma$ has the same sign as σ_{12} and $\cos 2\gamma$ has the same sign as $(\sigma_1^2 - \sigma_2^2)$.

Referring to Figure 2.6, we denote by u and v a rotated coordinate system, the axes of which coincide with the semimajor and semiminor axes of the ellipse. The system u, v is obtained from the original system x, y by a simple rotation through the angle γ. With respect to this new system, u, v, we may refer two new random variables $\tilde{u}$, $\tilde{v}$ with the marginal standard deviations $\bar{\sigma}_1$, $\bar{\sigma}_2$. From the foregoing discussion and Figure 2.6 it becomes clear that $\bar{\sigma}_1 = a$, $\bar{\sigma}_2 = b$, and that the new random variables $\tilde{u}$, $\tilde{v}$ are independent (uncorrelated), that is, $\bar{\sigma}_{12} = 0$. Consequently, through that transformation from x, y to u, v, it was possible to replace a pair of correlated random variables by another pair that is not correlated. This possibility is in fact general; it is always possible to replace a set (of any dimension) of correlated random variables by another set (of the same dimension) of uncorrelated random variables. The transformation matrix can be constructed by determining the normalized eigenvectors of the covariance matrix of the original set of random variables and using them as columns. This is equivalent to diagonalizing the original covariance matrix since the resulting covariance matrix of the new uncorrelated set of random variables is always in diagonal form. The elements of this diagonal matrix are the eigenvalues of the original covariance matrix. This concept is explained further in Appendix A (Section A.6).

Considering further the case of no correlation ($\rho = 0$), the density function of equation (2.70) can be written in the form,

$$\begin{aligned} f(x, y) &= \frac{1}{2\pi\sigma_1\sigma_2} \exp\left\{-\frac{1}{2}\left[\frac{(x-\mu_1)^2}{\sigma_1^2} + \frac{(y-\mu_2)^2}{\sigma_2^2}\right]\right\} \\ &= \left\{\frac{1}{\sigma_1\sqrt{2\pi}} \exp\left[-\frac{(x-\mu_1)^2}{2\sigma_1^2}\right]\right\} \\ &\quad \times \left\{\frac{1}{\sigma_2\sqrt{2\pi}} \exp\left[-\frac{(y-\mu_2)^2}{2\sigma_2^2}\right]\right\} \\ &= f_1(x) \cdot f_2(y) \end{aligned}$$

where $f_1(x)$ and $f_2(y)$ are one-dimensional normal density functions with means μ_1 and μ_2 and variances σ_1^2 and σ_2^2, respectively. Thus for no correlation, or when $\rho = 0$, the joint density function $f(x, y)$ can be written as the product of the marginal density functions for x and y. Referring to equation (2.21) we can directly conclude that for a normal distribution, the case of no correlation is identical to the case of statistical independence. Perhaps it is this fact that sometimes leads to the use of the terms "uncorrelated" and "independent" synonymously. This is obviously true *only* when the random variables in question *are normally distributed.*

Having defined the standard ellipse [see equation (2.75)], it is now of interest to find out the probability that the random vector $(\tilde{x}, \tilde{y})$ takes values within the ellipse [applying in principle equation (2.12)]. The derivation of that probability is simplified if we work with $(\tilde{u}, \tilde{v})$, instead of $(\tilde{x}, \tilde{y})$, as they are uncorrelated and lie along the principal axes of the ellipse. If the normalized variables $\tilde{u}/\bar{\sigma}_1$ and $\tilde{v}/\bar{\sigma}_2$ are used, the sum of their squares has, according to equation (2.59), a χ^2 distribution with two degrees of freedom. The general expression for the probability of being within or on an ellipse with axes $k\bar{\sigma}_1$ and $k\bar{\sigma}_2$ is

$$P\left\{\left[\frac{u^2}{\bar{\sigma}_1^2} + \frac{v^2}{\bar{\sigma}_2^2}\right] < k^2\right\} = P\{\chi_2^2 < k^2\} = 1 - \alpha \tag{2.80}$$

For the standard ellipse $k = 1$, and the value of $(1 - \alpha)$ is 0.3935, so the probability of falling inside the standard ellipse is 0.3935. In order to establish confidence regions (see Section 11.5.2), we select the α level and compute the multiplier k. For example, for $\alpha = 0.05$

$$P\{\chi_2^2 < \chi_{0.05,\,2}\} = P\{\chi_2^2 < 5.99\} = 0.95$$

from which $k = \sqrt{5.99} = 2.447$. Consequently, the 95% ellipse has semiaxes of $a = 2.447\bar{\sigma}_1$ and $b = 2.447\bar{\sigma}_2$. Table 2.1 gives other typical values.

TABLE 2.1

P	0.394	0.500	0.900	0.950	0.990
k	1.000	1.177	2.146	2.447	3.035

The reader should note that for each one-dimensional marginal normal distribution, the probability that each variable ($\tilde{x}$ or $\tilde{y}$) lies in the region within plus and minus one standard deviation ($\pm\sigma_1$ or σ_2) is 0.683 [see equation (2.54)]. By contrast the probability for the joint event, which is falling within the standard ellipse, is considerably less, being only 0.394.

2.6.2. Radial Standard Deviation in Two Dimensions In many cases we are interested in one measure (for example, standard deviation) instead of the two, σ_x and σ_y, in the two-dimensional estimation. If $\tilde{x}$, $\tilde{y}$ are two *independent* random variables, a new variable $\tilde{r} = \sqrt{\tilde{x}^2 + \tilde{y}^2}$ can be defined and its distribution derived from the joint distribution of $\tilde{x}$ and $\tilde{y}$. The value r that $\tilde{r}$ takes can be determined given the selected values for $\tilde{x}$ and $\tilde{y}$. For example, considering the case of standard ellipse for which the probability is 0.394, the value of r obtained from $P[\tilde{r} < r] = 0.394$ is called "circular standard error" and is given the symbol σ_c. In another case $P[\tilde{r} < r] = 0.500$, r is called the "circular probable error" (or CPE). The values of both σ_c and the circular probable error depend on the relative magnitudes of σ_x and σ_y. For practical reasons we may use the following approximation for σ_c:

$$\sigma_c \cong \tfrac{1}{2}(\sigma_x + \sigma_y) \tag{2.81}$$

particularly for $0.5 \leq (\sigma_y/\sigma_x) \leq 1.0$ (with $\sigma_x \geq \sigma_y$). Similarly, an approximation for the circular probable error is

$$\text{CPE} \cong 0.589(\sigma_x + \sigma_y) \tag{2.82}$$

especially for $0.2 \leq (\sigma_y/\sigma_x) \leq 1.0$ (when $\sigma_x \geq \sigma_y$).

Another circular measure of precision is termed the "mean square positional error" (MSPE) evaluated from

$$\text{MSPE} = \sqrt{\sigma_x^2 + \sigma_y^2} \tag{2.83}$$

$P[\tilde{r} < \text{MSPE}]$ varies with the ratio (σ_y/σ_x). When $\sigma_x = \sigma_y$, then $\text{MSPE} = \sqrt{2}\,\sigma_x$ and $P[\tilde{r} \leq \text{MSPE}] = 0.632$. This precision index (the mean square positional error) is not recommended for use because of the variation in probability.

2.6.3. Ellipsoids of Constant Probability Referring to equation (2.69), the function $(\mathbf{x} - \boldsymbol{\mu}_x)^t\boldsymbol{\Sigma}^{-1}(\mathbf{x} - \boldsymbol{\mu}_x) = k^2$ (which is a positive definite quadratic form) represents a family of hyperellipsoids of constant probability. The case of three dimensions is important because it is often encountered when applying statistical concepts to the determination of point coordinates (x, y, z) through photogrammetric or geodetic techniques. In this case the ellipsoid equation becomes (assuming for simplicity $\boldsymbol{\mu}_x = \mathbf{o}$)

$$h(x, y, z) = [x \quad y \quad z]\boldsymbol{\Sigma}^{-1}\begin{bmatrix} x \\ y \\ z \end{bmatrix} = k^2 \tag{2.84}$$

and when $k = 1$, it is called the *standard ellipsoid*. The semiaxes of the ellipsoid (a, b, c) are determined by diagonalizing $\boldsymbol{\Sigma}$ by writing

$$\begin{bmatrix} \sigma_u^2 & & \\ & \sigma_v^2 & \\ & & \sigma_w^2 \end{bmatrix} = \begin{bmatrix} a^2 & & \\ & b^2 & \\ & & c^2 \end{bmatrix} = \begin{bmatrix} \lambda_1 & & \\ & \lambda_2 & \\ & & \lambda_3 \end{bmatrix} = \mathbf{T}^t\boldsymbol{\Sigma}\mathbf{T} \tag{2.85}$$

where $\mathbf{T}$ is an orthogonal matrix whose columns are the normalized eigenvectors of $\boldsymbol{\Sigma}$; λ_1, λ_2, and λ_3 are the eigenvalues of $\boldsymbol{\Sigma}$; and the u, v, w is a rotated coordinate system such that the random variables in the directions of its axes are uncorrelated. In a manner similar to the two-dimensional case, the statistical significance of the ellipsoids may be derived. Thus for the probability of the point falling inside or on the ellipsoid defined by $a = k\sigma_u$, $b = k\sigma_v$, $c = k\sigma_w$ the expression is

$$P\left[\left(\frac{u^2}{\sigma_u^2} + \frac{v^2}{\sigma_v^2} + \frac{w^2}{\sigma_w^2}\right) < k^2\right] = P[\chi_3^2 < k^2] = 1 - \alpha$$

For the standard ellipsoid $(1 - \alpha) = 0.199$, which is obtained from χ^2 with three degrees of freedom. Furthermore, Table 2.2 gives different values of the probability and k.

TABLE 2.2

P	0.199	0.500	0.900	0.950	0.990
k	1.000	1.538	2.500	2.700	3.368

The reader should note that for one dimension ($n = 1$) the probability for the standard region is 0.683, for two dimensions it is 0.394, and for three dimensions it is 0.199. Table 2.3 gives the values for several other dimensions.

TABLE 2.3

n	1	2	3	4	5	6
P	0.683	0.394	0.199	0.090	0.037	0.014

When seeking a single measure for three-dimensional cases, the spherical standard error σ_s is defined as $P[\tilde{s} < \sigma_s] = 0.199$ and is approximated by

$$\sigma_s \cong \tfrac{1}{3}(\sigma_x + \sigma_y + \sigma_z) \tag{2.86}$$

for $0.35 \leq (\sigma_z/\sigma_x) \leq 1.0 (\sigma_x \geq \sigma_y \geq \sigma_z)$. Similarly, spherical probable error (SPE) is given by $P[\tilde{s} < \text{SPE}] = 0.500$ and is approximated by

$$\text{SPE} \cong 0.513(\sigma_x + \sigma_y + \sigma_z) \tag{2.87}$$

for $0.35 \leq (\sigma_z/\sigma_x) \leq 1.0$.

Finally, the mean radial spherical error (MRSE) is

$$\text{MRSE} = \sqrt{\sigma_x^2 + \sigma_y^2 + \sigma_z^2} \tag{2.88}$$

and if $\sigma_x = \sigma_y = \sigma_z$, then

$$\text{MRSE} = \sqrt{3}\,\sigma_x \qquad \text{and} \qquad P[\tilde{s} \leq \text{MRSE}] = 0.608$$

Since the probability varies with the different values of σ_x, σ_y, and σ_z, the mean radial spherical error is not suitable for use in practice.

Summarizing the discussion of the properties of two- and more-dimensional normal distributions, the basic concept of correlation should be emphasized. Correlation describes a certain dependence between x and y values of the random variables $\tilde{x}$ and $\tilde{y}$. The dependence is not strict (functional), however. It holds strictly (functionally) only for the expectations of both random variables. Their mutual dependence is given by regression lines. There are always two of them, since the expectation of y as a function of x is different from the expectation of x as a function of y.

Thus, in general terms, correlation for actual values of x and y expresses only a trend relationship, which is not strictly valid for individual pairs of values. The higher the correlation, however, the closer would be the tendency to have the regression relationship realized.

2.7. SAMPLING, ESTIMATION, AND CONFIDENCE MEASURES

In statistics a physical situation or process is assessed conceptually by a (hypothetical) functional model. The elements of that model are considered as random variables, associated with joint probability distributions. Statistical events (for instance, measurements) are *considered* realizations of the random variables. The realizations, whether repeated or not, are always limited in extent. They are called *samples* from the probability distributions.

From the sample values, the probability distributions of the random variables, or at least some of their parameters, are to be assessed or estimated. The method of least squares, for instance, is a method for estimating expectations of random variables from samples (measurements).

Related to estimation is the task of determining the reliability or accuracy with which the estimates are obtained (confidence measures). Furthermore, statistical testing treats the question of whether the results of estimations are in agreement with the initial assumptions (hypotheses).

Those concepts are increasingly linked to the application of the least squares method. Therefore introductory essentials to these topics are presented next.

2.7.1. Sampling Any set of measurements is considered the realization of the sample $(x_1, x_2, \ldots, x_n)$ of the random vector $\tilde{\mathbf{x}}$. The purpose of sampling is (in this context) to derive estimates for the probability distributions or some of their parameters. These distributions may be of the random variables concerned, or of some other random variables derived from the ones to which the measurements refer directly.

When repeated observations are taken, some factors that are commonly used to assess the sample data are as follows:

1. Frequency distributions.
2. Sample statistics for location (mean, median, mode, or midrange).
3. Sample statistics for dispersion (variances and covariances).
4. Other sample moments.

FREQUENCY DISTRIBUTIONS Establishing frequency distributions makes sense only if the size of the sample is relatively large, in particular for multidimensional frequency distributions. They are obtained by classifying data according to some appropriate subdivision and counting the events of each class, thus yielding class frequencies. In practice, relative frequencies are generally used. The table of class frequencies gives the frequency distribution, which can be graphically displayed by either a bar diagram or more frequently by a histogram.

BAR DIAGRAM AND HISTOGRAM As an example of a one-dimensional case, we shall consider measuring a distance (which is approximately 50 meters), say, 100 times, and recording the values to the nearest 4 mm. We shall regard any measurement whose last two digits fall between, say, 62 and 66 as 64 mm, that between 66 and 70 as 68 mm, and so on. (The actual values of the length would be 49,964 mm and 49,968 mm, but the digits 499 are the same for all readings and will be dropped). We may represent this sample of measurements by a bar diagram using the values 64, 68, 72, and so on, as the points along the horizontal axis. The number of measurements belonging to each of these values and the ratio between such numbers and the total number of measurements (100) which is the relative frequency are given in Table 2.4. The graph itself is shown in Figure 2.7, where $f(x)$ represents the values of the relative frequency of occurrence of a certain measurement value. In this presentation the approximation was made to take the middle value of an interval as representing a certain value of the observation. If we eliminate this approximation, we can replace each of the vertical lines by a rectangle that covers the appropriate interval and whose height is proportional to the relative frequency. In this case the probability of a certain *interval*, which is estimated as the ratio of measurements within the interval and the total number of measurements, is equal to the *area* of the rectangle constructed over it. In this manner the sum of the areas of all the rectangles will equal to 1, thus satisfying the requirement of having all probabilities adding up to 1. Such a frequency distribution is called a *histogram*. The

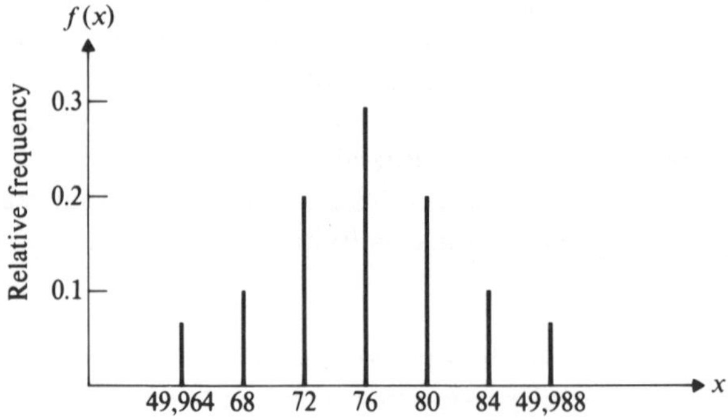

Figure 2.7

TABLE 2.4

Nominal Values	64	68	72	76	80	84	88	
Interval Range	62–66	66–70	70–74	74–78	78–82	82–86	86–90	
Number in Each Interval	6	10	20	28	20	10	6	$\Sigma = 100$
Relative Frequency	0.06	0.10	0.20	0.28	0.20	0.10	0.06	$\Sigma = 1.0$
Height of Rectangles in Histogram	0.015	0.025	0.05	0.07	0.05	0.025	0.015	

heights of the rectangles are given in the fifth line of Table 2.4 (which are one fourth of the values in the fourth row because the interval width is 4 mm) and the histogram itself is depicted in Figure 2.8.

BIDIRECTIONAL HISTOGRAM (STEREOGRAM) As an example for a two-dimensional case, we shall consider an example of a target chart grouping. Every point on the target chart represents one shot and its position may be determined by a pair of cartesian coordinates x and y with origin at the center of the target board. The behavior of the position of the points follows the laws of probability, in this case, *a bivariate* distribution with $\tilde{x}$ and $\tilde{y}$ as the random variables or variates. We shall assume that one person has taken 400 shots at the practice board. A grid made up of 3-cm squares is then superimposed on the target sheet and the number of points in each square are counted and entered in the upper half of the square as shown in Table 2.5. This table represents a two-dimensional frequency distribution of the rifle shots.

Table 2.5 can also serve as the base for constructing a stereogram (a bidirectional histogram) by erecting square columns over each square. The volume of each column is equal to the relative frequency of the number of

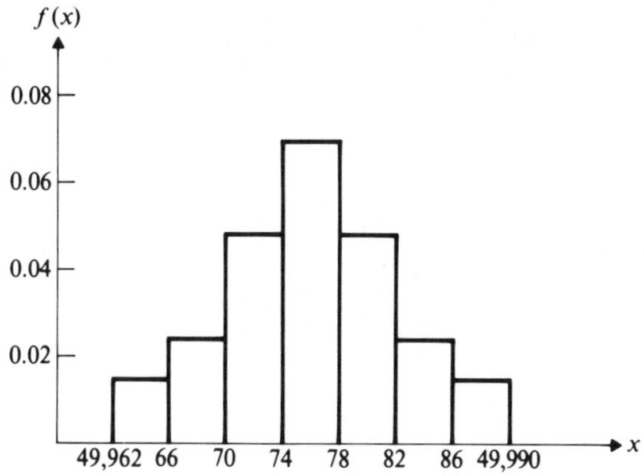

Figure 2.8

TABLE 2.5

The y-direction (cm)	The x direction (cm) −18	−15	−12	−9	−6	−3	0	3	6	9	12	15	18	Marginal y frequency
18							1 0.01			1 0.02				2
15							2 0.03	1 0.02	1 0.02	1 0.02				5
12				1 0.07		1 0.02	3 0.04	2 0.03	2 0.04	1 0.02	1 0.03		1 0.12	12
9		1 0.17	1 0.08		1 0.04	2 0.05	4 0.05	3 0.05	5 0.09	2 0.04	1 0.03	1 0.06		21
6			1 0.08	2 0.12	3 0.11	4 0.10	7 0.09	7 0.11	8 0.15	5 0.10	3 0.10	1 0.06	1 0.13	42
3	1 1.0	1 0.17	2 0.15	3 0.20	7 0.25	7 0.17	11 0.14	9 0.15	11 0.20	7 0.15	4 0.14	3 0.17	1 0.12	67
0		2 0.32	4 0.31	3 0.20	7 0.25	13 0.32	21 0.28	19 0.31	15 0.27	13 0.27	9 0.30	6 0.35	2 0.25	114
−3		1 0.17	3 0.23	3 0.20	6 0.20	5 0.12	10 0.13	7 0.11	5 0.09	6 0.13	4 0.14	2 0.12	2 0.25	54
−6		1 0.17	2 0.15	1 0.07	3 0.11	4 0.10	7 0.09	6 0.10	3 0.06	5 0.10	3 0.10	2 0.12	1 0.13	38
−9				1 0.07	1 0.04	2 0.05	5 0.07	4 0.06	2 0.04	4 0.08	3 0.10	1 0.06		23
−12				1 0.07		2 0.05	4 0.05	2 0.03	1 0.02	2 0.04	1 0.03	1 0.06		14
−15						1 0.02	1 0.01	2 0.03	1 0.02	1 0.02				6
−18							1 0.01				1 0.03			2
Marginal x frequency	1	6	13	15	28	41	77	62	54	48	30	17	8	Σ 400
Marginal x relative frequency	.002	.015	.033	.038	.070	.103	.192	.155	.135	.120	.075	.042	.020	Σ 1.000
Σ of $P(y\|x)$	1.00	1.00	1.00	1.00	1.00	1.00	1.00	1.00	1.00	0.99	1.00	1.00	1.00	

shots on which the column is erected. For example, the volume of the column over the square centered about (6, 6) is equal to $8/400 = 0.02$, where 8 is the number of shots in the square and 400 is the total number of shots. Similarly, the relative frequency and hence the volume of the column over $(-3, 6)$ is $4/400 = 0.01$. The relative frequency represents the probability of occurrence and therefore $P(6, 6) = 0.02$ and $P(-3, 6) = 0.01$.

In the foregoing discussion consideration was given to the probability of shots falling in a particular square on the target board, that is, the probability involving both random variables $\tilde{x}$ and $\tilde{y}$. Suppose, however, that we are interested in the aiming ability of the individual in the x direction only, without regard to the y direction. In other words, we wish to find out how the shots are distributed about the center of the target to the right and to the left of it. This is then a case of one random variable, $\tilde{x}$, and the values in the third row from the bottom in Table 2.5 show the frequency of shots relative to the x intervals of 3 cm each. These values are simply obtained by summing up the values in each column. Similarly, the distributions in the y direction only are obtained by adding up the numbers along each row and are given in the last column of Table 2.5. These two sets of numbers are called *marginal frequencies* as each represents the frequencies along one direction or margin. If the marginal frequencies are divided by the total number of shots, we would get the marginal relative frequencies. This is shown in the second to last row of Table 2.5 for the x direction. From these relative frequencies we would then also talk about *marginal probability* as being the probability that *only one random* variable takes on a certain value, or falls within a certain interval, without regard to other random variables. Thus the probability that the x coordinate of the shots falls within the interval $+7.5$ to $+10.5$ cm is 0.120, whereas it is 0.192 for the interval -1.5 to $+1.5$. The same can be computed for the marginal probabilities of the y coordinates. These ideas demonstrate the theoretical concepts given in Section 2.3.1.

Another univariate probability distribution is also possible for one variable as related to a particular interval along the other variable. As an example, the distribution of y given a value $x = -6$ is represented by the numbers (relative frequencies) 0.04, 0.11, 0.25, 0.25, 0.20, 0.11, and 0.04 in the column under $x = -6$. Similarly, each horizontal row in Table 2.5 gives the distribution of x for a given interval of y. Such distributions are called *conditional frequencies* or *conditional distributions.* If we are interested in the conditional distribution of x given intervals of y, we should study the relative frequencies along different rows. It can be seen that the relative frequencies differ from one row to another. This means that as one variable changes (in this case y), the conditional distribution of the other (x) also changes. Such an influence between the two random variables is an indication of *correlation.* As was said previously, statistical correlation or dependence should not be confused with functional (algebraic) dependence. If the latter existed between two variables, the value of one would completely determine the other. The example under consideration shows clearly that x and y are not functionally related; given x, the variate y may take several values, and vice versa.

Correlation between random variables may be strong or weak. Strong correlation would be indicated by the fact that relative frequencies of the conditional distribution of one random variable differ appreciably as the value of the other variable changes. On the other hand weak correlation would be reflected by having conditional relative frequencies that vary little when the other variable takes on different values. In the limit when correlation is zero, relative frequencies of the conditional distribution would be the same regardless of the value of the second variable. Such a case would be reflected in Table 2.5 by having equal relative frequencies from row to row and from column to column. This discussion of correlation and conditional distribution using numerical values from Table 2.5 amplifies on the theoretical treatment of Sections 2.3.2 and 2.3.3.

From frequency distributions, conclusions may be drawn as to the associated probability distribution. In applications to engineering, the assessment of probability distributions through frequency distributions is not often done simply because large sample size is not practical. In most cases the type of probability distribution associated with the random variable at hand is assumed to be known a priori. It is then some parameters of the distribution that are estimated from the sample. For such parameter estimation the sample size need not be as large as for estimation of the distribution itself.

The most common sample statistics used to estimate distribution parameters are discussed next.

Sample Statistics for Position Measures

Sample Mean: From a sample of a size n $(x_1, x_2, \ldots, x_n)$ of the stochastic variable $\tilde{x}$, the sample mean is calculated as

$$\bar{x} = \frac{1}{n} \sum_{i=1}^{n} x_i \tag{2.89}$$

which is the same as the arithmetic mean. Since $\tilde{x}$ is a random variable and its expected value is the population mean, that is, $E(\tilde{x}) = \mu$, the expectation of the mean is

$$E(\bar{x}) = E\left[\frac{1}{n} \sum_{i} x_i\right] = \frac{1}{n} E(x_1 + x_2 + \cdots + x_n)$$

$$= \frac{1}{n}[E(x_1) + E(x_2) + \cdots + E(x_n)] = \frac{1}{n} \cdot n\mu.$$

or $$E(\bar{x}) = \mu. \tag{2.90}$$

Thus the arithmetic mean of a set of independent observations is an estimate of the mean of the population describing the random variable for which observations are made.

The Median: A sample median is obtained by arranging the values in the sample in their order of magnitude. The median is the value in the middle if the number of observations n is odd, and it is the mean of the middle two values if n is even. There will therefore be an equal number of observations of larger magnitude as well as of smaller magnitude than the median.

It should be noted that for a continuous distribution the mean and median will be equal if the distribution is symmetric. This will also be nearly so for a sample of large size and approximately symmetrical values.

The Mode: The sample mode is the value that occurs most often in the sample. Therefore if a histogram is constructed, it is the value at which the highest rectangle is located.

Midrange: If the observation of smallest magnitude is subtracted from that of the largest magnitude, we obtain a value that is called the *range*. The value of the observation that is midway along the range is called the *midrange* and may be used, though infrequently, as a measure of position for a given sample. It is simply the arithmetic mean of the largest and smallest observations.

Sample Statistics for Dispersion Measures

The Range: The simplest dispersion measure is the range as defined in the preceding section. It is not, however, as indicative a measure as are others.

The Mean (or Average) Deviation The mean (or average) deviation is another measure of dispersion that is suited for coarse estimation and has been conventionally called the *average error*. It is the arithmetic mean of the absolute values of the deviations from any measure of position (mostly the mean). Thus the mean deviation from the mean for a sample of n observations would be given by

$$\text{mean deviation} = \frac{1}{n} \sum_{i=1}^{n} |(x_i - \bar{x})| \tag{2.91}$$

where $\bar{x}$ is the arithmetic mean [equation (2.89)].

Variance and Standard Deviation The mean deviation, though useful in certain cases, does not reflect the dispersion or scatter of the measured values as does the standard deviation that was defined previously as the square root of the variance. The variance of a sample is defined as (see also Section 2.5.2)

$$s_x^2 = \frac{1}{n-1} \sum_{i=1}^{n} (x_i - \bar{x})^2 \tag{2.92}$$

where $\bar{x}$ is the sample mean and n is the sample size. The reason for using $(n - 1)$ instead of n in equation (2.92) can be justified if it is shown that the expectation of s_x^2 is equal to the population variance. First

$$\begin{aligned} \sum_{i=1}^{n} (\tilde{x}_i - \bar{x})^2 &= \sum [(\tilde{x}_i - \mu) - (\bar{x} - \mu)]^2 \\ &= \sum (\tilde{x}_i - \mu)^2 + n(\bar{x} - \mu)^2 - 2(\bar{x} - \mu) \sum (\tilde{x}_i - \mu) \\ &= \sum (\tilde{x}_i - \mu)^2 + n(\bar{x} - \mu)^2 - 2(\bar{x} - \mu)(n\bar{x} - n\mu) \end{aligned}$$

or

$$\sum (\tilde{x}_i - \bar{x})^2 = \sum (\tilde{x}_i - \mu)^2 - n(\bar{x} - \mu)^2 \tag{2.93}$$

The expectation of the last term in equation (2.93) is

$$\begin{aligned} E[n(\bar{x} - \mu)^2] &= nE\left[\frac{\sum \tilde{x}_i}{n} - \mu\right]^2 = nE\left[\frac{(\sum \tilde{x}_i - n\mu)^2}{n^2}\right] \\ &= \frac{1}{n} E[\sum (\tilde{x}_i - \mu)^2] = \frac{1}{n} \sum E(\tilde{x}_i - \mu)^2 \end{aligned}$$

or

$$E[n(\bar{x} - \mu)^2] = \frac{n\sigma^2}{n} = \sigma^2 \tag{2.94}$$

Now, taking the expectation of equation (2.92),

$$\begin{aligned} (n-1)E(s_x^2) &= \sum E(\tilde{x}_i - \mu)^2 - \sigma^2 = n\sigma^2 - \sigma^2 \\ E(s_x^2) &= \sigma^2 \end{aligned} \tag{2.95}$$

Equation (2.95) is a very important result as regards to estimating parameters through calculation of statistics. It expresses the fact that the statistic s_x^2, as computed by equation (2.92), has a probability distribution whose expected value or mean is σ^2. If we used n instead of $(n - 1)$ in equation (2.92), we would find that the expected value is $[(n - 1)/n]\sigma^2$ instead of σ^2. In such a case the sample variance computed using n would be a *biased* estimate because its expectation is different from the population parameter, as will be discussed in Section 2.7.2.

SAMPLE COVARIANCE Given a set of n pairs of values $(x_1, y_1), (x_2, y_2), \ldots, (x_n, y_n)$ of the random vector $(\tilde{x}, \tilde{y})$, we can compute in addition to the marginal variances $s_{\tilde{x}}^2$ and $s_{\tilde{y}}^2$, the sample covariance as

$$s_{xy} = \frac{1}{n-1} \sum_{i=1}^{n} (x_i - \bar{x})(y_i - \bar{y}) \tag{2.96}$$

where $\bar{x}$ and $\bar{y}$ are the sample means of $\tilde{x}$ and $\tilde{y}$.

2.7.2. Estimation Drawing inferences regarding parameters of probability distributions on the basis of sample statistics is referred to as *estimation.* The sample statistic used for estimating the corresponding parameter is called an *estimator*, whereas the computed value is referred to as an *estimate* or, more precisely, a *point estimate.* These terms refer equally to single as well as multiple dimension cases.

As examples of estimators, the sample mean $\bar{x}$, (2.89), the sample variance $s_{\tilde{x}}^2$, (2.92), and the sample covariance s_{xy}, (2.96), are estimators for the distribution parameters μ, $\sigma_{\tilde{x}}^2$, and σ_{xy}, respectively. From the statistical point of view all values derived from observations are estimates for the distribution parameters in question. Because estimates can be obtained in different ways, certain criteria are needed to judge the quality of estimation. There are four different criteria.

1. *Consistence:* An estimator is called "consistent" if for $n \to \infty$ the probability for the estimator $\hat{p}$ to approach the parameter p converges toward 1. Thus for any small $\varepsilon > 0$,

$$\lim_{n \to \infty} P(|\hat{p} - p)| < \varepsilon) = 1 \tag{2.97}$$

The estimator is said to converge in probability to the parameter. As an example, the sample mean $\bar{x}$ given by equation (2.89) is a consistent estimator of the distribution mean μ_x of the random variable $\tilde{x}$.

2. *Unbiased Estimation:* Although consistence is a property related to the limit case of $n \to \infty$ it does not say anything about the property for small sample size. Among several possibilities, an estimator should not be biased. That means the expectation of the sample statistic $\hat{p}$ should be identical to the parameter p itself for any sample size n,

$$E(\hat{p}) = p \tag{2.98}$$

If this property holds only for $n \to \infty$, the estimator is said to be asymptotically unbiased.

As an example, the sample variance $s_{\tilde{x}}^2$ of equation (2.92) is an unbiased estimator of the population variance $\sigma_{\tilde{x}}^2$. As was shown in the preceding section if instead of using $(n-1)$ we used n in equation (2.92), the computed sample variance would be biased (too small). It is, however, asymptotically unbiased since

$$\lim_{n\to\infty} \frac{1}{n} \to \lim_{n\to\infty} \frac{1}{n-1}$$

Figure 2.9 shows the distribution of three different estimators of p. The first two, $\hat{p}_1$ and $\hat{p}_2$, are unbiased but the third is biased. The bias is defined as the difference between the value of the parameter and the expectation of the estimator,

$$\text{bias} = E(\hat{p}) - p \tag{2.99}$$

The bias is equivalent to the term "systematic effect" or "systematic error" as conventionally used. Just as increasing the number of observations will not alleviate systematic effects, neither will it eliminate the bias.

3. *Minimum Variance:* Although unbiasedness involves the relationship between the expectation of the estimator and the parameter, the variances of the estimates from an estimator offer another criterion for estimator selection. In Figure 2.9, we can see that the variance of $\hat{p}_1$ is smaller than that for $\hat{p}_2$, but the variance of $\hat{p}_3$ is smallest of the three. Hence, if we were to choose minimum variance as a criterion, $\hat{p}_3$ would be selected. However, realizing that $\hat{p}_3$ is biased, we must then weigh the benefits of unbiasedness against that of minimum variance.

Accuracy and Precision Figure 2.9 can be used to illustrate the related concepts of accuracy and precision. Precision may be defined as the degree of conformity among a set of observations of the same random variable. The spread (or dispersion) of the probability distribution is an indication of precision. Therefore in Figure 2.9 $\hat{p}_2$ is least precise and $\hat{p}_3$ is most precise.

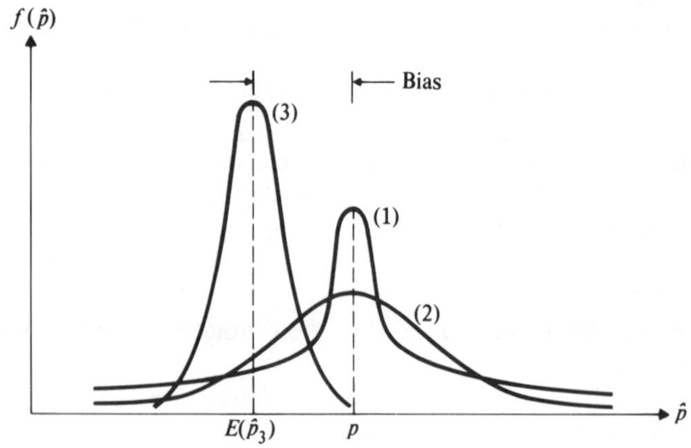

Figure 2.9. Biased Estimator

Accuracy, on the other hand, may be defined as the extent to which an estimate approaches its parameter. (In conventional terms, it was considered as the degree of closeness to the "true" value.)

In Figure 2.9, both $\hat{p}_1$ and $\hat{p}_2$ are equally accurate but neither is as precise as $\hat{p}_3$. By contrast, $\hat{p}_3$ is least accurate although it is the most precise.

The difference between precision and accuracy lies in the possible presence of bias or "systematic error." Although precision includes only random effects, accuracy encompasses both random and systematic effects. A measure of accuracy (proposed by Gauss) is the "mean square error" (MSE) given by

$$\text{MSE} = m^2 = E[(\hat{p} - E(\hat{p}))^2] \tag{2.100}$$

which can be shown to reduce to

$$\text{MSE} = m^2 = \sigma_{\hat{p}}^2 + (\text{bias})^2 \tag{2.101}$$

Although mean square error may be used for comparing relative accuracies of estimates that differ substantially in bias and precision, as do those in Figure 2.9, it is really not the most appropriate measure. Instead, we should assess accuracy on the basis of both the measure of precision and the bias.

4. *Efficiency and Sufficiency:* For unbiased estimators, that with minimum variance is called "efficient estimator." Thus if var $\hat{p}_1 <$ var $\hat{p}_2$ and $\hat{p}_1$ and $\hat{p}_2$ are estimators of the same parameter p, then $\hat{p}_1$ is efficient. The efficiency of $\hat{p}_2$ is equal to the ratio var $\hat{p}_1$/var $\hat{p}_2$.

To demonstrate the above concepts, let us consider the random variable $\tilde{x}$, whose expectation is

$$E(\tilde{x}) = \mu = \text{the mean of the distribution}$$

Obviously any observation x in a set of, say, n independent observations would be an *unbiased* estimator of μ. The arithmetic mean $\bar{x}$ is also an estimator of μ which is also *unbiased* as proved in equation (2.90). Thus on the basis of unbiasedness, both estimators are equally acceptable. However, the variance of x is obviously that of the population or σ^2. The variance of the mean $\sigma_{\bar{x}}^2$, on the other hand, may be evaluated from the basic definition as

$$\sigma_{\bar{x}}^2 = E[(\bar{x} - E(\bar{x}))^2] = E(\bar{x} - \mu)^2$$

which can be readily deduced from equation (2.94) as

$$\sigma_{\bar{x}}^2 = \frac{\sigma^2}{n} \tag{2.102}$$

Since $\sigma_{\bar{x}}^2 < \sigma^2$, the mean $\bar{x}$ satisfies the criterion of minimum variance better than an individual observation x. As a matter of fact $\bar{x}$ is also an efficient estimator since the efficiency of x as an estimator relative to $\bar{x}$ is $\sigma_{\bar{x}}^2/\sigma^2 = 1/n$, which can be a very small number. It can also be shown that $\bar{x}$ is a consistent estimator.

An estimator is called sufficient if it uses all the information about the parameter that is contained in the sample. The formulation of this property is too involved and will not therefore be considered here.

METHODS OF ESTIMATION There are many ways of estimating a parameter. Estimators that have the four properties mentioned above are called "best" estimators. Some of these properties may be valid only asymptotically, however. In addition, these estimators may be called best asymptotically normally distributed with $n \to \infty$.

In practical applications of statistical estimation from samples it is not necessary to evaluate all criteria to determine which one to apply in a particular estimation. Instead, methods for estimation are applied for which the criteria, or some of them, hold in general. A few methods are briefly mentioned here.

The *moment method* takes the kth sample moment,

$$m_k = \frac{1}{n} \sum x_i^k \tag{2.103}$$

as an estimate for the kth moment of the probability distribution.

The sample mean is one example. However, when equation (2.103) is applied to the second central moment, it yields only asymptotically unbiased estimators for the variance and covariance.

The *maximum likelihood method* is an estimation method that is widely used in statistics. It is used for estimating parameters such that the sample values take the highest probability. If the random variables x_i are independent and have the same distribution (density function $f(x_i)$), then the sample vector $(x_1, x_2, \ldots, x_n)$ has the joint density function

$$L(x_1, \ldots, x_n; p_1, \ldots, p_m) = \prod^{n} f(x_i; p_1, \ldots, p_m) \tag{2.104}$$

according to equation (2.22). Here, $p_1, \ldots, p_m$ are the parameters to be estimated. Equation (2.104) is called the likelihood function. The density $f(x_i)$ are functions of the unknown parameters $p_1, \ldots, p_m$, which in turn are related to the sample values $p_i = g_i(x_i, \ldots, x_n)$. It seems plausible to determine estimators $\hat{p}_i$ of p_i such that the joint probability density L [equation (2.104)] becomes a maximum. This leads to the system

$$\frac{\partial L}{\partial p} = 0 \qquad \text{or} \qquad \frac{\partial \ln L}{\partial p} = 0 \tag{2.105}$$

The solution from equation (2.105) yields estimators of the parameters p_i and these are called maximum likelihood estimators. Note that this method of estimation requires the knowledge of the distribution function of the random variable in question.

As an example, suppose that we would like to estimate the mean of a normal distribution using the data from the sample $x_1, x_2, \ldots, x_n$. Since the density function for a normal distribution is

$$f(x) = \frac{1}{\sigma\sqrt{2\pi}} e^{-1/2\left(\frac{x-\mu}{\sigma}\right)^2}$$

the likelihood function is given by

$$L = \left[\frac{1}{\sigma\sqrt{2\pi}} e^{-1/2\,\frac{(x_1-\mu)^2}{\sigma^2}}\right] \cdot \left[\frac{1}{\sigma\sqrt{2\pi}} e^{-1/2\,\frac{(x_2-\mu)^2}{\sigma^2}}\right] \cdots \left[\frac{1}{\sigma\sqrt{2\pi}} e^{-1/2\,\frac{(x_n-\mu)^2}{\sigma^2}}\right]$$

or

$$L = \frac{1}{(2\pi\sigma^2)^{n/2}} e^{-1/2 \sum_i \frac{(x_i-\mu)^2}{\sigma^2}}$$

To get the first derivative of L it would be easier if we work with the natural logarithm. Thus

$$\ln L = -\frac{n}{2} \ln (2\pi\sigma^2) - \frac{1}{2\sigma^2} \sum (x_i - \mu)^2$$

which when differentiated with respect to μ, gives

$$\frac{d(\ln L)}{d\mu} = \frac{-2}{2\sigma^2} \sum (x_i - \mu)(-1) = 0$$

for maximum

$$\sum_i (x_i - \mu) = \sum_i x_i - n\mu = 0$$

or

$$\mu = \frac{1}{n} \sum_i x_i$$

This shows that the maximum likelihood estimator of the mean of the normal distribution is the sample mean $\bar{x}$ as previously defined.

It is interesting to note that it was possible to estimate μ without encountering the second normal distribution parameter σ^2 (variance). Perhaps it is worthwhile to seek an estimator for σ^2 using the maximum likelihood function. Differentiating $\ln L$ with respect to σ^2 yields

$$\frac{\partial(\ln L)}{\partial(\sigma^2)} = -\frac{n\pi}{2\pi\sigma^2} + \frac{1}{2\sigma^4} \sum_i (x_i - \mu)^2 = 0$$

for maximum

$$\frac{1}{\sigma^2}\sum (x_i - \mu)^2 = n$$

or

$$\sigma^2 = \frac{1}{n}\sum_i (x_i - \mu)^2$$

which is not the same as s_x^2. In fact σ^2 is a biased estimate, which happens to be true about many maximum likelihood estimates. It is, however, consistent and asymptotically unbiased.

The *least squares* method is a very widely used procedure for estimating parameters of distributions from samples. Since it is the major topic of this book, its different techniques will be extensively treated in Parts II and III.

Least squares has been used in different branches of science and engineering for over a century and a half. Legendre published his own version of least squares in 1805 only to discover that Gauss had been using the same method since 1795. In addition to these two famed men, Laplace also holds a place of prominence with regard to the history of least squares and the theory of errors. Although both Gauss and Legendre were mainly interested in such sciences as astronomy and geodesy and the treatment of observations therein, Laplace had a purely theoretical interest. The latter's contributions were therefore important events in the history of calculus of probability and the theory of random errors.

Unlike the maximum likelihood method of estimation, the method of least squares does not require the knowledge of the distribution from which the observations are drawn for the purpose of parameter estimation. However, the construction of confidence intervals or the testing of hypotheses regarding the estimated parameters (which are given in the following sections) will require the knowledge of the distribution.

If the random variables to which the observations refer are normally distributed, the least squares method will give identical results to those from the maximum likelihood method. For instance, if $\mathbf{v}$ is the vector of observational residuals (which is equal to a posteriori observational estimate $\hat{\boldsymbol{l}}$, minus the a priori given estimate $\boldsymbol{l}$), for which $E(\mathbf{v}) = \mathbf{0}$, and which is assumed to be normally distributed, and $\boldsymbol{\Sigma}$ is the covariance matrix of the observations, then

$$\begin{aligned} f(\mathbf{v}) &= C \exp\left[-\tfrac{1}{2}(\mathbf{v} - E(\mathbf{v}))^t\boldsymbol{\Sigma}^{-1}(\mathbf{v} - E(\mathbf{v}))\right] \\ &= C \exp\left[-\tfrac{1}{2}\mathbf{v}^t\boldsymbol{\Sigma}^{-1}\mathbf{v}\right] \end{aligned}$$

The least squares criterion is $\mathbf{v}^t\boldsymbol{\Sigma}^{-1}\mathbf{v} \to$ minimum, as will be given in detail in Part II. It is clear, then, that minimizing $\mathbf{v}^t\boldsymbol{\Sigma}^{-1}\mathbf{v}$ would maximize $f(\mathbf{v})$, which is equivalent to yielding a maximum likelihood estimation.

With linear functions the estimated parameters (in particular the estimated expectations) are consistent, unbiased, efficient, sufficient, and have the minimum variance property, especially when there are no systematic effects in the observations.

2.7.3. Confidence Intervals The estimation of means, variances, and covariances of random variables from sample data is referred to as *point estimation*, because it results in one value for each parameter in question. By contrast to point estimation, establishing confidence intervals from sampling is referred to as *interval estimation*.

After having performed a point estimation—for instance, having estimated the coordinates of one intersected point—the question remains as to how much the deviation of the estimate is likely to be from the still unknown parameter. In other words, we would like to have an indication of how good the estimation is and how much it can be relied on. Here an absolute answer is not possible because sampling never leads to the true, theoretical distribution or its parameters. It is only possible to estimate probabilities with which the true value of the parameter in question is likely to be within a certain interval around the estimate. Such probabilities can be determined if the cumulative distribution function $F(x)$ of the random variable is given.

The probability that a random variable $\tilde{x}$ takes values within the boundaries x_1 and x_2 is given, according to equation (2.5), by

$$P(x_1 < \tilde{x} < x_2) = F(x_2) - F(x_1) = \int_{x_1}^{x_2} f(x)\, dx$$

By analogy to this, the probability statement for a *confidence interval* of a parameter p, the estimate of which is $\hat{p}$, is

$$P(p_1 < \tilde{p} < p_2) = 1 - \alpha \tag{2.106}$$

Here $(1 - \alpha)$ is called the *confidence level* or degree of confidence which is conventionally taken to be 90%, 95%, or 99%. The values p_1 and p_2 are the lower and upper confidence limits for the parameter p. Equation (2.106) defines the confidence interval for the parameter p as the interval around the estimate $\hat{p}$, such that the probability that this interval includes the (unknown) value of the parameter is $(1 - \alpha)$. The probability that the parameter does not fall in a given interval is the value α. The width of a confidence interval decreases as the degrees of freedom increase and as the level of probability associated with it decreases.

In constructing confidence intervals, it is essential to use suitable random variables whose values are determined by the sample data as well as by the parameters, but whose distributions do not involve the parameters in question. Examples are given in the following paragraphs.

CONFIDENCE INTERVALS FOR MEANS We shall consider a normally distributed random variable $\tilde{x}$, with unknown expectation μ and known variance σ^2. We will let the sample mean be $\bar{x}$, computed from a sample of size n. Then the estimator function $(\bar{x} - \mu)/(\sigma/\sqrt{n})$ will be normally distributed with mean 0 and variance 1. We choose from the normal distribution function a value $z_{\alpha/2}$ such that the probability of the standard normal random variable to be outside the interval $(\pm z_{\alpha/2})$ is α (see Figure 2.10). The probability statement for the confidence interval, which is symmetrical here, is then

$$P\left\{-z_{\alpha/2} < \frac{\bar{x} - \mu}{\sigma/\sqrt{n}} < z_{\alpha/2}\right\} = 1 - \alpha \tag{2.107a}$$

or

$$P\left\{\bar{x} - z_{\alpha/2} \cdot \frac{\sigma}{\sqrt{n}} < \mu < \bar{x} + z_{\alpha/2} \cdot \frac{\sigma}{\sqrt{n}}\right\} = 1 - \alpha \tag{2.107b}$$

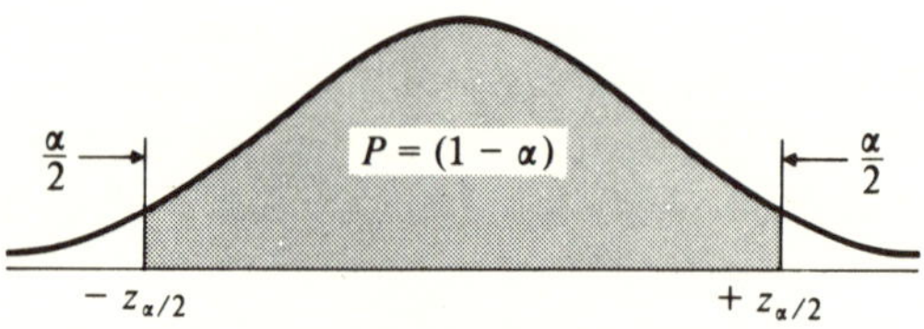

Figure 2.10

Since for $\alpha = 0.05$, $z_{\alpha/2} = 1.96$, we may write

$$P[\bar{x} - 1.96\sigma/\sqrt{n} < \mu < \bar{x} + 1.96\sigma/\sqrt{n}] = 0.95$$

This was an example of a two-sided confidence interval. On the other hand, for a one-sided confidence interval we write

$$P\left\{\mu < \bar{x} + z_\alpha\left(\frac{\sigma}{\sqrt{n}}\right)\right\} = 1 - \alpha \tag{2.108}$$

Example 2.3. Suppose a distance is measured $n = 8$ times with the mean $\bar{x} = 10.1$ cm. We assume that the variance of the normal population is known to be $\sigma^2 = 0.10$ cm^2. Then the 95% confidence interval on μ (which is unknown) according to equation (2.107b) is

$$P\left\{10.1 - 1.96\sqrt{\frac{0.10}{8}} < \mu < 10.1 + 1.96\sqrt{\frac{0.10}{8}}\right\} = 0.95$$

or

$$P[9.88 \text{ cm} < \mu < 10.32 \text{ cm}] = 0.95$$

for a two-sided confidence interval. For a one-sided interval, equation (2.108) may be applied,

$$P\left\{\mu < 10.1 + 1.645\sqrt{\frac{0.10}{8}}\right\} = 0.95$$

or

$$P[\mu < 10.28 \text{ cm}] = 0.95$$

Let us now consider the case in which the standard deviation of the distribution σ is not known and has to be replaced by the standard deviation s, as estimated from the sample [according to equation (2.92)].

Then the estimator $(\bar{x} - \mu)/(s/\sqrt{n})$ has a t distribution (see Section 2.5.2) with $(n - 1)$ degrees of freedom. The probability statement for the confidence interval of the random variable of the estimator will be

$$P\left\{-t_{\alpha/2,\,n-1} < \frac{\bar{x} - \mu}{s/\sqrt{n}} < t_{\alpha/2,\,n-1}\right\} = 1 - \alpha \tag{2.109}$$

or

$$P\left\{\bar{x} - t_{\alpha/2,\,n-1} \cdot \frac{s}{\sqrt{n}} < \mu < \bar{x} + t_{\alpha/2,\,n-1} \cdot \frac{s}{\sqrt{n}}\right\} = 1 - \alpha \tag{2.110}$$

For $\alpha = 0.05((1 - \alpha) = 0.95)$ we obtain from Appendix Table D.2 of t distribution for $(n - 1) = 7$, for instance, $t_{\alpha/2} = 2.365$, and $z_{\alpha/2} = 1.96$. The confidence interval is wider in this case compared to the case of known σ, as illustrated in the following example. As the degrees of freedom increase (that is, the larger the sample size), the difference between the values of t and z decreases, all other factors remaining the same. In practice, for a sample size of 30 or larger, the t distribution is often approximated by the standard normal distribution.

Example 2.4. Considering the preceding example of the measured distance, $n = 8$ times, $\bar{x} = 10.1$ cm, let $s^2 = 0.10$ cm^2 be the *sample* variance. The 95% confidence interval on μ is, according to equation (2.110),

$$P\left\{10.1 - 2.365\sqrt{\frac{0.10}{8}} < \mu < 10.1 + 2.365\sqrt{\frac{0.10}{8}}\right\} = 0.95$$

or

$$P[9.84 \text{ cm} < \mu < 10.36 \text{ cm}] = 0.95$$

Confidence Intervals for Variances and Covariances Suppose we are given a random sample of variance s^2 from a normal population with variance σ^2. The random variable ms^2/σ^2, where m is the number of degrees of freedom used in computing s^2, has a χ^2 distribution with m degrees of freedom. For the confidence level $(1 - \alpha)$ this random variable will take a value between $\chi^2_{1-\alpha/2,\, m}$ and $\chi^2_{\alpha/2,\, m}$ with the probability $(1 - \alpha)$. Thus

$$P\left\{\chi^2_{1-\alpha/2,\, m} < \frac{ms^2}{\sigma^2} < \chi^2_{\alpha/2,\, m}\right\} = 1 - \alpha \tag{2.111}$$

or

$$P\left\{\frac{ms^2}{\chi^2_{\alpha/2,\, m}} < \sigma^2 < \frac{ms^2}{\chi^2_{1-\alpha/2,\, m}}\right\} = 1 - \alpha \tag{2.112}$$

For instance, with $m = 10$ and $\alpha = 0.05$ we get from Appendix Table D.3 $\chi^2_{0.025,\, 10} = 20.48$ and $\chi^2_{0.975,\, 10} = 3.25$. This interval is two sided but is asymmetric. For the one-sided interval we have $\chi^2_{0.05,\, 10} = 18.31$.

Example 2.5. Suppose, as in preceding example, $n = 8$, $\bar{x} = 10.1$ cm, and $s_x^2 = 0.10$ cm^2. The 99% confidence interval on the population variance σ^2 may be established using equation (2.112). Here $m = n - 1 = 7$ degrees of freedom. Thus

$$P\left\{\frac{(7)(0.10)}{20.28} < \sigma^2 < \frac{(7)(0.10)}{0.99}\right\} = 0.99$$

or

$$P[0.035 \text{ cm}^2 < \sigma^2 < 0.707 \text{ cm}^2] = 0.99$$

which states that the probability is 0.99 that the distribution variance lies between 0.035 cm^2 and 0.707 cm^2.

For a confidence interval on the standard deviation σ, we have merely to take the square root of each side in the interval on the variance. Thus in this example

$$P[0.186 \text{ cm} < \sigma < 0.841 \text{ cm}] = 0.99.$$

CONFIDENCE INTERVALS FOR RATIOS OF VARIANCES If we are given two independent random samples of sizes n_1 and n_2 from normal populations with variances σ_1^2 and σ_2^2, respectively, then each of the random variables $m_1 s_1^2/\sigma_1^2$ and $m_2 s_2^2/\sigma_2^2$ with $m_1 = n_1 - 1$ and $m_2 = n_2 - 1$ has a χ^2 distribution with m_1 and m_2 degrees of freedom, respectively. The ratio

$$F = \frac{s_1^2/\sigma_1^2}{s_2^2/\sigma_2^2} \tag{2.113}$$

has an F distribution with m_1, m_2 degrees of freedom. For the confidence level $(1 - \alpha)$ the random variable F in equation (2.113) will take a value between $F_{1-\alpha/2,\, m_1,\, m_2}$ and $F_{\alpha/2,\, m_1,\, m_2}$ with the probability $(1 - \alpha)$. Thus

$$P\left\{F_{1-\alpha/2,\, m_1,\, m_2} < \frac{s_1^2/\sigma_1^2}{s_2^2/\sigma_2^2} < F_{\alpha/2,\, m_1,\, m_2}\right\} = 1 - \alpha \tag{2.114}$$

or

$$P\left\{\frac{s_2^2}{s_1^2}F_{1-\alpha/2,\,m_1,\,m_2} < \frac{\sigma_2^2}{\sigma_1^2} < \frac{s_2^2}{s_1^2}F_{\alpha/2,\,m_1,\,m_2}\right\} = 1-\alpha \tag{2.114a}$$

Because of the way the F tables are typically presented (see Appendix D) the following equality is often applied:

$$F_{1-\alpha/2,\,m_1,\,m_2} = \frac{1}{F_{\alpha/2,\,m_2,\,m_1}} \tag{2.115}$$

For instance, with $m_1 = 10$, $m_2 = 60$, and $\alpha = 0.05$,

$$F_{0.975,\,10,\,60} = \frac{1}{F_{0.025,\,60,\,10}} = \frac{1}{3.20} = 0.312$$

$$F_{0.025,\,10,\,60} = 2.27$$

which gives the following confidence interval for σ_2^2/σ_1^2:

$$0.312\frac{s_2^2}{s_1^2} < \frac{\sigma_2^2}{\sigma_1^2} < 2.27\frac{s_2^2}{s_1^2}$$

2.8. STATISTICAL TESTS

Statistical tests are increasingly applied in engineering and science and in combination with the least squares method. They are often used to compare results with previous ones or with given standards. In testing, one seeks a judgment as to whether some estimator function is consistent with the assumption (hypothesis) that the sample was drawn from a population with specified parameter values, such as normal distribution with a given standard deviation. Questions of this kind are the subject of hypothesis testing.

A hypothesis is a statement (explicit or implicit) about the probability distribution of a random variable. If the hypothesis covers the complete set of parameters for a distribution, we speak of a simple hypothesis. Conversely, a hypothesis is said to be composite if it covers only a number of distribution parameters, leaving others unspecified.

The general procedure of statistical testing always refers to a null hypothesis H_0—that is, the set of distribution parameters with respect to which the sample estimates at hand are to be compared. The intention is to formulate a statement as to whether the sample parameters are in sufficient agreement with H_0. In other words the result of the test is a statement whether, according to the available evidence, the null hypothesis can be considered acceptable or not.

A more specific task of a test is to make a decision between a null hypothesis H_0 and one or several alternative hypotheses. The alternative hypothesis can be simple, that is, specific or composite.

When accepting or rejecting a null hypothesis, there are four possible cases: (a) H_0 is true and is accepted; (b) H_0 is true but is rejected; (c) H_0 is false and is rejected; (d) H_0 is false and is accepted. Cases (a) and (c) are obviously correct decisions, and cases (b) and (d) are wrong decisions. Wrongly rejecting a true hypothesis is referred to as committing a type I error. The probability for type I error is designated by α, and is called the "significance level of the test." It is conventionally taken to be 5%, 2%, or 1%, as α has to be small in order that a test be useful. Case (d), which is the wrong acceptance of a false hypothesis, leads to committing a so-called type II error. The probability for type II error is designated β, and $(1 - \beta)$ is called the "power of a test."

Figure 2.11 demonstrates the relation between α and β. Let $D_1 = f(\hat{p} \mid H_0)$ be the conditional probability density function for the estimate of a parameter p when the null hypothesis H_0 is true; and let $D_2 = f(\hat{p} \mid H_1)$ be the

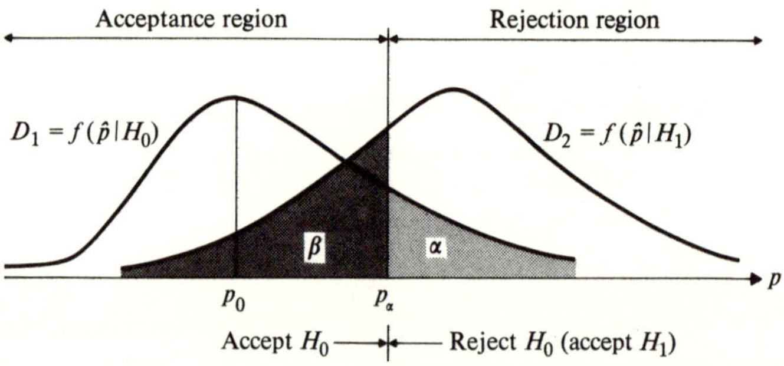

Figure 2.11

conditional probability density function for p in case the alternative hypothesis H_1 is true. We now consider the value p_α which marks the limit of sample values up to which H_0 is accepted and beyond which H_0 is rejected or H_1 is accepted. Thus H_0 is accepted if $\hat{p} < p_\alpha$, and H_0 is rejected or H_1 is accepted if $\hat{p} > p_\alpha$. The area α indicates the significance level of the test. It shows the probability that the sample parameter $(\hat{p} \mid H_0)$ will exceed p_α in case H_0 is true. It shows, in other words, the probability of rejecting H_0 when it is true (type I error). Conversely, the area β shows the probability of accepting H_0 when H_1 is true (type II error). Figure 2.11 illustrates that for a certain alternative hypothesis H_1 it is not possible to make both α and β arbitrarily small. Decreasing the probability for a type I error increases the probability for a type II error and vice versa. Balancing type I versus type II errors depends on the purpose of the test.

Figure 2.11 represents the so-called one-tail or one-sided test. In this case H_0 is $p = p_0$ whereas the alternative hypothesis is $H_1: p > p_0$. Another one-sided test is depicted in Figure 2.12 for which $H_0: p = p_0$ versus

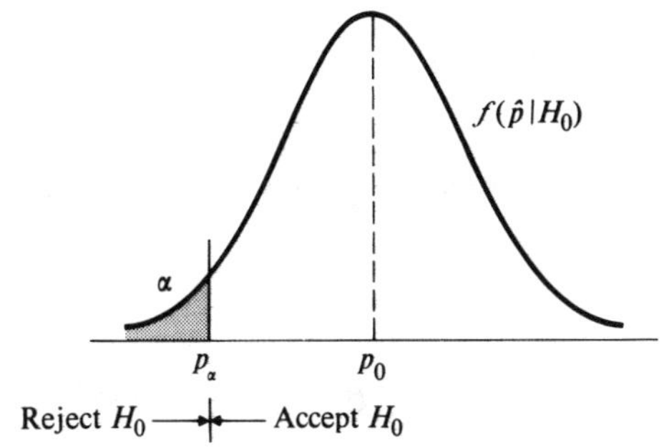

Figure 2.12

H_1: $p < p_0$. The two-tail or two-sided test is demonstrated in Figure 2.13. In this type of test the null hypothesis is H_0: $p = p_0$ whereas the alternative hypothesis is simply H_1: $p \neq p_0$. In the following sections hypothesis tests for different parameter estimates are given together with numerical examples.

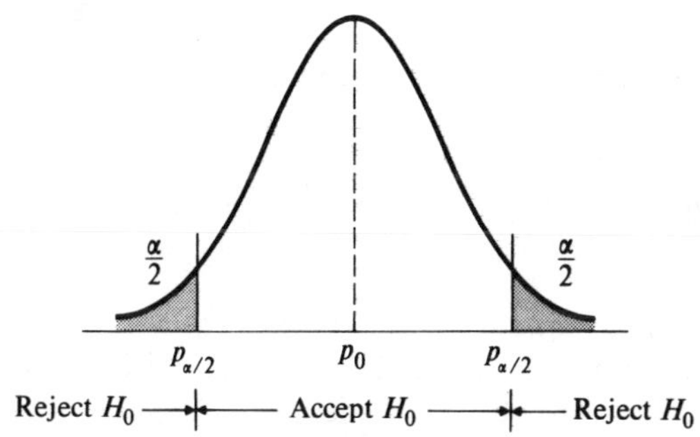

Figure 2.13

2.8.1. Test on Sample Means for Known σ In this case we are given a sample of size n with either the values x_i (from which the sample mean $\bar{x}$ may be computed) or the sample mean $\bar{x}$, as well as the normal population standard deviation σ. The null hypothesis is H_0: $\mu = \mu_0$, which seeks to test that the population mean is equal to some a priori known value μ_0. There are three possible alternative hypotheses: $\mu < \mu_0$, or $\mu > \mu_0$, or the two-tail case $\mu \neq \mu_0$. In order to perform any of the three possible hypothesis tests, the standardized normal random variable z is computed from

$$z = \frac{\bar{x} - \mu_0}{\sigma/\sqrt{n}} \tag{2.116}$$

If z_α and $z_{\alpha/2}$ are values from the normal table (Appendix Table D.1) with the level of significance being α, then the three possibilities are as follows:

1. H_0: $\mu = \mu_0$; H_1: $\mu < \mu_0$; reject H_0 when $z < -z_\alpha$ because

$$P\left\{z = \frac{\bar{x} - \mu_0}{\sigma/\sqrt{n}} < -z_\alpha\right\} = \alpha$$

2. H_0: $\mu = \mu_0$; H_1: $\mu > \mu_0$; reject H_0 when $z > z_\alpha$ because

$$P\left\{z = \frac{\bar{x} - \mu_0}{\sigma/\sqrt{n}} > z_\alpha\right\} = \alpha$$

3. H_0: $\mu = \mu_0$; H_1: $\mu \neq \mu_0$; reject H_0 when $z < -z_{\alpha/2}$ or $z > z_{\alpha/2}$ because

$$P\left\{-z_{\alpha/2} < \frac{\bar{x} - \mu_0}{\sigma/\sqrt{n}} < z_{\alpha/2}\right\} = 1 - \alpha$$

As examples, different values from Table D.1 are $z_{0.005} = 2.58$, $z_{0.01} = 2.33$, $z_{0.025} = 1.96$, $z_{0.05} = 1.64$.

2.8.2. Test on Sample Mean for Unknown σ When σ is unknown, the random variable

$$t = \frac{\bar{x} - \mu_0}{s/\sqrt{n}} \tag{2.117}$$

is used in place of z of equation (2.116). All three tests can be carried out as in the preceding section except that z_α and $z_{\alpha/2}$ are replaced by $t_{\alpha,m}$ and $t_{\alpha/2,m}$, respectively, where $m = n - 1$ degrees of freedom.

We shall consider the problem of comparing two sample means where one is concerned with hypotheses regarding differences of means, and shall assume that one is dealing with independent random samples of sizes n_1 and n_2 from two normal populations having the means μ_1 and μ_2 and the known variances σ_1^2 and σ_2^2. It is desired to test the null hypothesis $\mu_1 - \mu_2 = \delta$, where δ is a given constant, against one of the three alternatives $\mu_1 - \mu_2 \neq \delta$, $\mu_1 - \mu_2 > \delta$, or $\mu_1 - \mu_2 < \delta$. The test can be applied, based on the difference of the sample means $\bar{x}_1 - \bar{x}_2$, and using the standardized normal random variable

$$z = \frac{\bar{x}_1 - \bar{x}_2 - \delta}{\sqrt{\sigma_1^2/n_1 + \sigma_2^2/n_2}} \tag{2.118}$$

The rejection regions for the three tests are the same as those given in detail in Section (2.8.1.).

When the variances σ_1^2 and σ_2^2 are unknown but assumed equal, they are replaced by sample variances. In this case the random variable

$$t = \frac{\bar{x}_1 - \bar{x}_2 - \delta}{\sqrt{\dfrac{(n_1 - 1)s_1^2 + (n_2 - 1)s_2^2}{n_1 + n_2 - 2}}\sqrt{\dfrac{1}{n_1} + \dfrac{1}{n_2}}} \tag{2.119}$$

is used for a test of the null hypothesis $\mu_1 - \mu_2 = \delta$ against the same alternative hypotheses given above. Here t is a random variable having a t distribution with $m = (n_1 + n_2 - 2)$ degrees of freedom. The same three possible hypothesis tests can be performed using $t_{\alpha,\, m}$ and $t_{\alpha/2,\, m}$ in place of z_α and $z_{\alpha/2}$, respectively.

2.8.3. Tests Concerning Variances Let us consider a random sample of size n from a normal population. We may test the null hypothesis $\sigma^2 = \sigma_0^2$ against the alternatives, $\sigma^2 \neq \sigma_0^2$, $\sigma^2 > \sigma_0^2$, and $\sigma^2 < \sigma_0^2$. In these tests, σ_0^2 is a specified value for the variance. Here the χ^2 test is used, based on the sample variance s^2, with the variable

$$\chi_m^2 = \frac{ms^2}{\sigma_0^2} \tag{2.120}$$

computed from the sample data, where m is the degrees of freedom. If α is the level of significance, the three tests are as follows:

1. H_0: $\sigma^2 = \sigma_0^2$; H_1: $\sigma^2 \neq \sigma_0^2$; reject H_0 when $\chi_m^2 < \chi_{1-\alpha/2,\, m}$ or $\chi_m^2 > \chi_{\alpha/2,\, m}^2$ because

   $$P\left\{\chi_{1-\alpha/2,\, m}^2 < \frac{ms^2}{\sigma_0^2} < \chi_{\alpha/2,\, m}^2\right\} = 1 - \alpha$$

   where m = degrees of freedom.
2. H_0: $\sigma^2 = \sigma_0^2$; H_1: $\sigma^2 < \sigma_0^2$; reject H_0 when $\chi_m^2 < \chi_{1-\alpha,\, m}^2$ because

   $$P\left\{\frac{ms^2}{\sigma_0^2} < \chi_{1-\alpha,\, m}^2\right\} = \alpha$$

3. H_0: $\sigma^2 = \sigma_0^2$; H_1: $\sigma^2 > \sigma_0^2$; reject H_0 when $\chi_m^2 > \chi_{\alpha,\, m}^2$ because

   $$P\left\{\frac{ms^2}{\sigma_0^2} > \chi_{\alpha,\, m}^2\right\} = \alpha$$

Example 2.6. Given a sample of size $n = 12$ with sample mean $\bar{x} = 10.0$ and sample variance $s^2 = 0.07$, test the hypothesis that the population variance $\sigma^2 = 0.10$ against the alternative that $\sigma^2 < 0.10$ for a level of significance of 0.05.

Solution:

$$\chi_m^2 = \frac{ms^2}{\sigma_0^2} = \frac{(n-1)s^2}{\sigma_0^2} = \frac{(11)(0.07)}{0.10} = 7.70$$

We would reject the null hypothesis of $\sigma^2 = 0.10$ if χ_m^2 computed above is less than $\chi_{1-\alpha,m}^2$. Since $\chi_{0.95,11}^2 = 4.57$ is not larger than $\chi_m^2 = 7.70$, the null hypothesis is not rejected.

Another problem concerns the test on equality of the variances of two populations on the basis of the ratio between sample variances. Given two independent random samples of sizes n_1 and n_2 from two normal populations with the variances σ_1^2 and σ_2^2, the null hypothesis is H_0: $\sigma_1^2 = \sigma_2^2$.

If the sample variances are s_1^2 and s_2^2, with degrees of freedom $m_1 = n_1 - 1$ and $m_2 = n_2 - 1$, respectively, the random variable used in the test is

$$F_{m_1, m_2} = \frac{s_1^2}{s_2^2} \tag{2.121}$$

The three hypothesis tests possible are, in this case, as follows:

1. H_0: $\sigma_1^2 = \sigma_2^2$; H_1: $\sigma_1^2 < \sigma_2^2$; reject the null hypothesis if $F_{m_1,m_2} < F_{1-\alpha,m_1,m_2}$, because

$$P\left\{\frac{s_1^2}{s_2^2} < F_{1-\alpha,m_1,m_2}\right\} = \alpha$$

where α is the level of significance.

2. H_0: $\sigma_1^2 = \sigma_2^2$; H_1: $\sigma_1^2 > \sigma_2^2$; reject H_0 if $F_{m_1,m_2} > F_{\alpha,m_1,m_2}$ because

$$P\left\{\frac{s_1^2}{s_2^2} > F_{\alpha,m_1,m_2}\right\} = \alpha$$

3. H_0: $\sigma_1^2 = \sigma_2^2$; H_1: $\sigma_1^2 \neq \sigma_2^2$; reject H_0 if $F_{m_1,m_2} > F_{\alpha/2,m_1,m_2}$ when $s_1^2 > s_2^2$, which refers to the right boundary only. The rejection at the lower boundary is when $F_{m_1,m_2} < F_{1-\alpha/2,m_1,m_2}$.

Since $s_1^2 > s_2^2$, the value $F_{m_1,m_2} > 1$. Furthermore, $F_{1-\alpha/2,m_1,m_2} = 1/F_{\alpha/2,m_2,m_1}$ and $F_{\alpha/2,m_2,m_1} > 1$ (all entries in Table D.4 for the area under the F density function are larger than 1), then $F_{1-\alpha/2,m_1,m_2} < 1$. Henceforth F_{m_1,m_2} will always be larger than $F_{1-\alpha/2,m_1,m_2}$ and the test at the lower boundary is unnecessary.

Example 2.7. Given the following two sample data sets

$n_1 = 8$ $\quad \bar{x}_1 = 10.1$ $\quad s_1^2 = 0.10$

$n_2 = 12$ $\quad \bar{x}_2 = 10.0$ $\quad s_2^2 = 0.07$

Test the hypothesis that $\sigma_1^2 = \sigma_2^2$ against the alternative that $\sigma_1^2 \neq \sigma_2^2$ using the level of significance of 0.05.

Solution:

$$m_1 = 8 - 1 = 7 \qquad m_2 = 12 - 1 = 11$$

$$F_{7,11} = \frac{s_1^2}{s_2^2} = \frac{0.10}{0.07} = 1.429 \qquad (\text{note } s_1^2 > s_2^2)$$

We would reject the null hypothesis that $\sigma_1^2 = \sigma_2^2$ if F_{m_1, m_2} computed above is larger than $F_{\alpha/2, m_1, m_2}$. Since $\alpha/2 = 0.025$, and $F_{0.025, 7, 11} = 3.76$ is larger than $F_{7, 11} = 1.429$, then we cannot reject the null hypothesis.

There are of course a number of other tests that may be used in connection with adjustment problems such as (a) test for goodness of fit of frequency distribution on probability distribution, (b) test for correlation coefficients (see Section 11.5.3), and (c) test for regression coefficients. For such tests the reader should consult statistical literature. (See the Bibliography at the end of this book.)

3

Error Properties of Observations

3.1. INTRODUCTION

Modern information theory regards observations as signals the statistical properties of which are classified as having deterministic and stochastic components.

Until recently it was usual to speak of either the theory of observations or the theory of errors. Most of the foundation of present knowledge on the subject was laid down based on the idea of errors. The term "error" is actually somewhat misleading and seems to enjoy less and less usage. Instead, one often speaks in general of the statistical properties of observations.

Although the conventional concept of error should be avoided, the term still continues to be used to some extent. In order to avoid difficulties for the many who are still familiar with it, it will be occasionally used. In addition, an attempt will be made, whenever possible, to relate it to the modern concepts.

Classical ideas usually present errors as being of three types: random errors, systematic errors, and blunders.

3.2. RANDOM ERRORS

From the statistical standpoint, observations are considered to be samples from probability distributions of random variables. In the classical theory the variability due to sampling from probability distribution was what led to the idea of *observational errors*.

The term "observational error" or "random error," as it is sometimes called, is very often restricted to a normal probability distribution. This is due to the fact that repeated observations usually display a normal frequency distribution. In principle, however, the term "random" is not necessarily restricted to normal distribution. Furthermore, it is not limited to one-dimensional distributions but holds equally for multidimensional distributions.

The error properties of observations are equivalent to statistical properties of sampling. Although in statistical sampling emphasis is placed on the independence of sampling, this feature is normally not considered a serious problem with observations, because the population from which the observations are drawn is hypothetically considered continuous and infinite. Independent sampling essentially means that observations should be set up such that the results of previous observations should not influence the results of the following observations. This is to be clearly distinguished from possible correlation of subsequent observations, perhaps because of common influence of a certain physical parameter (such as refraction). The two more essential parameters of a distribution are

μ_x = expectation or mean that is a location parameter

σ_x = standard deviation

or

σ_x^2 = variance, or the second moment about μ_x
(the parameter of dispersion)

In case of a normal distribution the two parameters μ_x and σ_x determine it completely.

When observations $x_i (i = 1, \ldots, n)$ are made (or drawn), estimates for μ_x and σ_x are computed from them. Of direct importance is the sample mean $\bar{x}$ [see equation (2.89)] which represents an estimate for μ_x. In case of one observation only, $\bar{x} = x$ is considered an estimate for μ_x. The standard deviation σ_x of the random variable $\tilde{x}$ is often supposed to be known from previous experiments with similar equipment under similar conditions, and so on. If sufficient repetitions are available, then from the observations an estimate for the variance σ_x^2 can be obtained by computing the sample variance s_x^2 using equation (2.92).

The more general case refers to a multidimensional set of observations. Let us assume that there are several random variables $\tilde{x}_1, \tilde{x}_2, \ldots, \tilde{x}_m$ with a joint probability distribution. The parameters of interest are therefore

$\mu_1, \mu_2, \ldots, \mu_m$	the m expectations or means for the m random variables $\tilde{x}_i$
$\sigma_1^2, \sigma_2^2, \ldots, \sigma_m^2$	the variances of the marginal distributions for $\tilde{x}_1, \tilde{x}_2, \ldots, \tilde{x}_m$
$\sigma_{12}, \sigma_{13}, \ldots, \sigma_{1m}, \sigma_{23}, \ldots, \sigma_{m-1,m}$	covariances between $\tilde{x}_i$ and $\tilde{x}_j$ in all combinations

In case of a multidimensional normal distribution, the above parameters determine it completely.

Considering the general case of multiple observations for m-dimensional distribution, we get the following observational set:

$$
\begin{array}{l}
x_{11}, x_{12}, \ldots, x_{1n} \\
x_{21}, x_{22}, \ldots, x_{2n} \\
\quad \ldots \quad \ldots \\
x_{m1}, x_{m2}, \ldots, x_{mn}
\end{array}
$$

From such observations the sample means $\bar{x}_1, \bar{x}_2, \ldots, \bar{x}_m$ may be computed as estimates for the population means $\mu_1, \mu_2, \ldots, \mu_m$

$$
\begin{aligned}
\bar{x}_1 &= \frac{1}{n} \sum_{i=1}^{n} x_{1i} \\
\bar{x}_2 &= \frac{1}{n} \sum_{i=1}^{n} x_{2i} \\
&\ldots \quad \ldots \\
\bar{x}_m &= \frac{1}{n} \sum_{i=1}^{n} x_{mi}
\end{aligned}
\tag{3.1}
$$

In practice we usually assume that the other parameters of the distribution ($\sigma_1^2, \ldots, \sigma_m^2, \sigma_{12}, \ldots, \sigma_{1m}, \ldots, \sigma_{m-1,m}$) are known. In particular it is often assumed in practice that random variables $\tilde{x}_i, \tilde{x}_j$ are uncorrelated (for all i and j). In case enough repeated observations are available, the variances and covariances of the multidimensional distribution may also be estimated by the sample variances and sample covariances:

$$s_1^2 = \frac{1}{n-1} \sum_{i=1}^{n} (x_{1i} - \bar{x}_1)^2$$

$$s_2^2 = \frac{1}{n-1} \sum_{i=1}^{n} (x_{2i} - \bar{x}_2)^2$$

$$\cdots \quad \cdots$$

$$s_m^2 = \frac{1}{n-1} \sum_{i=1}^{n} (x_{mi} - \bar{x}_m)^2 \tag{3.2}$$

and

$$s_{12} = \frac{1}{n-1} \sum_{i=1}^{n} (x_{1i} - \bar{x}_1)(x_{2i} - \bar{x}_2)$$

$$\vdots$$

$$s_{kj} = \frac{1}{n-1} \sum_{i=1}^{n} (x_{ki} - \bar{x}_k)(x_{ji} - \bar{x}_j)$$

$$\vdots$$

$$s_{m-1,m} = \frac{1}{n-1} \sum_{i=1}^{n} (x_{m-1,i} - \bar{x}_{m-1})(x_{mi} - \bar{x}_m) \tag{3.3}$$

Thus the primary purpose of observations is to get estimates for the expectations of random variables. When we talk of *observational errors* or *random errors of observations*, we refer to the basic and inherent property that the estimates of a random variable $\tilde{x}$ do not agree, in general, with its expectation. Thus an observational error may in this context be defined as

$$\varepsilon_{ji} = \mu_j - x_{ji} \tag{3.4}$$

It can also be extended to the sample mean,

$$\varepsilon_j = \mu_j - \bar{x}_j \tag{3.5}$$

From equation (3.4) we can readily show that the distribution of ε about zero, is identical to the distribution of $\tilde{x}$ about its expectation μ_x. This is actually a result of the simple fact, directly obtainable from equation (3.4), that $E(\varepsilon) = 0$.

Error properties of observations are in general considered established once the complete probability distribution of the random variables to which they refer is specified. In practice, however, this is not always possible and often restricts consideration only to variances and covariances.

In this conception the mean of a sample of size n can be regarded as a *sample of size one* on a new random variable with mean μ and variance of σ^2/n. This case sometimes causes differing opinions. In the context of our discussion it may be considered as tantamount to having either n repeated observations, that is, a sample of size n on one random variable; or having n samples, each of size one on n different independent variables whose means μ and variance σ^2 are equal.

Closely related to the topic of random errors are the concepts of precision, accuracy, and weights.

3.3. PRECISION, ACCURACY, COFACTORS, AND WEIGHTS

Observations, as well as results of adjustments, are often assessed in practice by such terms as "accuracy" and "precision." These terms are used to summerize statistical properties of random variables as introduced in Section 2.7.2.

Briefly, *accuracy* refers to the degree of closeness of an estimate to its parameter, whereas *precision* expresses the degree of closeness of observations to their mean. Therefore accuracy reflects the closeness of a location statistic to the value of the parameter for which it is an estimate, and precision is directly related to the dispersion of a distribution. In the one-dimensional case precision is taken to be represented by the value of the standard deviation σ.

The term "precision" is also applied to multidimensional distribution. The description of the precision properties of multidimensional random variables (random vector) is the complete set of second central moments, or the covariance matrix $\mathbf{\Sigma} = \{\sigma_{ij}\}$.

A term related to the trace of this covariance matrix is often taken as an indication of the average precision:

$$\text{average precision} = \sqrt{\frac{\text{tr}\,\mathbf{\Sigma}}{n}} = \frac{1}{\sqrt{n}}\sqrt{\sigma_1^2 + \sigma_2^2 + \cdots + \sigma_n^2} \tag{3.6}$$

The covariance matrix can be used to determine the error ellipsoids of an n-dimensional random variable as explained in Section 2.6.3.

In practical application of adjustment the variances and covariances are often replaced by what should be called relative variances and covariances. For these the terms "weight coefficient" or "cofactors" are in common use. The term "cofactor" is selected and the letters q for one element and $\mathbf{Q}$ for a matrix, are used as symbols for it.

A cofactor is related to a covariance by

$$q_{ij} = \frac{\sigma_{ij}}{\sigma_0^2} \qquad \text{or} \qquad \sigma_{ij} = q_{ij}\sigma_0^2 \tag{3.7}$$

The definition in equation (3.7) includes as a special case the relation between a cofactor and the variance:

$$q_i = \frac{\sigma_i^2}{\sigma_0^2} \qquad \text{or} \qquad \sigma_i^2 = q_i \sigma_0^2 \tag{3.8}$$

In equations (3.7) and (3.8) the term σ_0^2 is an arbitrary constant with arbitrary dimension, called the *reference variance*. (Its square root, σ_0, will be referred to as the "reference standard deviation.") It has also been called "variance factor" and the "reference variance associated with the weight unity" (see below).

Applying equations (3.7) and (3.8) to the covariance matrix, results in the cofactor matrix,

$$\mathbf{Q} = \{q_{ij}\} = \frac{1}{\sigma_0^2}\{\sigma_{ij}\} = \frac{1}{\sigma_0^2}\boldsymbol{\Sigma} \tag{3.9}$$

Because of the symmetry of the covariance matrix $\boldsymbol{\Sigma}$ (when referring to the elements of the same random vector), the cofactor matrix $\mathbf{Q}$ is also symmetric for this case. It must also have only positive diagonal elements.

With equation (3.7) we can always switch from cofactors to covariances and vice versa. The inverse of the cofactor matrix $\mathbf{Q}$ (when it is square and nonsingular) is called the *weight matrix* $\mathbf{W}$, thus

$$\mathbf{W} = \{w_{ij}\} = \mathbf{Q}^{-1} \tag{3.10}$$

Again, when referring to the elements of the same random vector, the weight matrix must also be symmetric.

The concept of weights has had extensive use in the classical theory of error and of adjustment. In fact, the term "weight" was used to express precision by way of inverse relationship. Thus high weight meant high precision, which in turn meant small standard deviation. However, this type of relationship is not true in general. In fact the use of the term "weight" relating to individual (diagonal) elements of $\mathbf{W}$ should be totally restricted to the noncorrelation case. Only then is its use safe.

For the case of *no correlation* the covariance and cofactor matrices will both be diagonal, with all off-diagonal elements being equal to zero. In this case the diagonal elements are then

$$w_{ii} = \frac{1}{q_{ii}}$$

or, using equation (3.8),

$$w_{ii} = \frac{\sigma_0^2}{\sigma_{ii}^2} = \frac{c}{\sigma_{ii}^2}$$

These relationships are used often in textbooks on surveying. It must be stressed that they hold *only in the uncorrelated case*. In the general case, the diagonal elements w_{ii} of a weight matrix are *not* just the reciprocals of the associated cofactors,

$$w_{ii} \neq \frac{1}{q_{ii}}$$

It should be noted that in the case of perfect correlation, that is, $\rho = \pm 1$, between variables it is still possible to establish a covariance matrix and a cofactor matrix. *It is not possible, however, to establish a weight matrix in this case*. In addition, since covariances can be derived from observations directly, according to Equation (3.3), whereas weights cannot, the concept of weight is considered as a secondary concept. Caution must therefore be exercised when using weights.

3.4. BLUNDERS

Under the broad concept of error properties of observations, the conventional theory of errors also includes, in addition to random errors, blunders, or gross errors, as well as systematic errors. As regards blunders, the observer may take the wrong reading of a scale or a dial, or if he reads the proper value, he may record the wrong one by, for example, transposing numbers. If the operation of collecting the observations is performed through an automatic recording technique, mistakes may still occur due to failure of equipment, although they may be less frequent in this case. Another way of causing the occurrence of blunders is by failure in technique as in the case of reading the fraction on a tape on the wrong side of the zero mark, or by selecting the wrong whole degree in the measurement of angles that are very close to an integer of degrees. Finally, a mistake may also occur due to misinterpretation, such as sighting to the wrong target or selecting the wrong image for a control point in an aerial photograph. In fact, this last group of possibilities contains the more common causes of mistakes.

From a statistical point of view, blunders are observations that cannot be considered as belonging to the same samples from the distribution in question. They should not therefore be used together with other observations. Consequently, measurements should be planned and observational procedures designed in such a way as to allow for detecting blunders so they may be rejected. In practice, there are a variety of ways that can be employed: taking multiple readings and checking for reasonable consistency; careful checking of both pointing and recording; using simple and quick techniques for verification, applying logic and common sense; checking and verifying the performance of equipment, particularly those with automatic readouts; repeating the experiment with perhaps slightly different techniques; increasing the redundancy of the observations used in a model;

in case of relatively complex models, applying simplified geometric or algebraic checks to detect the mistakes; and, finally, simply noting that mistakes are deviations of large magnitude, whic may lead directly to their detection. (However, blunders of moderate magnitude are hard to detect even with modern statistical techniques). For example,

1. Measuring an angle or a distance several times and computing the average. Any single measurement deviating from that average by an amount that is larger than a preset value can be assumed to contain a blunder.
2. Taking two readings by the transit differing by 180 degrees.
3. Realizing simple facts such as a tape is usually 100 feet long; vertical angles are normally between ± 90 degrees; when checking photocoordinates, the format is mostly 230 × 230 mm (for aerial photography).
4. Making a quick check on a spherical triangle using plane trigonometry.
5. Using solar time instead of siderial time to check astro observations.
6. Using small portable calculators or even slide rules to check traverses and other nets to a low number of significant digits.

Despite design precautions, some blunders may still remain. Their detection and rejection should be carried out according to principles of statistical testing (data editing). It is worth mentioning here, that in practice too many observations are often discarded, although careful testing may prove the rejection to be unjustified.

3.5 SYSTEMATIC EFFECTS (ERRORS)

The term "systematic error," though used extensively in the past, is gradually being reconsidered. The practitioner is well aware of the existence of systematic effects and usually strives to minimize their presence through instrument calibration, observational procedures, and data preprocessing for atmospheric and other effects. However, from the statistical point of view it is essential to note that systematic errors will affect repeated observations in much the same way, hence they cannot be detected by having repetitive measurements. The concept of considering observations as samples from random probability distributions shows us where to place the effects of systematic errors. Although the variability of sampling relates to the deviations of the means of a distribution, the systematic effects of observations (or biases) concern its location parameter (μ). Thus perhaps with some exaggeration we can say that when systematic errors are present, there is nothing wrong with the observations; it is the interpretation that is wrong. In other words it is the functional model that is not appropriate. For example, a triangle on the earth's surface may be treated by one of three functional models: plane, spherical, or ellipsoidal. The choice of one over the others may result in systematic errors. Taking the concept of systematic error away from the observation and attaching it to the functional model may seem to be of little practical importance, as this does not solve the problem of how to consider such effects. But it is of basic importance to understand that systematic errors indicate an inconsistency between the observations (or better, between the random variables to which the observations refer) and the functional model. The linking of the observations to the model, that is, assigning sample means (which are estimates for the expectations) to certain variables of the functional model, will be inconsistent when systematic errors are present.

Systematic effects take on different forms depending on the value and sign of each of the effects. If the value and sign remain the same all through the measuring process, we would have the so-called constant error. An example of this is making distance measurements with a tape that is either too short (or too long) by a constant value. All lengths measured by that tape will undergo the same systematic effect due to the tape alone. If the sign of the systematic effect changes, perhaps due to personal bias of an observer, the resulting systematic errors are often called "counteracting." For example, on an aerial photograph earth curvature and atmospheric refraction cause opposite displacements of image points. Thus the systematic effect due to the first counteracts that due to the second.

In photogrammetry, geodesy, and surveying, systematic errors occur due to physical causes, due to instrumental factors, and due to the observer's human limitations. Physical sources are numerous of which we mention temperature, humidity, and pressure changes. These will affect angle measurements and distance measurements either by tapes or electro-optical equipment, and wili cause the bending of photogrammetric light rays due to atmospheric refraction, to name a few. Instrumental factors are caused by either imperfections in construction or lack of adequate adjustment of equipment before their use in data acquisition. Examples include unequal graduations on linear and circular scales, lack of centering of different components of the instrument, compromise in optical design thus leaving certain amounts of distortions and aberration (for example, photogrammetric camera lenses), and physical limitations in machining parts such as straight ways and pitch of screws in precision photogrammetric equipment.

Although automation has been considered and in some cases introduced (with its own sources of systematic effects) to several tasks, the human observer remains an important element in the activities of photogrammetry, geodesy, and surveying. Of his natural senses, he relies most on his vision and hearing abilities, both of which have limitations, and vary due to circumstances and from one individual to another. Although some of the personal systematic errors are constant and some are counteracting, a lot more may be erratic.

As a matter of principle, the functional model should be set up such that systematic effects are accounted for. This can be done in several ways:

1. The effects of systematic errors can be determined separately and considered directly a priori to the adjustment.
2. The effects can be considered as unknown random variables to be assessed by the observations during the adjustment.
3. The effects can be treated as observations themselves, perhaps with a priori given numerical values, for instance, refraction coefficients.

It is interesting to note that systematic effects are often not taken into account by the functional model but rather by "correcting" the observations for them. In other words the observations are replaced by another set of values, which are then treated as observations and which are more consistent with the functional model. It is said that the functional model is a "computational model" onto which the observations are "reduced." Working with such a simplified computational model is preferred over the "best" functional model in the sense of a best description of the physical system.

As an example we consider the case in which image coordinates in an aerial photograph are corrected for lens distortion and refraction in order to maintain the fictitious functional model of a perspective image. Sometimes the image coordinates are even corrected for earth curvature in order to have a simpler computational system.

In order to eliminate the effects of systematic errors, they must first be detected. One should always strive to analyze the measuring process and the functional model in order to determine as well as possible the systematic effects. In some instances even the best modeling of the systematic errors leaves some effects in the observations. This is often detected by performing tests on the residuals from the adjustment (for example, using regression on the residuals from least squares computation). In such cases the functional model may be extended in such a way as to include some added parameters to absorb the effect of the remaining systematic errors. Alternatively, some interpolation treatment a posteriori to the adjustment may be performed on the results (see Chapter 14). Before closing this chapter, some examples may help in focusing the attention of the reader on how to detect and treat systematic errors.

Example 3.1. Let us consider the taking of aerial photography for the purpose of photogrammetric aerotriangulation. Object points on the surface of the earth emanate light rays that will eventually impinge on the photographic emulsion in the aerial camera. As these rays travel through the atmosphere, the air density changes (decreases) thus causing their refraction and bending before they reach the camera lens assembly. While passing through the lens, the rays undergo further deviation from straight line paths due to the radial and tangential distortions of the lens. Therefore the latent images registered on the negative material will be at positions different from those that would result from perfectly straight line rays (perspective image).

The negative material is brought back and then subjected to a variety of changeable conditions (stresses, temperature and humidity changes, and so on) during development and printing to make the photographic postive. Image positions undoubtedly change further during this process. What are now measured on a comparator are the images on the positive plate, which are obviously shifted from the perspective position. The magnitudes of these shifts depend on the amounts and types of all systematic effects, some of which have already been discussed.

Once the existence of systematic errors is ascertained, their effects are modeled functionally. When formulation is made to fit the systematic portion of the problem at hand, the correction procedure for those systematic effects is not difficult. For example, to compensate for radial lens distortion the image coordinates need to be shifted a certain amount which is a function of the radial distance from the principal point. Knowing the function and the radial distance for any one image point, the x and y shifts can be readily computed and (algebraically) added to the measured coordinates. Other systematic effects may be treated in the same manner and the total correction computed and applied to compensate for all effects, as far as possible.

Example 3.2. Example 3.1, from photogrammetry, exemplifies the procedure followed to trace in an orderly manner the sources of systematic effects. Examples of a variety of systematic errors also abound in geodetic and surveying operations. Let us consider the operation of *taping* for the determination of distances between points on the earth's surface. The length of a given tape may be physically different from the values indicated by the numbers written on its graduations due to some or all of the following factors:

1. The temperature changes between that used for tape standardization (calibration) and the temperature actually recorded in the field during observation.
2. The tension or pull applied to the tape during measurement is different from that used during calibration.
3. The method of tape support is different during measurement from that used during calibration.
4. The end points of the distance to be measured are at different elevation. In this case a correction is needed due to the fact that the slope distance would be measured instead of the horizontal distance.

Example 3.3. Electronic (and electro-optical) distance measuring techniques are also subject to a number of systematic effects whose sources and characteristics must be determined and alleviated. We mention a few of these sources here: There may be a change in the density of air through which the signal travels as it causes a change in the signal frequency (due to variations in wave propagation velocity); the instrument (and sometimes the reflector or remote unit) may not be properly centered on the end of the line to be measured; and the path of propagation of the signal may not conform to the straight-line assumption and may be bent due to environmental and other factors.

Example 3.4. Another instrument that is used extensively in surveying and geodesy to measure horizontal and vertical angles is the transit. The following are some of the sources of systematic effects that may be associated with such instruments:

1. Horizontal circle may be off center.
2. Graduations on either or both circles may not be uniform.
3. The horizontal axis of the telescope (about which it rotates) may not be perpendicular to the vertical axis of the instrument.
4. The longitudinal (or optical) axis of the telescope may not be normal to the horizontal axis of rotation. (In this case, the axis of the telescope would describe a cone instead of a plane as it rotates through a complete revolution.)
5. When the optical axis of the telescope is horizontal, the reading on the vertical circle is different from zero.
6. The telescope axis and the axis of the leveling bubble may not be parallel.
7. This source does not pertain to the transit but to the target that is used to sight on. If the natural illuminating conditions are such that part of the target is in the shadow, the observer will tend to center the transit's cross hair so as to bisect the illuminated portion of the target. This type of error is often referred to as the phase error.

We have by no means exhausted all systematic sources in the general field of geometronics. There are numerous others in photogrammetry (particularly instrumental errors in stereoplotters, comparators, and so on) and geodesy (spherical and spheroidal excess, gravimeter and other instrument errors, and timing and other errors in astrogeodetic work). Systematic errors, once found, can often be accounted for as follows:

1. Actual formulation and computation of corrections that are then applied to the raw observations.
2. Careful calibration and adjustment of equipment and measuring under the same conditions specified by calibration results.
3. Devising observational procedures that will result in the elimination of systematic errors that would otherwise occur.
4. Extending the functional model to include the effects of the systematic errors.

Finally, systematic effects may arise due to highly correlated random errors that are not accommodated in the stochastic model. In fact, particularly after an adjustment, residual systematic effects are often treated as correlated random errors and filtering techniques are used for their reduction. It is important to note that in the phase on adjustment techniques in this book we do not deal with systematic effects or biases. The assumption will be made that both the stochastic and functional models are appropriate.

4

Principle and Techniques of Propagation

4.1. INTRODUCTION

Propagation involves obtaining the stochastic characteristics of (functionally) dependent variables given the characteristics of the independent variables and the functional relationships relating the two sets of variables.

Let $\tilde{x}_i$ be a set of random variables with associated density function (n dimensional) $f(x_1, x_2, \ldots)$. Let $\tilde{y}_k$ be another set of random variables related to $\tilde{x}_i$ by functional relationships

$$\tilde{y}_k = u_k(\tilde{x}_1, \tilde{x}_2, \ldots)$$

Then the task of propagation is to determine the stochastic properties of $\tilde{y}_k$ from those of the $\tilde{x}_i$, in other words, to determine basically the density function $f(y_1, y_2, \ldots)$.

In practical application this general problem is often simplified. We can distinguish conventionally between three cases:

1. Propagation of means (expectations).
2. Propagation of random errors (variances and covariances).
3. Propagation of systematic errors.

Each of these cases will be treated separately, following the general case of distribution propagation. In all the cases, however, practical application is concerned with linear (or linearized) functional relationships only.

4.2. PROPAGATION OF DISTRIBUTIONS

Let $(\tilde{x}_1, \tilde{x}_2, \ldots, \tilde{x}_n)$ be a set of stochastic variables, the joint probability distribution of which is given by the density function $f_x(x_1, x_2, \ldots, x_n)$. From them a set of variables $(\tilde{y}_1, \tilde{y}_2, \ldots, \tilde{y}_m)$ is derived by given functional relationships. The joint probability distribution $f_y(y_1, y_2, \ldots, y_m)$ is to be derived using f_x and the given functions.

4.2.1. One-Dimensional Case For simplicity, consider first a one-dimensional case. Let us take a continuous, differentiable function $\tilde{y} = g(\tilde{x})$ with $f(x)$ being the density function of $\tilde{x}$. The inverse function $\tilde{x} = h(\tilde{y})$ is assumed to exist and to be also unique, continuous, and differentiable.

Each event $\tilde{x}_i$ can be transformed to an event $\tilde{y}_i = g(\tilde{x}_i)$ and, vice versa, $\tilde{x}_i = h(\tilde{y}_i)$. Therefore the probability of $\tilde{x}$ falling into the interval $x_2 - x_1$ is the same as $\tilde{y}$ falling into the associated interval $y_2 - y_1$.

$$P(x_1 \leq \tilde{x} \leq x_2) = \int_{x_1}^{x_2} f_x(x)\, dx = \int_{y_1}^{y_2} f_y(y)\, dy = P(y_1 \leq \tilde{y} \leq y_2)$$

Using the functional relationships, we can substitute

$$f_x(x)\, dx = \left[f_x(h(y)) \frac{\partial h(y)}{\partial y} \right] dy = [f_y(y)]\, dy$$

Thus the density function for the random variable $\tilde{y} = g(\tilde{x})$ is

$$f_y(y) = f_x(h(y)) \left| \frac{\partial h(y)}{\partial y} \right| \tag{4.1}$$

The absolute value used in equation (4.1) is taken in order to insure having the proper sign.

Example 4.1

$$y = x^2$$

$$g(x) = x^2; \qquad f(x) = f_x(x)$$

$$h(y) = x = +\sqrt{y} \qquad \text{(using only the positive value)}$$

$$\frac{\partial h(y)}{\partial y} = \frac{1}{2\sqrt{y}}$$

density function for y: $f_y(y) = f_x(y) \cdot \dfrac{1}{2\sqrt{y}}$. Thus, if

$$f_x(x) = e^{-x2/2}$$

then

$$f_y(y) = e^{-y/2}\left(\frac{1}{2\sqrt{y}}\right)$$

4.2.2. Multidimensional Case Let us consider the functions

$$\tilde{u}_i = g_i(\tilde{x}_1, \tilde{x}_2, \ldots) = g_i(\tilde{\mathbf{x}}) \tag{4.2}$$

of the random vector $\tilde{\mathbf{x}}$ and the inverse functions

$$\tilde{x}_j = h_j(\tilde{u}_1, \tilde{u}_2, \ldots) = h_j(\tilde{\mathbf{u}}) \tag{4.3}$$

If $f(x_1, x_2, \ldots)$ is the joint density function of the random vector $\tilde{\mathbf{x}}$, then

$$\bar{f}(u_1, u_2, \ldots) = f[h_1(u_1, u_2, \ldots), h_2(u_1, u_2, \ldots), \ldots] \cdot |\mathbf{J}| \tag{4.4}$$

is the joint density function of the derived random variables $\tilde{u}_1, \tilde{u}_2, \ldots$. Here, $|\mathbf{J}|$ is the determinant of the Jacobian matrix of the inverse transformations

$$\tilde{x}_j = h_j(\tilde{\mathbf{u}}) \qquad \text{or} \qquad \mathbf{J} = \frac{\partial \mathbf{x}}{\partial \mathbf{u}}$$

The difficulty with propagation of distributions is that we have to assume that the inverse function exists. But in the case of linear transformation of normally distributed variables the result will also be normally distributed variables. Therefore propagation is generally limited to linear or linearized functions.

4.3. PROPAGATION OF MEANS

4.3.1. General Means or expectations of distributions are a special class of parameters by which probability distributions or density functions are defined. Therefore the general equation (4.4) must contain the special relationship for the propagation of means. Instead of using equation (4.4), however, it is simple to derive the propagation of means directly from the concept of statistical expectation introduced in Chapter 2. Let $\tilde{x}_1, \tilde{x}_2, \ldots$ be a set of random variables with expectations (or means) $E(\tilde{x}_1), E(\tilde{x}_2), \ldots$. We let $\tilde{y}_1, \tilde{y}_2, \ldots$ be a set of random variables that are functions of $\tilde{x}_1, \tilde{x}_2, \ldots$.

$$\tilde{y}_i = g_i(\tilde{x}_1, \tilde{x}_2, \ldots) \tag{4.5}$$

The expectation of $\tilde{y}_i$ being the expectation of the function g_i is given by

$$E(\tilde{y}_i) = E(g_i(\tilde{x}_1, \tilde{x}_2, \ldots)) \tag{4.6}$$

Equation (4.6) holds in general, but it does not provide a simple computational technique, although its general statistical definition makes possible the computation of $E(\tilde{y}_i)$ from

$$E(\tilde{y}_i) = \iint_{-\infty}^{+\infty} \cdots \int g_i(\tilde{x}_1, \tilde{x}_2, \ldots) f(x_1, x_2, \ldots)\, dx_1\, dx_2, \ldots \tag{4.7}$$

Here, $f(x_1, x_2, \ldots)$ is the density function of $\tilde{x}_1, \tilde{x}_2, \ldots$.

4.3.2. Propagation of Means for Linear Functions For linear or linearized functions, however, the relationship for the propagation of expectations can be easily derived. In this case the general formula (4.7) simplifies considerably. For example, for $\tilde{y} = a\tilde{x}$ we get

$$E(\tilde{y}) = E(a\tilde{x}) = \int_{-\infty}^{\infty} axf(x)\,dx = a\int_{-\infty}^{\infty} xf(x)\,dx = aE(\tilde{x}) \tag{4.8}$$

Extending equation (4.8) to include several variables

$$\begin{aligned} \tilde{y}_i &= a_{0i} + a_{1i}\tilde{x}_1 + a_{2i}\tilde{x}_2 + a_{3i}\tilde{x}_3 + \cdots \\ E(\tilde{y}_i) &= E(a_{0i} + a_{1i}\tilde{x}_1 + a_{2i}\tilde{x}_2 + a_{3i}\tilde{x}_3 + \cdots) \\ &= E(a_{0i}) + E(a_{1i}\tilde{x}_1) + E(a_{2i}\tilde{x}_2) + E(a_{3i}\tilde{x}_3) + \cdots \\ &= a_{0i} + a_{1i}E(\tilde{x}_1) + a_{2i}E(\tilde{x}_2) + a_{3i}E(\tilde{x}_3) + \cdots \end{aligned} \tag{4.9}$$

or

$$\mu_{yi} = a_{0i} + a_{1i}\mu_{x1} + a_{2i}\mu_{x2} + a_{3i}\mu_{x3} + \cdots \tag{4.10}$$

Thus the linear relationship holds for the expectations or means as well. It is important to point out that this propagation of means for linear functions is independent of the probability distribution of the random variables $\tilde{x}_1, \tilde{x}_2$, ... Equation (4.10) combines the rules given in Section 2.4.1. When the relations are nonlinear, the probability distribution is needed.

Example 4.2. Given a normally distributed random variable $\tilde{x}$ with

$$f(x) = \frac{1}{\sigma\sqrt{2\pi}} \exp\left[-\frac{1}{2}\left(\frac{x-\mu}{\sigma}\right)^2\right]$$

that is, with a mean μ and a variance of σ^2. If $\tilde{y} = \tilde{x}^2$ is another random variable, evaluate its mean μ_y.

Solution

$$\begin{aligned} \mu_y &= E(\tilde{y}) = E(\tilde{x}^2) \\ &= \frac{1}{\sigma\sqrt{2\pi}} \int_{-\infty}^{\infty} x^2 \exp\left[-\frac{1}{2}\left(\frac{x-\mu}{\sigma}\right)^2\right] dx \end{aligned}$$

Let

$$\begin{aligned} z &= \frac{x-\mu}{\sigma\sqrt{2}} \\ z^2 &= \frac{1}{2}\left(\frac{x-\mu}{\sigma}\right)^2 \\ x &= \mu + \sigma\sqrt{2}\,z \\ dx &= \sigma\sqrt{2}\,dz \end{aligned}$$

then

$$\mu_y = \frac{1}{\sigma\sqrt{2\pi}} \int_{-\infty}^{\infty} (\mu + \sigma\sqrt{2}\,z)^2 e^{-z^2}(\sigma\sqrt{2})\,dz$$

$$= \frac{1}{\sqrt{\pi}}\left[\mu^2 \int_{-\infty}^{\infty} e^{-z^2}\,dz + 2\mu\sigma\sqrt{2}\int_{-\infty}^{\infty} ze^{-z^2}\,dz + 2\sigma^2 \int_{-\infty}^{\infty} z^2 e^{-z^2}\,dz\right]$$

$$= \frac{1}{\sqrt{\pi}}\left[\sqrt{\pi}\,\mu^2 + 0 + 2\sigma^2\frac{\sqrt{\pi}}{2}\right]$$

or

$$\mu_y = \mu^2 + \sigma^2$$

This result could have been directly obtained from equation (2.38a) in Chapter 2, which, by the way, does not require knowledge of the density function.

In case $\tilde{x}$ has a standardized distribution $N(0, 1)$, or $\mu = 0$ and $\sigma^2 = 1$, then

$$\mu_y = \frac{1}{\sqrt{2\pi}} \int_{-\infty}^{\infty} x^2 e^{-(1/2)x^2}\,dx$$

Let

$$t = \frac{x}{\sqrt{2}}$$

$$t^2 = \tfrac{1}{2}x^2$$

$$dx = \sqrt{2}\,dt$$

then

$$\mu_y = \frac{1}{\sqrt{2\pi}} \int_{-\infty}^{\infty} 2t^2 e^{-t^2}\sqrt{2}\,dt$$

$$= \frac{4}{\sqrt{\pi}} \int_{0}^{\infty} t^2 e^{-t^2}$$

or

$$\mu_y = \frac{4}{\sqrt{\pi}} \cdot \frac{\sqrt{\pi}}{4} = 1$$

This result shows that whereas $\mu = 0$ (for $\tilde{x}$), $\mu_y = 0 + \sigma^2 = 0 + 1$ for $\tilde{y}$.

4.4. PROPAGATION OF VARIANCES AND COVARIANCES

4.4.1. General The propagation of variances and covariances is commonly known in practice as propagation of errors. In its general form it may be extracted as the second central moments from equation (4.4). It can in general be formulated for nonlinear functions in the following manner.

We let $\tilde{x}_1$, $\tilde{x}_2$ be random variables, with expectation μ_1, μ_2 and density function $f(x_1, x_2)$. We consider the functions

$$\tilde{y}_1 = g_1(\tilde{x}_1, \tilde{x}_2) \qquad \text{and} \qquad \tilde{y}_2 = g_2(\tilde{x}_1, \tilde{x}_2) \tag{4.11}$$

Variances and covariances are defined as the expectations

$$\sigma_{x_1}^2 = E[(\tilde{x}_1 - \mu_1)^2] = \int_{-\infty}^{\infty} (x_1 - \mu_1)^2 f(x_1)\, dx_1 \tag{4.12}$$

$$\sigma_{x_2}^2 = E[(\tilde{x}_2 - \mu_2)^2] = \int_{-\infty}^{\infty} (x_2 - \mu_2)^2 f(x_2)\, dx_2 \tag{4.13}$$

$$\begin{aligned} \sigma_{x_1 x_2} &= E[(\tilde{x}_1 - \mu_1)(\tilde{x}_2 - \mu_2)] \\ &= \int_{-\infty}^{\infty} \int_{-\infty}^{\infty} (x_1 - \mu_1)(x_2 - \mu_2) f(x_1, x_2)\, dx_1\, dx_2 \end{aligned} \tag{4.14}$$

Accordingly we get for the variances and covariances of $\tilde{y}_1$, $\tilde{y}_2$

$$\sigma_{y_1}^2 = \int_{-\infty}^{\infty} \int_{-\infty}^{\infty} (g_1(x_1, x_2) - \mu_{y_1})^2 f(x_1, x_2)\, dx_1\, dx_2 \tag{4.15}$$

$$\sigma_{y_2}^2 = \int_{-\infty}^{\infty} \int_{-\infty}^{\infty} (g_2(x_1, x_2) - \mu_{y_2})^2 f(x_1, x_2)\, dx_1\, dx_2 \tag{4.16}$$

$$\begin{aligned} = \int_{-\infty}^{\infty} \int_{-\infty}^{\infty} & (g_1(x_1, x_2) - \mu_{y_1})(g_2(x_1, x_2) - \mu_{y_2}) \\ & \times f(x_1, x_2)\, dx_1\, dx_2 \end{aligned} \tag{4.17}$$

Formulas (4.15) to (4.17) are general for any two functions of two random variables. They can readily be extended to the case of n random variables.

Such general formulas are not used in practice, however. The integrals involved in these equations often present problems. Therefore the propagation of variances and covariances is simplified to linear (or linearized) functions.

4.4.2. Variance and Covariance Propagation for Linear Functions Let $\tilde{x}_1$, $\tilde{x}_2$ be the random variables, with expectations μ_1, μ_2 and joint density function $f(x_1, x_2)$. Consider the linear functions

$$\begin{aligned} \tilde{y}_1 &= a_0 + a_1 \tilde{x}_1 + a_2 \tilde{x}_2 \\ \tilde{y}_2 &= b_0 + b_1 \tilde{x}_1 + b_2 \tilde{x}_2 \end{aligned} \tag{4.18}$$

The variances and covariances of $\tilde{y}_1$, $\tilde{y}_2$ are derived by applying the definition,

$$\begin{aligned}
\sigma_{y1}^2 &= E(\tilde{y}_1 - \mu_{y_1})^2 = E(a_0 + a_1\tilde{x}_1 + a_2\tilde{x}_2 - a_0 - a_1\mu_1 - a_2\mu_2)^2 \\
&= E(a_1(\tilde{x}_1 - \mu_1) + a_2(\tilde{x}_2 - \mu_2))^2 \\
&= E(a_1^2(\tilde{x}_1 - \mu_1)^2 + a_2^2(\tilde{x}_2 - \mu_2)^2 + 2a_1a_2(\tilde{x}_1 - \mu_1)(\tilde{x}_2 - \mu_2)) \\
&= a_1^2E(\tilde{x}_1 - \mu_1)^2 + a_2^2E(\tilde{x}_2 - \mu_2)^2 + 2a_1a_2E(\tilde{x}_1 - \mu_1)(\tilde{x}_2 - \mu_2) \\
&= a_1^2\sigma_{x_1}^2 + a_2^2\sigma_{x_2}^2 + 2a_1a_2\sigma_{x_1x_2}
\end{aligned} \tag{4.19}$$

Similarly,

$$\sigma_{y2}^2 = b_1^2\sigma_{x_1}^2 + b_2^2\sigma_{x_2}^2 + 2b_1b_2\sigma_{x_1x_2} \tag{4.20}$$

For the covariance, we have

$$\begin{aligned}
\sigma_{y_1y_2} &= E(\tilde{y}_1 - \mu_{y_1})(\tilde{y}_2 - \mu_{y_2}) \\
&= E[(a_0 + a_1\tilde{x}_1 + a_2\tilde{x}_2 - a_0 - a_1\mu_1 - a_2\mu_2) \\
&\quad \times (b_0 + b_1\tilde{x}_1 + b_2\tilde{x}_2 - b_0 - b_1\mu_1 - b_2\mu_2)] \\
&= E[(a_1(\tilde{a}_1 - \mu_1) + a_2(\tilde{x}_2 - \mu_2)) \\
&\quad \times ((b_1(\tilde{x}_1 - \mu_1) + b_2(\tilde{x}_2 - \mu_2))] \\
&= E(a_1b_1(\tilde{x}_1 - \mu_1)^2 + 2(a_1b_2 + a_2b_1) \\
&\quad \times (\tilde{x}_1 - \mu_1)(\tilde{x}_2 - \mu_2) + a_2b_2(\tilde{x}_2 - \mu_2)^2) \\
&= a_1b_1E(\tilde{x}_1 - \mu_1)^2 + a_2b_2E(\tilde{x}_2 - \mu_2)^2 + 2(a_1b_2 + a_2b_1) \\
&\quad \times E(\tilde{x}_1 - \mu_1)(\tilde{x}_2 - \mu_2) \\
&= a_1b_1\sigma_{x_1}^2 + a_2b_2\sigma_{x_2}^2 + 2(a_1b_2 + a_2b_1)\sigma_{x_1x_2}
\end{aligned} \tag{4.21}$$

Again it is noticeable that the propagation of variances and covariances with linear functions is independent of the density functions. Equations (4.19) to (4.21) are valid for any probability distribution.

These equations were derived in a lengthy manner due to the application of the basic definitions of expectation. They can, however, be more concisely presented using matrices. Let

$$\tilde{\mathbf{y}} = [\tilde{y}_1\,\tilde{y}_2]^t \qquad \text{and} \qquad \tilde{\mathbf{x}} = [\tilde{x}_1\,\tilde{x}_2]^t$$

be the two random vectors involved. Equation (4.18) may be written in the form

$$\tilde{\mathbf{y}} = \mathbf{c} + \mathbf{C}\tilde{\mathbf{x}} \tag{4.22}$$

where

$$\mathbf{c} = [a_0 \quad b_0]^t$$

$$\mathbf{C} = \begin{bmatrix} a_1 & a_2 \\ b_1 & b_2 \end{bmatrix}$$

Further, let

$$\boldsymbol{\Sigma}_{xx} = \begin{bmatrix} \sigma_{x_1}^2 & \sigma_{x_1x_2} \\ \sigma_{x_1x_2} & \sigma_{x_2}^2 \end{bmatrix} \qquad \boldsymbol{\Sigma}_{yy} = \begin{bmatrix} \sigma_{y_1}^2 & \sigma_{y_1y_2} \\ \sigma_{y_1y_2} & \sigma_{y_2}^2 \end{bmatrix}$$

be the covariance matrices for the two random vectors $\tilde{\mathbf{x}}$ and $\tilde{\mathbf{y}}$. The three equations (4.19), (4.20), and (4.21) may all be combined in one

$$\boldsymbol{\Sigma}_{yy} = \mathbf{C}\boldsymbol{\Sigma}_{xx}\,\mathbf{C}^t \tag{4.23}$$

The (1, 1) terms in equation (4.23) give equation (4.19), the (1, 2) or (2, 1) terms give equation (4.20), and the (2, 2) terms give equation (4.21). The matrix **C** represents the Jacobian of $\tilde{\mathbf{y}}$ with respect to $\tilde{\mathbf{x}}$; thus equation (4.23) may be further generalized symbolically to read

$$\boldsymbol{\Sigma}_{yy} = \mathbf{J}_{yx}\,\boldsymbol{\Sigma}_{xx}\,\mathbf{J}_{yx}^t \tag{4.24}$$

with

$$\mathbf{J}_{yx} = \frac{\partial \tilde{\mathbf{y}}}{\partial \tilde{\mathbf{x}}} \tag{4.25}$$

If $\boldsymbol{\Sigma}_{xx}$ is a diagonal matrix, $\boldsymbol{\Sigma}_{yy}$ may still in general be a full matrix. Therefore even if the original variables are uncorrelated, the new ones are usually correlated.

4.4.3. Variance and Covariance Propagation for Nonlinear Functions Linear equations are not often encountered in practice, and therefore equations (4.18) are usually the result of linearizing functions that were originally nonlinear. For instance, if the orginal equations are those given by equation (4.11), their linearized form at the approximate (or initial) values x_1^0, x_2^0 (see Appendix B) will be identical to equation (4.18).

Of course, the linearized form is stopped after the zero and first-order terms neglecting the second and higher-order terms. In fact, the zero-order terms are not needed since $g_1(x_1^0, x_2^0)$ and $g_2(x_1^0, x_2^0)$, which correspond to a_0, b_0 of the directly linear form, would not appear in the forms for propagated $\sigma_{y_1}^2$, $\sigma_{y_1y_2}$, and $\sigma_{y_2}^2$. This leaves only the partial derivatives and thus makes equation (4.24) general, in as much as it can be applied to linear and nonlinear cases alike. Here, the Jacobian matrix **J** would include all four partial derivatives or

$$\mathbf{J} = \begin{bmatrix} \dfrac{\partial y_1}{\partial x_1} & \dfrac{\partial y_1}{\partial x_2} \\ \dfrac{\partial y_2}{\partial x_1} & \dfrac{\partial y_2}{\partial x_2} \end{bmatrix} \quad \text{evaluated at } x_1^0, x_2^0 \tag{4.26}$$

Although in practical applications linearized functions are used regularly for the propagation of variances and covariances, it should be pointed out that this is permitted if the range of dispersion in $\tilde{x}_1, \tilde{x}_2$ is small when linear approximation is compared to the curvature of the function in the neighborhood of x_1^0, x_2^0. In other words the function should be approximated well by its tangent within the region of interest—that is, the region of dispersion of the random variables.

From a formal point of view it should be noted that in linearization the properties of random variables change from the variables themselves (in the nonlinear form) to the increment,

$$x_i = x_{i0} + \Delta x_i \qquad y_i = y_{i0} + \Delta y_i$$

Thus the error properties—that is, the probability distribution—is now associated with Δx_i and Δy_i instead of x_i, y_i, respectively.

Within the range of linearity, the geometrical consideration confirms the result, obtained previously, that linear transformation of normally distributed variables $\tilde{x}_i$ also yields normally distributed variables $\tilde{y}_i$. It also allows us to assess the validity of the linearization (see Figure 4.1).

As a final comment, propagation rules for other moments could be derived in the same way as for the second moments in the case of linear

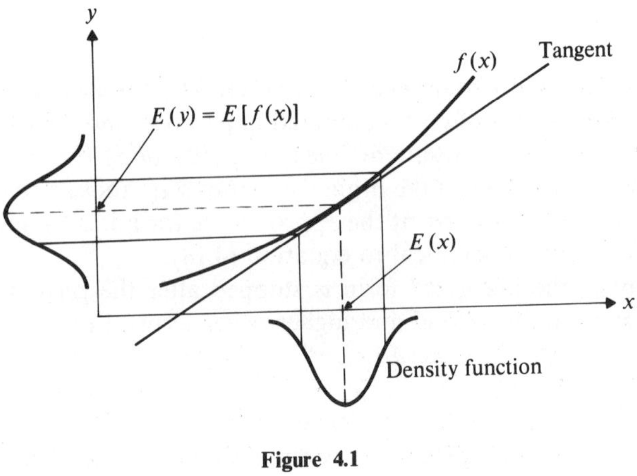

Figure 4.1

functions. In general, the propagation of moments is independent of the distribution with linear functions.

4.4.4. Alternate Derivation for Variance and Covariance Propagation A derivation, which is due to Brown,† is possible without the application of the concept of moments. It is based on the following alternate definitions of variance and covariance. If $\tilde{x}_i$ is a random variable taking many values x_{ik}, each of which having a random error dx_{ik} (as a deviation from the mean), then the variance σ_{xi}^2 of x_i is given by

$$\sigma_{xi}^2 = \lim_{m \to \infty} \frac{1}{m} \sum_{k=1}^{m} (dx_{ik})^2 \tag{4.27}$$

In a similar manner the covariance of two elements $\tilde{x}_i, \tilde{x}_j$ of a random vector $\tilde{\mathbf{x}}$ may be defined by

$$\sigma_{xixj} = \lim_{m \to \infty} \frac{1}{m} \sum_{k=1}^{m} (dx_{ik}\, dx_{jk}) \tag{4.28}$$

With these two basic definitions, the propagation of variances and covariances from one set of random variables $\tilde{\mathbf{x}}$ to another set $\tilde{\mathbf{y}}$ can be formulated. We let

$$\begin{aligned} \tilde{y}_1 &= y_1(\tilde{x}_1, \tilde{x}_2, \ldots, \tilde{x}_n) \\ &\vdots \\ \tilde{y}_q &= y_q(\tilde{x}_1, \tilde{x}_2, \ldots, \tilde{x}_n) \end{aligned} \tag{4.29}$$

be a set of q functions relating the variables $\tilde{\mathbf{x}}$ to $\tilde{\mathbf{y}}$. In order to propagate the "errors" $dx_{1k}, dx_{2k}, \ldots, dx_{nk}$ into all $\tilde{y}_i$, let us define the auxiliaries (partial derivatives),

$$y'_{ij} = \frac{\partial y_i}{\partial x_j} \qquad i = 1, 2, \ldots, q, \quad j = 1, 2, \ldots, n \tag{4.30}$$

Thus

$$\begin{aligned} dy_{1k} &= y'_{11}\, dx_{1k} + y'_{12}\, dx_{2k} + \cdots + y'_{1n}\, dx_{nk} \\ &\vdots \\ dy_{qk} &= y'_{q1}\, dx_{1k} + y'_{q2}\, dx_{2k} + \cdots + y'_{qn}\, dx_{nk} \end{aligned} \tag{4.31}$$

or in matrix form

$$d\mathbf{y}_k = \mathbf{Y}'\, d\mathbf{x}_k \tag{4.32}$$

† See bibliography at end of book.

where $d\mathbf{y}_k$ is a $q \times 1$ vector, and $d\mathbf{x}_k$ is a $n \times 1$ vector; $\mathbf{Y}'$ is a $q \times n$ matrix of partial derivatives which is the same as the Jacobian $\mathbf{J}_{yx}$. Hence, equation (4.32) may be written as

$$d\mathbf{y}_k = \mathbf{J}_{yx}\, d\mathbf{x}_k \tag{4.33}$$

If we choose the ith and rth elements to represent any two typical elements, then

$$\begin{aligned} dy_{ik} &= \mathbf{j}_i\, d\mathbf{x}_k \\ dy_{rk} &= \mathbf{j}_r\, d\mathbf{x}_k \end{aligned} \tag{4.34}$$

where $\mathbf{j}_i$, $\mathbf{j}_r$ are the ith and rth rows of $\mathbf{J}$. Next,

$$dy_{ik}\, dy_{rk} = \underset{1,\,n}{\mathbf{j}_i}\; \underset{n,\,1}{d\mathbf{x}_k}\; \underset{1,\,n}{\mathbf{j}_r}\; \underset{n,\,1}{d\mathbf{x}_k}$$

or

$$dy_{ik}\, dy_{rk} = \underset{1,\,n}{\mathbf{j}_i}\; \underset{n,\,1}{d\mathbf{x}_k}\; \underset{1,\,n}{(d\mathbf{x}_k)^t}\; \underset{n,\,1}{\mathbf{j}_r^t} \tag{4.35}$$

Let

$$\underset{n,\,n}{\mathbf{M}_k} = d\mathbf{x}_k \cdot (d\mathbf{x}_k)^t = \begin{bmatrix} (dx_1)^2 & (dx_1\, dx_2) & \cdots & (dx_1\, dx_n) \\ \vdots & \vdots & & \vdots \\ (dx_n\, dx_1) & (dx_n\, dx_2) & \cdots & (dx_n)^2 \end{bmatrix}_k \tag{4.36}$$

then

$$dy_{ik}\, dy_{rk} = \mathbf{j}_i\, \mathbf{M}_k\, \mathbf{j}_r^t \tag{4.37}$$

We let k run from 1 to m, take the sum, divide by m, and take the limit, or

$$\lim_{m\to\infty} \frac{1}{m} \sum_{k=1}^{m} dy_{ik}\, dy_{rk} = \mathbf{j}_i \left[\lim_{m\to\infty} \frac{1}{m} \sum_{k=1}^{m} \mathbf{M}_k \right] \mathbf{j}_r^t \tag{4.38}$$

In view of the definitions of equations (4.27) and (4.28), it is readily seen that equation (4.38) leads to

$$\sigma_{y_i y_r} = \mathbf{j}_i\, \boldsymbol{\Sigma}_{xx}\, \mathbf{j}_r^t \tag{4.39}$$

Equation (4.39) represents *one element* on the ith row and rth column of the total covariance matrix $\boldsymbol{\Sigma}_{yy}$. It should be rather straightforward to extend equation (4.39) to

$$\boldsymbol{\Sigma}_{yy} = \mathbf{J}_{yx}\, \boldsymbol{\Sigma}_{xx}\, \mathbf{J}_{yx}^{t} \tag{4.40}$$

which is identical to equation (4.24). Equation (4.40) is sufficiently general in that it concerns a vector $\tilde{\mathbf{y}}_{(q,\,1)}$ as a general (nonlinear) function of another vector $\tilde{\mathbf{x}}_{(n,\,1)}$ which is of different dimension. Since cofactor matrices are related to covariance matrices by only a scale factor, that is,

$$\mathbf{Q}_{xx} = \frac{1}{\sigma_0^2}\boldsymbol{\Sigma}_{xx} \qquad \text{and} \qquad \mathbf{Q}_{yy} = \frac{1}{\sigma_0^2}\boldsymbol{\Sigma}_{yy}$$

Then equation (4.40) may be written in terms of cofactor matrices

$$\mathbf{Q}_{yy} = \mathbf{J}_{yx}\,\mathbf{Q}_{xx}\,\mathbf{J}_{yx}^t \tag{4.41}$$

It is worth noting here that we may speak of covariance and cofactor propagation but *not of weight* propagation. This follows in view of the fact that earlier discussion showed that weight matrices may not even be defined. Furthermore, the Jacobian matrices are often rectangular and the inverse of equation (4.41) [or equation (4.40)] cannot be effected on individual matrices.

A further generalization of equation (4.41) is possible when multiple functions of different random vectors are involved. For example, let $\mathbf{x}_{(n,\,1)}$ and $\mathbf{t}_{(m,\,1)}$ be two correlated vectors with cofactor matrices $\mathbf{Q}_{xx}$, $\mathbf{Q}_{xt}$ and $\mathbf{Q}_{tt}$. Two other vectors $\mathbf{y}_{(q,\,1)}$ and $\mathbf{z}_{(p,\,1)}$ are functions of $\mathbf{x}$ and $\mathbf{t}$ ($\mathbf{y}$ is functionally independent of $\mathbf{t}$ and $\mathbf{z}$ of $\mathbf{x}$)

$$\begin{aligned} \mathbf{y} &= y(\mathbf{x}) \\ \mathbf{z} &= z(\mathbf{t}) \end{aligned} \tag{4.42}$$

from which the following Jacobians are derived

$$\begin{aligned} \mathbf{J}_{yx} &= \frac{\partial \mathbf{y}}{\partial \mathbf{x}} \\ \mathbf{J}_{zt} &= \frac{\partial \mathbf{z}}{\partial \mathbf{t}} \end{aligned} \tag{4.43}$$

Two expanded vectors

$$\mathbf{r} = \begin{bmatrix} \mathbf{y} \\ \mathbf{z} \end{bmatrix} \qquad \text{and} \qquad \mathbf{s} = \begin{bmatrix} \mathbf{x} \\ \mathbf{t} \end{bmatrix}$$

can be set up with the corresponding expanded Jacobian

$$\mathbf{J}_{rs} = \begin{bmatrix} \mathbf{J}_{yx} & \mathbf{0} \\ \mathbf{0} & \mathbf{J}_{zt} \end{bmatrix} \tag{4.44}$$

and equation (4.41) may be applied

$$\mathbf{Q}_{rr} = \mathbf{J}_{rs}\,\mathbf{Q}_{ss}\,\mathbf{J}_{rs}^t$$

or

$$\begin{bmatrix} \mathbf{Q}_{yy} & \mathbf{Q}_{yz} \\ \mathbf{Q}_{zy} & \mathbf{Q}_{zz} \end{bmatrix} = \begin{bmatrix} \mathbf{J}_{yx} & \mathbf{0} \\ \mathbf{0} & \mathbf{J}_{zt} \end{bmatrix} \begin{bmatrix} \mathbf{Q}_{xx} & \mathbf{Q}_{xt} \\ \mathbf{Q}_{tx} & \mathbf{Q}_{tt} \end{bmatrix} \begin{bmatrix} \mathbf{J}^t_{yx} & \mathbf{0} \\ \mathbf{0} & \mathbf{J}^t_{zt} \end{bmatrix}$$

$$= \begin{bmatrix} \mathbf{J}_{yx}\mathbf{Q}_{xx}\mathbf{J}^t_{yx} & \mathbf{J}_{yx}\mathbf{Q}_{xt}\mathbf{J}^t_{zt} \\ \mathbf{J}_{zt}\mathbf{Q}_{tx}\mathbf{J}^t_{yx} & \mathbf{J}_{zt}\mathbf{Q}_{tt}\mathbf{J}^t_{zt} \end{bmatrix}$$

from which we can write the following four general relationships

$$\mathbf{Q}_{yy} = \mathbf{J}_{yx}\mathbf{Q}_{xx}\mathbf{J}^t_{yx} \tag{4.45a}$$

$$\mathbf{Q}_{yz} = \mathbf{J}_{yx}\mathbf{Q}_{xt}\mathbf{J}^t_{zt} \tag{4.45b}$$

$$\mathbf{Q}_{zy} = \mathbf{J}_{zt}\mathbf{Q}_{tx}\mathbf{J}^t_{yx} = \mathbf{Q}^t_{yz} \tag{4.45c}$$

$$\mathbf{Q}_{zz} = \mathbf{J}_{zt}\mathbf{Q}_{tt}\mathbf{J}^t_{zt} \tag{4.45d}$$

The crosscofactor matrices are easily constructed as, for example,

$$\underset{n,\,m}{\mathbf{Q}_{xt}} = \begin{bmatrix} q_{x1t1} & q_{x1t2} & \cdots & q_{x1tm} \\ \vdots & \vdots & & \vdots \\ q_{xnt1} & q_{xnt2} & \cdots & q_{xntm} \end{bmatrix} \tag{4.46}$$

which directly leads to the fact that

$$\underset{n,\,m}{\mathbf{Q}_{tx}} = \mathbf{Q}^t_{xt}$$

4.4.5. Symbolic Multiplication In an effort to assist the practitioner, Tienstra† devised a scheme for the application of propagation rules that he called "symbolic multiplication." It is a strictly mnemotechnical rule that makes possible obtaining the elements of the covariance matrix separately. It is briefly given here for the sake of those readers consulting European literature.

In order to illustrate the scheme, we consider the following equations,

$$u = a_0 + a_1 x + a_2 y + a_3 z \tag{4.47}$$

$$v = b_0 + b_1 x + b_2 y + b_3 z$$

According to equation (4.41), the cofactors of u, v are

$$q_{uu} = a_1^2 q_{xx} + a_2^2 q_{yy} + a_3^2 q_{zz} + 2a_1 a_2 q_{xy} + 2a_1 a_3 q_{xz} + 2a_2 a_3 q_{yz}$$

$$q_{vv} = b_1^2 q_{xx} + b_2^2 q_{yy} + b_3^2 q_{zz} + 2b_1 b_2 q_{xy} + 2b_1 b_3 q_{xz} + 2b_2 b_3 q_{yz}$$

$$q_{uv} = a_1 b_1 q_{xx} + a_2 b_2 q_{yy} + a_3 b_3 q_{zz} + (a_1 b_2 + a_2 b_1) q_{xy} + (a_1 b_3 + a_3 b_1) q_{xz} + (a_2 b_3 + a_3 b_2) q_{yz} \tag{4.48}$$

† See bibliography at end of book.

The coefficients of the cofactors on the right-hand side of equation (4.48) can be obtained directly from corresponding multiplication of the original equations themselves [equations (4.47)] disregarding a_0, b_0. For instance, the product (uu) would be

$$(uu) = a_1^2(xx) + a_2^2(yy) + a_3^2(zz) + 2a_1a_2(xy) + 2a_1a_3(xz) + 2a_2a_3(yz) \tag{4.49}$$

The terms between parentheses refer to the indices of cofactors.

Comparing equation (4.49) to the first line of equation (4.48) shows that such a "multiplication" does in fact yield the coefficients of the cofactors. Thus with the understanding that products of random variables are shifted to indices of cofactors, any auto- or crosscofactor can be obtained by "symbolic multiplication" of the equations. It was devised before the extensive use of matrix algebra, and was helpful in particular when a single element of the total cofactor matrix is required. Recently the same scheme has also been applied to matrix equations. Of course, in this case the matrix sequence must be carefully observed, thus leading to transposition. For example, for

$$\mathbf{y} = \mathbf{Ax} + \mathbf{a}$$
$$\mathbf{z} = \mathbf{Bt} + \mathbf{b}$$

The crosscofactor matrix $\mathbf{Q}_{yz}$ may be obtained by "symbolic multiplication" of $\mathbf{y}$ and $\mathbf{z}$ (again disregarding the terms $\mathbf{a}$ and $\mathbf{b}$ which are irrelevant to propagation). Thus

$$\mathbf{yz}^{(t)} = (\mathbf{Ax})(\mathbf{Bt})^t = \mathbf{A}(\mathbf{xt}^{(t)})\mathbf{B}^t \tag{4.50}$$

The random vectors are transferred to indices for cofactor matrices and dropping the transpose symbol, or

$$\mathbf{Q}_{yz} = \mathbf{AQ}_{xt}\,\mathbf{B}^t \tag{4.51}$$

Noting that $\mathbf{J}_{yx} = \mathbf{A}$ and $\mathbf{J}_{zt} = \mathbf{B}$, equation (4.51) becomes identical to equation (4.45b).

4.4.6. Further Elaboration on Propagation Using Matrices Matrix formulation expounded in the preceding sections allows for the propagation of variances through several transformations. As an example, we consider the following three relations:

$$\begin{aligned}\mathbf{y} &= \mathbf{Ax} + \mathbf{a}\\ \mathbf{z} &= \mathbf{By} + \mathbf{b}\\ \mathbf{r} &= \mathbf{Cz} + \mathbf{c}\end{aligned} \tag{4.52}$$

Let the random vector $\tilde{\mathbf{x}}$ be known, with its cofactor matrix $\mathbf{Q}_{xx}$. Then there are two ways of obtaining the cofactor matrices of $\mathbf{z}$ and $\mathbf{r}$, known as "substitution" and "stepwise execution."

PROPAGATION THROUGH SUBSTITUTION We express **z** and **r** in terms of **x**, by substitution:

$$\begin{aligned}
\mathbf{y} &= \mathbf{Ax} + \mathbf{a} \\
\mathbf{z} &= \mathbf{By} + \mathbf{b} = \mathbf{B}(\mathbf{Ax} + \mathbf{a}) + \mathbf{b} = (\mathbf{BA})\mathbf{x} + (\mathbf{Ba} + \mathbf{b}) \\
\mathbf{r} &= \mathbf{Cz} + \mathbf{c} = \mathbf{C}(\mathbf{By} + \mathbf{b}) + \mathbf{c} = (\mathbf{CB})\mathbf{y} + (\mathbf{Cb} + \mathbf{c}) \\
&\qquad\qquad = (\mathbf{CB})(\mathbf{Ax} + \mathbf{a}) + (\mathbf{Cb} + \mathbf{c}) \\
&\qquad\qquad = (\mathbf{CBA})\mathbf{x} + (\mathbf{CBa} + \mathbf{Cb} + \mathbf{c})
\end{aligned} \tag{4.53}$$

Applying propagation of cofactor relations in equation (4.45) (or using symbolic multiplication) we get

$$\begin{aligned}
\mathbf{Q}_{yy} &= \mathbf{AQ}_{xx}\,\mathbf{A}^t \\
\mathbf{Q}_{zz} &= (\mathbf{BA})\mathbf{Q}_{xx}(\mathbf{BA})^t = \mathbf{BAQ}_{xx}\,\mathbf{A}^t\mathbf{B}^t \\
\mathbf{Q}_{rr} &= (\mathbf{CBA})\mathbf{Q}_{xx}(\mathbf{CBA})^t = \mathbf{CBAQ}_{xx}\,\mathbf{A}^t\mathbf{B}^t\mathbf{C}^t \\
\mathbf{Q}_{yz} &= \mathbf{AQ}_{xx}(\mathbf{BA})^t = \mathbf{AQ}_{xx}\,\mathbf{A}^t\mathbf{B}^t \\
\mathbf{Q}_{yr} &= \mathbf{AQ}_{xx}(\mathbf{CBA})^t = \mathbf{AQ}_{xx}\,\mathbf{A}^t\mathbf{B}^t\mathbf{C}^t \\
\mathbf{Q}_{zr} &= \mathbf{BAQ}_{xx}(\mathbf{CBA})^t = \mathbf{BAQ}_{xx}\,\mathbf{A}^t\mathbf{B}^t\mathbf{C}^t
\end{aligned} \tag{4.54}$$

STEPWISE PROPAGATION The same result in equation (4.54) can be obtained by applying propagation in steps as follows:

$$\begin{aligned}
\mathbf{y} &= \mathbf{Ax} + \mathbf{a} \qquad & \mathbf{Q}_{yy} &= \mathbf{AQ}_{xx}\,\mathbf{A}^t \\
\mathbf{z} &= \mathbf{By} + \mathbf{b} \qquad & \mathbf{Q}_{zz} &= \mathbf{BQ}_{yy}\,\mathbf{B}^t = \mathbf{BAQ}_{xx}\,\mathbf{A}^t\mathbf{B}^t \\
\mathbf{r} &= \mathbf{Cz} + \mathbf{c} \qquad & \mathbf{Q}_{rr} &= \mathbf{CQ}_{zz}\,\mathbf{C}^t = \mathbf{CBAQ}_{xx}\,\mathbf{A}^t\mathbf{B}^t\mathbf{C}^t \\
& & \mathbf{Q}_{yz} &= \mathbf{AQ}_{xy}\,\mathbf{B}^t \\
& & \mathbf{Q}_{yr} &= \mathbf{AQ}_{xz}\,\mathbf{C}^t \\
& & \mathbf{Q}_{zr} &= \mathbf{BQ}_{yz}\,\mathbf{C}^t
\end{aligned} \tag{4.55}$$

The last three (crosscofactor) relations of equation (4.55) do not correspond to those in equation (4.54), particularly because of the absence of the matrices $\mathbf{Q}_{xy}$, $\mathbf{Q}_{xz}$, and $\mathbf{Q}_{yz}$. However, these matrices can be derived if equations (4.52) are supplemented by simple identities. Thus

$$\begin{aligned}
\mathbf{x} &= \mathbf{Ix} \\
\mathbf{y} &= \mathbf{Ax} + \mathbf{a} \\
\mathbf{x} &= \mathbf{Ix} \\
\mathbf{z} &= \mathbf{By} + \mathbf{b} \\
\mathbf{y} &= \mathbf{Iy} \\
\mathbf{z} &= \mathbf{By} + \mathbf{b}
\end{aligned}$$

and

$$
\begin{aligned}
\mathbf{Q}_{xy} &= \mathbf{I}\mathbf{Q}_{xx}\,\mathbf{A}^t = \mathbf{Q}_{xx}\,\mathbf{A}^t \\
\mathbf{Q}_{xz} &= \mathbf{Q}_{xy}\,\mathbf{B}^t = \mathbf{Q}_{xx}\,\mathbf{A}^t\mathbf{B}^t \\
\mathbf{Q}_{yz} &= \mathbf{Q}_{yy}\,\mathbf{B}^t
\end{aligned} \tag{4.56}
$$

Substituting equation (4.56) in equation (4.55) leads directly to the equations in (4.54), thus ascertaining the equivalence of both procedures.

Example 4.3. Given

$$\begin{bmatrix} y_1 \\ y_2 \end{bmatrix} = \begin{bmatrix} 2 & 1 \\ 1 & 2 \end{bmatrix}\begin{bmatrix} x_1 \\ x_2 \end{bmatrix}; \qquad \boldsymbol{\Sigma}_{yy} = \begin{bmatrix} 3 & 1 \\ 1 & 3 \end{bmatrix}; \qquad z = x_1 + x_2$$

1. Compute σ_z^2, $\boldsymbol{\Sigma}_{zy}$, and $\boldsymbol{\Sigma}_{xy}$
2. Compute the elements of the standard ellipse for **x** and **z**.

Solution: 1. First

$$\begin{bmatrix} x_1 \\ x_2 \end{bmatrix} = \tfrac{1}{3}\begin{bmatrix} 2 & -1 \\ -1 & 2 \end{bmatrix}\begin{bmatrix} y_1 \\ y_2 \end{bmatrix}$$

and

$$
\begin{aligned}
\boldsymbol{\Sigma}_{xx} &= \mathbf{J}_{xy}\,\boldsymbol{\Sigma}_{yy}\,\mathbf{J}_{xy}^t = \tfrac{1}{9}\begin{bmatrix} 11 & -7 \\ -7 & 11 \end{bmatrix} \\
\sigma_z^2 &= \mathbf{J}_{zx}\,\boldsymbol{\Sigma}_{xx}\,\mathbf{J}_{zx}^t = \tfrac{8}{9} \\
\boldsymbol{\Sigma}_{zy} &= \mathbf{J}_{zx}\,\boldsymbol{\Sigma}_{xx}\,\mathbf{J}_{yx}^t = \tfrac{1}{3}[4 \quad 4] \\
\boldsymbol{\Sigma}_{xy} &= \mathbf{J}_{xy}\boldsymbol{\Sigma}_{yy}\,\mathbf{J}_{yy}^t = \tfrac{1}{3}\begin{bmatrix} 5 & -1 \\ -1 & 5 \end{bmatrix}
\end{aligned}
$$

2. (a) Standard ellipse for **x**: The characteristic equation is

$$\lambda^2 + \operatorname{tr}(\boldsymbol{\Sigma}_{xx})(-\lambda) + |\boldsymbol{\Sigma}_{xx}| = 0$$

or

$$\lambda^2 - \tfrac{22}{9}\lambda + \tfrac{8}{9} = 0$$

thus

$$\lambda_1 = 2 \qquad \text{and} \qquad \lambda_2 = \tfrac{4}{9}$$

For $\lambda_1 = 2$ the eigenvector is $(x_1, -x_1)$, and for $\lambda_2 = \frac{4}{9}$ the eigenvector is (x_1, x_1). Hence the semimajor axis of the standard ellipse is $a = \sqrt{\lambda_1} = \sqrt{2}$ and is oriented along a line making an angle $-45°$ with the x_1 axis. The semiminor axis is normal to it and is equal to $b = \sqrt{\lambda_2} = \frac{2}{3}$.

(b) Standard ellipse for z: Since this is a one-dimensional case, the ellipse degenerates into a line segment with $+\sigma_z = (2\sqrt{2})/3$ along the positive z axis from the estimate of the point, and $-\sigma_z$ in the opposite direction.

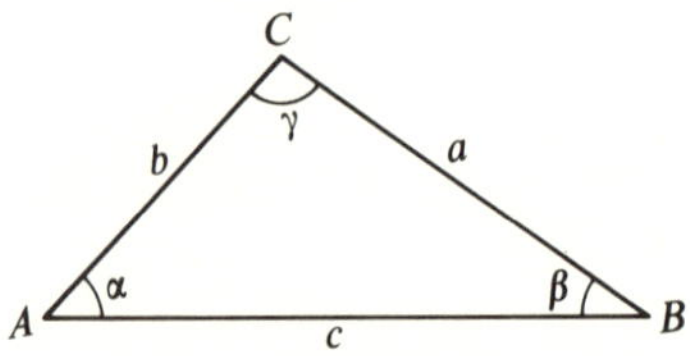

Figure 4.2

Example 4.4. Figure 4.2 shows a triangle for which

$$f = \alpha + \beta + \gamma - (\pi + \varepsilon)$$

represents the lack of closure, with ε being the spherical excess.
1. If s_α, s_β, s_γ are the standard deviations for the three angles, which are assumed to be uncorrelated, then

$$s_f^2 = [1 \quad 1 \quad 1]\begin{bmatrix} s_\alpha^2 & 0 & 0 \\ 0 & s_\beta^2 & 0 \\ 0 & 0 & s_\gamma^2 \end{bmatrix}\begin{bmatrix} 1 \\ 1 \\ 1 \end{bmatrix} = s_\alpha^2 + s_\beta^2 + s_\gamma^2$$

If, further, $s_\alpha = s_\beta = s_\gamma = s$, then

$$s_f^2 = 3s^2$$

2. Each angle may be considered as the difference between two directions that are uncorrelated and of equal precision, s_d^2. For example: $\alpha = d_1 - d_2$ and

$$s^2 = s_\alpha^2 = [1 \quad -1]\begin{bmatrix} s_d^2 & 0 \\ 0 & s_d^2 \end{bmatrix}\begin{bmatrix} 1 \\ -1 \end{bmatrix} = 2s_d^2$$

Thus

$$s_f^2 = 6s_d^2 \qquad \text{or} \qquad s_d^2 = \tfrac{1}{6}s_f^2$$

The latter relationship is often referred to as Ferrero's.

Example 4.5. In the plane triangle of Figure 4.2, the two angles α and β and the side c are measured and known to be uncorrelated. It is desired to derive the autocofactor of side a.

Solution:

$$a = \frac{c \sin \alpha}{\sin(\pi - \alpha - \beta)}$$

$$\mathbf{J} = \left[c\frac{(\cos\alpha \sin\gamma \ \sin\alpha\cos\gamma)}{\sin^2\gamma} \quad \frac{c\sin\alpha\cos\gamma}{\sin^2\gamma} \quad \frac{\sin\alpha}{\sin\gamma}\right]$$

$$= \left[(a\cot\alpha + a\cot\gamma) \quad a\cot\gamma \quad \frac{a}{c}\right]$$

$$q_{aa} = a^2(\cot\alpha + \cot\gamma)^2 q_{\alpha\alpha} + a^2\cot^2\gamma q_{\beta\beta} + \frac{a^2}{c^2}q_{cc}$$

Another procedure is as follows:

$$a = \frac{\sin \alpha}{\sin \gamma} \qquad \gamma = \pi - \alpha - \beta$$

$$\mathbf{J} = \begin{bmatrix} a \cot \alpha & -a \cot \gamma & \frac{a}{c} \end{bmatrix}$$

$$q_{aa} = a^2 \cot^2 \alpha q_{\alpha\alpha} + a^2 \cot^2 \gamma q_{\gamma\gamma} - 2a^2 \cot \alpha \cot \gamma q_{\alpha\gamma} + \frac{a^2}{c^2} q_{cc}$$

Since $q_{\gamma\gamma} = q_{\alpha\alpha} + q_{\beta\beta}$ and $q_{\alpha\gamma} = -q_{\alpha\alpha}$, then q_{aa} reduces to the same expression above.

4.5. PROPAGATION OF SYSTEMATIC AND "TRUE" ERRORS

In this section we shall discuss briefly the propagation of "known" terms through known functions. Let $\tilde{x}$, $\tilde{y}$ be random variables, for which the following function is given

$$r = g(x, y) \tag{4.57}$$

Let Δx, Δy be increments, the nature of which will be discussed below. Then we have directly for the propagation of the increments

$$r + \Delta r = g(x + \Delta x, y + \Delta y)$$

$$\Delta r = g(x + \Delta x, y + \Delta y) - g(x, y) \tag{4.58}$$

If the increments are small, the function can be expanded and only linear terms used. Thus

$$r + \Delta r = g(x, y) + \frac{\partial g}{\partial x} \Delta x + \frac{\partial g}{\partial y} \Delta y + \cdots$$

or

$$\Delta r = \frac{\partial g}{\partial x} \Delta x + \frac{\partial g}{\partial y} \Delta y \tag{4.59}$$

To this point, no probability considerations enter into the picture. It is treated as a calculus problem as far as the formula (4.59) is concerned. In practical applications, however, the increments Δx, Δy are in a way related to the error concept. They can be (a) corrections due to computational errors; (b) "true" errors that are assumed to be known; or (c) systematic errors.

It is essential that in all cases Δx, Δy are supposed to be known and not associated with random variables. This is, of course, a simplification. From a statistical point of view the case can be considered as a special case of propagation of means. The random variables $\tilde{x}$, $\tilde{y}$ are replaced by derived variables $u = x + \Delta x$, $v = y + \Delta y$, where Δx and Δy are also, in principle, random variables. We ask here for the expectation of $r = g(u, v)$ as compared to that of $r = g(x, y)$.

The above rule in equation (4.59) can be considered a limit case when Δx, Δy are considered as constants, their stochastic properties being reduced to zero. This propagation rule is from a theoretical point of view of very limited importance. However, it is useful in practice for studying certain effects, in particular in connection with simplified mathematical models. Systematic errors are, in principle, deficiencies of the mathematical model. Consequently, such rules can be used in studying the propagation of residual systematic effects.

Problems for Part I

1. From the bivariate normal density function for two correlated random variables $\tilde{x}$, $\tilde{y}$ [see equation (2.70)], derive the marginal distributions for each variable $\tilde{x}$ and $\tilde{y}$.

2. Given the bivariate normal density function $f(x, y)$ [equation (2.70)] derive the conditional density function $f(x \mid y)$.

3. From equation (2.69) give the (noncentral) joint normal density function for the variables $\tilde{x}$, $\tilde{y}$, $\tilde{z}$ such that there is no correlation between $\tilde{x}$ and $\tilde{z}$ ($\rho_{xz} = 0$) and between $\tilde{y}$ and $\tilde{z}$ ($\rho_{yz} = 0$). Show that the joint density can be written as the product of the bivariate density function $f(x, y)$ and the univariate density function $f(z)$.

4. Show that the mean is zero and the variance is equal to one for the standardized random variable

$$\tilde{z} = \frac{\tilde{x} - E(\tilde{x})}{\sigma_x}$$

5. Derive the mean and variance of the χ^2 distribution, given in equations (2.61) and (2.62).

6. Given the following 20 measurements of an angle in order of their observation:

(a) 9.3″	(b) 6.6″	(c) 7.9″	(d) 6.8″
(e) 6.1″	(f) 5.9″	(g) 6.1″	(h) 7.8″
(i) 7.2″	(j) 8.0″	(k) 7.1″	(l) 4.9″
(m) 5.7″	(n) 5.2″	(o) 6.9″	(p) 6.2″
(q) 10.1″	(r) 9.7″	(s) 4.1″	(t) 6.2″

Sketch a histogram of the measurements using an interval of 0.5″ and a lower bound of 4.0″ and compute the sample mean value, the sample standard deviation of a single measurement, and of the mean value.

7. Given a random variable $\tilde{y}$ with standard normal distribution, what is the probability that (a) $\tilde{y}$ assumes a value between 0.35 and 1.65; (b) $\tilde{y}$ assumes a value between -1.96 and 2.42.

8. The distribution of the diameters of the ball bearings in a certain shipment is approximately normal with mean 0.500 cm and standard deviation 0.010 cm. If a ball bearing is effective when its diameter lies between 0.490 and 0.515 cm, find the probability of obtaining an effective ball bearing from the shipment. (Use the standard normal tables.)

9. The following observations were recorded for the measurement of two angles, $\tilde{\alpha}$ and $\tilde{\beta}$, each measured 20 times.

OBSERVATION	$\tilde{\alpha}$	$\tilde{\beta}$
1	20° 05′ 17.9″	16° 42′ 29.8″
2	17.0	29.9
3	19.1	30.1
4	17.1	31.3
5	17.7	31.3
6	16.8	30.9
7	17.6	30.7
8	18.5	29.5
9	17.6	30.0
10	16.4	31.2
11	17.5	30.2
12	18.4	30.2
13	17.6	30.6
14	16.6	29.1
15	18.2	30.3
16	17.8	29.9
17	17.0	29.8
18	18.3	29.5
19	17.7	29.6
20	18.0	30.6

(a) Compute the sample variance and sample standard deviation for a single observation of each of the angles $\tilde{\alpha}$ and $\tilde{\beta}$. (b) Compute the sample covariance and estimate of the correlation coefficient between $\tilde{\alpha}$ and $\tilde{\beta}$. (Assume $\tilde{\alpha}$ and $\tilde{\beta}$ are the random variables of a bivariate distribution.) (c) Write out the variance–covariance matrix for the observations $\tilde{\alpha}$, $\tilde{\beta}$.

10. A distance x is measured nine times with sample mean of 190.09 meters and sample variance of the mean of 4 cm^2. Present the confidence intervals for μ_x and σ_x at the 95% and 99% levels.

11. A distance x is measured 30 times with sample mean of 190.09 meters and sample standard deviation of a single value of 6 cm. Present the confidence intervals for μ_x and σ_x at the 95% and 99% levels.

12. A sample of nine measurements of a distance yields a mean value of 8.4 cm and a standard deviation in the mean value of 0.4 mm. Test the hypothesis that the true length of the distance is 8.0 cm against the alternative that the true length is greater than 8.0 cm. Use a 5% level of significance.

13. The average of 100 measurements is 21.0 cm with a standard deviation in a single measurement of 1.5 cm. Can it be claimed that the true length of the measurement exceeds 20.0 cm? Use a 5% level of significance.

14. The following 15 measurements were observed and assumed to be from a normal distribution:

132.2	128.0	125.0	130.2	126.6
128.6	127.8	128.4	128.0	128.4
128.0	128.5	128.4	128.7	127.2

Test the following hypotheses at a 5% level of significance:

a. H_0: $\mu = 130.0$ against H_1: $\mu \neq 130.0$
b. H_0: $\mu = 130.0$ against H_1: $\mu < 130.0$
c. H_0: $\sigma^2 = 9.0$ against H_1: $\sigma^2 \neq 9.0$
d. H_0: $\sigma^2 = 5.0$ against H_1: $\sigma^2 < 5.0$

15. The following two sets of measurements were supposedly taken of the same physical quantity by identical techniques:

Sample I:	0.806	0.827	0.816	0.809	0.814
Sample II:	0.813	0.803	0.810	0.807	0.804

Test the following hypotheses at a 5% level of significance:

a. H_0: $\mu_{\text{I}} - \mu_{\text{II}} = 0$ against H_1: $\mu_{\text{I}} - \mu_{\text{II}} \neq 0$
b. H_0: $\sigma_{\text{I}}^2 = \sigma_{\text{II}}^2$ against H_1: $\sigma_{\text{I}}^2 \neq \sigma_{\text{II}}^2$

16. The random variable $\tilde{X}$ has mean μ_x. The random variable $\tilde{Y}$ has mean μ_y. If a and b are constants, determine the expected value of $\tilde{Z}$ in terms of a, b, μ_x, and μ_y if $\tilde{Z} = \tilde{X} - a\tilde{X} + \tilde{Y} - b\tilde{Y}$.

17. A sample of 12 measurements of an angle θ yields a mean value of 60 degrees. The standard deviation in the mean value of the angle is 30 seconds. Present a 95% confidence interval for μ_θ and σ_θ^2.

18. From two sets of measurements supposedly from the same population, the following data were recorded:

SET	n	$\bar{X}$	σ_X
I	8	10.1	0.10
II	12	10.0	0.07

At the 10% level of significance, test the hypothesis that the variances are equal against the alternative that they are not equal.

19. If $x_1 = [1 \quad 2]\begin{bmatrix} u_1 \\ u_2 \end{bmatrix}$, $x_2 = [2 \quad 3]\begin{bmatrix} u_1 \\ u_2 \end{bmatrix}$, and $\mathbf{\Sigma}_{uu} = \begin{bmatrix} 1 & 0 \\ 0 & 2 \end{bmatrix}$ is the covariance matrix of u_1 and u_2, compute the covariance matrix $\mathbf{\Sigma}_{zz}$ for

$$\mathbf{z} = \begin{bmatrix} z_1 \\ z_2 \end{bmatrix} = \begin{bmatrix} 6 & 2 \\ 2 & 3 \end{bmatrix}\begin{bmatrix} x_1 \\ x_2 \end{bmatrix}$$

20. Given the cartesian coordinates $x = 300$ mm with $\sigma_x = 0.04$ mm, and $y = 400$ mm with $\sigma_y = 0.05$ mm, where x and y are assumed uncorrelated, compute the polar coordinates r and θ and their covariance matrix.

21. $x_1 = [1 \quad 3]\begin{bmatrix} u_1 \\ u_2 \end{bmatrix}$, $x_2 = [1 \quad 4]\begin{bmatrix} u_1 \\ u_2 \end{bmatrix}$, $x_1^2 + x_2^2 = y^2$, and $\mathbf{\Sigma}_{uu} = \begin{bmatrix} 0.1 & 0 \\ 0 & 0.2 \end{bmatrix}$. Compute σ_y if $u_1 = 0$ and $u_2 = 1$.

22. $\mathbf{x} = [2 \quad 3]^t$, $\mathbf{\Sigma}_{xx} = \begin{bmatrix} 1 & -0.1 \\ -0.1 & 2 \end{bmatrix}$, $y = \mathbf{x}^t\mathbf{x}$. Compute σ_y^2.

23. $y = x_1 \cos\theta - x_2 \sin\theta$, $\theta = 45°$, $\mathbf{W}_{xx} = \begin{bmatrix} 5 & 1 \\ 1 & 4 \end{bmatrix}$. Compute the weight of y if $\sigma_0^2 = 1$.

24. $z = [1 \quad 2]^t$, $\mathbf{\Sigma}_{zz} = \begin{bmatrix} 2 & 0 \\ 0 & 3 \end{bmatrix}$, $u = 2\mathbf{z}^t\mathbf{A}\mathbf{z}$ for $\mathbf{A} = \begin{bmatrix} 1 & 1 \\ 1 & 2 \end{bmatrix}$. Compute σ_u^2.

25. The figure below shows a small triangulation chain, in which L is the measured baseline and $\alpha_1, \alpha_2, \alpha_3, \beta_1, \beta_2$, and β_3 are known angles. The exit base M may be computed from

$$M = L\frac{\sin\alpha_1 \sin\alpha_2 \sin\alpha_3}{\sin\beta_1 \sin\beta_2 \sin\beta_3}$$

Compute the value of M and its standard deviation if the measured quantities and their corresponding estimated accuracies are

$L = 5286.278$ meters; $\sigma_L = 12.7$ mm

ANGLE	DEGREES	MINUTES	SECONDS	σ IN SECONDS
α_1	53	56	41.3	7.6
β_1	74	47	40.2	10.5
α_2	50	42	26.3	6.8
β_2	81	32	56.3	2.3
α_3	63	50	46.6	3.7
β_3	69	22	06.5	5.3

Note: All observations are assumed to be uncorrelated.

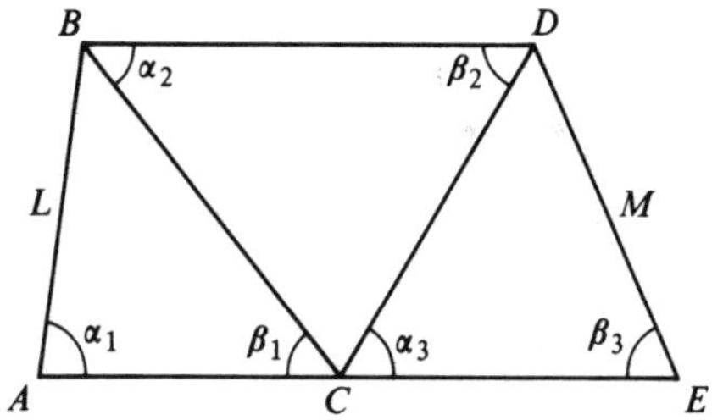

26. The following two functions x_1 and x_2 relate three observed correlated variates u_1, u_2, and u_3, where $u_1 = 3.0$, $u_2 = 4.0$, and $u_3 = 5.0$,

$$x_1 = u_1^2 + 3u_2^3 + u_3$$

$$x_2 = \frac{1}{u_1^2} + 4u_2 + \frac{1}{u_3}$$

Compute the covariance matrix of the functions x_1 and x_2 given that the weight matrix of the observations is

$$\mathbf{W}_{uu} = \begin{bmatrix} 2 & 0.5 & 0.6 \\ 0.5 & 1 & -0.7 \\ 0.6 & -0.7 & 3 \end{bmatrix}$$

and the reference variance is 0.8.

27. It is estimated that in 1 km of differential leveling, the standard deviation in elevation difference determination is 2 mm. The elevation of bench mark A is considered known at a standard deviation of 3 mm. How long a level line can be run from BM A to establish the elevation of point Q at a standard deviation of 1 cm?

28. It is estimated that each interior angle of a nine-sided closed traverse is measured with a standard deviation of 10 seconds. Within what range would you expect the sum of the measured interior angles of the traverse to fall, knowing that the sum of the interior angles of an n-sided polygon is $(n - 2)$ 180 degrees?

29. The upper base (UB), lower base (LB), and height (H) of a trapezoid are each measured n times and the respective standard deviation of a single measurement for each is determined (σ_{UB}, σ_{LB}, and σ_H). The mean value of the area of the trapezoid is determined from $\bar{A} = [(\overline{UB} + \overline{LB})/2]\bar{H}$. What is the standard deviation of $\bar{A}$ in terms of the variables identified? (Assume the measurements on UB, LB, and H are independent.)

30. σ_A, σ_B, σ_C, and σ_D are the standard deviations of A, B, C, and D, respectively. The measurements on A, B, C, and D are not independent, hence $\sigma_{AB} \neq \sigma_{AC} \neq \sigma_{AD} \neq \sigma_{BC} \neq \cdots \neq 0$. If $S = A + 2B - 3C + D$, what is the standard deviation of S?

31. In a triangle, the misclosure $w = a + b + c - (180° - e)$, where a, b, and c are the three measured angles with standard deviations σ_a, σ_b, and σ_c, respectively, and e is the spherical excess. (a) Determine the standard deviation of w, σ_w. (In terms of variables defined above.) (b) Determine σ_w if $\sigma_a = \sigma_b = \sigma_c = \sigma_A$. (c) If each angle is

determined as the difference of two directions as read from a horizontal circle of a theodolite, and if the standard deviation for a direction σ_D is assumed equal for all directions, represent σ_a, σ_b, σ_c, and σ_w in terms of σ_D.

32. In stadia distance measurement, the horizontal distance $D = KR \cos^2 \alpha$, where K is the stadia constant (normally 100), R is the level rod intercept (upper wire reading − lower wire reading), and α is the vertical inclination of the telescope above horizontal. At a sight distance of 300 feet, the wire readings on the level rod can be made with a standard deviation of approximately ± 0.01 feet. (a) Determine the standard deviation in D for a horizontal sight distance of 300 feet. (b) State or show why the effects of error in K and α are much less than those in the reading of the level rod.

33. In a subtense bar distance measurement process, the horizontal distance $D = (b/2) \cot (\alpha/2)$, where b is the length of the subtense bar, α is the horizontal angle measured between the targets designating the length of the subtense bar, and the line of sight is perfectly perpendicular to the subtense bar at its midpoint. If the length of the subtense bar is 2 meters $\pm$ 0.1 mm, determine the standard deviation that must be attained in the measurement of the angle α in order to measure a 100-foot line with a relative error of 1/5000. Repeat for a 200-foot line and a 300-foot line. (Note: For the purposes of this problem consider the relative error to be the ratio of the estimated *standard deviation* in the line length over the *line length*.)

34. The figure shows a scaled layout of a land area where all the angles in the figure are perfectly known to be right angles. You are given the following information:

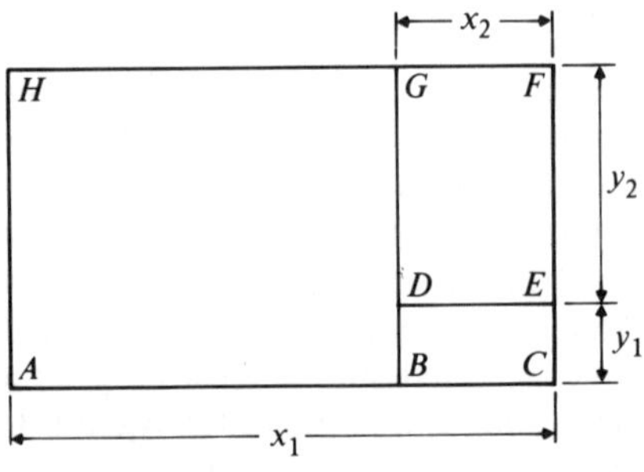

$x_1 = 7$ cm
$x_2 = 2$ cm

$$\Sigma_{xx} = \begin{bmatrix} 2 & 1 \\ 1 & 2 \end{bmatrix} \text{mm}$$

$y_1 = 1$ cm
$y_2 = 3$ cm

$$\Sigma_{yy} = \begin{bmatrix} 2 & 1 \\ 1 & 2 \end{bmatrix} \text{mm}$$

Assuming no correlation between the x_i and y_i measurements, compute the area of the rectangle *ABGH* and its variance.

35. An observable quantity l has been measured three times with assumed uncorrelated values l_1, l_2, and l_3 having standard deviation σ_1, σ_2, and σ_3, respectively. Using the *propagation* principle derive the *standard deviation* of the weighted mean of

these three observations. (Assume that the reference variance is unity.) The weighted mean $\bar{x}$, of any variable x, is given by

$$\bar{x} = \frac{(w_1 x_1 + w_2 x_2 + \cdots w_n x_n)}{w_1 + w_2 + \cdots w_n}$$

where x_i is the ith observation and w_i is the ith weight, *assuming no correlation.*

36. a. Sketch the error ellipse for

$$\boldsymbol{\Sigma}_{xx} = \begin{bmatrix} \frac{5}{2} & \frac{3}{2} \\ \frac{3}{2} & \frac{5}{2} \end{bmatrix}$$

b. Describe in your own words what the above ellipse means.

PART II

5

Introduction to Least Squares Adjustment

5.1. GENERAL

In Part I the reader was introduced to the basic concepts underlying observations, the mathematical model, and adjustment. It was pointed out that adjustment is in general meaningful only in those cases in which redundant observations are available. In a statistical sense, adjustment is a method of deriving estimates for stochastic variables and their distribution parameters from observed samples. Of the different adjustment methods *least squares* is by far the most common.

Since its first application to an astronomical problem by C. F. Gauss, least squares adjustment has been introduced and applied in a vast number of fields in science and engineering. Its practical importance has recently been enhanced by the introduction of electronic computers, by the formulation of its techniques in matrix notation, and by connecting its concept to statistics.

Before planning observations, a general functional model about the system to be assessed must be specified. Such a functional model, which refers to a finite closed system, is determined by a certain number of variables (parameters or observations or both) and the relationships between them.

There is always a *minimum number* of independent variables that uniquely determines a chosen model. It is important to note that the same physical object or set of events can be associated with different functional models. These models would have different elements depending on the particular choice made for the purpose or extent to which information is needed. Once that choice is made and the problem is specified, the *minimum number* of variables is consequently fixed—although the choice of actual variables still remains optional. In other words, we would determine the minimum number of variables without specifying the particular variables, as illustrated by the following examples.

1. The *shape of a plane triangle* (a particular geometric functional model) is uniquely determined by *two distinct* variables (minimum number). These may be chosen as *any* two angles of the three, or as the ratios of the three sides (optional choice of actual variables).
2. The *size and shape of a plane triangle* (another particular functional model) require a minimum of *three distinct* variables (which therefore cannot be all angles since a linear dimension is necessary for determining size). Several choices remain open, such as two angles and one side, two sides and one angle, or three sides.
3. If, in addition, the location and orientation with respect to a specified cartesian coordinate system of a plane triangle (yet another functional model) are of interest, a *total* of *six* variables becomes necessary. Obviously more choices also exist in this case.

As has been indicated above, once a model is selected it is expressed by a minimum number of distinct variables. This minimum number of variables will always be referred to by n_0. Unless the given observations are sufficient for determining the n_0 variables, the situation will obviously be deficient. We denote the number of observations by n. These observations must be functionally independent, that is, not one of the n observations can be derived from any or all remaining $(n - 1)$ observations. When n is larger than n_0, *redundancy* is said to exist and *adjustment* is needed in order to obtain a unique set of estimates for the model variables. The redundancy, which is denoted by r, is given by

$$r = n - n_0 \tag{5.1}$$

and is equal to the (*statistical*) *degrees of freedom.*

The redundancy r is meaningful only if the observations and the functional model are mutually consistent. The observations should be sufficient to determine the model, that is, they should refer to at least the minimum number of distinct random variables. Otherwise there will be a deficiency in the model.

A deficiency can exist in spite of apparent existence of redundancy as, for instance, when multiple observations are present (see the following example). In complex systems such as large photogrammetric blocks, deficiency may be encountered without the presence of multiple observations.

Example 5.1. Denote the three interior angles of a plane triangle by α, β, γ, and assume that determining the shape of the triangle is of interest. Therefore the functional model is defined as that of the shape of a triangle for which $n_0 = 2$, each referring to one distinct angle α, β, or γ. Now, there are several possibilities for actual observations.

1. Each angle is observed once, that is, α_1, β_1, γ_1. Thus $n = 3$ and $r = n - n_0 = 1$.
2. Each angle is observed twice, or α_1, α_2, β_1, β_2, γ_1, γ_2. Hence $n = 6$ and $r = 4$.
3. Angle α observed twice, and β, γ each observed once, or α_1, α_2, β, γ. Here $n = 4$ and $r = 2$.
4. Angle α is observed four times, α_1, α_2, α_3, α_4. In this case although $n = 4$, we actually have a deficient situation, because the observations do not refer to the minimum of $n_0 = 2$ distinct variables of the specified model. Therefore these observations are no longer adequate for the model and the shape of the triangle cannot be determined.

If we are asked to indicate what can be determined from these observations, the model must first be reduced to that of one angle α, for which $n_0 = 1$. In this case $r = 3$ and the adjustment would involve obtaining a unique estimate for the angle α from the four given observations.

Part (4) of the above example demonstrates that before we can evaluate the value of the redundancy r, a check must be made to ascertain that the total model selected is determinable from the given observations.

5.2. THE LEAST SQUARES PRINCIPLE

Because of the inherent stochastic properties of observations, redundant observations are not usually compatible with the functional model. Any sufficient subset of functionally (but not necessarily stochastically) independent observations can be used to assess the model. Due to random variability of the observational sampling, each minimum subset would yield a different result. For example, any two of three measured angles in a plane triangle suffice to determine its shape. However, for every possible pair of angles (for example, α and β; β and γ; or γ and α) a triangle shape results that is in general different from that obtained from another pair; that is, no unique result is possible from redundant observations unless an additional criterion is introduced. In this situation it is the basic principle of adjustment to derive a unique set of estimates for all the model variables, with certain optimum properties.

The original set of observations, which will be denoted by the vector $\boldsymbol{l}$ and which includes redundant observations and is inconsistent with the model, is replaced in the adjustment by another set of estimates $\hat{\boldsymbol{l}}$ which satisfies the model. This implies that the functional relationships comprising the model are accepted as superior to the given sample values of the observations. Consequently, after the adjustment is performed, model relationships remain unaltered and will be strictly satisfied by the newly estimated set of values $\hat{\boldsymbol{l}}$.

The set of estimates $\hat{\boldsymbol{l}}$ is in general different from the original set $\boldsymbol{l}$. The difference between the two sets is

$$\mathbf{v} = \hat{\boldsymbol{l}} - \boldsymbol{l} \tag{5.2}$$

which has been termed in the classical theory as either "corrections" or "residuals." Neither of the two terms is descriptive or precise, and other suggested terms such as "alterations" or "shifts" are equally so. For lack of a better term, the term "residuals" is selected as adequate, and will be used throughout this book.

The vector of residuals $\mathbf{v}$ plays an important role after the adjustment process. It is often possible to analyze the elements of $\mathbf{v}$ in order to test the adequacy of the model. In certain instances this analysis may even lead to discarding a chosen model completely and to remodeling the problem.

Due to redundancy, there would be an infinite number of estimates for $\mathbf{v}$, or $\hat{\boldsymbol{l}}$, which would satisfy the model. Amongst all the possibilities there exists one set of estimates that, in addition to being consistent with the model, satisfies another criterion commonly referred to as the *least squares principle*. This principle endeavors to ascertain that the new estimates $\hat{\boldsymbol{l}}$ are as close as possible to the sample values of the observations $\boldsymbol{l}$ taking their stochastic properties also into consideration. Such a criterion is obviously plausible since the available sample of observations is the best that we have, although it is superseded by the model functions. Therefore any variations in $\boldsymbol{l}$ necessitated by the existence of inconsistencies with the model due to redundancy must be as small as possible. In the limit when $r = 0$, $\hat{\boldsymbol{l}} = \boldsymbol{l}$ and the residuals will all be zero.

The least squares principle states that

$$\phi = \mathbf{v}^t\mathbf{W}\mathbf{v} \rightarrow \text{minimum} \tag{5.3}$$

where $\mathbf{W}$ is the weight matrix of the observations. It should be denoted by $\mathbf{W}_{ll} = \mathbf{Q}_{ll}^{-1}$ with $\mathbf{Q}_{ll}$ being the cofactor matrix. However, since they will be used extensively, neither $\mathbf{Q}$ nor $\mathbf{W}$ will have a subscript for reasons of simplification. In the exceptional cases when ambiguity may arise, the subscripts will be restored. The weight matrix $\mathbf{W}$ is square and of order equal to n, the number of observations. Its elements reflect the stochastic properties, such as variation and correlation, of all the observations as discussed in Part I.

The minimum criterion given above is the most general expression. Several special cases may be derived from it by considering weight matrices of special structure. The first specialization is when the observations are considered as uncorrelated. The weight matrix **W** would be a diagonal matrix, and the minimum principle becomes

$$\phi = \sum_{i=1}^{n} (w_i v_i^2) \rightarrow \text{minimum} \tag{5.4}$$

where w_i is the ith diagonal element of **W** and v_i is the residual associated with the corresponding ith observation.

A yet simpler case is when the observations are taken as of equal precision in addition to being uncorrelated. Then **W** becomes the identity matrix and

$$\phi = \sum_{i=1}^{n} (v_i^2) \rightarrow \text{minimum} \tag{5.5}$$

will be the form of the minimum principle. This last case is the oldest and most classical, and possibly the one that gave rise to the name "least squares," since in this case we seek the "least" of the sum of the squares of the residuals.

It is important to note that the application of least squares principle given above *does not require a priori knowledge of the distribution associated with the observations.* All that is necessary is to have either **W** or **Q** defined and known. In the past it has been erroneously stated that least squares adjustment requires normal distribution. Perhaps the reason for this incorrect assertion is that when the observations are normally distributed, the least squares estimates will have some special properties, such as being identical to those from the method of maximum likelihood.

Least squares has been by far the predominant technique of data adjustment in photogrammetry, geodesy, surveying, and many other fields. The reason for such extensive use lies in the fact that it yields a computational algorithm for a unique answer even in very complicated cases. This was essential from the beginning, as early attempts at adjustment of geodetic nets in the nineteenth century failed mainly due to the lack of a suitable algorithm. The advantage of such an algorithm has been enhanced lately by the adoption of matrix notation and the use of electronic computers for performing the computations.

5.3. THE TECHNIQUES OF LEAST SQUARES

The mathematical model represents the point of beginning at which the basic philosophy of the adjustment task is established. Once this has been accomplished, the model remains in the background and consideration is then given to the practical and computational aspects of selecting a particular technique of least squares. The relationship between these phases is shown schematically in Figure 5.1. The least squares computations yield updated estimates of all model variables as well as their respective cofactor or covariance matrices. After the computational algorithm is applied to the given data, another step is necessary where the results are statistically evaluated. This is again a judgment operation which lies on the same side as the model. In fact, depending on the degree of sophistication, the results of the statistical evaluation could lead to remodeling of the adjustment task if the original model is found to be inadequate.

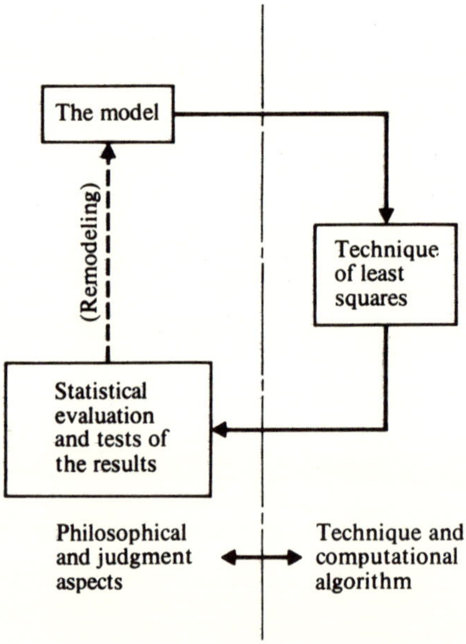

Figure 5.1

Although for a specified model and a given set of data the least squares yields unique results, there are several techniques that can be employed. It should be emphasized that whatever technique is used, the final answers are always the same.

In addition to the observations, the model may also include other variables and numerical constants. The family of those other variables, which are also stochastic, will be termed "parameters" to distinguish them from the observations with given a priori sample values. Invariably these parameters have unknown values at the beginning and estimates are therefore derived for them in the adjustment. They represent functional unknowns, although after the adjustment they are treated as stochastic variables undergoing statistical testing just like the variables representing the observations.

The number of parameters carried in the adjustment will be designated by the letter u (for unknowns). The vector $\Delta_{u,1}$ will be used to represent those parameters. Once the functional and stochastic models are established, the least squares technique or algorithm operates with a set of mathematical functions or equations. These equations describe the functional model of the adjustment problem. For example, the equation for the closure of a plane triangle and the relationship equating the final values of two observations on a distance or angle express the functional models of the respective problems. We shall distinguish between two types of equations: condition equations and constraint equations. Any equation that includes one or more *observations* will be called a *condition equation.* Consequently, every adjustment problem will entail condition equations. The total number of these equations or conditions in the adjustment will always be denoted by c. By contrast, those equations that *do not* include any observations will be called *constraint equations, parameter constraints,* or simply *constraints.* By definition, these equations will be functions of only parameters and constants. If the same vector of parameters Δ appears in both the condition equations and constraint equations, the number of the latter (constraints) will be given the letter s. In some cases, in formulating the constraints it may be convenient to include in addition to Δ another vector of *added parameters* Δ'. In those cases s' will represent the number of constraints and q the number of added parameters.

To recapitulate then, Table 5.1 gives the names of the different symbols that have been introduced.

TABLE 5.1

SYMBOL	EXPLANATION
l	Vector of sample values of the observational variables $\tilde{l}$ which will also be denoted by l for simplification
$\hat{l}$	Estimated ("corrected") observations
$\mathbf{v}$	Vector of residuals
Δ	(Unknown) parameters
Δ'	Added parameters (appearing in constraint equations)
n_0	Minimum number of distinct variables expressing a specified model
n	Number of given observations (estimates)
r	The redundancy or (statistical) degrees of freedom
u	Number of (unknown) parameters (in Δ)
q	Number of added parameters (in Δ')
c	Number of condition equations
s	Number of constraints (in terms of only Δ)
s'	Number of constraints with added parameters Δ'
$\mathbf{Q}$	Cofactor matrix of the observations l (replacing $\mathbf{Q}_{ll}$)
$\mathbf{W}$	Weight matrix of the observations l (replacing $\mathbf{W}_{ll}$)

There are two groups of adjustment techniques that will be treated as one encompassing technique at the end. The two groups are entitled, (a) *least squares adjustment with conditions only*, and (b) *least squares adjustment with conditions and constraints*. These two groups of techniques arise mainly because of making a distinction between observations and parameters (hence conditions and constraints). Recently, it was found that this distinction can be eliminated by considering all model variables as observations, thus leading to only one inclusive method called (c) *a unified least squares adjustment*.

Since many practical adjustment problems can be treated by a technique from the first two groups, they will therefore be developed in detail in the remainder of Part II. Then the unified approach will be introduced in Part III.

The schematic of Figure 5.1 showed that we should, for practical reasons, differentiate clearly between the setting up of the model and the least squares adjustment technique. It must be stressed that the observations, or rather the variables they refer to, including additional parameters associated with them (calibration parameters, orientation parameters, and so on) are part of the total system. The eventual merits of an adjustment are mostly decided by the choice of the model at the beginning.

For a long time the importance of the model concept has not been realized, because in conventional applications the model is given a priori and therefore it is not problematic. But even if no specific reference to the model is made when performing an adjustment, it is nevertheless fixed by implication. Many questions about least squares adjustment, and also practical difficulties, refer in fact to the problem of adequate modeling.

Although the techniques of least squares adjustment are mainly concerned with obtaining the least squares estimates for a given model, the real problem is often to refine or change the model if preliminary adjustments turn out to be unsatisfactory.

5.4. LINEAR AND NONLINEAR FUNCTIONS IN THE MODEL

The conditional as well as the constraint equations involved in an adjustment problem can, in general, be nonlinear. However, least squares treatments are generally performed with linear functions, since it is rather difficult and often impractical, at least at present, to seek a least squares solution of nonlinear equations. Consequently, whenever the equations in the model are originally nonlinear, some means of linearization must be used to get linear equations. Series expansions, and Taylor's series in particular, are often used for the purpose, where only the zero and first-order terms are used and all other higher-order terms are neglected.

When applying a series expansion, a set of approximate values for the unknowns in the equations must be chosen. The choice of those approximations is an important aspect of solving the problem at hand. Unfortunately there is no concrete and unique way of choosing approximations that can be applied to all adjustment problems. Sometimes experience is relied on, and in other situations some computational shortcuts may be employed. In all cases the attempt should always be made to obtain close approximations that can be obtained by using relatively simple and uninvolved techniques.

To demonstrate how linearization is performed, let any set of m nonlinear functions be denoted by

$$F(\mathbf{x}) = \mathbf{0} \tag{5.6}$$

where $\mathbf{x}_{p,1}$ is a vector of unknown variables (which in the context of adjustment may be observations, parameters, and so on). If $\mathbf{x}^0_{p,1}$ denotes the vector of approximate values of the variables, the zero and first-order terms of a series expansion would be

$$F(\mathbf{x}^0) + \left.\frac{\partial F(\mathbf{x})}{\partial \mathbf{x}}\right|_{\mathbf{x}=\mathbf{x}^0} \cdot \Delta\mathbf{x} = \mathbf{0} \tag{5.7}$$

The set of partial derivatives of the functions with respect to the elements of the vector of variables is a rectangular matrix $\mathbf{U}$ of dimensions $m \times p$. The vector $\Delta\mathbf{x}$ is a $p \times 1$ vector of corrections to the approximations which replaces the vector of unknowns $\mathbf{x}$. The result of applying the series expansion is that the nonlinear equation (5.6) becomes a set of linear equations of the general form

$$\underset{m,p}{\mathbf{U}}\ \underset{p,1}{\Delta\mathbf{x}} = \underset{m,1}{\mathbf{u}} \tag{5.8}$$

where $\mathbf{u} = -F(\mathbf{x}^0)$.

After least squares adjustment, we get as a solution the vector $\Delta\mathbf{x}$. If the original approximation vector $\mathbf{x}^0$ was sufficiently close for equation (5.7) to be an adequate substitute for equation (5.6), that is, for second- and higher-order terms of the series to be in fact negligible, then the final least squares estimate is $(\mathbf{x}^0 + \Delta\mathbf{x})$. Often, however, $\mathbf{x}^0$ is not that close an approximation and adding $\Delta\mathbf{x}$ to $\mathbf{x}^0$ yields *only an improved* approximation. The updated vector of approximations must now be used again to formulate equation (5.8) or equation (5.7) and least squares is used to obtain an updated vector $\Delta\mathbf{x}$ the elements of which are, in general, smaller than those of the first one. The process of relinearization at an updated vector of approximation continues until the last value of $\Delta\mathbf{x}$ is insignificantly small and the iterative procedure terminates. The final estimate $\hat{\mathbf{x}}$ will be the sum of the original approximation $\mathbf{x}^0$ and *all* the correction vectors $\Delta\mathbf{x}$. (Or equivalently the last updated approximation vector plus the last correction vector.)

One must note that the method of least squares is applied only within each iteration and on a linear (linearized) set of equations. It does not really have anything to do with the linearization process. The linearization is only a scheme to treat a problem of nonlinear estimation which although possible in principle, is not normally solved directly by least squares.

6

Adjustment with Conditions Only—General Case

Adjustment of Observations and Functionally Independent Parameters

6.1. INTRODUCTION

In this group of techniques the parameters, if they exist, are functionally independent. The existence of the parameters in the adjustment is relatively arbitrary and will depend mostly on the type of problem to be solved. In principle, we can perform the adjustment using only the observations. However, it may be considerably more convenient to include a parameter vector. This is particularly true when the parameters themselves are the variables of interest.

When starting an adjustment, the first thing to be specified is the model, particularly the functional model, which will specify n_0. Then the given n observations are examined with respect to the model in order to ascertain that there is no deficiency. Thus the redundancy $r = n - n_0$ can be computed. This redundancy, or degrees of freedom, may be interpreted to mean that among the n observations there exist r functions (conditions) that must be satisfied. This can be explained by considering that when $r = 0$, then $n = n_0$ and the observations fit the model perfectly. When n exceeds n_0 by one, that is, $r = 1$, one function must be written relating the n observations together. The sum of the three angles of a plane triangle is such a function, for example, involving one redundancy. This analysis can obviously be extended to include all r degrees of freedom in the adjustment.

Now we consider the existence of parameters in addition to observations. If we have to write r conditions when there are no parameters, then $r + 1$ conditions need to be written if one (unknown) parameter is added. This added condition is necessary to allow for the determination of that extra unknown, and to *retain the same number of degrees of freedom* r in the system. Thus if there are u unknown independent parameters in the adjustment, the number of conditions will, in general, be

$$c = r + u \tag{6.1}$$

This is regarded as the general case within the group of techniques containing independent parameters. Special cases are designated according to the number of parameters u carried in the adjustment. The lower limit occurs obviously when $u = 0$. The upper limit occurs when $u = n_0$, since in this case $c = n$. If u is taken larger than n_0, then the number of conditions would exceed the number of observations, which is not possible unless the parameters are not independent. Hence, the following two inequalities must be satisfied

$$r \le c \le n \tag{6.2}$$

$$0 \le u \le n_0 \tag{6.3}$$

For the present case the number of parameters u may take any value larger than zero but less than n_0. The chosen parameters may be only those of direct interest, or those of interest plus others. While we are analyzing the adjustment problem, the number of parameters u can be composed not only of parameters of direct interest but also of other parameters that make the writing of the condition equations either easier or less complicated.

As soon as these u parameters are specified, a total of $c = r + u$ *independent* condition equations must be set up between the n observations and u parameters. These equations may be linear or linearized as pointed out in the preceding chapter. We let the condition equations take the general form

$$\mathbf{A}(\boldsymbol{l} + \mathbf{v}) + \mathbf{B}\boldsymbol{\Delta} = \mathbf{d} \tag{6.4}$$

in which **A** and **B** are coefficient matrices, **d** is a column vector of constants, and the remaining symbols have been previously defined (see Table 5.1). Rearranging equation (6.4) leads to the following form which will be used as a standard throughout this book

$$\mathbf{A}\mathbf{v} + \mathbf{B}\boldsymbol{\Delta} = \mathbf{f} \tag{6.5}$$

with

$$\mathbf{f} = \mathbf{d} - \mathbf{A}\boldsymbol{l} \tag{6.6}$$

In view of the inequalities of equations (6.2) and (6.3) and other explanatory remarks, the following are the characteristics of the matrices in equations (6.5) and (6.6):

$$\begin{aligned} \mathbf{A} &= c \times n \text{ matrix} \qquad \text{rank } (\mathbf{A}) = c \\ \mathbf{v} &= n \times 1 \text{ vector} \\ \mathbf{B} &= c \times u \text{ matrix} \qquad \text{rank } (\mathbf{B}) = u \\ \mathbf{\Delta} &= u \times 1 \text{ vector} \\ \mathbf{f} &= c \times 1 \text{ vector} \end{aligned} \tag{6.7}$$

There are three other different cases that may be considered as special examples of the present case. These are when $u = 0$, giving the lower limit of $c = r$ and no parameters; when $u = n_0$, giving the upper limit when $c = n$; and a further simplification of the case of $c = n$, with the matrix **A** being the identity matrix. Those will be dealt with in later sections.

6.2. DERIVATION

Equation (6.5) is the fundamental form of condition equations for the adjustment of observations and independent parameters combined. It represents c linear equations in $(n + u)$ unknowns, which are the elements of the two vectors **v** and **Δ**. Since from equations (6.1) and (6.2) c is less than $(n + u)$, there exist many solutions for equation (6.5). A unique least squares solution is obtained by adding the basic criterion of equation (5.3), or

$$\phi = \mathbf{v}^t\mathbf{W}\mathbf{v} \rightarrow \text{minimum}$$

To enforce this criterion and at the same time have a solution for equation (6.5), the method of constrained minima by Lagrange multipliers is used. Thus if $\mathbf{k}_{c,\,1}$ represents the yet unknown Lagrange multipliers, then we should seek the minimum for the following function, noting that the quantity between parenthesis is zero when equation (6.5) is satisfied.

$$\phi' = \mathbf{v}^t\mathbf{W}\mathbf{v} - 2\mathbf{k}^t(\mathbf{A}\mathbf{v} + \mathbf{B}\mathbf{\Delta} - \mathbf{f}) \tag{6.8}$$

To minimize ϕ', its partial derivatives with respect to **v** and to **Δ** are equated to zero. Consequently, according to the rules in Section A8 (Appendix A), we get

$$\frac{\partial \phi'}{\partial \mathbf{v}} = 2\mathbf{v}^t\mathbf{W} - 2\mathbf{k}^t\mathbf{A} = \mathbf{0}^t$$

and

$$\frac{\partial \phi'}{\partial \mathbf{\Delta}} = -2\mathbf{k}^t\mathbf{B} = \mathbf{0}^t$$

which are in row form. Transposing and rearranging yields

$$-\underset{n,n}{\mathbf{W}}\ \underset{n,1}{\mathbf{v}} + \underset{n,c}{\mathbf{A}^t}\ \underset{c,1}{\mathbf{k}} = \mathbf{0} \tag{6.9}$$

$$\underset{u,c}{\mathbf{B}^t}\ \underset{c,1}{\mathbf{k}} = \mathbf{0} \tag{6.10}$$

realizing of course that $\mathbf{W}$ is a symmetric matrix.

It is important to pause here and consider the number of unknowns versus the number of linear equations. There is a total of $(n + u + c)$ unknowns. Correspondingly, there are c linear equations in equation (6.5), n equations in equation (6.9), and u equations in equation (6.10). In matrix form the total system is

$$\begin{array}{c} \\ n \\ c \\ u \end{array} \begin{array}{c} \begin{array}{ccc} n & c & u \end{array} \\ \begin{bmatrix} -\mathbf{W} & \mathbf{A}^t & \mathbf{0} \\ \mathbf{A} & \mathbf{0} & \mathbf{B} \\ \mathbf{0} & \mathbf{B}^t & \mathbf{0} \end{bmatrix} \end{array} \begin{bmatrix} \mathbf{v} \\ \mathbf{k} \\ \boldsymbol{\Delta} \end{bmatrix} = \begin{bmatrix} \mathbf{0} \\ \mathbf{f} \\ \mathbf{0} \end{bmatrix} \tag{6.11}$$

This system of equations has usually been referred to as *the total system of normal equations*. The matrix of coefficients is a square symmetric matrix of order $(n + c + u)$, which is always *nonsingular* (that is, its rank is equal to its order), unless the model is improperly constructed. In view of this fact the least squares problem will be solved by inverting the system of equation (6.11), or

$$\begin{bmatrix} \mathbf{v} \\ \mathbf{k} \\ \boldsymbol{\Delta} \end{bmatrix} = \begin{bmatrix} -\mathbf{W} & \mathbf{A}^t & \mathbf{0} \\ \mathbf{A} & \mathbf{0} & \mathbf{B} \\ \mathbf{0} & \mathbf{B}^t & \mathbf{0} \end{bmatrix}^{-1} \begin{bmatrix} \mathbf{0} \\ \mathbf{f} \\ \mathbf{0} \end{bmatrix} \tag{6.12}$$

The existence of zero matrices along the main diagonal of the coefficient matrix should not be disturbing as long as the matrix is nonsingular and because the zero matrices can be avoided simply by rearranging the equations.

For relatively small problems in which the *total* system of normal equations is limited in number, equation (6.12) may be used and the problem may be directly solved in this manner. However, many practical problems are not that small and the resulting total normal equations constitute a rather large system for which a direct solution may not be convenient, or practical. Furthermore, we may not be interested in both $\mathbf{v}$ and $\boldsymbol{\Delta}$ but only in one of them, and we rarely need $\mathbf{k}$ for its own sake. Henceforth, an alternative scheme may be desirable. Fortunately, the system of equation (6.11) contains many zero submatrices and a solution by partitioning is relatively simple. From equation (6.9),

$$\mathbf{v} = \mathbf{W}^{-1}\mathbf{A}^t\mathbf{k} = \mathbf{Q}\mathbf{A}^t\mathbf{k} \tag{6.13}$$

and substituting in equation (6.5) gives

$$\mathbf{AQA^t k} + \mathbf{B\Delta} = \mathbf{f} \tag{6.14}$$

Let

$$\boldsymbol{l}_e = \mathbf{A}\boldsymbol{l} \tag{6.15}$$

represent an "equivalent" set of observations. Applying the propagation rule, the cofactor matrix $\mathbf{Q}_e$ for $\boldsymbol{l}_e$ is

$$\mathbf{Q}_e = \mathbf{AQA^t} \tag{6.16}$$

which when used in equation (6.14) yields

$$\mathbf{Q}_e \mathbf{k} + \mathbf{B\Delta} = \mathbf{f}$$

solving for $\mathbf{k}$

$$\mathbf{k} = \mathbf{Q}_e^{-1}(-\mathbf{B\Delta} + \mathbf{f}) = \mathbf{W}_e(-\mathbf{B\Delta} + \mathbf{f}) \tag{6.17}$$

where

$$\mathbf{W}_e = \mathbf{Q}_e^{-1} = (\mathbf{AQA^t})^{-1} \tag{6.18}$$

substituting equation (6.17) into (6.10) and reducing

$$(\mathbf{B^t W}_e \mathbf{B})\mathbf{\Delta} = (\mathbf{B^t W}_e \mathbf{f}) \tag{6.19}$$

or

$$[\mathbf{B^t}(\mathbf{AQA^t})^{-1}\mathbf{B}]\mathbf{\Delta} = [\mathbf{B^t}(\mathbf{AQA^t})^{-1}\mathbf{f}] \tag{6.20}$$

Equation (6.19) or (6.20) represents a set of u equations in u unknown parameters (the elements of $\mathbf{\Delta}$) which are termed *partially reduced normal equations.* With the auxiliaries

$$\mathbf{N} = \mathbf{B^t W}_e \mathbf{B} = \mathbf{B^t}(\mathbf{AQA^t})^{-1}\mathbf{B} \tag{6.21}$$

$$\mathbf{t} = \mathbf{B^t W}_e \mathbf{f} = \mathbf{B^t}(\mathbf{AQA^t})^{-1}\mathbf{f} \tag{6.22}$$

a more compact form of equations (6.19) and (6.20) will be

$$\underset{u,u}{\mathbf{N}}\ \underset{u,1}{\mathbf{\Delta}} = \underset{u,1}{\mathbf{t}} \tag{6.23}$$

The vector $\mathbf{\Delta}$ may be obtained from equation (6.23) by direct inversion such that

$$\mathbf{\Delta} = \mathbf{N}^{-1}\mathbf{t} \tag{6.24}$$

In the derivation above, several inverses were taken. First, $\mathbf{W}^{-1}$ in equation (6.13) is all right since $\mathbf{W}$ is nonsingular because the observations are functionally independent. Then $\mathbf{Q}_e^{-1}$ in equation (6.17) is allowable because the rank of $\mathbf{Q}_e$ is equal to the rank of $\mathbf{A}$, or c, which is equal to its order. Finally, $\mathbf{N}^{-1}$ in equation (6.24) is also possible because $\mathbf{N}$ has a rank and order that are equal ($= u$). Having the value of the parameters $\mathbf{\Delta}$, we can compute the vector **k** from equation (6.17) and substitute it into equation (6.13) to evaluate the vector of residuals **v**. With this vector the quadratic form of the residuals may be computed and an unbiased estimate of the reference variance obtained from (a derivation and discussion of this aspect is given in Section 11.3)

$$\hat{\sigma}_0^2 = \frac{\mathbf{v}^t\mathbf{W}\mathbf{v}}{r} \tag{6.25}$$

where r is the number of degrees of freedom (or redundancy) as defined earlier and is equal to the number of condition equations c minus the number of unknown parameters u.

Finally, if the least squares estimate of the observations (which satisfy the model) is also desired, it may be obtained by adding the residuals to the given observations.

6.2.1. A Computational Check The equality

$$\mathbf{a} = \mathbf{B}^t\mathbf{W}_e\,\mathbf{A}\mathbf{v} = \mathbf{B}^t(\mathbf{A}\mathbf{Q}\mathbf{A}^t)^{-1}\mathbf{A}\mathbf{v} = \mathbf{0} \tag{6.26}$$

may be used for checking the correctness of the computation, particularly for smaller problems. Equation (6.26) may be verified by first substituting for **v** from equation (6.13); then for **k** from equation (6.17), and finally referring to equation (6.12). Thus

$$\begin{aligned}
\mathbf{a} &= \mathbf{B}^t(\mathbf{A}\mathbf{Q}\mathbf{A}^t)^{-1}\mathbf{A}\mathbf{Q}\mathbf{A}^t\mathbf{k} \\
&= \mathbf{B}^t(\mathbf{A}\mathbf{Q}\mathbf{A}^t)^{-1}(-\mathbf{B}\boldsymbol{\Delta} + \mathbf{f}) \\
&= -\mathbf{B}^t(\mathbf{A}\mathbf{Q}\mathbf{A}^t)^{-1}\mathbf{B}\boldsymbol{\Delta} + \mathbf{B}^t(\mathbf{A}\mathbf{Q}\mathbf{A}^t)^{-1}\mathbf{f} \\
&= \mathbf{0}
\end{aligned}$$

6.2.2. Computation of Quadratic Form Equation (6.25) shows that in order to compute the a posteriori estimate of the reference variance, $\hat{\sigma}_0^2$, we must first compute the quadratic form ($\mathbf{v}^t\mathbf{W}\mathbf{v}$). To do this we recall equation (6.13),

$$\mathbf{v} = \mathbf{Q}\mathbf{A}^t\mathbf{k}$$

and from equation (6.17)

$$\mathbf{v} = \mathbf{Q}\mathbf{A}^t\mathbf{W}_e(-\mathbf{B}\boldsymbol{\Delta} + \mathbf{f})$$

Thus

$$\mathbf{v}^t\mathbf{W}\mathbf{v} = (-\boldsymbol{\Delta}^t\mathbf{B}^t + \mathbf{f}^t)\mathbf{W}_e\,\mathbf{A}\mathbf{Q}\mathbf{W}\mathbf{Q}\mathbf{A}^t\mathbf{W}_e(-\mathbf{B}\boldsymbol{\Delta} + \mathbf{f})$$

or

$$\begin{aligned}
\mathbf{v}^t\mathbf{W}\mathbf{v} &= \mathbf{f}^t\mathbf{W}_e\,\mathbf{f} - 2\mathbf{f}^t\mathbf{W}_e\,\mathbf{B}\boldsymbol{\Delta} + \boldsymbol{\Delta}^t\mathbf{B}^t\mathbf{W}_e\,\mathbf{B}\boldsymbol{\Delta} \\
&= \mathbf{f}^t\mathbf{W}_e\,\mathbf{f} - 2\mathbf{t}^t\boldsymbol{\Delta} + \boldsymbol{\Delta}^t\mathbf{N}\boldsymbol{\Delta} \\
&= \mathbf{f}^t\mathbf{W}_e\,\mathbf{f} - 2\boldsymbol{\Delta}^t\mathbf{t} + \boldsymbol{\Delta}^t\mathbf{t}
\end{aligned}$$

or

$$\mathbf{v}^t\mathbf{W}\mathbf{v} = \mathbf{f}^t\mathbf{W}_e\,\mathbf{f} - \boldsymbol{\Delta}^t\mathbf{t} \tag{6.27}$$

In a nonlinear problem (see Section 11.1) if the iterative process has been carried out sufficiently, such that the last correction vector $\boldsymbol{\Delta}$ is essentially zero, we can disregard the last term of the right-hand side of equation (6.27) and use the following relation for computing the quadratic form

$$\mathbf{v}^t\mathbf{W}\mathbf{v} = \mathbf{f}^t\mathbf{W}_e\,\mathbf{f} = \mathbf{f}^t(\mathbf{A}\mathbf{Q}\mathbf{A}^t)^{-1}\mathbf{f} \tag{6.28}$$

6.2.3. Precision Estimation A second, and quite important, part of adjustment is the determination of the precision of the quantities that have been estimated in the adjustment. Such precision estimation is in the form of cofactor matrices, and it may also, of course, be in the form of covariance matrices. These matrices can be derived from the least squares solution by simply applying the rules of propagation of Chapter 4. For completeness, all matrices are given here.

$$\begin{aligned}
\boldsymbol{l} &= \mathbf{I}\boldsymbol{l}\\
\mathbf{f} &= \mathbf{d} - \mathbf{A}\boldsymbol{l}\\
\boldsymbol{\Delta} &= \mathbf{N}^{-1}\mathbf{B}^t\mathbf{W}_e\,\mathbf{f} = \mathbf{N}^{-1}\mathbf{B}^t\mathbf{W}_e(\mathbf{d} - \mathbf{A}\boldsymbol{l})\\
\mathbf{k} &= \mathbf{W}_e[-\mathbf{B}\boldsymbol{\Delta} + \mathbf{f}] = \mathbf{W}_e(-\mathbf{B}\mathbf{N}^{-1}\mathbf{B}^t\mathbf{W}_e + \mathbf{I})(\mathbf{d} - \mathbf{A}\boldsymbol{l})\\
\mathbf{v} &= \mathbf{Q}\mathbf{A}^t\mathbf{k} = \mathbf{Q}\mathbf{A}^t\mathbf{W}_e(-\mathbf{B}\mathbf{N}^{-1}\mathbf{B}^t\mathbf{W}_e + \mathbf{I})(\mathbf{d} - \mathbf{A}\boldsymbol{l})\\
\hat{\boldsymbol{l}} &= \boldsymbol{l} + \mathbf{v}
\end{aligned}$$

Autocofactor Matrices

$$\mathbf{Q}_{ll} = \mathbf{Q}$$

$$\mathbf{Q}_{ff} = (-\mathbf{A})\mathbf{Q}(-\mathbf{A})^t = \mathbf{Q}_e$$

$$\begin{aligned}
\mathbf{Q}_{\Delta\Delta} &= (-\mathbf{N}^{-1}\mathbf{B}^t\mathbf{W}_e\,\mathbf{A})\mathbf{Q}(-\mathbf{A}^t\mathbf{W}_e\,\mathbf{B}\mathbf{N}^{-1})\\
&= \mathbf{N}^{-1}\mathbf{B}^t\mathbf{W}_e\,\mathbf{B}\mathbf{N}^{-1} = \mathbf{N}^{-1}
\end{aligned} \tag{6.29}$$

$$\begin{aligned}
\mathbf{Q}_{kk} &= [-\mathbf{W}_e(-\mathbf{B}\mathbf{N}^{-1}\mathbf{B}^t\mathbf{W}_e + \mathbf{I})\mathbf{A}]\mathbf{Q}[-\mathbf{A}^t(-\mathbf{W}_e\,\mathbf{B}\mathbf{N}^{-1}\mathbf{B}^t + \mathbf{I})\mathbf{W}_e]\\
&= (-\mathbf{W}_e\,\mathbf{B}\mathbf{N}^{-1}\mathbf{B}^t\mathbf{W}_e + \mathbf{W}_e)\mathbf{Q}_e(-\mathbf{W}_e\,\mathbf{B}\mathbf{N}^{-1}\mathbf{B}^t + \mathbf{I})\mathbf{W}_e\\
&= (-\mathbf{W}_e\mathbf{B}\mathbf{N}^{-1}\mathbf{B}^t + \mathbf{I})^2\mathbf{W}_e = \mathbf{W}_e - \mathbf{W}_e\,\mathbf{B}\mathbf{N}^{-1}\mathbf{B}^t\mathbf{W}_e \qquad \text{(idempotent)}
\end{aligned}$$

$$\begin{aligned}
\mathbf{Q}_{vv} &= \mathbf{Q}\mathbf{A}^t(\mathbf{W}_e - \mathbf{W}_e\,\mathbf{B}\mathbf{N}^{-1}\mathbf{B}^t\mathbf{W}_e)\mathbf{A}\mathbf{Q}\\
&= \mathbf{Q}\mathbf{A}^t\mathbf{W}_e\,\mathbf{A}\mathbf{Q} - \mathbf{Q}\mathbf{A}^t\mathbf{W}_e\,\mathbf{B}\mathbf{Q}_{\Delta\Delta}\,\mathbf{B}^t\mathbf{W}_e\,\mathbf{A}\mathbf{Q}
\end{aligned} \tag{6.30}$$

$$\mathbf{Q}_{\hat{l}\hat{l}} = \mathbf{Q}_{ll} + \mathbf{Q}_{lv} + \mathbf{Q}_{vl} + \mathbf{Q}_{vv} = \mathbf{Q} - \mathbf{Q}_{vv} \tag{6.31}$$

Crosscofactor Matrices

$$
\begin{aligned}
\mathbf{Q}_{lf} &= -\mathbf{Q}\mathbf{A}^t \\
\mathbf{Q}_{l\Delta} &= -\mathbf{Q}\mathbf{A}^t\mathbf{W}_e\,\mathbf{B}\mathbf{N}^{-1} \\
\mathbf{Q}_{lk} &= -\mathbf{Q}\mathbf{A}^t(\mathbf{W}_e - \mathbf{W}_e\,\mathbf{B}\mathbf{N}^{-1}\mathbf{B}^t\mathbf{W}_e) \\
\mathbf{Q}_{lv} &= -\mathbf{Q}\mathbf{A}^t(\mathbf{W}_e - \mathbf{W}_e\,\mathbf{B}\mathbf{N}^{-1}\mathbf{B}^t\mathbf{W}_e)\mathbf{A}\mathbf{Q} = -\mathbf{Q}_{vv} \\
\mathbf{Q}_{l\hat{l}} &= \mathbf{Q}_{ll} + \mathbf{Q}_{lv} = \mathbf{Q} - \mathbf{Q}_{vv} = \mathbf{Q}_{\hat{l}\hat{l}} \\
\mathbf{Q}_{f\Delta} &= +\mathbf{A}\mathbf{Q}\mathbf{A}^t\mathbf{W}_e\,\mathbf{B}\mathbf{N}^{-1} = \mathbf{B}\mathbf{N}^{-1} \\
\mathbf{Q}_{fk} &= \mathbf{Q}_e(\mathbf{W}_e - \mathbf{W}_e\,\mathbf{B}\mathbf{N}^{-1}\mathbf{B}^t\mathbf{W}_e) = (\mathbf{I} - \mathbf{B}\mathbf{N}^{-1}\mathbf{B}^t\mathbf{W}_e) \\
\mathbf{Q}_{fv} &= \mathbf{Q}_e(\mathbf{I} - \mathbf{W}_e\,\mathbf{B}\mathbf{N}^{-1}\mathbf{B}^t)\mathbf{W}_e\,\mathbf{A}\mathbf{Q} = (\mathbf{I} - \mathbf{B}\mathbf{N}^{-1}\mathbf{B}^t\mathbf{W}_e)\mathbf{A}\mathbf{Q} \\
\mathbf{Q}_{f\hat{l}} &= \mathbf{Q}_{fl} + \mathbf{Q}_{fv} = -\mathbf{A}\mathbf{Q} + \mathbf{A}\mathbf{Q} - \mathbf{B}\mathbf{N}^{-1}\mathbf{B}^t\mathbf{W}_e\,\mathbf{A}\mathbf{Q} \\
&= -\mathbf{B}\mathbf{N}^{-1}\mathbf{B}^t\mathbf{W}_e\,\mathbf{A}\mathbf{Q} \\
\mathbf{Q}_{\Delta k} &= (\mathbf{N}^{-1}\mathbf{B}^t\mathbf{W}_e)\mathbf{Q}_{ff}(-\mathbf{W}_e\,\mathbf{B}\mathbf{N}^{-1}\mathbf{B}^t\mathbf{W}_e + \mathbf{W}_e)^t \\
&= \mathbf{N}^{-1}\mathbf{B}^t\mathbf{W}_e\,\mathbf{Q}_e\,\mathbf{W}_e - \mathbf{N}^{-1}\mathbf{B}^t\mathbf{W}_e\,\mathbf{Q}_e\,\mathbf{W}_e\,\mathbf{B}\mathbf{N}^{-1}\mathbf{B}^t\mathbf{W}_e \\
&= \mathbf{N}^{-1}\mathbf{B}^t\mathbf{W}_e - \mathbf{N}^{-1}\mathbf{B}^t\mathbf{W}_e = \mathbf{0}
\end{aligned}
\tag{6.32}
$$

or

$$
\begin{aligned}
\mathbf{Q}_{\Delta k} &= \mathbf{Q}_{\Delta\Delta}(-\mathbf{W}_e\,\mathbf{B})^t + \mathbf{Q}_{\Delta f}\,\mathbf{W}_e^t \\
&= -\mathbf{N}^{-1}\mathbf{B}^t\mathbf{W}_e + \mathbf{N}^{-1}\mathbf{B}^t\mathbf{W}_e = \mathbf{0} \\
\mathbf{Q}_{\Delta v} &= \mathbf{Q}_{\Delta k}(\mathbf{Q}\mathbf{A}^t)^t = \mathbf{0} \\
\mathbf{Q}_{\Delta\hat{l}} &= \mathbf{Q}_{\Delta l} + \mathbf{Q}_{\Delta v} = -\mathbf{N}^{-1}\mathbf{B}^t\mathbf{W}_e\,\mathbf{A}\mathbf{Q} \\
\mathbf{Q}_{kv} &= \mathbf{Q}_{kk}(\mathbf{Q}\mathbf{A}^t)^t = (\mathbf{W}_e - \mathbf{W}_e\,\mathbf{B}\mathbf{N}^{-1}\mathbf{B}^t\mathbf{W}_e)\mathbf{A}\mathbf{Q} \\
&= \mathbf{W}_e\,\mathbf{A}\mathbf{Q} - \mathbf{W}_e\,\mathbf{B}\mathbf{N}^{-1}\mathbf{B}^t\mathbf{W}_e\,\mathbf{A}\mathbf{Q} \\
&= \mathbf{Q}_{kl} \\
\mathbf{Q}_{k\hat{l}} &= \mathbf{Q}_{kl} + \mathbf{Q}_{kv} = \mathbf{Q}_{kl} - \mathbf{Q}_{kl} = \mathbf{0} \\
\mathbf{Q}_{v\hat{l}} &= \mathbf{Q}_{vl} + \mathbf{Q}_{vv} = -\mathbf{Q}_{vv} + \mathbf{Q}_{vv} = \mathbf{0}
\end{aligned}
\tag{6.33}
$$

Of the cofactor matrices, equation (6.29) gives a relation that is of fundamental importance in least squares estimation. It states that the cofactor matrix of the parameters is simply the inverse of the partially reduced normal equation coefficient matrix. Thus as far as the precision of the unknown parameters is concerned, it is obtained as a by-product of the least squares solution, particularly if such a solution was performed by inverting the coefficient matrix in equation (6.24). If $\mathbf{x}^0$ is the vector of approximations, which is a numerical constant as far as propagation is concerned, then $\hat{\mathbf{x}} = \mathbf{x}^0 + \boldsymbol{\Delta}$ for nonlinear systems, and

$$\mathbf{Q}_{\hat{x}\hat{x}} = \mathbf{Q}_{\Delta\Delta} = \mathbf{N}^{-1} \tag{6.34}$$

In certain situations it is possible, and even necessary, to obtain a solution by stopping the elimination procedure after replacing $\mathbf{v}$. This means that the following system is solved directly.

$$\begin{bmatrix} \mathbf{Q}_e & \mathbf{B} \\ \mathbf{B}^t & \mathbf{0} \end{bmatrix} \begin{bmatrix} \mathbf{k} \\ \boldsymbol{\Delta} \end{bmatrix} = \begin{bmatrix} \mathbf{f} \\ \mathbf{0} \end{bmatrix} \tag{6.35}$$

from which

$$\begin{bmatrix} \mathbf{k} \\ \boldsymbol{\Delta} \end{bmatrix} = \begin{bmatrix} \mathbf{Q}_e & \mathbf{B} \\ \mathbf{B}^t & \mathbf{0} \end{bmatrix}^{-1} \begin{bmatrix} \mathbf{f} \\ \mathbf{0} \end{bmatrix} \tag{6.36}$$

The desired vector $\boldsymbol{\Delta}$ can be taken as the appropriate subvector. However, it is now an important question to find out how $\mathbf{Q}_{\Delta\Delta}$ can be evaluated. In order to do this, we seek the inverse in equation (6.36) by partitioning (see Section A3.8, Appendix A.) Thus if

$$\begin{bmatrix} \mathbf{Q}_e & \mathbf{B} \\ \mathbf{B}^t & \mathbf{0} \end{bmatrix}^{-1} = \begin{bmatrix} \boldsymbol{\alpha} & \boldsymbol{\beta} \\ \boldsymbol{\beta}^t & \boldsymbol{\gamma} \end{bmatrix} \tag{6.37}$$

and, with $\mathbf{N} = \mathbf{B}^t\mathbf{W}_e\mathbf{B}$, from equation (6.21), we write

$$\begin{aligned} \boldsymbol{\alpha} &= \mathbf{W}_e - \mathbf{W}_e\mathbf{B}\mathbf{N}^{-1}\mathbf{B}^t\mathbf{W}_e \\ \boldsymbol{\beta} &= \mathbf{W}_e\mathbf{B}\mathbf{N}^{-1} \\ \boldsymbol{\gamma} &= -\mathbf{N}^{-1} \end{aligned} \tag{6.38}$$

or

$$\begin{bmatrix} \mathbf{Q}_e & \mathbf{B} \\ \mathbf{B}^t & \mathbf{0} \end{bmatrix}^{-1} = \begin{bmatrix} (\mathbf{W}_e - \mathbf{W}_e\mathbf{B}\mathbf{N}^{-1}\mathbf{B}^t\mathbf{W}_e) & \mathbf{W}_e\mathbf{B}\mathbf{N}^{-1} \\ \mathbf{N}^{-1}\mathbf{B}^t\mathbf{W}_e & -\mathbf{N}^{-1} \end{bmatrix} \tag{6.39}$$

Although $\mathbf{Q}_{kk} = (\mathbf{W}_e - \mathbf{W}_e\mathbf{B}\mathbf{N}^{-1}\mathbf{B}^t\mathbf{W}_e)$, the other two submatrices are not $\mathbf{Q}_{k\Delta}$ and $\mathbf{Q}_{\Delta\Delta}$, respectively. As derived above $\mathbf{Q}_{k\Delta} = \mathbf{0}$ and $\mathbf{Q}_{\Delta\Delta} = \mathbf{N}^{-1}$ and *not* $-\mathbf{N}^{-1}$. Consequently

$$\begin{bmatrix} \mathbf{Q}_e & B \\ \mathbf{B}^t & \mathbf{0} \end{bmatrix}^{-1} \neq \begin{bmatrix} \mathbf{Q}_{kk} & Q_{k\Delta} \\ \mathbf{Q}_{\Delta k} & \mathbf{Q}_{\Delta\Delta} \end{bmatrix} \tag{6.40}$$

But the cofactor matrix of interest, $\mathbf{Q}_{\Delta\Delta}$, can in fact be extracted from the inverse in equation (6.36) by taking the negative of the diagonal submatrix corresponding to $\boldsymbol{\Delta}$, or

$$\mathbf{Q}_{\Delta\Delta} = -\boldsymbol{\gamma} \tag{6.41}$$

6.2.4. Functional Dependence and a Posteriori Cofactor Matrices It has been indicated so far that, a priori, both the observations and parameters are functionally independent. Functional independence of the n observations was said to mean that no one observation can be deduced or computed from the remaining $(n - 1)$ observations. Also the difference between functional independence and stochastical independence or lack of correlation was stressed. Although the n observations are functionally independent, they may or may not be correlated. The difference between these two concepts can be related to the covariance or cofactor matrix associated with the vector of stochastic variables under consideration. If these variables are *functionally independent*, then the *rank of the covariance* or cofactor matrix is equal to its order, thus implying nonsingularity. On the other hand if the variables are *stochastically independent* or uncorrelated, then all elements off the main diagonal of the covariance, or cofactor, matrix must be zero. This establishes the fundamental difference between the two concepts. (In addition, perfect stochastic correlation is equivalent to linear functional dependence.)

The a priori vectors $\mathbf{l}$ and $\mathbf{\Delta}$ are functionally independent and thus $\mathbf{Q}$ and $\mathbf{Q}_{\Delta\Delta}$ are both nonsingular, the first having a rank of n and the second a rank of u. Other vectors computed from the adjustment include the residuals $\mathbf{v}$ and the estimated observations $\hat{\mathbf{l}}$ with corresponding cofactor matrices $\mathbf{Q}_{vv}$ and $\mathbf{Q}_{\hat{l}\hat{l}}$. The elements of both of these vectors are functionally dependent with the result that both $\mathbf{Q}_{vv}$ and $\mathbf{Q}_{\hat{l}\hat{l}}$ are singular matrices. Perhaps it is easier to show why the elements of $\hat{\mathbf{l}}$ are functionally dependent. After the adjustment, $\hat{\mathbf{l}}$ satisfies the condition equations. Thus since there are c conditions, then at least c estimated observations (that is, c elements of $\hat{\mathbf{l}}$) can be computed from the remaining $(n - c)$ elements. Consequently, the rank of $\mathbf{Q}_{\hat{l}\hat{l}}$ is less than its order. In fact,

$$\text{rank}\,(\mathbf{Q}_{vv}) = \text{redundancy} = r \tag{6.42}$$

$$\begin{aligned}\text{rank}\,(\mathbf{Q}_{\hat{l}\hat{l}}) &= (n - r)\\ &= \text{the minimum number of variables}\\ &\quad\ \text{specifying the model, } n_0\end{aligned} \tag{6.43}$$

Equations (6.42) and (6.43) are derived in Appendix C for this as well as other cases of adjustment.

Example 6.1

Given: Figure 6.1 shows a much simplified problem of an object point which is photographed by three terrestrial cameras. All three camera stations are assumed to lie on the same line which is taken to be the X_1 axis of the object coordinate system. The X_2 axis is taken to coincide with the optical axis of the first camera, S_1. The camera axes are horizontal and parallel. All interior orientations of the cameras are

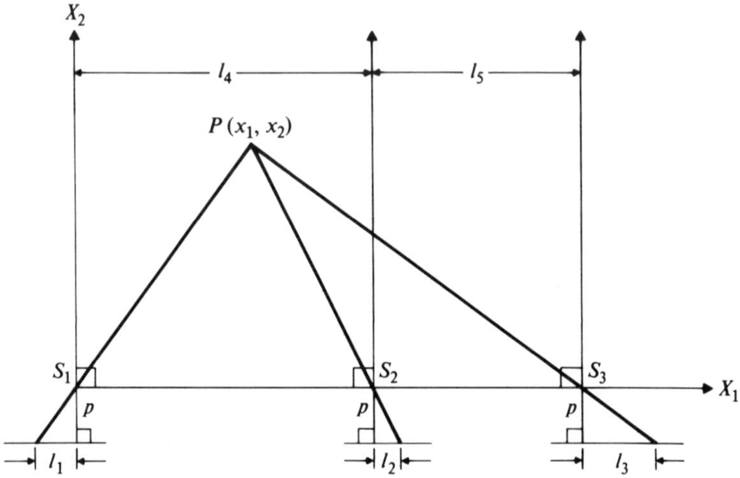

Figure 6.1. (Not to scale).

assumed known and without errors with principal distance $p = 100$ mm (as a constant value). The observations $\boldsymbol{l}$ are those five distances denoted by that symbol in Figure 6.1. Their absolute values (as distances and not coordinates) as well as their standard deviations are given in Table 6.1. All those observations are assumed, for simplicity, to be in the X_1X_2 plane. We shall assume further that no correlation exists between all the observations.

TABLE 6.1

OBSERVATION	VALUE	STANDARD DEVIATION
l_1	16.5 mm	0.10 mm
l_2	3.8 mm	0.10 mm
l_3	20.4 mm	0.10 mm
l_4	10.0 m	0.05 m
l_5	8.0 m	0.05 m

Required: We require the least squares estimate of the coordinates (X_1, X_2) of point P, and the a posteriori estimates of the reference variance, the cofactor matrices, and the covariance matrices.

Solution: The first step is to analyze the mathematical model of the problem. This includes all the information that in this case specifies the geometry of the problem and all the assumptions inherent in that information, whether explicit or implied. To begin with, the object coordinate system is fixed, the camera axes are parallel to each other and normal to the X_1 axis, and the profiles of all three photographs are parallel to the X_1 axis and at the equal distance of $p = 100$ mm from it. The given observations fix, with respect to the framework given, *three* points: S_2, S_3, and P. Next, we need to determine the minimum number of variables, n_0. Since both S_2 and S_3 lie on the X_1 axis, by fundamental assumption of the given model, only two parameters need to be determined to fix their positions. Added to those are the two that would determine point P, which makes a total of *four* variables as the necessary minimum number for determining the total geometry of the problem. Since n is equal to *five* observations, then there is *1 degree of freedom*, or $r = 1$.

In the statement of the problem it was indicated that two parameters, the pair of coordinates of point P, are desired. Thus $u = 2$ and the number of independent condition equations is $c = r + u = 3$ [see equation (6.1)]. Consulting Figure 22.1 one equation may be written for each ray from P to a camera station. (A sort of simplified version of the collinearity condition equation, as it is known in photogrammetry). Thus

$$\frac{l_1}{p} = \frac{x_1}{x_2}, \qquad \frac{l_2}{p} = \frac{l_4 - x_1}{x_2}, \qquad \text{and} \qquad \frac{l_3}{p} = \frac{l_4 + l_5 - x_1}{x_2}$$

or

$$\begin{aligned} f_1 &= l_1 x_2 - p x_1 &&= 0 \\ f_2 &= l_2 x_2 - p(l_4 - x_1) &&= 0 \\ f_3 &= l_3 x_2 - p(l_4 + l_5 - x_1) &&= 0 \end{aligned}$$

In these equations l_i refer to the *variables* representing the observations and not the numerical values of the observations. Linearization by Taylor series according to Section 11.1.2 leads to the form $\mathbf{Av} + \mathbf{B\Delta} = \mathbf{f}$ where

$$\underset{3,5}{\mathbf{A}} = \frac{\partial f_i}{\partial l_j} = \begin{bmatrix} x_2^0 & 0 & 0 & 0 & 0 \\ 0 & x_2^0 & 0 & -p & 0 \\ 0 & 0 & x_2^0 & -p & -p \end{bmatrix} \qquad \underset{3,2}{\mathbf{B}} = \frac{\partial f_i}{\partial x_j} = \begin{bmatrix} -p & l_1 \\ p & l_2 \\ p & l_3 \end{bmatrix}$$

$$\underset{3,1}{\mathbf{f}} = \begin{bmatrix} p x_1^0 - l_1 x_2^0 \\ p(l_4 - x_1^0) - l_2 x_2^0 \\ p(l_4 + l_5 - x_1^0) - l_3 x_2^0 \end{bmatrix}$$

At this early stage, $\Delta \boldsymbol{l}$ in equations (11.5) and (11.6) reduces to $\mathbf{v}$ because the approximations for the observational variables will be taken equal to the numerical values of the observations, and no iterations on the observations are effected. Approximations for x_1 and x_2, however, are needed and may be evaluated from the following geometric relationships:

$$x_2^0 = \frac{l_4 p}{l_1 + l_2} = \frac{10(\text{m}) \times 100(\text{mm})}{20.3(\text{mm})} \simeq 50 \text{ m}$$

$$x_1^0 = \frac{l_1 x_2^0}{p} = \frac{16.5(\text{mm}) \times 50(\text{m})}{100(\text{mm})} \simeq 8 \text{ m}$$

The stochastic model in the problem is specified by the variances of the observations and the fact that there are no correlations. Assuming that the *reference variance is unity*, the covariance matrix is used as the cofactor matrix, or

$$\underset{5,5}{\mathbf{Q}} = \begin{bmatrix} 0.01\ \text{mm}^2 & 0 & 0 & 0 & 0 \\ & 0.01\ \text{mm}^2 & 0 & 0 & 0 \\ & & 0.01\ \text{mm}^2 & 0 & 0 \\ \text{symmetric} & & & 0.0025\ \text{m}^2 & 0 \\ & & & & 0.0025\ \text{m}^2 \end{bmatrix}$$

Using the numerical values of the observations and the approximations of the parameters computed above, the elements of **A**, **B**, and **f** may be computed as

$$\mathbf{A} = \begin{bmatrix} 50(\text{m}) & 0 & 0 & 0 & 0 \\ 0 & 50(\text{m}) & 0 & -100(\text{mm}) & 0 \\ 0 & 0 & 50(\text{m}) & -100 & -100 \end{bmatrix}$$

$$B = \begin{bmatrix} -100 & 16.5 \\ 100 & 3.8 \\ 100 & 20.4 \end{bmatrix} (\text{mm})$$

$$f = \begin{bmatrix} -25 \\ 10 \\ -20 \end{bmatrix} (\text{mm m})$$

Then

$$\mathbf{Q}_e = (\mathbf{AQA}^t) = 25 \begin{bmatrix} 1 & 0 & 0 \\ 0 & 2 & 1 \\ 0 & 1 & 3 \end{bmatrix} (\text{mm}^2\ \text{m}^2)$$

$$\mathbf{W}_e = \mathbf{Q}_e^{-1} = \tfrac{1}{125} \begin{bmatrix} 5 & 0 & 0 \\ 0 & 3 & -1 \\ 0 & -1 & 2 \end{bmatrix} (\text{mm}^{-2}\ \text{m}^{-2})$$

and

$$\mathbf{N} = \mathbf{B}^t\mathbf{W}_e\,\mathbf{B} = \tfrac{1}{125} \begin{bmatrix} 8 \times 10^4 & -5.45 \times 10^3 \\ -5.45 \times 10^3 & 2.082 \times 10^3 \end{bmatrix} (\text{m}^{-2})$$

$$\mathbf{t} = \mathbf{B}^t\mathbf{W}_e\,\mathbf{f} = \tfrac{1}{125} \begin{bmatrix} 1.25 \times 10^4 \\ -2.893 \times 10^3 \end{bmatrix} (\text{m}^{-1})$$

$$\mathbf{N}^{-1} = \frac{125}{13684.55 \times 10^4} \begin{bmatrix} 2.082 \times 10^3 & 5.45 \times 10^3 \\ 5.45 \times 10^3 & 8 \times 10^4 \end{bmatrix} (\text{m}^2)$$

and, finally,

$$\Delta = \mathbf{N}^{-1}\mathbf{t} = \begin{vmatrix} 0.075(\text{m}) \\ -1.193(\text{m}) \end{vmatrix}$$

Adding $\mathbf{\Delta}$ to the approximations yields

$$x_1 = 8.075(\text{m}) \qquad \text{and} \qquad x_2 = 48.807(\text{m})$$

The two values computed here will be considered the final estimates for the coordinates of point P. It should be mentioned, however, that the problem is nonlinear and more consideration should be given. Since the concern at this stage is mainly with linear (or linearized) conditions, there will be no elaboration on this aspect of nonlinearity now. Suffice it to say that if the solution is properly iterated to account for the neglected higher-order terms, the final answer would have been $x_1 = 8.0749(\text{m})$ and $x_2 = 48.8093(\text{m})$. More discussion of this subject will be given in later chapters and examples.

In order to compute the vector of residuals, we first compute $\mathbf{k}$

$$\mathbf{k} = \mathbf{W}_e(-\mathbf{B\Delta} + \mathbf{f}) = \tfrac{1}{125}[10.95 \quad 24.16 \quad -13.32]^t$$

Then

$$\mathbf{v} = \mathbf{QA}^t\mathbf{K} = [0.04\text{ mm} \quad 0.10\text{ mm} \quad -0.05\text{ mm} \quad -0.022\text{ m} \quad 0.027\text{ m}]^t$$

The values in $\mathbf{v}$ were truncated to two decimals in millimetres (mm) and three decimals in metres (m), as indicated by the given data and variances.

Using this vector of residuals, the vector of estimated observations may be computed from $\hat{\boldsymbol{l}} = \boldsymbol{l} + \mathbf{v}$, or

$$\hat{\boldsymbol{l}} = [16.54\text{ mm} \quad 3.90\text{ mm} \quad 20.35\text{ mm} \quad 9.978\text{ m} \quad 8.027\text{ m}]^t$$

As a *computational check*, the expression of equation (6.26) may be evaluated to

$$\mathbf{a} = \mathbf{B}^t(\mathbf{AQA}^t)^{-1}\mathbf{Av} = 10^{-9}[6.17 \quad 6.10]^t$$

which is practically a zero vector as expected. The cofactor matrices are

$$\mathbf{Q}_{\Delta\Delta} = \mathbf{N}^{-1} = 10^{-3}\begin{bmatrix} 1.90 & 4.98 \\ 4.98 & 73.08 \end{bmatrix}(\text{m}^2) \qquad \text{rank} = 2$$

$$\mathbf{Q}_{ll} = 10^{-3}\begin{bmatrix} 8.993 & -2.238 & 1.231 & 0.503 & -0.616 \\ & 5.025 & 2.737 & 1.119 & -1.369 \\ & & 8.494 & -0.616 & 0.753 \\ & & & 2.248 & 0.308 \\ \text{symmetric} & & & & 2.124 \end{bmatrix} \qquad \text{rank} = 4 = n_0$$

The reference variance is computed, realizing that here there is 1 degree of freedom,

$$\hat{\sigma}_0^2 = 1.9367 \qquad \text{and} \qquad \hat{\sigma}_0 = 1.39$$

It is important to note that the above values were computed after the first iteration of a nonlinear problem. Although this is, in the case of this example, adequate, here are the corresponding values after the last (fifth) iteration for the sake of comparison with other methods later on (the subscript f is used to signify final),

$$\mathbf{Q}_{\Delta\Delta_{(f)}} = 10^{-3}\begin{bmatrix} 1.86 & 5.03 \\ 5.03 & 71.33 \end{bmatrix}(\mathrm{m}^2)$$

$$\mathbf{Q}_{ll_{(f)}} = 10^{-3}\begin{bmatrix} 9.024 & -2.189 & 1.213 & 0.500 & -0.621 \\ & 5.090 & 2.721 & 1.121 & -1.394 \\ & & 8.492 & -0.621 & 0.772 \\ & & & 2.244 & 0.318 \\ \text{symmetric} & & & & 2.104 \end{bmatrix}$$

The reference variance is also recomputed after the last iteration to give

$$\hat{\sigma}^2_{0_{(f)}} = 1.9634 \qquad \text{and} \qquad \hat{\sigma}_{0_{(f)}} = 1.40$$

(For a posteriori statistical analysis on $\hat{\sigma}_0^2$ and $\mathbf{Q}_{\Delta\Delta}$ refer to Examples 11.4 and 11.6).

Example 6.2

Given: The two-parameter transformation, rotation and scale (see Chapter 8, for more detail), between two coordinate systems $\mathbf{x}$ and $\mathbf{y}$ is represented by

$$y_{1i} = ax_{1i} - bx_{2i}$$

$$y_{2i} = bx_{1i} + ax_{2i}$$

where $(x_1, x_2)_i$ are the $\mathbf{x}$ coordinates of any point i, and $(y_1, y_2)_i$ are its corresponding $\mathbf{y}$ coordinates, and (a, b) are the transformation parameters. In order to estimate a and b, Table 6.2 gives the three points of known coordinates in both systems.

TABLE 6.2

i	x_{1i}	x_{2i}	y_{1i}	y_{2i}
1	0.0	1.0	−2.1	1.1
2	1.0	0.0	1.0	2.0
3	1.0	1.0	−0.9	2.8

For all three given points, the cofactor matrix is equal to the covariance matrix for the pair of coordinates in the $\mathbf{x}$ coordinate system and is equal to

$$\mathbf{Q}_{xx} = \mathbf{\Sigma}_{xx} = \begin{bmatrix} 0.01 & 0 \\ 0 & 0.01 \end{bmatrix} = 0.01\mathbf{I}_2$$

All y coordinates are to be considered as constants insofar as the adjustment is concerned.

Required: We require the least squares estimates of the transformation parameters a and b and the a posteriori estimates of the reference variance and covariance matrices.

Solution: Similar to Example 6.1 the functional model of this example involves plane geometry, which is demonstrated schematically in Figure 6.2. As implied from the given information, we are interested in transforming from the **x** system to the **y** system. Therefore the **y** coordinates of the points are those desired. Having three points of known coordinates in both systems, a total of six observations, that is, $n = 6$, are available. The minimum number of variables necessary for a unique determination of a and b is $n_0 = 2$. Consequently the degrees of freedom or redundancy is $r = 4$. For the purposes of this example there are two unknown parameters, or $u = 2$, and the number of condition equations is $c = 6$. The stochastic model is relatively simple since the cofactor matrices for the **x** coordinates of all points are the same. Then the total cofactor matrix of the observations is

$$\underset{6,6}{\mathbf{Q}} = \begin{bmatrix} \mathbf{Q}_{xx} & \mathbf{0} & \mathbf{0} \\ & \mathbf{Q}_{xx} & \mathbf{0} \\ \text{symmetric} & & \mathbf{Q}_{xx} \end{bmatrix} = 0.01\mathbf{I}_6$$

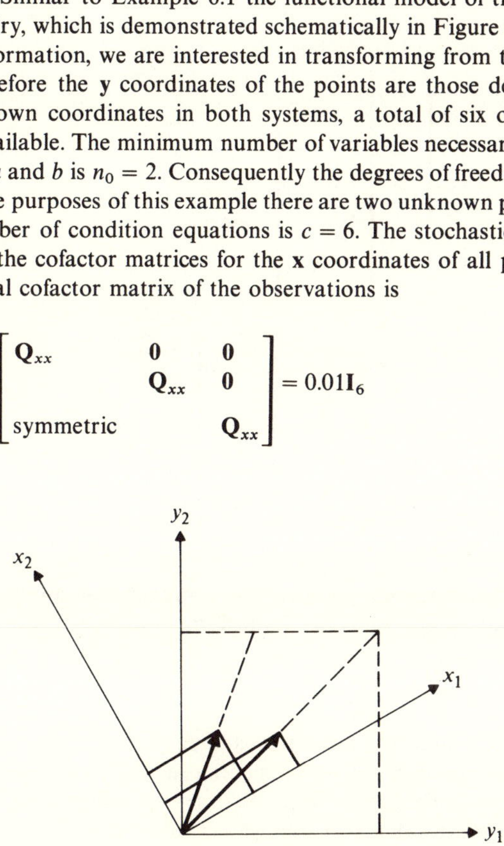

Figure 6.2

The six condition equations may be obtained by writing the pair of transformation equations for each of the three given points (a more general algorithm is given in Chapter 8),

$$x_{1i}a - x_{2i}b - y_{1i} = 0$$

$$x_{2i}a + x_{1i}b - y_{2i} = 0 \qquad i = 1, 2, 3$$

Linearization yields

$$\underset{6,6}{\mathbf{A}}\ \underset{6,6}{\mathbf{v}} + \underset{6,2}{\mathbf{B}}\ \underset{2,1}{\mathbf{\Delta}} = \underset{6,1}{\mathbf{f}}$$

where **A** is a block diagonal matrix of 2 × 2 submatrices along the main diagonal (see Section A9, Appendix A). If the approximations† a^0, b^0 are evaluated by simple computation to be 1 and 2, respectively, each submatrix on the diagonal of **A** will be

$$\underset{2,2}{\mathbf{A}_i} = \begin{bmatrix} a^0 & -b^0 \\ b^0 & a^0 \end{bmatrix} = \begin{bmatrix} 1 & -2 \\ 2 & 1 \end{bmatrix}$$

Similarly, the matrices **B** and **f** are evaluated as

$$\underset{6,2}{\mathbf{B}} = \begin{bmatrix} x_{11} & -x_{21} \\ x_{21} & x_{11} \\ x_{12} & -x_{22} \\ x_{22} & x_{12} \\ x_{13} & -x_{23} \\ x_{23} & x_{13} \end{bmatrix} = \begin{bmatrix} 0 & -1 \\ 1 & 0 \\ 1 & 0 \\ 0 & 1 \\ 1 & -1 \\ 1 & 1 \end{bmatrix}$$

$$\underset{6,1}{\mathbf{f}} = \begin{bmatrix} y_{11} - x_{11}a^0 + x_{21}b^0 \\ y_{21} - x_{21}a^0 - x_{11}b^0 \\ y_{12} - x_{12}a^0 + x_{22}b^0 \\ y_{22} - x_{22}a^0 - x_{12}b^0 \\ y_{13} - x_{13}a^0 + x_{23}b^0 \\ y_{23} - x_{23}a^0 - x_{13}b^0 \end{bmatrix} = \begin{bmatrix} -0.1 \\ 0.1 \\ 0 \\ 0 \\ 0.1 \\ -0.2 \end{bmatrix}$$

noting that x_{ij} denotes the x_i coordinate of point j, and the same applies for y_{ij}. A set of two reduced normal equations may be formed from the given condition equations the solution of which is

$$\boldsymbol{\Delta} = [0.00 \quad -0.05]^t$$

Thus the estimates of a and b, obtained by adding $\boldsymbol{\Delta}$ to the approximations, are $\hat{a} = 1.00$ and $\hat{b} = 1.95$. Again it should be mentioned that for the purposes of this chapter we shall stop after the first iteration. In order to evaluate **a** for the computational check of equation (6.26), first the numerical values are

$$\mathbf{v} = [0.01 \quad 0.08 \quad 0.02 \quad 0.01 \quad -0.05 \quad -0.05]^t$$

$$\mathbf{Q} = 0.01\mathbf{I}_6$$

$$\mathbf{A} = \begin{bmatrix} 1 & -2 & 0 & 0 & 0 & 0 \\ 2 & 1 & 0 & 0 & 0 & 0 \\ 0 & 0 & 1 & -2 & 0 & 0 \\ 0 & 0 & 2 & 1 & 0 & 0 \\ 0 & 0 & 0 & 0 & 1 & -2 \\ 0 & 0 & 0 & 0 & 2 & 1 \end{bmatrix} \qquad \mathbf{B} = \begin{bmatrix} 0 & 1 & 1 & 0 & 1 & 1 \\ -1 & 0 & 0 & 1 & -1 & 1 \end{bmatrix}^t$$

† No iterations on the observations is done here; therefore their approximations are the same as their a priori estimates given in the data table.

Then

$$\mathbf{a} = \begin{bmatrix} 1.2 \times 10^{-10} \\ -5.8 \times 10^{-11} \end{bmatrix} \cong \mathbf{0}$$

Next, the cofactor matrices are

$$\mathbf{Q}_{\Delta\Delta} = \mathbf{N}^{-1} = 10^{-2}\begin{bmatrix} 1.25 & 0 \\ 0 & 1.25 \end{bmatrix} \qquad \text{rank} = 2$$

$$\mathbf{Q}_{ll} = 10^{-3}\begin{bmatrix} 2.5 & 0 & 0 & -2.5 & 2.5 & -2.5 \\ & 2.5 & 2.5 & 0 & 2.5 & 2.5 \\ & & 2.5 & 0 & 2.5 & 2.5 \\ & & & 2.5 & -2.5 & 2.5 \\ & & & & 5.0 & 0 \\ \text{symmetric} & & & & & 5.0 \end{bmatrix}$$

and with 4 degrees of freedom, $\hat{\sigma}_0^2 = 0.3123$ and $\hat{\sigma}_0 = 0.56$. Again, for the sake of later checking, the final cofactor matrices are

$$\mathbf{Q}_{\Delta\Delta_{(f)}} = 10^{-2}\begin{bmatrix} 1.212 & 0 \\ 0 & 1.212 \end{bmatrix}$$

$$\mathbf{Q}_{ll_{(f)}} = 10^{-3}\begin{bmatrix} 2.916 & 0 & 0.052 & -2.750 & 2.579 & -2.538 \\ & 2.916 & 2.750 & 0.052 & 2.538 & 2.579 \\ & & 2.595 & 0 & 2.439 & 2.387 \\ & & & 2.595 & -2.387 & 2.439 \\ & & & & 4.489 & 0 \\ \text{symmetric} & & & & & 4.489 \end{bmatrix}$$

$$\hat{\sigma}^2_{0_{(f)}} = 0.3114 \qquad \text{and} \qquad \hat{\sigma}_{0_{(f)}} = 0.56$$

The a posteriori reference variance value may be statistically tested against the a priori value according to Section 11.5.1. The a priori value is $\sigma_0^2 = 1.0000$ (with infinite degrees of freedom), whereas the a posteriori estimate is $\hat{\sigma}_0^2 = 0.3114$ with $r = 4$ degrees of freedom. Hence, according to Section 2.8.3

$$\chi_m^2 = \frac{ms^2}{\sigma_0^2} \qquad \text{or} \qquad \chi_4^2 = \frac{(4)(0.3114)}{1.0000} = 1.2446$$

and H_0: $\sigma^2 = \sigma_0^2$ versus H_1: $\sigma^2 > \sigma_0^2$. The null hypothesis H_0 is rejected when $\chi_4^2 > \chi_{\alpha,4}^2$. At the level of significance $\alpha = 0.05$, Table D3 lists $\chi^2_{0.05,4} = 9.49$ which is larger than χ_4^2 and therefore H_0 cannot be rejected. Hence the a posteriori estimate of the reference variance is not statistically significantly different from its a priori value.

It should be noted that the differences between the results after the first and last iteration are not significant from the practical standpoint. However, differences can exist and there may be situations in which they are significant and therefore nonlinear problems should be appropriately iterated.

6.3. ADJUSTMENT WITH MAXIMUM NUMBER OF INDEPENDENT PARAMETERS (ADJUSTMENT WITH n CONDITION EQUATIONS)

We now discuss a special case that represents the upper limit and for which $u = n_0$ and $c = n$. This leads to the same form of conditions but with **A** being an $n \times n$ nonsingular matrix, and **B** an $n \times u$ matrix with a rank of u. All relationships developed for the general case apply directly here while taking the change in dimensions of **A** and **B** into account. For example, $\mathbf{Q}_e$ may be computed from equation (6.16) but will be of order n; no change occurs in **N**, **t**, and $\boldsymbol{\Delta}$ as they may be obtained from equations (6.21), (6.22), and (6.24), respectively; and **k** can be computed from equation (6.17) noting that it is an $n \times 1$ vector in this case. Without undue repetition, all relations for precision estimation given in Section 6.2.3 also apply here.

There are two possibilities for using this present case of adjustment. The first is when all $u = n_0$ parameters are of interest and need to be estimated. The second is when those parameters of interest are less than n_0, but we wish to write all possible conditions to avoid errors in constructing the model. The Example 6.3 is a useful illustration of the case.

Example 6.3. Refer to Example 6.1, Figure 6.1, dealing with determination of the coordinates of an object point P, from data obtained from three terrestrial photographs. In example 6.1 it was determined that $n_0 = 4$ and $r = 1$. To demonstrate the case of $c = n$ condition equations, the four variables used to determine n_0 are used as unknown parameters, namely,

x_1 as the x_1 coordinate of point P

x_2 as the x_2 coordinate of point P

x_3 as the x_1 coordinate of point S_2

x_4 as the x_1 coordinate of point S_3

With $u = 4$ the number of condition equations will be equal to $n = 5$,

$$
\begin{aligned}
&l_1 x_2 - p x_1 &&= 0\\
&l_2 x_2 - p(l_4 - x_1) &&= 0\\
&l_3 x_2 - p(x_4 - x_1) &&= 0\\
&l_4 - x_3 &&= 0\\
&l_4 + l_5 - x_4 &&= 0
\end{aligned}
$$

in which l_i refer to the variables representing the observations. Linearization by Taylor series (without iteration on the observations) gives

$$\underset{5,\,5}{\mathbf{A}} = \begin{bmatrix} x_2^0 & 0 & 0 & 0 & 0 \\ 0 & x_2^0 & 0 & -p & 0 \\ 0 & 0 & x_2^0 & 0 & 0 \\ 0 & 0 & 0 & 1 & 0 \\ 0 & 0 & 0 & 1 & 1 \end{bmatrix} \qquad \underset{5,\,4}{\mathbf{B}} = \begin{bmatrix} -p & l_1 & 0 & 0 \\ p & l_2 & 0 & 0 \\ p & l_3 & 0 & -p \\ 0 & 0 & -1 & 0 \\ 0 & 0 & 0 & -1 \end{bmatrix}$$

$$\underset{5,\,1}{\mathbf{f}} = \begin{bmatrix} px_1^0 - l_1 x_2^0 \\ p(l_4 - x_1^0) - l_2 x_2^0 \\ p(x_4^0 - x_1^0) - l_3 x_2^0 \\ x_3^0 - l_4 \\ x_4^0 - l_4 - l_5 \end{bmatrix}$$

Take the approximations as $x_1^0 = 8$ m, $x_2^0 = 50$ m, $x_3^0 = 10$ m, $x_4^0 = 18$ m, and the same $\mathbf{Q}$ matrix given in Example 6.1. The $\mathbf{A}$ and $\mathbf{B}$ matrices are very similar to those computed previously noting that they are augmented by simple numbers and there is no need to rewrite them. The vector $\mathbf{f}$, however, will be

$$\mathbf{f} = [-25 \quad 10 \quad -20 \quad 0 \quad 0]^t$$

For the sake of variation use the partially reduced system of normal equations given in equation (6.35),

$$\begin{bmatrix} \mathbf{Q}_e & \mathbf{B} \\ \mathbf{B}^t & \mathbf{0} \end{bmatrix} \begin{bmatrix} \mathbf{k} \\ \mathbf{\Delta} \end{bmatrix} = \begin{bmatrix} \mathbf{f} \\ \mathbf{0} \end{bmatrix} \qquad \text{or} \qquad \underset{9,\,9}{\mathbf{M}} \begin{bmatrix} \mathbf{k} \\ \mathbf{\Delta} \end{bmatrix} = \begin{bmatrix} \mathbf{f} \\ \mathbf{0} \end{bmatrix}$$

where

$$\mathbf{Q}_e = \begin{bmatrix} 25.0 & 0 & 0 & 0 & 0 \\ & 25.0 & 0 & 0 & 0 \\ & & 25.0 & 0 & 0 \\ & & & 0.0025 & 0.0025 \\ \text{symmetric} & & & & 0.0050 \end{bmatrix} \qquad \text{and}$$

$$\mathbf{B} = \begin{bmatrix} -100 & 16.5 & 0 & 0 \\ 100 & 3.8 & 0 & 0 \\ 100 & 20.4 & 0 & -100 \\ 0 & 0 & -1 & 0 \\ 0 & 0 & 0 & -1 \end{bmatrix}$$

The inverse of the 9 × 9 matrix $\mathbf{M}$, is computed as

$$\mathbf{M}^{-1} = \begin{bmatrix} 0.0040 & 0.0090 & -0.0049 & -0.8952 & 0.4925 & -0.0043 & 0.0283 & -0.0010 & 0.0002 \\ & 0.0199 & -0.0109 & -1.990 & 1.0948 & 0.0027 & 0.0027 & -0.0022 & -0.0003 \\ & & 0.0060 & 1.0948 & -0.6023 & 0.0030 & 0.0256 & 0.0012 & -0.0003 \\ & & & 198.9996 & -109.4767 & -0.2684 & -0.2704 & -0.7762 & -0.0499 \\ & & & & 60.2270 & -0.2995 & -2.5613 & -0.1231 & -0.9726 \\ & & & & & -0.0019 & -0.0050 & -0.0014 & -0.0022 \\ & & & & & \text{II} & -0.0731 & -0.0071 & -0.0135 \\ & & & & & & & -0.0022 & -0.0026 \\ \text{symmetric} & & & & & \text{I} & & \text{III} & -0.0050 \end{bmatrix}$$

The solution of the normal equations would give a 9 × 1 vector containing both the Lagrange multipliers **k** and **Δ**. The latter is of more interest to our present problem, and thus

$$\mathbf{\Delta} = [0.075 \quad -1.193 \quad -0.022 \quad 0.005]^t \text{ (m)}$$

Adding these values to the specified approximations gives the estimates of the four parameters

$$\hat{x}_1 = 8.075 \text{ m} = x_1 \text{ coordinate of point } P$$

$$\hat{x}_2 = 48.807 \text{ m} = x_2 \text{ coordinate of point } P$$

$$\hat{x}_3 = 9.978 \text{ m} = x_1 \text{ coordinate of point } S_2$$

$$\hat{x}_4 = 18.005 \text{ m} = x_1 \text{ coordinate point } S_3$$

It is clear that $\hat{x}_1$ and $\hat{x}_2$ are identical to the parameters obtained in Example 6.1. To check $\hat{x}_3$ and $\hat{x}_4$, we should realize by consulting Figure 6.1 that these two parameters are closely related to two of the observations and their determination is tantamount to computing estimated observations. For example, $\hat{x}_3$ is nothing but $\hat{l}_4$, and the answers from here and from Example 6.1 show that they are indeed equal. The estimate $\hat{x}_4$, however, should be equal to $\hat{l}_4 + \hat{l}_5$ or $(9.978 + 8.027) = 18.005$ m, which is identical to the value computed above.

If we are also interested in the a posteriori cofactor matrices, they can be obtained in the usual manner. For example, the cofactor matrix for all four parameters is the negative of the 4 × 4 submatrix designated (I) in $\mathbf{M}^{-1}$ [see Equation (6.38)]. Within it, the negative of submatrix (II) is the cofactor matrix for the first two parameters that were used in Example 6.1. Considering that all these answers are taken after the first iteration, only the negative of submatrix (II) agrees quite well with $\mathbf{Q}_{\Delta\Delta}$ computed in Example 6.1.

The negative of submatrix (III) in $\mathbf{M}^{-1}$ represents the cofactor matrix of the two parameters $\hat{x}_3$ and $\hat{x}_4$. It can be verified by recognizing that $\hat{x}_3 = \hat{l}_4$ and $\hat{x}_4 = \hat{l}_4 + \hat{l}_5$ or

$$\begin{bmatrix} \hat{x}_3 \\ \hat{x}_4 \end{bmatrix} = \begin{bmatrix} 1 & 0 \\ 1 & 1 \end{bmatrix} \begin{bmatrix} \hat{l}_4 \\ \hat{l}_5 \end{bmatrix}$$

and that from Example 6.1 the cofactor matrix for $\hat{l}_4$, $\hat{l}_5$ is

$$\begin{bmatrix} 0.00224 & 0.00032 \\ 0.00032 & 0.00211 \end{bmatrix}$$

Thus

$$\begin{bmatrix} 1 & 0 \\ 1 & 1 \end{bmatrix} \begin{bmatrix} 0.00224 & 0.00032 \\ 0.00032 & 0.00211 \end{bmatrix} \begin{bmatrix} 1 & 1 \\ 0 & 1 \end{bmatrix} = \begin{bmatrix} 0.0022 & 0.0026 \\ 0.0026 & 0.0050 \end{bmatrix}$$

which is identical to the negative of submatrix (III).

To recapitulate, then, the cofactor matrix of all four parameters is

$$\mathbf{Q}_{\Delta\Delta} = 10^{-3} \begin{bmatrix} 1.9 & 5.0 & 1.4 & 2.2 \\ & 73.1 & 7.1 & 13.5 \\ & & 2.2 & 2.6 \\ \text{symmetric} & & & 5.0 \end{bmatrix}$$

and $\mathbf{Q}_{ll}$ when computed turn out to be identical to that evaluated in Example 6.1. Finally, the estimate of the reference variance is $\hat{\sigma}_0^2 = 1.9727$.

It should therefore be noted that the choice of four parameters here as compared to two parameters in Example 6.1 has no effect on the evaluation of the remaining variables, $\mathbf{v}$, $\mathbf{Q}_{vv}$, ..., and so on.

For the sake of completeness the final cofactor matrix for the parameters after the last iteration is

$$\mathbf{Q}_{\Delta\Delta_{(f)}} = 10^{-3} \begin{bmatrix} 1.862 & 5.029 & 1.420 & 2.170 \\ & 71.332 & 7.106 & 1.347 \\ & & 2.244 & 2.562 \\ \text{symmetric} & & & 4.985 \end{bmatrix}$$

The value of $\mathbf{Q}_{ll_{(f)}}$ was also computed and found to be identical to that given in Example 6.1. Finally, the last value of the reference variance is

$$\hat{\sigma}^2_{0_{(f)}} = 1.9634$$

6.4. GEOMETRIC INTERPRETATION OF THE LEAST SQUARES PRINCIPLE

Although the foundation and derivation of the least squares principle are best done starting from the minimum variance property, a geometrical demonstration of its meaning can give the reader a better appreciation and familiarity with the method. One interpretation is expressed by Brown (see Bibliography):

> Consider an n-dimensional coordinate system with orthogonal axes $v_1, v_2, \ldots, v_n$ (the residuals). Then the quadratic form $\phi = \mathbf{v}^t\mathbf{\Sigma}^{-1}\mathbf{v}$, being positive definite, will represent an n-dimensional ellipsoid. The ellipsoid is centered at the origin. If $\mathbf{\Sigma}$ is diagonal, the axes of the ellipsoid will coincide with the coordinate axes, while for $\mathbf{\Sigma}$ nondiagonal the ellipsoid will be in a tilted orientation. It is clear that by a rotational coordinate transformation, a tilted ellipsoid can be reoriented into a standard position. Such a transformation is specified by $\mathbf{v} = \mathbf{R}\hat{\mathbf{v}}$ where $\mathbf{R}$ is an $n \times n$ matrix whose rows (or columns) are composed of the normalized characteristic vectors (eigenvectors) of $\mathbf{\Sigma}$. Thus a problem involving correlated observations can be reduced to one involving derived observations which are uncorrelated. The dimensions of the hyperellipsoid are, of course, unaffected by a rotation. In fact, the lengths of the axes are directly proportional to the square roots of the characteristic roots (eigenvalues) of $\mathbf{\Sigma}$. The constant of proportionality which determines their absolute dimensions is simply $\phi^{1/2}$. It thus follows that the volume of the ellipsoid is directly proportional to $\phi^{n/2}$. Therefore, minimizing ϕ is equivalent to minimizing the volume of the ellipsoid, it being understood, naturally, that the condition equations must be satisfied by some point on the ellipsoid. To simplify matters we may assume that any parameters have been eliminated from the linearized condition equations, leaving r relations between the residuals alone. Each condition equation then represents a hyperplane, and the residuals must lie on the intersection of the r hyperplanes. Now consider the family of hyperellipsoids defined by varying ϕ. The orientation and relative dimensions of such ellipsoids will be constant, and all will be centered at the origin. We may think of the family as being formed by the balloonlike expansion of an initial infinitesimal ellipsoid. Let the ellipsoid expand until it becomes tangent to the

intersection of the condition equations hyperplanes. For this point all the condition equations are satisfied and the volume of the ellipsoid (and consequently ϕ) is obviously minimum. Hence the coordinates of the point of tangency give the most probable residuals. . . ."

To demonstrate the statements quoted above we consider the simple case of two observations with the residuals v_1 and v_2.

A two-dimensional coordinate system as shown in Figure 6.3 would represent the space expressed by these two residuals. The condition that may relate v_1 to v_2 can be of the general form $a_0 + a_1 v_1 + a_2 v_2 = 0$ and is depicted by a straight line in the figure. The quadratic form $\phi = \mathbf{v}^t\mathbf{W}\mathbf{v}$ represents in this case a family of ellipses. The dotted ellipse in Figure 6.3, which is tangent to the line of condition, corresponds to the minimum value for ϕ. The point of tangency yields the least squares estimates of v_1 and v_2.

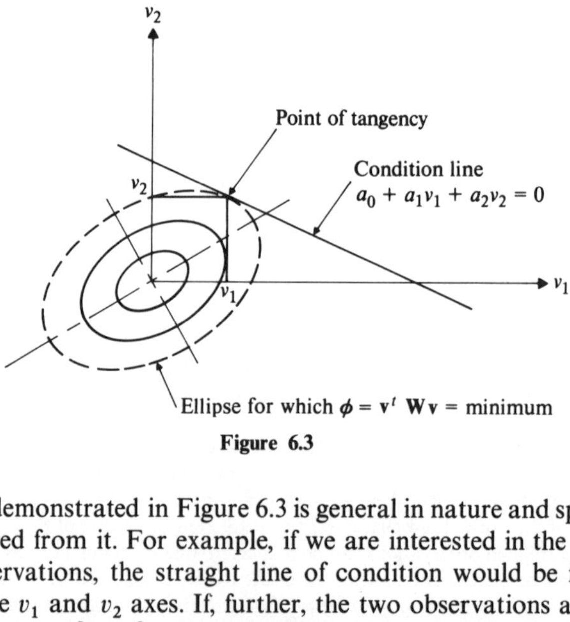

Figure 6.3

The case demonstrated in Figure 6.3 is general in nature and special cases may be derived from it. For example, if we are interested in the equality of the two observations, the straight line of condition would be inclined 45 degrees to the v_1 and v_2 axes. If, further, the two observations are of equal precision (that is, $\sigma_1^2 = \sigma_2^2$), then the ellipse will become a circle. In such a case the point of tangency will be at the foot of the normal from the origin to the line with the obvious result that the estimates of v_1 and v_2 will be equal.

A slightly different interpretation is possible if the observations themselves, and not the residuals, are used as the axes of the coordinate system. As an example consider that the three angles α, β, and γ of a plane triangle are measured with equal precision. With a redundancy of $r = 1$, the well-known condition $\alpha + \beta + \gamma - \pi = 0$ should be satisfied.

The condition represents a plane E_1 in the three-dimensional vector space α, β, γ in Figure 6.4. The a priori values of the observations α, β, γ determine the point P which would not necessarily lie on the plane. Adjustment of the

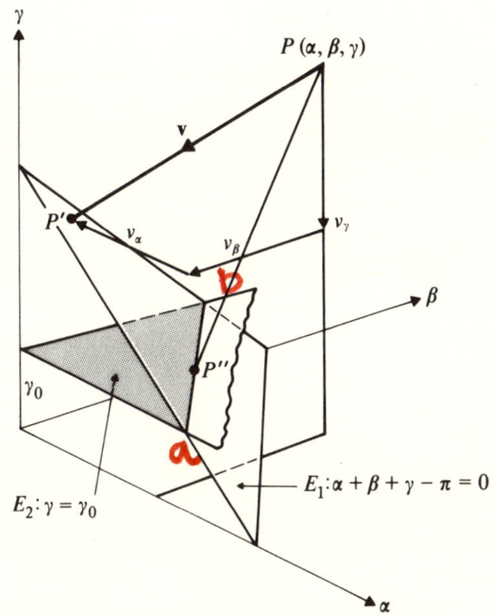

Figure 6.4

given data would replace P by P' such that the least squares principle is satisfied, making the vector **v** as short as possible, which means making the line $\overline{PP'}$ normal to the plane E_1. This implies directly that the least squares solution would give

$$v_\alpha = v_\beta = v_\gamma = \tfrac{1}{3}(\pi - \alpha - \beta - \gamma)$$

because the plane E_1 is equally inclined to all three axes. All three residuals will have the same sign, being positive if P lies below the plane and negative if it lies above it.

Assume, now, that for one reason or another we want to impose the constraint that the angle γ should have a certain fixed and known value γ_0 in addition to the given data. This additional condition refers to a geometrical model having redundancy $r = 2$, since with γ given, only one observation would suffice to fix the shape of the triangle. The condition $\gamma = \gamma_0$ in the three-dimensional vector space is a plane parallel to the α, β plane. The adjustment in this case has to provide a point P'' that would satisfy both conditions. Thus the point P'' must lie on the intersection of the two planes or on the line ab in Figure 6.4. It will be fixed on this line of intersection by the least squares principle, which would minimize the distance $\overline{PP''}$.

In case the observations α, β, γ are given different a priori weights or are considered correlated, the geometrical explanation of the case remains unaltered if the vector space is spanned by a skew axis system, according to the weight relation.

6.5. SUMMARY OF EQUATIONS

This section is included as an easy reference for those readers who have become familiar with the concepts and wish to extract equations quickly for use in actual adjustment problems. It is a summary of the basic symbols and equations, with reference to the numbers used in the text. Although concise, it is intended to be self-sufficient.

n = number of observations in the vectors $\mathbf{l}$ (a priori observations); $\mathbf{v}$ (residuals); and $\hat{\mathbf{l}}$ (estimated observations)
r = redundancy $= n - n_0$ (where n_0 is the minimum specifying the model)
u = number of parameters in $\mathbf{\Delta}(0 < u \leq n_0)$
$c = r + u$ = number of condition equations in

$$\mathbf{Av} + \mathbf{B\Delta} = \mathbf{d} - \mathbf{Al} = \mathbf{f} \qquad (6.5, 6.6)$$

$\mathbf{A} = c \times n$ matrix of coefficients; rank $(\mathbf{A}) = c$ (upper limit, $\mathbf{A}$ is $n \times n$ with rank n)
$\mathbf{B} = c \times u$ matrix of coefficients; rank $(B) = u$ (upper limit, $\mathbf{B}$ is $n \times u$ with rank $= u$)
$\mathbf{d}$ and $\mathbf{f} = c \times 1$ vectors of constants (upper limit, dimension of $n \times 1$)

$$\mathbf{Q}_e = \mathbf{AQA}^t \qquad (6.16)$$

$$\underset{u,u}{\mathbf{N}} = \mathbf{B}^t\mathbf{W}_e\mathbf{B} = \text{partially reduced normal equations coefficient matrix} \qquad (6.21)$$

$$\mathbf{t} = \mathbf{B}^t\mathbf{W}_e\mathbf{f} = \text{partially reduced normal equations constant vector} \qquad (6.22)$$

$$\underset{u,1}{\mathbf{\Delta}} = \mathbf{N}^{-1}\mathbf{t} = \text{estimate of parameters}$$ (In the event that the conditions are originally nonlinear, then $\mathbf{x} = \mathbf{x}^0 + \mathbf{\Delta}$, $\mathbf{x}^0$ is a vector of approximations.) (6.24)

$$\mathbf{v} = \mathbf{QA}^t\mathbf{W}_e(-\mathbf{B\Delta} + \mathbf{f}) \qquad (6.13, 6.17)$$

$$\hat{\mathbf{l}} = \mathbf{l} + \mathbf{v} = \text{estimated observations} \qquad (5.2)$$

$$\hat{\sigma}_0^2 = \mathbf{v}^t\mathbf{Wv}/r = (\mathbf{f}^t\mathbf{W}_e\mathbf{f} - \mathbf{\Delta}^t\mathbf{t})/r \qquad (6.25, 6.27)$$

$\mathbf{Q}_{\Delta\Delta} = \mathbf{N}^{-1}$ = cofactor matrix of the parameters (rank $\mathbf{Q}_{\Delta\Delta} = u$) For nonlinear conditions, $\mathbf{Q}_{\hat{x}\hat{x}}$ is also equal to $\mathbf{N}^{-1}$ (6.29)

$$\mathbf{Q}_{vv} = \mathbf{QA}^t(\mathbf{W}_e - \mathbf{W}_e\mathbf{BN}^{-1}\mathbf{B}^t\mathbf{W}_e)\mathbf{AQ}$$

$$= \mathbf{QA}^t\mathbf{W}_e\mathbf{AQ} - \mathbf{QA}^t\mathbf{W}_e\mathbf{BQ}_{\Delta\Delta}\mathbf{B}^t\mathbf{W}_e\mathbf{AQ} \qquad (6.30)$$

= cofactor matrix of the residuals (rank $(\mathbf{Q}_{vv}) = r$) (6.42)

$$\mathbf{Q}_{\hat{l}\hat{l}} = \mathbf{Q} - \mathbf{Q}_{\mathrm{vv}} \tag{6.31}$$

$= \text{cofactor matrix of the estimated observations (rank } (\mathbf{Q}_{\hat{l}\hat{l}}) = n - r = n_0)$ (6.43)

$$\mathbf{Q}_{\Delta l} = \mathbf{Q}_{\Delta \hat{l}} = -\mathbf{N}^{-1}\mathbf{B}^t\mathbf{W}_e\,\mathbf{AQ} = \mathbf{Q}^t_{\hat{l}\Delta} = \mathbf{Q}^t_{l\Delta} \qquad (6.32, 6.33)$$

7

Adjustment with Conditions Only—Special Cases

Adjustment of Functionally Independent Parameters and Observations

7.1. INTRODUCTION

The method of least squares has been conventionally used with two cases that are simpler than, and special cases of, the general case given in the preceding chapter. In this chapter these two cases will be treated and their relationships to the general case shown. The choice from the three techniques will depend mostly on the mathematical model of the problem to be solved, and to a lesser degree on other factors. Such factors may include the type of computational facilities available, the size of the problem, and the preference of the individual involved.

The first of the two special cases involves situations in which only observations appear in the condition equations, thus $u = 0$. The number of condition equations would be equal to the redundancy. Therefore this method is often used for simple geometric problems, such as the adjustment of a plane triangle, a small level net, and the like. The solution of the normal equations, whose number is equal to the number of conditions, yields values for Lagrange multipliers that may be used in turn to compute the vector of residuals, $\mathbf{v}$. By adding $\mathbf{v}$ to the given observations $\boldsymbol{l}$, we compute the estimated observations, $\hat{l}$. Should we be interested in another set of variables which are related to the elements of $\hat{l}$ by specified functions, we compute such variables *after* the adjustment is completed. The cofactor, or covariance, matrix of these variables is also computed after the adjustment, using $\mathbf{Q}_{\hat{l}\hat{l}}$ and the given functions and applying the principle of propagation. Thus the above two operations are considered outside the process of least squares adjustment.

The second special case is more like the general case given in the preceding chapter. In certain simple circumstances it is possible to formulate the condition equations in such a way that each condition equation contains *only one observation.* This implies that parameters are present, that the number of conditions is equal to the number of observations, and that u is at its upper limit of n_0. Because of having only one observation per condition equation, the **A** matrix reduces to the identity matrix. The number of reduced normal equations is less than the number of condition equations in this case, as it is equal to the number of parameters to be determined. Consequently, their solution yields the values of the parameters and their cofactor matrix directly. If we are also interested in the estimated observations $\hat{l}$, they can still be obtained from the adjustment.

Each of these two cases will now be discussed in detail in separate sections, with numerical examples worked out by both techniques for the sake of comparison.

7.2. ADJUSTMENT OF OBSERVATIONS ONLY

Adjustment of observations only is the technique conventionally referred to as the method of "adjustment by conditions." But since we have utilized the term "condition equations" in a much more general sense, we shall avoid this term. We will simply refer to the technique as "adjustment of observations only."

7.2.1. Formulation Since there are no parameters, the linear condition equations will then simply be

$$\mathbf{A}(\boldsymbol{l} + \mathbf{v}) = \mathbf{d} \tag{7.1}$$

or

$$\mathbf{A}\mathbf{v} = \mathbf{d} - \mathbf{A}\boldsymbol{l} = \mathbf{f} \tag{7.2}$$

in which **d** and **f** are constant vectors.

The application of the least squares in this case follows exactly the same derivation steps as those given in the preceding chapter. Therefore for the sake of brevity, such a derivation is left as an exercise for the reader and only the summary results will be given here. The total system of normal equations may be written directly from equation (6.11) by enforcing $\mathbf{B} = \mathbf{0}$, thus

$$\begin{bmatrix} -\mathbf{W} & \mathbf{A}^t \\ \mathbf{A} & \mathbf{0} \end{bmatrix} \begin{bmatrix} \mathbf{v} \\ \mathbf{k} \end{bmatrix} = \begin{bmatrix} \mathbf{0} \\ \mathbf{f} \end{bmatrix} \tag{7.3}$$

Solution of equation (7.3) for the two vectors $\mathbf{k}$ and $\mathbf{v}$ yields

$$\mathbf{k} = \mathbf{W}_e\,\mathbf{f} = \mathbf{Q}_e^{-1}\mathbf{f} = (\mathbf{AQA}^t)^{-1}\mathbf{f} \tag{7.4}$$

$$\mathbf{v} = \mathbf{QA}^t\mathbf{k} = \mathbf{QA}^t\mathbf{W}_e\,\mathbf{f} \tag{7.5}$$

The matrix $\mathbf{Q}_e$ is the partially reduced normal equations coefficient matrix in this case. The unknown vector in these equations is obviously the vector of Lagrange multipliers, $\mathbf{k}$.

The a priori estimate of the reference variance may be computed from equation (6.25). The quadratic form can be evaluated directly using $\mathbf{v}$ and $\mathbf{W}$, or using equation (6.28) or its equivalent in view of equation (7.4),

$$\mathbf{v}^t\mathbf{W}\mathbf{v} = \mathbf{f}^t\mathbf{k} = \mathbf{k}^t\mathbf{f} \tag{7.6}$$

7.2.2. Estimation of Precision In order to apply the principle of propagation to obtain autocofactor and crosscofactor matrices, the pertinant relations are first assembled as follows:

$$\boldsymbol{l} = \boldsymbol{l} \qquad \text{denote } \mathbf{Q}_{ll} \text{ by } \mathbf{Q}$$

$$\mathbf{f} = \mathbf{d} - \mathbf{A}\boldsymbol{l}$$

$$\mathbf{k} = (\mathbf{AQA}^t)^{-1}\mathbf{f} = \mathbf{Q}_e^{-1}\mathbf{f} = \mathbf{W}_e\,\mathbf{d} - \mathbf{W}_e\,\mathbf{A}\boldsymbol{l} = -\mathbf{W}_e\mathbf{A}\boldsymbol{l} + \mathbf{k}_0$$

$$\mathbf{v} = \mathbf{QA}^t\mathbf{k} = -\mathbf{QA}^t\mathbf{W}_e\,\mathbf{A}\boldsymbol{l} + \mathbf{v}_0$$

$$\hat{\boldsymbol{l}} = \boldsymbol{l} + \mathbf{v} = (\mathbf{I} - \mathbf{QA}^t\mathbf{W}_e\,\mathbf{A})\boldsymbol{l} + \mathbf{v}_0$$

Using the method of substitution we get the following autocofactor and crosscofactor matrices.

Autocofactor Matrices

$$\mathbf{Q}_{ff} = (-\mathbf{A})\mathbf{Q}(-\mathbf{A})^t = \mathbf{AQA}^t = \mathbf{Q}_e$$

$$\mathbf{Q}_{kk} = \mathbf{W}_e\,\mathbf{AQA}^t\mathbf{W}_e = \mathbf{W}_e\,\mathbf{Q}_e\,\mathbf{W}_e = \mathbf{W}_e$$

$$\mathbf{Q}_{vv} = \mathbf{QA}^t\mathbf{W}_e\,\mathbf{AQA}^t\mathbf{W}_e\,\mathbf{AQ} = \mathbf{QA}^t\mathbf{W}_e\,\mathbf{AQ}$$

$$\begin{aligned} \mathbf{Q}_{\hat{l}\hat{l}} &= (\mathbf{I} - \mathbf{QA}^t\mathbf{W}_e\,\mathbf{A})\mathbf{Q}(\mathbf{I} - \mathbf{A}^t\mathbf{W}_e\,\mathbf{AQ}) \\ &= (\mathbf{I} - \mathbf{QA}^t\mathbf{W}_e\,\mathbf{A})^2\mathbf{Q} \end{aligned}$$

(Employing the property of idempotent matrices, see Appendix A.)

$$\begin{aligned} \mathbf{Q}_{\hat{l}\hat{l}} &= \mathbf{Q} - \mathbf{QA}^t\mathbf{W}_e\,\mathbf{AQ} \\ &= \mathbf{Q} - \mathbf{Q}_{vv} \end{aligned}$$

Crosscofactor Matrices

$$\mathbf{Q}_{lf} = (\mathbf{I})\mathbf{Q}(-\mathbf{A})^t = -\mathbf{Q}\mathbf{A}^t$$

$$\mathbf{Q}_{lk} = (\mathbf{I})\mathbf{Q}(-\mathbf{W}_e\,\mathbf{A})^t = -\mathbf{Q}\mathbf{A}^t\mathbf{W}_e$$

$$\mathbf{Q}_{lv} = (\mathbf{I})\mathbf{Q}(-\mathbf{Q}\mathbf{A}^t\mathbf{W}_e\,\mathbf{A})^t = -\mathbf{Q}\mathbf{A}^t\mathbf{W}_e\,\mathbf{A}\mathbf{Q} = -\mathbf{Q}_{vv}$$

$$\mathbf{Q}_{l\hat{l}} = (\mathbf{I})\mathbf{Q}(\mathbf{I} - \mathbf{Q}\mathbf{A}^t\mathbf{W}_e\,\mathbf{A})^t = \mathbf{Q} - \mathbf{Q}\mathbf{A}^t\mathbf{W}_e\,\mathbf{A}\mathbf{Q} = \mathbf{Q}_{\hat{l}\hat{l}}$$

$$\mathbf{Q}_{fk} = (-\mathbf{A})\mathbf{Q}(-\mathbf{W}_e\,\mathbf{A})^t = \mathbf{A}\mathbf{Q}\mathbf{A}^t\mathbf{W}_e = \mathbf{Q}_e\,\mathbf{W}_e = \mathbf{I}$$

$$\mathbf{Q}_{fv} = (-\mathbf{A})\mathbf{Q}(-\mathbf{Q}\mathbf{A}^t\mathbf{W}_e\,\mathbf{A})^t = \mathbf{A}\mathbf{Q}\mathbf{A}^t\mathbf{W}_e\,\mathbf{A}\mathbf{Q} = \mathbf{A}\mathbf{Q}$$

$$\mathbf{Q}_{f\hat{l}} = (-\mathbf{A})\mathbf{Q}(\mathbf{I} - \mathbf{Q}\mathbf{A}^t\mathbf{W}_e\,\mathbf{A})^t = -\mathbf{A}\mathbf{Q} + \mathbf{A}\mathbf{Q} = \mathbf{0}$$

$$\mathbf{Q}_{kv} = (-\mathbf{W}_e\,\mathbf{A})\mathbf{Q}(-\mathbf{Q}\mathbf{A}^t\mathbf{W}_e\,\mathbf{A})^t = \mathbf{W}_e\,\mathbf{A}\mathbf{Q}\mathbf{A}^t\mathbf{W}_e\,\mathbf{A}\mathbf{Q} = \mathbf{W}_e\,\mathbf{A}\mathbf{Q}$$

$$\begin{aligned}\mathbf{Q}_{k\hat{l}} &= (-\mathbf{W}_e\,\mathbf{A})\mathbf{Q}(\mathbf{I} - \mathbf{Q}\mathbf{A}^t\mathbf{W}_e\,\mathbf{A})^t \\ &= -\mathbf{W}_e\,\mathbf{A}\mathbf{Q} + \mathbf{W}_e\,\mathbf{A}\mathbf{Q}\mathbf{A}^t\mathbf{W}_e\,\mathbf{A}\mathbf{Q} = \mathbf{0}\end{aligned}$$

$$\begin{aligned}\mathbf{Q}_{v\hat{l}} &= (-\mathbf{Q}\mathbf{A}^t\mathbf{W}_e\,\mathbf{A})\mathbf{Q}(\mathbf{I} - \mathbf{Q}\mathbf{A}^t\mathbf{W}_e\,\mathbf{A})^t \\ &= -\mathbf{Q}\mathbf{A}^t\mathbf{W}_e\,\mathbf{A}\mathbf{Q} + \mathbf{Q}\mathbf{A}^t\mathbf{W}_e\,\mathbf{A}\mathbf{Q}\mathbf{A}^t\mathbf{W}_e\,\mathbf{A}\mathbf{Q} = \mathbf{0}\end{aligned}$$

Using stepwise propagation we get the same results as follows

Autocofactor Matrices

$$\mathbf{Q}_{ff} = \mathbf{A}\mathbf{Q}\mathbf{A}^t = \mathbf{Q}_e$$

$$\mathbf{Q}_{kk} = \mathbf{W}_e\,\mathbf{Q}_{ff}\,\mathbf{W}_e = \mathbf{W}_e$$

$$\mathbf{Q}_{vv} = (\mathbf{Q}\mathbf{A}^t)\mathbf{Q}_{kk}(\mathbf{Q}\mathbf{A}^t)^t = \mathbf{Q}\mathbf{A}^t\mathbf{W}_e\,\mathbf{A}\mathbf{Q} \tag{7.7}$$

$$\mathbf{Q}_{\hat{l}\hat{l}} = \mathbf{Q} + \mathbf{Q}_{vv} + \mathbf{Q}_{vl} + \mathbf{Q}_{lv}$$

which when using $\mathbf{Q}_{vv}$ and $\mathbf{Q}_{lv}(= \mathbf{Q}_{vl}^t)$ given next, becomes

$$\mathbf{Q}_{\hat{l}\hat{l}} = \mathbf{Q} - \mathbf{Q}\mathbf{A}^t\mathbf{W}_e\,\mathbf{A}\mathbf{Q} = \mathbf{Q} - \mathbf{Q}_{vv} \tag{6.31}$$

Crosscofactor Matrices

$$\mathbf{Q}_{lf} = -\mathbf{Q}\mathbf{A}^t$$

$$\mathbf{Q}_{lk} = \mathbf{Q}_{lf}\,\mathbf{W}_e = -\mathbf{Q}\mathbf{A}^t\mathbf{W}_e$$

$$\mathbf{Q}_{lv} = \mathbf{Q}_{lk}(\mathbf{Q}\mathbf{A}^t)^t = -\mathbf{Q}\mathbf{A}^t\mathbf{W}_e\,\mathbf{A}\mathbf{Q} = -\mathbf{Q}_{vv}$$

$$\mathbf{Q}_{l\hat{l}} = \mathbf{Q} + \mathbf{Q}_{lv} = \mathbf{Q} - \mathbf{Q}_{vv} = \mathbf{Q}_{\hat{l}\hat{l}}$$

$$\mathbf{Q}_{fk} = (-\mathbf{A})\mathbf{Q}_{lf}\,\mathbf{W}_e = \mathbf{A}\mathbf{Q}\mathbf{A}^t\mathbf{W}_e = \mathbf{I}$$

$$\mathbf{Q}_{fv} = (-\mathbf{A})\mathbf{Q}_{lk}(\mathbf{Q}\mathbf{A}^t)^t = \mathbf{A}\mathbf{Q}\mathbf{A}^t\mathbf{W}_e\,\mathbf{A}\mathbf{Q} = \mathbf{A}\mathbf{Q}$$

$$\mathbf{Q}_{f\hat{l}} = (-\mathbf{A})(\mathbf{Q} + \mathbf{Q}_{lv}) = -\mathbf{A}\mathbf{Q} + \mathbf{A}\mathbf{Q}\mathbf{A}^t\mathbf{W}_e\,\mathbf{A}\mathbf{Q} = \mathbf{0}$$

or

$$= \mathbf{Q}_{fl} + \mathbf{Q}_{fv} = -\mathbf{AQ} + \mathbf{AQ} = \mathbf{0}$$

$$\mathbf{Q}_{kv} = (\mathbf{W}_e)\mathbf{Q}_{fk}(\mathbf{AQ}) = \mathbf{W}_e\,\mathbf{AQ}$$

$$\mathbf{Q}_{k\hat{l}} = \mathbf{W}_e(\mathbf{Q}_{fl} + \mathbf{Q}_{fv}) = \mathbf{W}_e(-\mathbf{AQ} + \mathbf{AQ}) = \mathbf{0}$$

or

$$= \mathbf{Q}_{kl} + \mathbf{Q}_{kv} = -\mathbf{W}_e\,\mathbf{AQ} + \mathbf{W}_e\,\mathbf{AQ} = \mathbf{0}$$

$$\mathbf{Q}_{v\hat{l}} = \mathbf{Q}_{vl} + \mathbf{Q}_{vv} = -\mathbf{Q}_{vv} + \mathbf{Q}_{vv} = \mathbf{0}$$

As shown in Appendix C, the ranks of $\mathbf{Q}_{vv}$ and $\mathbf{Q}_{\hat{l}\hat{l}}$ are

$$\text{rank}(\mathbf{Q}_{vv}) = r \tag{6.42}$$

$$\text{rank}(\mathbf{Q}_{\hat{l}\hat{l}}) = n - r = n_0 \tag{6.43}$$

Example 7.1. Consider a simple plane triangle with the following observed angles

Angle A: $40°19'02'' = l_1$

Angle B: $70°30'01'' = l_2$

Angle C: $69°11'00'' = l_3$

All angles were measured with equal precision and are assumed to be uncorrelated. It is required to compute the least squares estimates of these three angles, their cofactor matrix and its rank, and an estimate of the reference variance.

Solution: There is one redundant observation and therefore 1 degree of freedom. Thus with $n = 3$ and $c = r = 1$, the condition among the observations is

$$l_1 + l_2 + l_3 - \pi = 0$$

or

$$[1 \quad 1 \quad 1]\mathbf{v} = \pi - l_1 - l_2 - l_3 = -3''$$

From the simple stochastic model given, $\mathbf{Q} = \mathbf{I}$, compute

$$\mathbf{Q}_e = \mathbf{AQA}^t = 3 \qquad \mathbf{k} = \mathbf{W}_e\,\mathbf{f} = -1''$$

$$\mathbf{v} = \mathbf{QA}^t\mathbf{k} = \begin{bmatrix} 1 \\ 1 \\ 1 \end{bmatrix}(-1'') = \begin{bmatrix} -1'' \\ -1'' \\ -1'' \end{bmatrix}$$

Finally, $\hat{\boldsymbol{l}} = \boldsymbol{l} + \mathbf{v}$, or

$\hat{l}_1$ = angle $A = 40°19'01''$

$\hat{l}_2$ = angle $B = 70°30'00''$

$\hat{l}_3$ = angle $C = 69°10'59''$

It can be easily ascertained that $\hat{l}_1 + \hat{l}_2 + \hat{l}_3 - \pi = 0$, which shows that the condition is exactly satisfied after the adjustment. This is equivalent to saying that $\hat{\mathbf{l}}$ is consistent with the model.

To compute the cofactor matrix of $\hat{\mathbf{l}}$, first compute the cofactor matrix of the residuals, $\mathbf{Q}_{vv}$, using equation (7.7),

$$\mathbf{Q}_{vv} = \frac{1}{3}\begin{bmatrix} 1 & 1 & 1 \\ 1 & 1 & 1 \\ 1 & 1 & 1 \end{bmatrix}$$

Although this is a 3×3 matrix, its rank can be easily computed to be 1 (which is equal to r). Next, $\mathbf{Q}_{\hat{l}\hat{l}}$ is computed from equation (6.31) as

$$\mathbf{Q}_{\hat{l}\hat{l}} = \frac{1}{3}\begin{bmatrix} 2 & -1 & -1 \\ -1 & 2 & -1 \\ -1 & -1 & 2 \end{bmatrix}$$

It can be readily shown that the rank of $\mathbf{Q}_{\hat{l}\hat{l}}$ is 2 which is equal to n_0.

Finally, to compute $\hat{\sigma}_0^2$, first $\mathbf{v}^t\mathbf{W}\mathbf{v} = 3\ \text{sec}^2$; and with $r = 1$, then $\hat{\sigma}_0^2 = 3\ \text{sec}^2$. An alternative would be to use equation (7.6), which would give the same value.

Example 7.2. Refer to Example 6.1 and solve the same problem using the technique developed in this section.

Consulting Figure 6.1 and the data given in Example 6.1 for the basic information on the problem, the redundancy is $r = 1$. The stochastic model is expressed by the a priori cofactor matrix,

$$\mathbf{Q} = \text{diag.}\ \{0.01, 0.01, 0.01, 0.0025, 0.0025\}$$

The first three terms are in square millimeteres and the last two in square metres. In the present case of adjustment, given 1 degree of freedom, one condition equation in terms of *only* the observations is needed. Such a condition equation is not easily obtainable from the geometry of the problem. Therefore first write the three condition equations in Example 6.1 and then eliminate the parameters x_1 and x_2. (It can be argued that if such three condition equations are written first, then adjustment using the technique discussed in the past chapter should be applied. This would be, in fact, the direct way to solve the problem. Elimination of the parameters is done here mainly to provide an exercise for the present technique.)

Using the first two equations solve for x_1 and x_2

$$x_1 = \frac{l_1 l_4}{l_1 + l_2} \quad \text{and} \quad x_2 = \frac{p l_4}{l_1 + l_2}$$

substituting into the third equation leads to

$$-l_1 l_5 - l_2 l_4 - l_2 l_5 + l_3 l_4 = 0$$

where l_i refers to the observational variables.

This is the one condition equation in terms of the five observations. Linearization by Taylor series to the form $\mathbf{A}\mathbf{v} = \mathbf{f}$ gives

$$\begin{aligned} \mathbf{A} &= [-l_5 \quad -(l_4 + l_5) \quad l_4 \quad (l_3 - l_2) \quad -(l_1 + l_2)] \\ &= [-8 \quad -18 \quad 10 \quad 16.6 \quad -20.3] \\ \mathbf{f} &= l_1 l_5 + l_2 l_4 + l_2 l_5 - l_3 l_4 = -3.6\ (\text{mm m}) \end{aligned}$$

With **Q** given above, the normal equation matrix,

$$\mathbf{Q}_e = 6.6\ (\text{mm}^2\ \text{m}^2)$$

the Lagrange multiplier is

$$\mathbf{k} = -(6.6)^{-1}3.6 = -0.56\ (\text{mm}^{-1}\ \text{m}^{-1})$$

and the vector of observational residuals

$$\mathbf{v} = \begin{bmatrix} 0.04\ (\text{mm}) \\ 0.10\ (\text{mm}) \\ -0.05\ (\text{mm}) \\ -0.022\ (\text{m}) \\ 0.027\ (\text{m}) \end{bmatrix}$$

Using these values, the final estimates of the observations will be

$$\hat{l}_1 = 16.54\ (\text{mm}), \qquad \hat{l}_2 = 3.90\ (\text{mm}), \qquad \hat{l}_3 = 20.35\ (\text{mm})$$

$$\hat{l}_4 = 9.978\ (\text{m}) \qquad \text{and} \qquad \hat{l}_5 = 8.027\ (\text{m})$$

These values are identical to those computed in Example 6.1. In that example, the coordinates x_1 and x_2 of point P were the quantities of interest. Their estimates are

$$\hat{x}_1 = \frac{\hat{l}_1 \hat{l}_4}{\hat{l}_1 + \hat{l}_2} = \frac{16.54\ (\text{mm})\ 9.978\ (\text{m})}{(16.54 + 3.90)\ (\text{mm})} = 8.07\ (\text{m})$$

$$\hat{x}_2 = \frac{p\hat{l}_4}{(\hat{l}_1 + \hat{l}_2)} = \frac{100\ (\text{mm})\ 9.978\ (\text{m})}{20.44\ (\text{mm})} = 48.82$$

These two values are nearly the same as the answers obtained in Example 6.1. Actually, we would obtain exactly the same answers if both cases were iterated properly. The a posteriori cofactor matrix of the estimated observations is

$$\mathbf{Q}_{ll} = 10^{-3}\begin{bmatrix} 9.03 & -2.18 & 1.21 & 0.50 & -0.62 \\ & 5.09 & 2.73 & 1.13 & -1.38 \\ & & 8.48 & -0.63 & 0.77 \\ & & & 2.24 & 0.32 \\ \text{symmetric} & & & & 2.11 \end{bmatrix}$$

It can be shown that the rank of this matrix is $(n - r) = (5 - 1) = 4$.

To compute $\mathbf{Q}_{\hat{x}\hat{x}}$ we apply propagation principle on the formulas

$$\hat{\mathbf{x}} = \begin{bmatrix} \hat{x}_1 \\ \hat{x}_2 \end{bmatrix} = \begin{bmatrix} \dfrac{\hat{l}_1 \hat{l}_4}{\hat{l}_1 + \hat{l}_2} \\[2ex] \dfrac{p\hat{l}_4}{\hat{l}_1 + \hat{l}_2} \end{bmatrix} = F(\hat{l}')$$

$$\underset{2,3}{\mathbf{J}} = \frac{\partial \hat{\mathbf{x}}}{\partial \hat{l}'} = \begin{bmatrix} \left(\dfrac{\hat{x}_1}{\hat{l}_1} - \dfrac{\hat{x}_1}{\hat{l}_1 + \hat{l}_2}\right) & -\dfrac{\hat{x}_1}{\hat{l}_1 \hat{l}_2} & \dfrac{\hat{x}_1}{\hat{l}_4} \\[2ex] \dfrac{-\hat{x}_2}{\hat{l}_1 + \hat{l}_2} & -\dfrac{\hat{x}_2}{\hat{l}_1 + \hat{l}_2} & \dfrac{\hat{x}_2}{p} \end{bmatrix}$$

or

$$\mathbf{J} = \begin{bmatrix} 0.0931 & -0.3950 & 0.8093 \\ -2.3877 & -2.3877 & 4.8919 \end{bmatrix}$$

Because the two values $\hat{x}_1$ and $\hat{x}_2$ are computed from only a subvector $\hat{l}'$ of the total vector $\hat{l}$, the pertinent cofactor matrix $\mathbf{Q}_{l'l'}$ is constructed from $\mathbf{Q}_{ll}$ by deleting the latter's third and fifth rows and columns. Thus

$$\mathbf{Q}_{l'l'} = 10^{-3}\begin{bmatrix} 9.03 & -2.18 & 0.50 \\ & 5.09 & 1.13 \\ \text{symmetric} & & 2.24 \end{bmatrix}$$

The cofactor matrix of the derived parameters is therefore

$$\mathbf{Q}_{\hat{x}\hat{x}} = \mathbf{J}\mathbf{Q}_{l'l'}\,\mathbf{J}^t = 10^{-3}\begin{bmatrix} 1.851 & 4.968 \\ 4.968 & 71.006 \end{bmatrix} \qquad \text{rank} = 2$$

Finally, the estimate of the reference variance is $\hat{\sigma}_0^2 = 2.016$.

For reference, the values after the final iteration (f) are

$$\mathbf{Q}_{ll(f)} = 10^{-3}\begin{bmatrix} 9.024 & -2.189 & 1.213 & 0.500 & -0.621 \\ & 5.090 & 2.721 & 1.121 & -1.394 \\ & & 8.492 & -0.621 & 0.772 \\ & & & 2.244 & 0.318 \\ \text{symmetric} & & & & 2.104 \end{bmatrix}$$

$$\mathbf{Q}_{\hat{x}\hat{x}(f)} = 10^{-3}\begin{bmatrix} 1.86 & 5.03 \\ 5.03 & 71.33 \end{bmatrix}$$

and

$$\hat{\sigma}_{0(f)}^2 = 1.9633$$

which agree well with those computed in Example 6.1.

Example 7.3. Consider the problem of Example 6.2, the two-parameter transformation, for adjustment with observations only.

In Example 6.2 the redundancy was evaluated as $r = 4$ for the given six observations. Inspection of the problem does not lead, at least not directly, to a simple way of writing four equations in terms of the observations. In fact this type of problem (transformation) is rather awkward to solve using the technique of adjustment of observations only.

Writing a pair of transformation equations for any of the three given points, solve for a and b in terms of observations and constants. For example, using the pair for the first point,

$$y_{11} = ax_{11} - bx_{21}$$

$$y_{21} = bx_{11} + ax_{21}$$

then

$$a = \frac{x_{11}y_{11} + x_{21}y_{21}}{x_{11}^2 + x_{21}^2} = \alpha \qquad b = \frac{x_{11}y_{21} - x_{21}y_{11}}{x_{11}^2 + x_{21}^2} = \beta$$

Substituting these in the transformation equations gives

$$y_{1i} = \alpha x_{1i} - \beta x_{2i}$$

$$y_{2i} = \alpha x_{2i} + \beta x_{1i}$$

or

$$f_{1i} = (x_{11} y_{11} + x_{21} y_{21})x_{1i} - (x_{11} y_{21} - x_{21} y_{11})x_{2i} - y_{1i}(x_{11}^2 + x_{21}^2) = 0$$

$$f_{2i} = (x_{11} y_{11} + x_{21} y_{21})x_{2i} + (x_{11} y_{21} - x_{21} y_{11})x_{1i} - y_{2i}(x_{11}^2 + x_{21}^2) = 0$$

with $i = 2, 3$.

These are four condition equations involving the six observations only. From the given data, the stochastic model is represented by the cofactor matrix of the observations $\mathbf{Q} = 0.01\ \mathbf{I}_6$.

Linearization: The condition equations may now be linearized to the form,

$$\underset{4,6}{\mathbf{A}}\ \underset{6,1}{\mathbf{v}} = \underset{4,1}{\mathbf{f}}$$

where

$$\mathbf{v} = [v_{x_{11}} v_{x_{21}} v_{x_{12}} v_{x_{22}} v_{x_{13}} v_{x_{23}}]^t$$

and

$$a_{11} = \frac{\partial f_{12}}{\partial x_{11}} = y_{11} x_{12} - y_{21} x_{22} - 2y_{12} x_{11}$$

$$a_{12} = \frac{\partial f_{12}}{\partial x_{21}} = y_{21} x_{12} + y_{11} x_{22} - 2y_{12} x_{21}$$

$$a_{13} = \frac{\partial f_{12}}{\partial x_{12}} = x_{11} y_{11} + x_{21} y_{21}$$

$$a_{14} = \frac{\partial f_{12}}{\partial x_{22}} = -x_{11} y_{21} + x_{21} y_{11}$$

$$a_{15} = \frac{\partial f_{12}}{\partial x_{13}} = 0$$

$$a_{16} = \frac{\partial f_{12}}{\partial x_{23}} = 0$$

$$a_{21} = \frac{\partial f_{22}}{\partial x_{11}} = y_{11} x_{22} + y_{21} x_{12} - 2y_{22} x_{11}$$

$$a_{22} = \frac{\partial f_{22}}{\partial x_{21}} = y_{21} x_{22} - y_{11} x_{12} - 2y_{22} x_{21}$$

$$a_{23} = \frac{\partial f_{22}}{\partial x_{12}} = x_{11} y_{21} - x_{21} y_{11}$$

$$a_{24} = \frac{\partial f_{22}}{\partial x_{22}} = x_{11}y_{11} + x_{21}y_{21}$$

$$a_{25} = \frac{\partial f_{22}}{\partial x_{13}} = 0$$

$$a_{26} = \frac{\partial f_{22}}{\partial x_{23}} = 0$$

$$a_{31} = \frac{\partial f_{13}}{\partial x_{11}} = y_{11}x_{13} - y_{21}x_{23} - 2y_{13}x_{11}$$

$$a_{32} = \frac{\partial f_{13}}{\partial x_{21}} = y_{21}x_{13} + y_{11}x_{23} - 2y_{13}x_{21}$$

$$a_{33} = \frac{\partial f_{13}}{\partial x_{12}} = 0$$

$$a_{34} = \frac{\partial f_{13}}{\partial x_{22}} = 0$$

$$a_{35} = \frac{\partial f_{13}}{\partial x_{13}} = a_{13}$$

$$a_{36} = \frac{\partial f_{13}}{\partial x_{23}} = a_{14}$$

$$a_{41} = \frac{\partial f_{23}}{\partial x_{11}} = y_{11}x_{23} + y_{21}x_{13} - 2y_{23}x_{11}$$

$$a_{42} = \frac{\partial f_{23}}{\partial x_{21}} = y_{21}x_{23} - y_{11}x_{13} - 2y_{23}x_{21}$$

$$a_{43} = \frac{\partial f_{23}}{\partial x_{12}} = 0$$

$$a_{44} = \frac{\partial f_{23}}{\partial x_{22}} = 0$$

$$a_{45} = \frac{\partial f_{23}}{\partial x_{13}} = a_{23}$$

$$a_{46} = \frac{\partial f_{23}}{\partial x_{23}} = a_{24}$$

The elements of the vector $\mathbf{f}_{4,1}$ is the evaluation of the four equations with the given observations. The following are the numerical values:

$$\underset{4,6}{\mathbf{A}} = \begin{bmatrix} -2.1 & -0.9 & 1.1 & -2.1 & 0.0 & 0.0 \\ 1.1 & -1.9 & 2.1 & 1.1 & 0.0 & 0.0 \\ -3.2 & 0.8 & 0.0 & 0.0 & 1.1 & -2.1 \\ -1.0 & -2.4 & 0.0 & 0.0 & 2.1 & 1.1 \end{bmatrix}$$

$$\underset{4,1}{\mathbf{f}} = \begin{bmatrix} 0.10 \\ 0.10 \\ -0.10 \\ 0.40 \end{bmatrix}$$

$$\mathbf{Q}_e = \begin{bmatrix} 10.84 & -0.60 & 6.0 & 4.26 \\ & 10.44 & -5.04 & 3.46 \\ & & 16.50 & 1.28 \\ \text{symmetric} & & & 12.38 \end{bmatrix}$$

$$\mathbf{W}_e = \begin{bmatrix} 13.36 & -0.074 & -4.56 & -4.10 \\ & 13.00 & 4.31 & -4.05 \\ & & 9.08 & -0.57 \\ \text{symmetric} & & & 10.68 \end{bmatrix}$$

$$\mathbf{k} = \mathbf{W}_e\,\mathbf{f} = \begin{bmatrix} -0.143 \\ 0.760 \\ 1.163 \\ -3.514 \end{bmatrix}$$

$$\mathbf{v} = \begin{bmatrix} 0.0093 \\ 0.0805 \\ 0.0144 \\ 0.0114 \\ -0.0610 \\ -0.0631 \end{bmatrix} \qquad \begin{aligned} \hat{x}_{11} &= 0.0093 \\ \hat{x}_{21} &= 1.0805 \\ \hat{x}_{12} &= 1.0144 \\ \hat{x}_{22} &= 0.0114 \\ \hat{x}_{13} &= 0.9390 \\ \hat{x}_{23} &= 0.9369 \end{aligned}$$

Using the estimated observations compute the transformation parameters

$$\hat{a} = \frac{\hat{x}_{11}\hat{y}_{11} + \hat{x}_{21}\hat{y}_{21}}{\hat{x}_{11}^2 + \hat{x}_{21}^2} = 1.00$$

$$\hat{b} = \frac{\hat{x}_{11}\hat{y}_{21} - \hat{x}_{21}\hat{y}_{11}}{\hat{x}_{11}^2 + \hat{x}_{21}^2} = 1.95$$

which are identical to those computed in Example 6.2. It should be obvious that for this type of problem, and for transformation problems in general, the general case of adjustment of observations and parameters combined given in the preceding chapter is more appropriate. No special effort is necessary in formulating the condition equations. Furthermore, for this particular example the general technique gave rise to only two reduced normal equations, a number that remains fixed even when the number of observations is increased. In addition, the 2×2 matrix of the inverse of the normal equations coefficient matrix is automatically the cofactor matrix of the parameters. By contrast, adjustment of observations only requires special effort to formulate the condition equations, and the number of these equations is dependent on the number of observations. Furthermore, both the parameters and their cofactor matrix require further computation after obtaining the estimated observations and their cofactor matrix from the adjustment.

For the sake of completeness, $\mathbf{Q}_{ll}$ and $\hat{\sigma}_0^2$ are

$$\mathbf{Q}_{ll} = 10^{-3}\begin{bmatrix} 2.501 & 0.055 & 0.049 & -2.499 & 2.541 & -2.459 \\ & 3.420 & 3.030 & 0.067 & 2.586 & 2.574 \\ & & 2.685 & 0.059 & 2.292 & 2.281 \\ & & & 2.501 & -2.449 & 2.550 \\ & & & & 4.456 & -0.554 \\ \text{symmetric} & & & & & 4.437 \end{bmatrix}$$

$$\hat{\sigma}_0^2 = 0.2656$$

7.2.3. Further Examples on Adjustment of Observations Only

Example 7.4. *Adjustment of a Level Net*

Given: Figure 7.1 shows a small level net involving eight lines, four points (B, C, D, and E), and a known elevation of the starting point A of 800.000 metres. The observations, which are the differences in elevation along the indicated lines, are given in Table 7.1. These observations (the elevation differences) are assumed to be uncorrelated. However, the weight of each elevation difference in square metres has

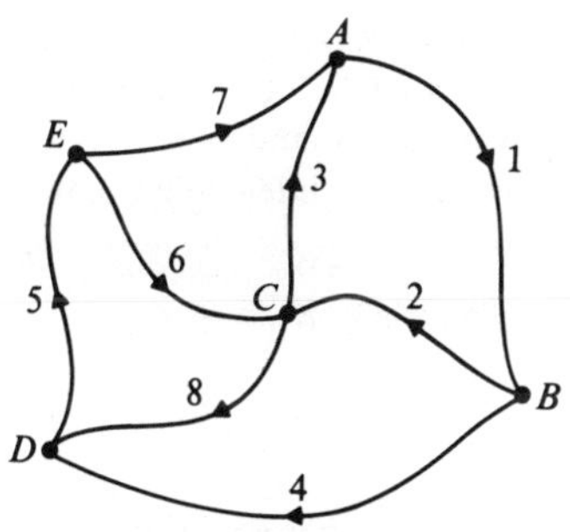

Figure 7-1

TABLE 7-1

LINE NUMBER	DISTANCE (km)	ELEVATION DIFFERENCE (m)
1	18.1	+25.42
2	9.4	+10.34
3	14.2	−35.20
4	17.6	−15.54
5	13.5	+21.32
6	9.9	+ 4.82
7	13.8	−31.02
8	14.0	−26.11

been found empirically to be equal to the reciprocal of the corresponding distance taken in 10 km units. Thus, for example, the weight for the first elevation difference (line 1 between points A and B) is $w_1 = (1/1.8)\ \text{m}^{-2}$.

Required: Compute the least squares estimates of the elevations of points B, C, D, and E, and their precision.

Solution: The functional model of the problem is concerned with a simple geometric case of vertical distances between points on the earth's surface within a relatively small area. *Four variables* are needed to determine uniquely the elevations of the four points in the problem. Therefore with eight observations, the redundancy is $r = 4$.

There are two possible adjustment techniques. In the first, four condition equations involving only observations are written and values for all eight estimated observations, $\hat{l}$, are obtained. Using these and simple geometric relations, we would then compute the required elevations of the four points. Also, using $\mathbf{Q}_{ll}$ from the adjustment and applying the rules of propagation, the cofactor matrix of the computed elevations can be evaluated. In the second method the elevations of points B, C, D, and E are introduced as unknown parameters and their values as well as the corresponding cofactor matrix are computed directly from the adjustment. In the present example we shall consider the first case of adjustment with observations only.

Going around a "loop," *in one direction*, the added elevation differences must be zero. Using the four obvious loops (ABC, ACE, BDC, and DEC), and noting that a value is positive along the given arrow and negative when opposite to the given arrow, we may write the following four independent condition equations:

$$
\begin{aligned}
l_1 + l_2 + l_3 \qquad\qquad\qquad\qquad &= 0 \\
- l_3 \qquad - l_6 + l_7 \qquad &= 0 \\
- l_2 \qquad + l_4 \qquad\qquad - l_8 &= 0 \\
l_5 + l_6 \qquad + l_8 &= 0
\end{aligned}
$$

It should be pointed out that several other equations may be written. It is therefore important not only to have the right number of equations, but also that they be independent. For example, the sum of the first two equations would correspond to the loop $ABCE$, but it is obviously a dependent equation.

Rewriting these equations in the matrix form $\mathbf{Av} = \mathbf{f}$, thus

$$
\begin{bmatrix}
1 & 1 & 1 & 0 & 0 & 0 & 0 & 0 \\
0 & 0 & -1 & 0 & 0 & -1 & 1 & 0 \\
0 & -1 & 0 & 1 & 0 & 0 & 0 & -1 \\
0 & 0 & 0 & 0 & 1 & 1 & 0 & 1
\end{bmatrix}
\begin{bmatrix} v_1 \\ v_2 \\ v_3 \\ v_4 \\ v_5 \\ v_6 \\ v_7 \\ v_8 \end{bmatrix}
=
\begin{bmatrix} -l_1 - l_2 - l_3 \\ l_3 + l_6 - l_7 \\ l_2 - l_4 + l_8 \\ -l_5 - l_6 - l_8 \end{bmatrix}
=
\begin{bmatrix} -0.56\ \text{m} \\ +0.64\ \text{m} \\ -0.23\ \text{m} \\ -0.03\ \text{m} \end{bmatrix}
$$

From the information given in the problem, the weight matrix of the given observations is the following diagonal matrix:

$$
\mathbf{W} = \text{diag.}\left\{\frac{1}{1.81}, \frac{1}{0.94}, \frac{1}{1.42}, \frac{1}{1.76}, \frac{1}{1.35}, \frac{1}{0.99}, \frac{1}{1.38}, \frac{1}{1.40}\right\}(\text{m}^{-2})
$$

and the cofactor matrix is

$$\mathbf{Q} = \text{diag. } \{1.81, 0.94, 1.42, 1.76, 1.35, 0.99, 1.38, 1.40\}(\text{m}^2)$$

Using the above values for **A** and **Q**, the reduced normal equations coefficient matrix is

$$\mathbf{Q}_e = \begin{bmatrix} 4.17 & -1.42 & -0.94 & 0 \\ -1.42 & 3.79 & 0 & -0.99 \\ -0.94 & 0 & 4.10 & -1.40 \\ 0 & -0.99 & -1.40 & 3.74 \end{bmatrix} (\text{m}^2)$$

and its inverse

$$\mathbf{W}_e = \begin{bmatrix} 0.3065 & 0.1333 & 0.0944 & 0.0706 \\ 0.1333 & 0.3445 & 0.0707 & 0.1177 \\ 0.0944 & 0.0707 & 0.3118 & 0.1354 \\ 0.7060 & 0.1177 & 0.1354 & 0.3492 \end{bmatrix} (\text{m}^{-2})$$

The vector of Lagrange multipliers is computed as

$$\mathbf{k} = [-0.1102 \quad 0.1261 \quad -0.0834 \quad -0.0059]^t(\text{m}^{-1})$$

from which, the vector of residuals is computed to be

$$\mathbf{v} = [-0.199 \quad -0.025 \quad -0.335 \quad -0.147 \quad -0.008 \quad -0.131 \quad 0.174 \quad 0.109]^t(\text{m})$$

The estimated observations $\hat{l}$ are obtained by adding the vector of residuals $\mathbf{v}$ to the given observations l.

$$\hat{l} = [25.220 \quad 10.315 \quad -35.535 \quad -15.685 \quad 21.312 \quad 4.689 \quad -30.846 \quad -26.001]^t(\text{m})$$

To compute the elevations of points, B, C, D, and E, denote them by x_1, x_2, x_3, and x_4, respectively, and the elevation of point A by x_0. Thus, from Figure 7.1,

$$x_1 = x_0 + \hat{l}_1 \qquad = 825.220 \text{ m} = \text{elev. of point } B$$

$$x_2 = x_0 - \hat{l}_3 \qquad = 835.535 \text{ m} = \text{elev. of point } C$$

$$x_3 = x_0 + \hat{l}_1 + \hat{l}_4 = 809.535 \text{ m} = \text{elev. of point } D$$

$$x_4 = x_0 - \hat{l}_7 \qquad = 830.846 \text{ m} = \text{elev. of point } E$$

In order to compute the cofactor matrix for these elevations, first compute $\mathbf{Q}_{ll}$

$$\mathbf{Q}_{ll} = \begin{bmatrix} 0.806 & -0.361 & -0.445 & -0.301 & -0.173 & 0.112 & -0.333 & 0.061 \\ & 0.560 & -0.200 & 0.360 & 0.082 & 0.118 & -0.081 & -0.201 \\ & & 0.645 & -0.059 & 0.090 & -0.231 & 0.414 & 0.141 \\ & & & 0.794 & -0.322 & -0.113 & -0.172 & -0.434 \\ & & & & 0.714 & -0.309 & -0.219 & -0.404 \\ & & & & & 0.541 & 0.310 & -0.091 \\ & & & & & & 0.724 & -0.091 \\ \text{symmetric} & & & & & & & 0.635 \end{bmatrix}$$

Referring to the vector of four elevations by $\mathbf{x}$, we can write $\mathbf{x} = \mathbf{J}\hat{\mathbf{l}}$ where

$$\mathbf{J} = \begin{bmatrix} 1 & 0 & 0 & 0 & 0 & 0 & 0 & 0 \\ 0 & 0 & -1 & 0 & 0 & 0 & 0 & 0 \\ 1 & 0 & 0 & 1 & 0 & 0 & 0 & 0 \\ 0 & 0 & 0 & 0 & 0 & 0 & -1 & 0 \end{bmatrix}$$

and through propagation compute

$$\mathbf{Q}_{xx} = \begin{bmatrix} 0.806 & 0.445 & 0.505 & 0.333 \\ & 0.645 & 0.504 & 0.414 \\ & & 0.999 & 0.505 \\ \text{symmetric} & & & 0.724 \end{bmatrix} (\text{m}^2)$$

If we were interested in the covariance matrix of the elevations, instead of the cofactor matrix which gives only relative values, we would need to scale the latter by the reference variance $\hat{\sigma}_0^2$. Since the a priori precision estimation was only in the form of an empirically derived weight matrix, we can only use the a posteriori estimate of the reference variance (that is, in this case there is no possibility of performing a statistical test on $\hat{\sigma}_0^2$).

The computation of the reference variance leads to

$$\hat{\sigma}_0^2 = 0.0404$$

Thus the covariance matrix of the elevations is

$$\hat{\Sigma}_{xx} = \hat{\sigma}_0^2\, \mathbf{Q}_{xx} = \begin{bmatrix} 0.0326 & 0.0180 & 0.0204 & 0.0135 \\ & 0.0261 & 0.0204 & 0.0167 \\ & & 0.0404 & 0.0204 \\ \text{symmetric} & & & 0.0293 \end{bmatrix} (\text{m}^2)$$

One last remark regarding the computation of $\mathbf{Q}_{xx}$ is worth mentioning here. Examining the matrix $\mathbf{J}$, we see that it contains four zero columns, the second, fifth, sixth, and eighth. This is because the computation of $\mathbf{x}$ requires the use of $\hat{l}_1$, $\hat{l}_3$, $\hat{l}_4$, and $\hat{l}_7$ only. Therefore if the computations are performed by simple means (by hand, desk calculator, or a small desk computer), it would be economical to shorten the operation of matrix multiplication in the following way. First compact $\mathbf{J}$ to a size 4×4, thus

$$\begin{array}{cccc} 1 & 3 & 4 & 7 \end{array} \quad \text{columns of } \mathbf{J}$$

$$\underset{4,4}{\mathbf{J}'} = \begin{bmatrix} 1 & 0 & 0 & 0 \\ 0 & -1 & 0 & 0 \\ 1 & 0 & 1 & 0 \\ 0 & 0 & 0 & -1 \end{bmatrix}$$

noting that the columns now correspond to $\hat{l}_1$, $\hat{l}_3$, $\hat{l}_4$, and $\hat{l}_7$, respectively. Using this fact, the submatrix of $\mathbf{Q}_{l'l'}$ which is needed for computing $\mathbf{Q}_{xx}$ would be a 4×4 matrix and is constructed using elements pertinent to the above four quantities, or

$$\underset{4,4}{\mathbf{Q}_{l'l'}} = \begin{bmatrix} 0.806 & -0.445 & -0.301 & -0.333 \\ & 0.645 & -0.059 & 0.414 \\ & & 0.794 & -0.172 \\ \text{symmetric} & & & 0.724 \end{bmatrix}$$

Using $\mathbf{J}'$ and $\mathbf{Q}_{l'l'}$ into the propagation formula, we would get exactly the same $\mathbf{Q}_{xx}$ as computed above. It must be emphasized, however, that if the problem is being solved by a standardized adjustment program, the above modification need not be applied.

Example 7.5. *Quadrilateral Adjustment by Angles and by Directions.* There has been some discussion in the literature regarding the use of angles versus the use of directions when adjusting triangulation nets. Although the method of observing angles may be different and preferable over that for observing directions, the concern here is not with particular surveying techniques. What should be demonstrated is that as far as adjustment computation is concerned, the two techniques of using angles and directions are identical; that is, starting with the same numerical data, we would arrive at exactly the same answer if we construct corresponding mathematical models properly and carry out the computations correctly. In this example we shall ascertain this fact by a numerical example concerning the adjustment of a single quadrilateral.

Figure 7.2 shows a quadrilateral (in a plane) *ABCD* with the two diagonals *AC* and *BD* connected. It shall be assumed at the beginning that the area covered by the quadrilateral is sufficiently small to allow working with plane trigonometry.

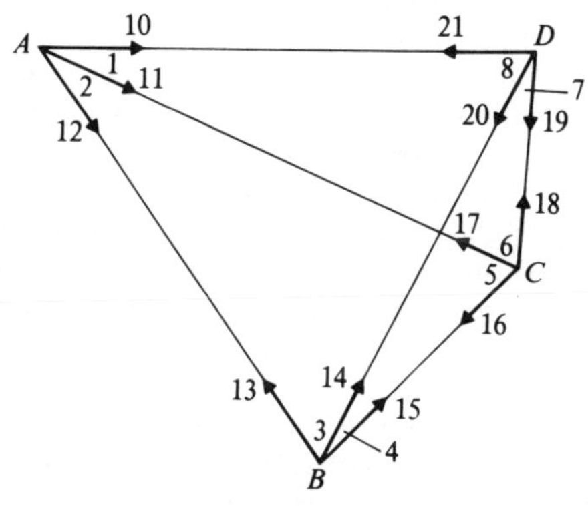

Figure 7-2

At each of the four vertices a theodolite was set up and readings along all three lines were taken. These will be considered here as directions. Any angle at one of the vertices will be simply the difference between the two directions which form that angle. Under these assumptions, then, the given observed data is listed in Table 7.2. For simplicity, all the directions in the table will be assumed as uncorrelated and of equal precision. The corresponding observed angles are fixed as in Table 7.3 once the directions are given.

Adjustment Using Directions: The fundamental observation is assumed to be a single direction which is one reading on a horizontal circle. Such a reading is actually meaningless in itself since no useful information can be derived from a *single* direction. Unless *at least* two directions are designated, we would actually establish nothing by an individual direction. Therefore with a set of directions from one point a

TABLE 7-2

DIRECTION, d	VALUE
10	00°00′00″00
11	22°01′42″51
12	38°46′13″71
13	00°00′00″00
14	57°08′57″10
15	76°42′11″23
16	00°00′00″00
17	86°33′13″45
18	145°19′49″38
19	00°00′00″00
20	15°06′52″28
21	99°11′42″94

TABLE 7-3

ANGLE l	VALUE
1	22°01′42″51
2	16°44′31″20
3	57°08′57″10
4	19°33′14″13
5	86°33′13″45
6	58°46′35″93
7	15°06′52″28
8	84°04′50″66

reference is attached that may be the zero direction and that may or may not be known. To analyze the given information, which is composed of four sets of three directions each, begin by determining the minimum number of observations that are necessary to define the quadrilateral. Starting with point A, use directions 10, 11, and 12. Since the size of the quadrilateral is immaterial as the problem involves only its shape, choose point C anywhere along the direction 11. From C, use directions 16, 17, and 18 to complete the quadrilateral. At this point we would suspect that six observations have been used to determine the required figure leaving six extra observed directions. *This does not mean, however, that there are 6 degrees of freedom* ($r = 6$). As explained above, a direction taken at a point does not provide a useful and independent piece of information unless it is referred to another from the same point. Consequently, for each of the two sets of directions observed at points B and D, there exists an inherent parameter representing the zero direction. Since these two zero directions were not designated, they are to be considered as unknown variables of the model. Hence, six extra observations minus two parameters lead to *only 4 degrees of freedom*, or a redundancy of $r = 4$.

In this problem the interest is in showing that the use of either directions or angles leads to the same adjustment results, and therefore the technique of adjustment to be chosen is not of primary importance. In the present case we shall be comparing the estimated observations and their cofactor matrix, and thus the case of adjustment involving observations only using $c = r = 4$ independent condition equations is appropriate.

The reader who is familiar with surveying techniques will recognize that there are, for the present case, *only three independent angular conditions* in a quadrilateral and *one* so-called *side condition.* The side condition guarantees that one side computed from two different sets of triangles must be the same. It is actually the one condition which ascertains that we do have a quadrilateral (otherwise, we would have adjacent triangles). Designating directions by d (as used in Table 7.2) the following are four independent condition equations:

$$\left.\begin{aligned} -d_{10} + d_{12} - d_{13} + d_{14} - d_{20} + d_{21} &= 0 \\ -d_{14} + d_{15} - d_{16} + d_{18} - d_{19} + d_{20} &= 0 \\ -d_{10} + d_{11} - d_{17} + d_{18} - d_{19} + d_{21} &= 0 \end{aligned}\right\} \quad \text{angle conditions}$$

$$\frac{\sin (d_{11} - d_{10}) \sin (d_{20} - d_{19}) \sin (d_{17} - d_{16}) \sin (d_{14} - d_{13})}{\sin (d_{12} - d_{11}) \sin (d_{21} - d_{20}) \sin (d_{18} - d_{17}) \sin (d_{15} - d_{14})} - 1 = 0$$

side condition

Linearization by Taylor series: Let us denote these four condition equations by $\dot{f}_1$, $\dot{f}_2$, $\dot{f}_3$, and $\dot{f}_4$. Then

$$\begin{bmatrix} -1 & 0 & 1 & -1 & 1 & 0 & 0 & 0 & 0 & 0 & -1 & 1 \\ 0 & 0 & 0 & 0 & -1 & 1 & -1 & 0 & 1 & -1 & 1 & 0 \\ -1 & 1 & 0 & 0 & 0 & 0 & 0 & -1 & 1 & -1 & 0 & 1 \\ a_{41} & a_{42} & a_{43} & a_{44} & a_{45} & a_{46} & a_{47} & a_{48} & a_{49} & a_{4,10} & a_{4,11} & a_{4,12} \end{bmatrix}$$

$$\times \begin{bmatrix} v_{10} \\ v_{11} \\ v_{12} \\ \vdots \\ v_{21} \end{bmatrix} = \begin{bmatrix} -\dot{f}_1 \\ -\dot{f}_2 \\ -\dot{f}_3 \\ -\dot{f}_4 \end{bmatrix}$$

which is of the form

$$\underset{4,12}{\mathbf{A}} \; \underset{12,1}{\mathbf{v}} = \underset{4,1}{\mathbf{f}}$$

where

Value (sec)

$$f_1 = +d_{10} - d_{12} + d_{13} - d_{14} + d_{20} - d_{21} \qquad = -1.4700$$

$$f_2 = +d_{14} - d_{15} + d_{16} - d_{18} + d_{19} - d_{20} \qquad = 4.2100$$

$$f_3 = +d_{10} - d_{11} + d_{17} - d_{18} + d_{19} - d_{21} \qquad = -1.3800$$

$$f_4 = 1 - \frac{\sin (d_{11} - d_{10}) \sin (d_{20} - d_{19}) \sin (d_{17} - d_{16}) \sin (d_{14} - d_{13})}{\sin (d_{12} - d_{11}) \sin (d_{21} - d_{20}) \sin (d_{18} - d_{17}) \sin (d_{15} - d_{14})} = -6.2670$$

If the fourth equation is

$$\dot{f}_4 = \frac{U}{V} - 1$$

then the elements of the fourth row of the **A** matrix can be evaluated as

$$a_{4,1} = -(U/V) \cot (d_{11} - d_{10}) \qquad = -2.47162$$

$$a_{4,2} = (U/V)[\cot (d_{11} - d_{10}) + \cot (d_{12} - d_{11})] = 5.79604$$

$$a_{4,3} = -(U/V) \cot (d_{12} - d_{11}) \qquad = -3.32242$$

$$a_{4,4} = -(U/V) \cot (d_{14} - d_{13}) \qquad = -0.64573$$

$$a_{4,5} = (U/V)[\cot (d_{14} - d_{13}) + \cot (d_{15} - d_{14})] = 3.46131$$

$$a_{4,6} = -(U/V)\cot(d_{15} - d_{14}) = -2.81557$$

$$a_{4,7} = -(U/V)\cot(d_{17} - d_{16}) = -0.06022$$

$$a_{4,8} = (U/V)[\cot(d_{17} - d_{16}) + \cot(d_{18} - d_{17})] = 0.66642$$

$$a_{4,9} = -(U/V)\cot(d_{18} - d_{17}) = -0.60620$$

$$a_{4,10} = -(U/V)\cot(d_{20} - d_{19}) = -3.70255$$

$$a_{4,11} = (U/V)[\cot(d_{20} - d_{19}) + \cot(d_{21} - d_{20})] = -3.80623$$

$$a_{4,12} = -(U/V)\cot(d_{21} - d_{20}) = -0.10368$$

Since the observed directions were given as uncorrelated and of equal precision, then

$$\mathbf{Q}_e = \begin{bmatrix} 6.0 & -2.0 & 2.0 & -0.6557 \\ & 6.0 & 2.0 & 0.6859 \\ & & 6.01 & 10.5939 \\ \text{symmetric} & & & 100.1021 \end{bmatrix} (\text{sec of arc})^2$$

and

$$\mathbf{Q}_e^{-1} = \mathbf{W}_e = \begin{bmatrix} 0.2773 & 0.1490 & -0.1765 & 0.0195 \\ & 0.2712 & -0.1703 & 0.0171 \\ & & 0.3471 & -0.0367 \\ \text{symmetric} & & & 0.0139 \end{bmatrix} (\text{sec of arc})^{-2}$$

The vector of four Lagrange multipliers is

$$\mathbf{k} = [0.3414 \quad 1.0502 \quad -0.7066 \quad 0.0072]^t (\text{sec of arc})^{-1}$$

and then the vector of residuals $\mathbf{v}$, and estimated directions $\hat{\mathbf{d}}$, are

$$\mathbf{v} = \begin{bmatrix} 0.3474 \\ -0.6648 \\ 0.3174 \\ -0.3460 \\ -0.6838 \\ 1.0299 \\ -1.0506 \\ 0.7114 \\ 0.3392 \\ -0.3703 \\ 0.7362 \\ -0.3660 \end{bmatrix} (\text{sec}) \qquad \hat{\mathbf{d}} = \begin{bmatrix} \hat{d}_{10} \\ \hat{d}_{11} \\ \hat{d}_{12} \\ \hat{d}_{13} \\ \hat{d}_{14} \\ \hat{d}_{15} \\ \hat{d}_{16} \\ \hat{d}_{17} \\ \hat{d}_{18} \\ \hat{d}_{19} \\ \hat{d}_{20} \\ \hat{d}_{21} \end{bmatrix} = \begin{bmatrix} 0^\circ & 0' & 0''.3474 \\ 22 & 1 & 41.8452 \\ 38 & 46 & 14.0274 \\ 359 & 59 & 59.6540 \\ 57 & 8 & 56.4162 \\ 76 & 42 & 12.2599 \\ 359 & 59 & 58.9494 \\ 86 & 33 & 14.1614 \\ 145 & 19 & 49.7192 \\ 359 & 59 & 59.6297 \\ 15 & 6 & 53.0162 \\ 99 & 11 & 42.5740 \end{bmatrix}$$

From the above directions the following eight angles of the quadrilateral are computed (see Figure 7.2):

$$\hat{l}_1 = \hat{d}_{11} - \hat{d}_{10} = 22^\circ 1' 41''.4978$$

$$\hat{l}_2 = \hat{d}_{12} - \hat{d}_{11} = 16^\circ 44' 32''.1822$$

$$\hat{l}_3 = \hat{d}_{14} - \hat{d}_{13} = 57^\circ 8' 56''.7622$$

$$\hat{l}_4 = \hat{d}_{15} - \hat{d}_{14} = 19°33'15''.8437$$

$$\hat{l}_5 = \hat{d}_{17} - \hat{d}_{16} = 86°33'15''.2120$$

$$\hat{l}_6 = \hat{d}_{18} - \hat{d}_{17} = 58°46'35''.5578$$

$$\hat{l}_7 = \hat{d}_{20} - \hat{d}_{19} = 15°6'53''.3865$$

$$\hat{l}_8 = \hat{d}_{21} - \hat{d}_{20} = 84°4'49''.5578$$

Values for these eight angles will be computed in the second part of the problem directly from a least squares adjustment and compared to those given above. To complete the comparison, we also check on the a posteriori cofactor matrix $\mathbf{Q}_{aa}$.

$$\mathbf{Q}_{aa} = \begin{bmatrix}
0.7290 & 0.1788 & 0.0922 & -0.1600 & -0.1870 & -0.0270 & -0.0221 & -0.0685 & 0.0906 & -0.1641 & -0.0629 & 0.2270 \\
 & 0.6121 & 0.2091 & -0.0354 & -0.1588 & 0.1942 & -0.0683 & 0.1051 & -0.0368 & 0.2253 & -0.1592 & -0.0661 \\
 & & 0.6987 & 0.1953 & -0.0281 & -0.1672 & 0.0904 & -0.0366 & -0.1538 & -0.0120 & 0.2221 & -0.1609 \\
 & & & 0.6917 & 0.2282 & 0.0800 & -0.1618 & 0.2192 & -0.0573 & -0.0652 & -0.0215 & 0.0867 \\
 & & & & 0.5671 & 0.2047 & -0.0597 & -0.1669 & 0.2266 & -0.0095 & 0.0666 & -0.0572 \\
 & & & & & 0.7153 & 0.2216 & -0.0523 & -0.1693 & 0.0747 & -0.0451 & -0.0295 \\
 & & & & & & 0.7267 & 0.1845 & 0.0887 & -0.1662 & 0.1904 & -0.0242 \\
 & & & & & & & 0.5978 & 0.2177 & -0.0196 & -0.1673 & 0.1869 \\
 & & & & & & & & 0.6936 & 0.1858 & -0.0231 & -0.1627 \\
\text{symmetric} & & & & & & & & & 0.6770 & 0.2408 & 0.8210 \\
 & & & & & & & & & & 0.5662 & 0.1930 \\
 & & & & & & & & & & & 0.7248
\end{bmatrix}$$

The quadratic form may also be computed as $\mathbf{v}^t\mathbf{W}_{dd}\mathbf{v} = 4.8492$ (sec of arc)^2.

In order to compute the cofactor matrix of the eight angles first write

$$\begin{bmatrix} \hat{l}_1 \\ \hat{l}_2 \\ \hat{l}_3 \\ \hat{l}_4 \\ \hat{l}_5 \\ \hat{l}_6 \\ \hat{l}_7 \\ \hat{l}_8 \end{bmatrix} = \begin{bmatrix}
-1 & 1 & 0 & 0 & 0 & 0 & 0 & 0 & 0 & 0 & 0 & 0 \\
0 & -1 & 1 & 0 & 0 & 0 & 0 & 0 & 0 & 0 & 0 & 0 \\
0 & 0 & 0 & -1 & 1 & 0 & 0 & 0 & 0 & 0 & 0 & 0 \\
0 & 0 & 0 & 0 & -1 & 1 & 0 & 0 & 0 & 0 & 0 & 0 \\
0 & 0 & 0 & 0 & 0 & 0 & -1 & 1 & 0 & 0 & 0 & 0 \\
0 & 0 & 0 & 0 & 0 & 0 & 0 & -1 & 1 & 0 & 0 & 0 \\
0 & 0 & 0 & 0 & 0 & 0 & 0 & 0 & 0 & -1 & 1 & 0 \\
0 & 0 & 0 & 0 & 0 & 0 & 0 & 0 & 0 & 0 & -1 & 1
\end{bmatrix} \begin{bmatrix} \hat{d}_{10} \\ \hat{d}_{11} \\ \vdots \\ \hat{d}_{21} \end{bmatrix}$$

or $\hat{\mathbf{l}} = \mathbf{F}\hat{\mathbf{d}}$ and then apply propagation to compute $\mathbf{Q}_{ll}$.

$$\mathbf{Q}_{ll} = \begin{bmatrix}
0.9835 & -0.3164 & -0.4704 & 0.5669 & 0.2198 & -0.3009 & -0.4858 & -0.1968 \\
 & 0.8926 & -0.1000 & -0.4920 & -0.3005 & 0.1247 & 0.6679 & -0.4762 \\
 & & 0.8024 & -0.2143 & -0.4881 & 0.6700 & 0.0324 & -0.2320 \\
 & & & 0.8731 & -0.1668 & -0.5104 & -0.1959 & 0.1394 \\
 & & & & 0.9554 & -0.2843 & -0.5044 & -0.5688 \\
\text{symmetric} & & & & & 0.8559 & -0.0612 & -0.4937 \\
 & & & & & & 0.7615 & -0.2145 \\
 & & & & & & & 0.9050
\end{bmatrix}$$

Adjustment Using Angles: It is of fundamental importance here to recognize that although the directions are the original uncorrelated observations, the angles that are computationally derived from them need not be uncorrelated. As a matter of fact, since $\boldsymbol{l} = \mathbf{Fd}$, the a priori cofactor matrix of the angles is $\mathbf{Q}_{ll} = \mathbf{F}\mathbf{Q}_{dd}\mathbf{F}^t = \mathbf{F}\mathbf{F}^t$ since $\mathbf{Q}_{dd} = \mathbf{I}$.

This is a basic point that involves choosing the correct stochastic model to fit the variations in the functional model. Thus

$$\mathbf{Q}_{ll} = \begin{bmatrix} 2 & -1 & 0 & 0 & 0 & 0 & 0 & 0 \\ & 2 & 0 & 0 & 0 & 0 & 0 & 0 \\ & & 2 & -1 & 0 & 0 & 0 & 0 \\ & & & 2 & 0 & 0 & 0 & 0 \\ & & & & 2 & -1 & 0 & 0 \\ & & & & & 2 & 0 & 0 \\ & & & & & & 2 & -1 \\ \text{symmetric} & & & & & & & 2 \end{bmatrix}$$

This matrix indicates a fact that should be obvious from Figure 7.2; that is, angles 1 and 2 are correlated, angles 3 and 4 are correlated, and so on, but each pair is not correlated with any other. This is because each pair of angles is derived from the directions measured at one point. Consequently, it would be an inaccurate stochastic model if we were to use the identity matrix instead of the given $\mathbf{Q}_{ll}$ above.

To formulate the number of condition equations we have only to realize that it takes a minimum of four angles to construct the shape of a unique quadrilateral thus leaving a redundancy of *four*. Four independent condition equations (three angular and one side) are

$$f_1 = l_1 + l_2 + l_3 + l_8 - \pi = 0$$

$$f_2 = l_4 + l_5 + l_6 + l_7 - \pi = 0$$

$$f_3 = l_1 + l_6 + l_7 + l_8 - \pi = 0$$

$$f_4 = \frac{\sin(l_1)\sin(l_3)\sin(l_5)\sin(l_7)}{\sin(l_2)\sin(l_4)\sin(l_6)\sin(l_8)} - 1 = \frac{Y}{Z} - 1 = 0$$

Linearization

$$\begin{bmatrix} 1 & 1 & 1 & 0 & 0 & 0 & 0 & 1 \\ 0 & 0 & 0 & 1 & 1 & 1 & 1 & 0 \\ 1 & 0 & 0 & 0 & 0 & 1 & 1 & 1 \\ a_{41} & a_{42} & a_{43} & a_{44} & a_{45} & a_{46} & a_{47} & a_{48} \end{bmatrix} \begin{bmatrix} v_1 \\ v_2 \\ \vdots \\ v_8 \end{bmatrix} = \begin{bmatrix} -f_1 \\ -f_2 \\ -f_3 \\ -f_4 \end{bmatrix} = \begin{bmatrix} 0.3414 \\ 1.0502 \\ -0.7066 \\ 0.0072 \end{bmatrix} (\text{sec of arc})^{-1}$$

where

$$a_{41} = (Y/Z)\cot(l_1) = 2.4716$$

$$a_{42} = -(Y/Z)\cot(l_2) = -3.3244$$

$$a_{43} = (Y/Z)\cot(l_3) = 0.6457$$

$$a_{44} = -(Y/Z)\cot(l_4) = -2.8156$$

$$a_{45} = (Y/Z)\cot(l_5) = 0.0602$$

$$a_{46} = -(Y/Z)\cot(l_6) = -0.6062$$

$$a_{47} = (Y/Z)\cot(l_7) = 3.7025$$

$$a_{48} = -(Y/Z)\cot(l_8) = 0.1037$$

The reduced normal equations coefficient matrix is

$$\mathbf{Q}_e = \begin{bmatrix} 6.0 & -2.0 & 2.0 & -0.6557 \\ & 6.0 & 2.0 & 0.6859 \\ & & 6.0 & 10.5939 \\ \text{symmetric} & & & 100.1021 \end{bmatrix} (\text{sec of arc})^2$$

and its inverse

$$\mathbf{W}_e = \begin{bmatrix} 0.2773 & 0.1490 & -0.1765 & 0.0195 \\ & 0.2712 & -0.1703 & 0.0171 \\ & & 0.3471 & -0.0367 \\ \text{symmetric} & & & 0.0139 \end{bmatrix} (\text{sec of arc})^{-2}$$

The vector of Lagrange multipliers is

$$\mathbf{k} = [0.3414 \quad 1.0502 \quad -0.7066 \quad 0.0072]^t (\text{sec of arc})^{-1}$$

The vectors of residuals $\mathbf{v}$ and estimated observations $\hat{\boldsymbol{l}}$ are

$$\underset{8,1}{\mathbf{v}} = \begin{bmatrix} -1.0122 \\ 0.9822 \\ -0.3378 \\ 1.7137 \\ 1.7620 \\ -0.3722 \\ 1.1065 \\ -1.1022 \end{bmatrix} (\text{sec}) \quad \text{and} \quad \begin{aligned} \hat{l}_1 &= 22°1'41''4978 \\ \hat{l}_2 &= 16°44'32''1822 \\ \hat{l}_3 &= 57°8'56''7622 \\ \hat{l}_4 &= 19°33'15''8437 \\ \hat{l}_5 &= 86°33'15''2120 \\ \hat{l}_6 &= 58°46'35''5578 \\ \hat{l}_7 &= 15°6'53''3865 \\ \hat{l}_8 &= 84°4'49''5578 \end{aligned}$$

Finally, the cofactor matrix of the estimated angles is

$$\mathbf{Q}_{\hat{l}\hat{l}} = \begin{bmatrix} 0.9835 & -0.3164 & -0.4704 & 0.5669 & 0.2198 & -0.3009 & -0.4858 & -0.1968 \\ & 0.8926 & -0.1000 & -0.4920 & -0.3005 & 0.1247 & 0.6679 & 0.4762 \\ & & 0.8024 & -0.2143 & -0.4881 & 0.6700 & 0.0324 & -0.2320 \\ & & & 0.8731 & -0.1668 & -0.5104 & -0.1959 & 0.1394 \\ & & & & 0.9554 & -0.2843 & -0.5044 & 0.5688 \\ & & & & & 0.8559 & -0.0612 & -0.4937 \\ & & & & & & 0.7615 & -0.2145 \\ \text{symmetric} & & & & & & & 0.9050 \end{bmatrix}$$

and the computed quadratic form $\mathbf{v}^t \mathbf{W}_{ll} \mathbf{v} = 4.8492$.

It should be clear now that the results obtained above match identically with those computed from an adjustment using observed directions.

With $r = 4$, the a posteriori estimate of the reference variance is

$$\hat{\sigma}_0^2 = \frac{4.8492}{4} = 1.2123$$

Using this value the "standard deviation-correlation" matrix† for $\hat{l}$ in seconds is

$$\mathbf{M}_{\sigma\rho} = \begin{bmatrix} 1.092 & -0.338 & -0.529 & 0.612 & 0.227 & -0.328 & -0.561 & -0.209 \\ & 1.040 & -0.118 & -0.557 & -0.325 & 0.143 & 0.810 & -0.530 \\ & & 0.986 & -0.256 & -0.557 & 0.808 & 0.041 & -0.272 \\ & & & 1.029 & -0.183 & -0.590 & -0.240 & 0.157 \\ & & & & 1.076 & -0.314 & -0.591 & 0.612 \\ & & & & & 1.019 & -0.076 & -0.561 \\ & & & & & & 0.961 & -0.258 \\ \text{symmetric} & & & & & & & 1.047 \end{bmatrix}$$

Difficulties may arise when the a priori correlation between angles is neglected, if such angles are computed from measured directions. To demonstrate this point, the same problem was solved one more time except that instead of using the correct cofactor matrix used above, $\mathbf{Q}_{ll} = 2\mathbf{I}$ was used. The reason for the multiplier of 2 should be obvious since each angle is compiled from two directions each having a cofactor of unity, thus it will have a cofactor of two. The results of this nonrigorous adjustment are

$$\mathbf{v} = \begin{bmatrix} -1.02 \\ 0.26 \\ 0.38 \\ 1.69 \\ 1.78 \\ 0.30 \\ 0.43 \\ -1.10 \end{bmatrix} \text{(sec)} \quad \text{and} \quad \begin{aligned} \hat{l}_1 &= 22°1'41''.49 \\ \hat{l}_2 &= 16°44'31''.46 \\ \hat{l}_3 &= 57°8'57''.48 \\ \hat{l}_4 &= 19°33'15''.82 \\ \hat{l}_5 &= 86°33'15''.23 \\ \hat{l}_6 &= 58°46'36''.23 \\ \hat{l}_7 &= 15°6'52''.71 \\ \hat{l}_8 &= 84°4'49''.56 \end{aligned}$$

Comparing the set of residuals computed here and those computed using the correct adjustment we find some differences. The variation is both in the magnitude and sign of the corrections. The maximum difference in the values of the estimated angles is 0.72 seconds. However, this value is within the reference standard deviation and may be tolerated.

This example represents a case of "approximate adjustment," inasmuch as a simplified stochastic model was used. Such simplified computational methods have been, and still are, used in practice. They lose importance, however, with the development of computer techniques which increasingly make the use of simplified procedures unwarranted. The example further demonstrates that the estimated observations and the estimates of the parameters can be close to the proper least squares estimates. By contrast to estimated observations and parameters, however, the a posteriori cofactor and covariance matrices computed on the basis of a priori simplified matrices are usually not close to those obtained from proper least squares.

† The "standard deviation-correlation" matrix is used frequently in statistics and is given here for the sake of introducing the reader to another useful concept. Its elements on the main diagonal are the standard deviations, whereas those off the diagonal are the correlation coefficients.

Because this is quite frequent in practice, a general warning is appropriate against relying on such a posteriori matrices. Instead, new matrices should be computed on the basis of the nonsimplified cofactors of the original observations.

Let $\mathbf{Q}$ be the original rigorous a priori cofactor matrix of the observations and $\bar{\mathbf{Q}}$ the simplified matrix. Let $\mathbf{s}$ represent the set of variables estimated using $\mathbf{Q}$. Instead of using the a posteriori cofactor matrix for the estimates $\hat{\mathbf{s}}$ as computed with $\bar{\mathbf{Q}}$, we should apply the principle of propagation to get a $\mathbf{Q}_{\hat{s}\hat{s}}$ using the original $\mathbf{Q}$.

7.3. ADJUSTMENT OF INDIRECT OBSERVATIONS

Adjustment of indirect observations is the second of the two classical cases of least squares adjustment. Here the adjustment is performed with both observations and parameters but with the restriction that each condition equation contains only *one observation.* Therefore the number of condition equations is the same as the number of observations, or $c = n$.

7.3.1. Formulation Because of the restriction on this case, the general linear functional equations would be of the form

$$\boldsymbol{l} + \mathbf{v} + \mathbf{B\Delta} = \mathbf{d} \tag{7.8}$$

or

$$\mathbf{v} + \mathbf{B\Delta} = (-\boldsymbol{l} + \mathbf{d}) = \mathbf{f} \tag{7.9}$$

where $\mathbf{d}$ and $\mathbf{f}$ are constant vectors.

Since the residual vector $\mathbf{v}$ is explicitly separable in equation (7.9), the application of the least squares principle can be done more directly than in the cases discussed previously. For this reason, and as a further demonstration of the minimum criterion, we give the relatively short development here, although the results can be obtained directly from the equations obtained in Chapter 6. Thus the scalar to be minimized is

$$\begin{aligned}\phi &= \mathbf{v}^t\mathbf{W}\mathbf{v}\\ &= (\mathbf{f} - \mathbf{B\Delta})^t\mathbf{W}(\mathbf{f} - \mathbf{B\Delta})\\ &= (\mathbf{f}^t - \mathbf{\Delta}^t\mathbf{B}^t)(\mathbf{W}\mathbf{f} - \mathbf{W}\mathbf{B\Delta})\\ &= \mathbf{\Delta}^t\mathbf{B}^t\mathbf{W}\mathbf{B\Delta} - \mathbf{\Delta}^t\mathbf{B}^t\mathbf{W}\mathbf{f} + \mathbf{f}^t\mathbf{W}\mathbf{f} - \mathbf{f}^t\mathbf{W}\mathbf{B\Delta}\end{aligned}$$

Realizing that the second and last terms on the right-hand side are scalars (as well as all other terms) then

$$\phi = \mathbf{\Delta}^t\mathbf{B}^t\mathbf{W}\mathbf{B\Delta} - 2\mathbf{f}^t\mathbf{W}\mathbf{B\Delta} + \mathbf{f}^t\mathbf{W}\mathbf{f} \tag{7.10}$$

The free variable in equation (7.10) is the vector of parameters $\boldsymbol{\Delta}$, thus for ϕ to be a minimum $\partial\phi/\partial\boldsymbol{\Delta}$ must be zero, or

$$\frac{\partial\phi}{\partial\boldsymbol{\Delta}} = 2\boldsymbol{\Delta}^t\mathbf{B}^t\mathbf{W}\mathbf{B} - 2\mathbf{f}^t\mathbf{W}\mathbf{B} = \mathbf{0}^t \tag{7.11}$$

recognizing that the last term in equation (7.10) is a constant insofar as the partial differentiation is concerned. Transposing equation (7.11) and rearranging

$$(\mathbf{B}^t\mathbf{W}\mathbf{B})\boldsymbol{\Delta} = \mathbf{B}^t\mathbf{W}\mathbf{f} \tag{7.12}$$

Equation (7.12) is the u reduced normal equations in the u unknown parameters $\boldsymbol{\Delta}$. With the auxiliaries used previously [see equations (6.21) and (6.22)], then

$$\mathbf{N}\boldsymbol{\Delta} = \mathbf{t} \tag{7.13}$$

or

$$\boldsymbol{\Delta} = \mathbf{N}^{-1}\mathbf{t} \tag{7.14}$$

It is important to note here that all the relations developed in this section are essentially the same as those derived in Section 6.2. Having obtained $\boldsymbol{\Delta}$, the estimated observations $\hat{\boldsymbol{l}}$ can be directly obtained from

$$\hat{\boldsymbol{l}} = \boldsymbol{l} + \mathbf{f} - \mathbf{B}\boldsymbol{\Delta} \tag{7.15}$$

7.3.2. Computational Check and Reference Variance The computational check in this case is

$$\mathbf{a} = \mathbf{B}^t\mathbf{W}\mathbf{A}\mathbf{v} = \mathbf{0} \tag{7.16}$$

The proof of equation (7.16) is exactly the same as that given previously (see Section 6.2.1).

The a posteriori estimate of the reference variance $\hat{\sigma}_0^2$ is given by equation (6.25). The quadratic form $\mathbf{v}^t\mathbf{W}\mathbf{v}$ may be evaluated directly using a $\mathbf{v}$ computed from equation (7.9), or from

$$\mathbf{v}^t\mathbf{W}\mathbf{v} = \mathbf{f}^t\mathbf{W}\mathbf{f} - \boldsymbol{\Delta}^t\mathbf{t} \tag{7.17}$$

7.3.3. Estimation of Precision The autocofactor and crosscofactor matrices are

$$\boldsymbol{l} = \mathbf{I}\boldsymbol{l}$$

$$\mathbf{f} = -\boldsymbol{l} + \mathbf{d}$$

$$\boldsymbol{\Delta} = \mathbf{N}^{-1}\mathbf{B}^t\mathbf{W}\mathbf{f} = -\mathbf{N}^{-1}\mathbf{B}^t\mathbf{W}\boldsymbol{l} + \boldsymbol{\Delta}_0$$

$$\mathbf{v} = \mathbf{f} - \mathbf{B}\boldsymbol{\Delta} = \mathbf{f} - \mathbf{B}\mathbf{N}^{-1}\mathbf{B}^t\mathbf{W}\mathbf{f}$$

$$\hat{\boldsymbol{l}} = \boldsymbol{l} + \mathbf{v}$$

Autocofactor Matrices

$$\mathbf{Q}_{ll} = \mathbf{Q}$$

$$\mathbf{Q}_{ff} = \mathbf{Q}$$

$$\mathbf{Q}_{\Delta\Delta} = \mathbf{N}^{-1}\mathbf{B}^t\mathbf{W}\mathbf{Q}_{ff}\,\mathbf{W}\mathbf{B}\mathbf{N}^{-1} = \mathbf{N}^{-1} \tag{7.18}$$

$$\mathbf{Q}_{vv} = \mathbf{Q} - \mathbf{B}\mathbf{N}^{-1}\mathbf{B}^t \tag{7.18a}$$

$$\mathbf{Q}_{\hat{l}\hat{l}} = \mathbf{Q} - \mathbf{Q}_{vv} \tag{7.19}$$

Crosscofactor Matrices

$$\mathbf{Q}_{lf} = -\mathbf{Q}$$

$$\mathbf{Q}_{l\Delta} = -\mathbf{Q}(\mathbf{W}\mathbf{B}\mathbf{N}^{-1}) = -\mathbf{B}\mathbf{N}^{-1}$$

$$\mathbf{Q}_{lv} = \mathbf{Q}_{lf} - \mathbf{Q}_{lf}(\mathbf{W}\mathbf{B}\mathbf{N}^{-1}\mathbf{B}^t) = -\mathbf{Q} + \mathbf{B}\mathbf{N}^{-1}\mathbf{B}^t = -\mathbf{Q}_{vv}$$

$$\mathbf{Q}_{l\hat{l}} = \mathbf{Q} + \mathbf{Q}_{lv} = \mathbf{Q} - \mathbf{Q}_{vv} = \mathbf{Q}_{\hat{l}\hat{l}}$$

$$\mathbf{Q}_{f\Delta} = \mathbf{Q}(\mathbf{W}\mathbf{B}\mathbf{N}^{-1}) = \mathbf{B}\mathbf{N}^{-1}$$

$$\mathbf{Q}_{fv} = \mathbf{Q}_{ff} - \mathbf{Q}_{f\Delta}\mathbf{B}^t = \mathbf{Q} - \mathbf{B}\mathbf{N}^{-1}\mathbf{B}^t$$

$$\mathbf{Q}_{f\hat{l}} = \mathbf{Q}_{fl} + \mathbf{Q}_{fv} = -\mathbf{Q} + \mathbf{Q} - \mathbf{B}\mathbf{N}^{-1}\mathbf{B}^t = -\mathbf{B}\mathbf{N}^{-1}\mathbf{B}^t$$

$$\mathbf{Q}_{\Delta v} = \mathbf{Q}_{\Delta f} - \mathbf{Q}_{\Delta\Delta}\mathbf{B}^t = \mathbf{N}^{-1}\mathbf{B}^t - \mathbf{N}^{-1}\mathbf{B}^t = \mathbf{0}$$

$$\mathbf{Q}_{\Delta\hat{l}} = \mathbf{Q}_{\Delta l} + \mathbf{Q}_{\Delta v} = -\mathbf{N}^{-1}\mathbf{B}^t$$

$$\mathbf{Q}_{v\hat{l}} = \mathbf{Q}_{vl} + \mathbf{Q}_{vv} = -\mathbf{Q}_{vv} + \mathbf{Q}_{vv} = \mathbf{0}$$

Example 7.6. Reconsider the case of the plane triangle given in Example 7.1. Although it was worked out in the most straightforward manner there, we shall solve it again here using the technique of adjustment of indirect observations.

Consider the two angles l_1 and l_2 also as unknown parameters x_1 and x_2, respectively. The three condition equations will therefore be

$$l_1 - x_1 = 0$$

$$l_2 - x_2 = 0$$

$$l_3 + x_1 + x_2 - \pi = 0$$

or

$$\mathbf{v} + \begin{bmatrix} -1 & 0 \\ 0 & -1 \\ 1 & 1 \end{bmatrix} \begin{bmatrix} x_1 \\ x_2 \end{bmatrix} = \begin{bmatrix} -l_1 \\ -l_2 \\ \pi - l_3 \end{bmatrix} = \begin{bmatrix} -40°19'02'' \\ -70°30'01'' \\ +110°49'00'' \end{bmatrix}$$

With $\mathbf{W} = \mathbf{Q}^{-1} = \mathbf{I}$, the reduced normal equations may be computed as

$$\mathbf{N} = \mathbf{B}^t\mathbf{W}\mathbf{B} = \mathbf{B}^t\mathbf{B} = \begin{bmatrix} 2 & 1 \\ 1 & 2 \end{bmatrix}$$

$$\mathbf{t} = \mathbf{B}^t\mathbf{W}\mathbf{f} = \mathbf{B}^t\mathbf{f} = \begin{bmatrix} 151°08'02'' \\ 181°19'01'' \end{bmatrix}$$

$$\mathbf{N}^{-1} = \tfrac{1}{3}\begin{bmatrix} 2 & -1 \\ -1 & 2 \end{bmatrix}$$

$$\boldsymbol{\Delta} = \hat{\mathbf{x}} = \mathbf{N}^{-1}\mathbf{t} = \begin{bmatrix} 40°19'01'' \\ 70°30'00'' \end{bmatrix}$$

These two values are the same as the values of $\hat{l}_1$ and $\hat{l}_2$ obtained in Example 7.1. The cofactor matrix of the parameters is

$$\mathbf{Q}_{\Delta\Delta} = \mathbf{N}^{-1} = \tfrac{1}{3}\begin{bmatrix} 2 & -1 \\ -1 & 2 \end{bmatrix}$$

which is also the submatrix of $\mathbf{Q}_{ll}$ corresponding to $\hat{l}_1$ and $\hat{l}_2$ in that example. To complete the problem we compute

$$\mathbf{v} = \mathbf{f} - \mathbf{B}\boldsymbol{\Delta} = \begin{bmatrix} -40 & 19 & 02 \\ -70 & 30 & 01 \\ 110 & 49 & 00 \end{bmatrix} - \begin{bmatrix} -40 & 19 & 01 \\ -70 & 30 & 00 \\ 110 & 49 & 01 \end{bmatrix} = \begin{bmatrix} -1'' \\ -1'' \\ -1'' \end{bmatrix}$$

and

$$\hat{l}_1 = 40°19'01'' \qquad \hat{l}_2 = 70°30'00'' \qquad \hat{l}_3 = 69°10'59''$$

$$\mathbf{Q}_{ll} = \mathbf{B}^t\mathbf{N}^{-1}\mathbf{B} = \tfrac{1}{3}\begin{bmatrix} 2 & -1 & -1 \\ & 2 & -1 \\ \text{symmetric} & & 2 \end{bmatrix}$$

which is the same as the matrix $\mathbf{Q}_{ll}$ computed in Example 7.1.

Another verification for the computations is possible when applying equation (7.16).

$$\mathbf{a} = \mathbf{B}^t\mathbf{W}\mathbf{v} = \mathbf{B}^t\mathbf{v} = \begin{bmatrix} -1 & 0 & 1 \\ 0 & -1 & 1 \end{bmatrix}\begin{bmatrix} -1 \\ -1 \\ -1 \end{bmatrix} = \begin{bmatrix} 0 \\ 0 \end{bmatrix} = \mathbf{0}$$

For the quadratic form we use the computed $\mathbf{v}$ directly, or

$$\mathbf{v}^t\mathbf{W}\mathbf{v} = [-1 \quad -1 \quad -1]\begin{bmatrix} -1 \\ -1 \\ -1 \end{bmatrix} = 3 \text{ sec}^2$$

then

$$\hat{\sigma}_0^2 = 3 \text{ sec}^2$$

Example 7.7. Consider the problem of Example 6.1 according to the technique of indirect observations. As indicated above, as many condition equations as there are observations (or five equations) are needed for this problem. From Figure 6.1 we can readily write the following five equations with the parameters being (a) x_1 is the x_1 coordinate of point P; (b) x_2 is the x_2 coordinate of point P; (c) x_3 is the distance $S_1 S_2$; and (d) x_4 is the distance $S_2 S_3$.

$$l_1 - \frac{p x_1}{x_2} = 0 \qquad l_2 - \frac{p(x_3 - x_1)}{x_2} = 0 \qquad l_3 - \frac{p(x_3 + x_4 - x_1)}{x_2} = 0$$

$$l_4 - x_3 = 0 \qquad l_5 - x_4 = 0$$

l_i represent observational variables. Linearization leads to

$$\underset{5,4}{\mathbf{B}} = \begin{bmatrix} -p/x_2^0 & p x_1^0/x_2^{02} & 0 & 0 \\ p/x_2^0 & p(x_3^0 - x_1^0)/x_2^{02} & -p/x_2^0 & 0 \\ p/x_2^0 & p(x_4^0 - x_1^0)/x_2^{02} & -p/x_2^0 & -p/x_2^0 \\ 0 & 0 & -1 & 0 \\ 0 & 0 & 0 & -1 \end{bmatrix}$$

Approximations may be computed as follows:

$$x_1^0 = \frac{l_1 l_4}{(l_1 + l_2)} = 8.128 \text{ m}$$

$$x_2^0 = \frac{p l_4}{(l_1 + l_2)} = 49.261 \text{ m}$$

$$x_3^0 = l_4 = 10.0 \text{ m}$$

$$x_4^0 = l_5 = 8.0 \text{ m}$$

with $p = 100$ the numerical value of $\mathbf{B}$ becomes

$$\mathbf{B} = \begin{bmatrix} -2.03 & 0.335 & 0 & 0 \\ 2.03 & 0.077 & -2.03 & 0 \\ 2.03 & 0.407 & -2.03 & -2.03 \\ 0 & 0 & -1 & 0 \\ 0 & 0 & 0 & -1 \end{bmatrix} (\text{mm m}^{-1})$$

With the covariance matrix used before, the weight matrix may be obtained from

$$\frac{1}{\sigma_0^2}\mathbf{W} = \mathbf{\Sigma}^{-1} = \begin{bmatrix} 100 & & & & \\ & 100 & & & \\ & & 100 & & \\ & & & 400 & \\ & & & & 400 \end{bmatrix} = 100 \begin{bmatrix} 1 & & & & \\ & 1 & & & \\ & & 1 & & \\ & & & 4 & \\ & & & & 4 \end{bmatrix}$$

From which the a priori value of σ_0^2 is 0.01 and the weight matrix is

$$\mathbf{W} = \text{diag.}\ \{1, 1, 1, 4, 4\}$$

Then

$$\mathbf{N} = \begin{bmatrix} 12.363 & 0.303 & -8.242 & -4.121 \\ & 0.284 & -0.982 & -0.826 \\ & & 12.242 & 4.121 \\ \text{symmetric} & & & 8.121 \end{bmatrix}$$

In order to compute **t**, first evaluate **f** as follows:

$$f_1 = \frac{px_1^0}{x_2^0} - l_1 = 0$$

$$f_2 = \frac{p(x_3^0 - x_1^0)}{x_2^0} - l_2 = -0.175 \text{ mm}$$

$$f_3 = \frac{p(x_3^0 + x_4^0 - x_1^0)}{x_2^0} - l_3 = -0.360 \text{ mm}$$

$$f_4 = x_3^0 - l_4 = 0$$

$$f_5 = x_4^0 - l_5 = 0$$

Then

$$\mathbf{t} = \begin{bmatrix} -1.065 \\ -1.160 \\ +1.065 \\ +0.831 \end{bmatrix}$$

With

$$\mathbf{N}^{-1} = \begin{bmatrix} 0.189 & 0.513 & 0.143 & 0.075 \\ & 7.334 & 0.718 & 0.642 \\ & & 0.224 & 0.031 \\ \text{symmetric} & & & 0.211 \end{bmatrix}$$

then the solution is

$$\mathbf{\Delta} = [-0.054 \quad -0.455 \quad -0.022 \quad 0.028]^t \text{ (mm)}$$

leading to the parameter values

$$\hat{x}_1 = 8.074 \text{ m} \qquad \hat{x}_2 = 48.806 \text{ m}$$

$$\hat{x}_3 = 9.978 \text{ m} \qquad \hat{x}_4 = 8.028 \text{ m}$$

which agree well with the values computed previously.

This example brings up one last remark. In order that we may use this case of adjustment of indirect observations, it is necessary that the *maximum number of independent parameters* involved in the model be carried in the adjustment. Although this may be desirable when all such parameters are of interest, it is not convenient or useful to have to do this when only a subset is needed. This example illustrates the point since we were only interested in two parameters, x_1 and x_2, but had to solve for four.

Example 7.8. In this example, solve the two-parameter transformation problem of Example 6.2 using the technique of adjustment of indirect observations. The transformation equations were given as

$$y_{1i} = ax_{1i} - bx_{2i}$$

$$y_{2i} = bx_{1i} + ax_{2i}$$

or in matrix form

$$\begin{bmatrix} y_1 \\ y_2 \end{bmatrix}_i = \begin{bmatrix} a & -b \\ b & a \end{bmatrix} \begin{bmatrix} x_1 \\ x_2 \end{bmatrix}_i$$

Since the **x** coordinates are the observations, first write the inverse transformation, or

$$\begin{bmatrix} x_1 \\ x_2 \end{bmatrix}_i = \frac{1}{(a^2+b^2)} \begin{bmatrix} a & b \\ -b & a \end{bmatrix} \begin{bmatrix} y_1 \\ y_2 \end{bmatrix}_i$$

or

$$f_{1i} = x_{1i} - \frac{a}{(a^2+b^2)} y_{1i} - \frac{b}{(a^2+b^2)} y_{2i} = 0 \qquad \text{(a)}$$

$$f_{2i} = x_{2i} + \frac{b}{(a^2+b^2)} y_{1i} - \frac{a}{(a^2+b^2)} y_{2i} = 0$$

with $i = 1, 2, 3$, which are the six condition equations ($n = 6$). Linearization by Taylor series to the form $\mathbf{v} + \mathbf{B}\boldsymbol{\Delta} = \mathbf{f}$ yields

$$\underset{2,1}{\mathbf{v}_i} + \frac{1}{(a^2+b^2)^2} \begin{bmatrix} [(a^2-b^2)y_{1i} + 2aby_{2i}] & [(-a^2+b^2)y_{2i} + 2aby_{1i}] \\ [(a^2-b^2)y_{2i} - 2aby_{1i}] & [(a^2-b^2)y_{1i} + 2aby_{2i}] \end{bmatrix} \boldsymbol{\Delta}$$

$$= \begin{bmatrix} -f_{1i} \\ -f_{2i} \end{bmatrix}$$

with the given data

$$\mathbf{x}_1 = (0.0,\ 1.0),\ \mathbf{x}_2 = (1.0,\ 0.0),\ \mathbf{x}_3 = (1.0,\ 1.0)$$

$$\mathbf{y}_1 = (-2.1,\ 1.1),\ \mathbf{y}_2 = (1.0,\ 2.0),\ \mathbf{y}_3 = (-0.9,\ 2.8)$$

and the approximations $a^0 = 1$ and $b^0 = 2$, the numerical values of the matrices become

$$\mathbf{B} = \begin{bmatrix} 0.428 & -0.204 \\ 0.204 & 0.428 \\ 0.200 & 0.400 \\ -0.400 & 0.200 \\ 0.556 & 0.192 \\ -0.192 & 0.556 \end{bmatrix} \quad \text{and} \quad \mathbf{f} = \begin{bmatrix} 0.02 \\ 0.06 \\ 0.00 \\ 0.00 \\ -0.06 \\ -0.08 \end{bmatrix}$$

Remembering that the **x** coordinates are uncorrelated and of equal precision, for the sake of variation take an a priori $\sigma_0^2 = 0.01$ and thus a $\mathbf{W} = \mathbf{I}$. Then

$$\mathbf{N} = \begin{bmatrix} 0.7708 & 0.0 \\ 0.0 & 0.7708 \end{bmatrix} \qquad \mathbf{N}^{-1} = \begin{bmatrix} 1.2974 & 0.0 \\ 0.0 & 1.2974 \end{bmatrix}$$

$$\mathbf{\Delta} = \begin{bmatrix} 0.0036 \\ -0.0446 \end{bmatrix}$$

and

$$\hat{a} = 1.0036 \qquad \hat{b} = 1.9554$$

These answers agree with those obtained twice before. They, however, represent values obtained after the first iteration of a nonlinear problem. If we were to continue the iterations until the last $\mathbf{\Delta}$ is insignificantly small, we would get as a final answer $\hat{a} = 1.0031$ and $\hat{b} = 1.9561$.

We have solved the problem above in a straightforward manner without any attempt at simplification. It is possible, however, to eliminate the nonlinearity in the problem as well as simplify the algebraic operations by using two new parameters c and d, mainly because the **y** coordinates were given in the problem as constants. Referring to equation (a), make the substitution

$$c = \frac{a}{a^2 + b^2} \qquad d = \frac{b}{a^2 + b^2} \tag{b}$$

It is important to emphasize that reparameterization is possible provided we always make certain that the *number of the new parameters is the same as the old parameters, and that the new parameters are also functionally independent just as the old ones were.* In our present case both restrictions are met by the choice of c and d. Thus

$$\begin{aligned} f_{1i} &= x_{1i} - cy_{1i} - dy_{2i} = 0 \\ f_{2i} &= x_{2i} + dy_{1i} - cy_{2i} = 0 \end{aligned} \tag{c}$$

with $i = 1, 2, 3$; and then in the usual form $\mathbf{v} + \mathbf{B}\mathbf{\Delta} = f$, or

$$\mathbf{v}_i + \begin{bmatrix} -y_{1i} & -y_{2i} \\ -y_{2i} & y_{1i} \end{bmatrix} \begin{bmatrix} c \\ d \end{bmatrix} = \begin{bmatrix} -x_{1i} \\ -x_{2i} \end{bmatrix} \tag{d}$$

again with $i = 1, 2, 3$. With $\mathbf{Q} = \mathbf{I}_6$, then

$$\mathbf{N} = \begin{bmatrix} 19.27 & 0 \\ 0 & 19.27 \end{bmatrix} \qquad \mathbf{t} = \begin{bmatrix} 4.0 \\ 7.8 \end{bmatrix}$$

and

$$\mathbf{\Delta} = \begin{bmatrix} \hat{c} \\ \hat{d} \end{bmatrix} = \begin{bmatrix} 0.207576 \\ 0.404774 \end{bmatrix}$$

To get the estimates of the original parameters a and b we solve for them from (b), or

$$\hat{a} = \frac{\hat{c}}{\hat{c}^2 + \hat{d}^2} = 1.0031$$

$$\hat{b} = \frac{\hat{d}}{\hat{c}^2 + \hat{d}^2} = 1.9561$$

These values are identical to those obtained from the first solution, *but after* the problem has been iterated sufficiently. This should be clear to the reader since in the linear case we would obtain the exact solution directly, which in the nonlinear case is attainable only after the effects of linearization have been reduced to a negligible level.

Example 7.9. *Simple Mean, Weighted Mean, and Correlated Mean*

a. *Simple Mean.* Given n repeated measurements on a variate l, derive the relation for the best estimate if all the measurements are uncorrelated and of equal precision (weight).

Solution: Denoting the best estimate by x (a parameter), the n condition equations would be

$$l_1 + v_1 - x = 0$$

$$l_2 + v_2 - x = 0$$

$$\cdots\cdots\cdots\cdots$$

$$l_n + v_n - x = 0$$

or

$$\begin{bmatrix} v_1 \\ v_2 \\ \vdots \\ v_n \end{bmatrix} + \begin{bmatrix} -1 \\ -1 \\ \vdots \\ -1 \end{bmatrix} \underset{1,1}{\mathbf{x}} = \begin{bmatrix} -l_1 \\ -l_2 \\ \vdots \\ -l_n \end{bmatrix}$$

With $\mathbf{W} = \mathbf{Q}^{-1} = \mathbf{I}$, then

$$\mathbf{N} = n \qquad \mathbf{t} = \sum_i l_i$$

$$\mathbf{x} = \frac{1}{n} \sum_i l_i = \bar{l} = \text{the arithmetic mean} \tag{7.20}$$

$$\mathbf{Q}_{xx} = \frac{1}{n} = \text{cofactor matrix of arithmetic mean} \tag{7.21}$$

b. *Weighted Mean.* Given n uncorrelated observations on the same stochastic variable l, each of which having a weight w_i, derive the formula for the best estimate of that variable.

Solution: The condition equations are the same as those in (a) above, but the weight matrix **W** in this case is

$$\mathbf{W} = \text{diag.}\,[w_1, w_2, \ldots, w_n]$$

and

$$\mathbf{N} = \sum_{i=1}^{n} w_i$$

$$\mathbf{t} = \sum_{i=1}^{n} w_i l_i$$

$$\mathbf{x} = \frac{\sum_{i=1}^{n} w_i l_i}{\sum_{i=1}^{n} w_i} = \text{the weighted arithmetic mean} \tag{7.22}$$

$$\mathbf{Q}_{xx} = \left[\sum_{i=1}^{n} w_i\right]^{-1} \tag{7.23}$$

$$\mathbf{W}_{xx} = \sum_{i=1}^{n} w_i$$

which says that the weight of the weighted mean is equal to the sum of all weights of constituent observations.

c. *Correlated Mean.* Given n observations on the same random variable l that are correlated and of unequal precision, if the weight matrix of these observations is **W**, compute the best estimate of l.

Solution: The weight matrix may be written in expanded form as

$$\mathbf{W} = \begin{bmatrix} w_{11} & w_{12} & \cdots & w_{1n} \\ w_{21} & w_{22} & \cdots & w_{2n} \\ \vdots & \vdots & & \vdots \\ w_{n1} & w_{n2} & \cdots & w_{nn} \end{bmatrix}$$

The condition equations are as in (a), but the normal equations become

$$\mathbf{N} = \mathbf{B}^t\mathbf{W}\mathbf{B} = [-1 \quad -1 \quad \cdots \quad -1]\mathbf{W}\begin{bmatrix} -1 \\ -1 \\ \vdots \\ -1 \end{bmatrix}$$

$$= \sum_{i=1}^{n} \sum_{j=1}^{n} w_{ij}$$

and, similarly

$$t = \sum_{i=1}^{n} \sum_{j=1}^{n} w_{ij} l_i$$

The best estimate is therefore

$$\mathbf{x} = \frac{\sum_{i=1}^{n} \sum_{j=1}^{n} w_{ij} l_i}{\sum_{i=1}^{n} \sum_{j=1}^{n} w_{ij}} \tag{7.24}$$

and

$$\mathbf{Q}_{xx} = \left[\sum_{i=1}^{n}\sum_{j=1}^{n} w_{ij}\right]^{-1}$$

$$\mathbf{W}_{xx} = \sum_{i=1}^{n}\sum_{j=1}^{n} w_{ij} \tag{7.25}$$

which equals the sum of all the elements of the weight matrix of the given observational vector $\boldsymbol{l}$.

Example 7.10. *The Vector Mean.* The plane coordinates of a point have been independently determined twice by two methods with the following results. By the first method

$$\mathbf{y} = \begin{bmatrix} y_1 \\ y_2 \end{bmatrix} \qquad \text{with a covariance matrix } \underset{2,\,2}{\boldsymbol{\Sigma}_{yy}}$$

By the second method

$$\mathbf{z} = \begin{bmatrix} z_1 \\ z_2 \end{bmatrix} \qquad \text{with a covariance matrix } \underset{2,\,2}{\boldsymbol{\Sigma}_{zz}}$$

If there is no correlation between **y** and **z**, find by the method of least squares the best estimate of the coordinates.

Solution: Let the final estimate coordinate vector be denoted by the parameter vector

$$\mathbf{x} = \begin{bmatrix} x_1 \\ x_2 \end{bmatrix}$$

Realizing that the **y** and **z** are observational vectors, we can write the following pair of vector condition equations

$$\mathbf{y} + \mathbf{v}_y - \mathbf{x} = \mathbf{0}$$

$$\mathbf{z} + \mathbf{v}_z - \mathbf{x} = \mathbf{0}$$

or

$$\begin{bmatrix} \mathbf{v}_y \\ \mathbf{v}_z \end{bmatrix} + \begin{bmatrix} -\mathbf{I} \\ -\mathbf{I} \end{bmatrix} \mathbf{x} = \begin{bmatrix} -\mathbf{y} \\ -\mathbf{z} \end{bmatrix}$$

which is of the form $\mathbf{v} + \mathbf{B}\boldsymbol{\Delta} = \mathbf{f}$.

Since there is no correlation between the **y** and **z** observational vectors, the total covariance matrix of the observations can be constructed as

$$\underset{4,\,4}{\boldsymbol{\Sigma}} = \begin{bmatrix} \boldsymbol{\Sigma}_{yy} & \mathbf{0} \\ \mathbf{0} & \boldsymbol{\Sigma}_{zz} \end{bmatrix}$$

and hence

$$\mathbf{W} = \boldsymbol{\Sigma}^{-1} = \begin{bmatrix} \boldsymbol{\Sigma}_{yy}^{-1} & \mathbf{0} \\ \mathbf{0} & \boldsymbol{\Sigma}_{zz}^{-1} \end{bmatrix}$$

The normal equations are

$$\mathbf{N} = [-\mathbf{I} \quad -\mathbf{I}]\begin{bmatrix}\boldsymbol{\Sigma}_{yy}^{-1} & \mathbf{0} \\ \mathbf{0} & \boldsymbol{\Sigma}_{zz}^{-1}\end{bmatrix}\begin{bmatrix}-\mathbf{I} \\ -\mathbf{I}\end{bmatrix}$$

$$= (\boldsymbol{\Sigma}_{yy}^{-1} + \boldsymbol{\Sigma}_{zz}^{-1})$$

$$\mathbf{t} = [-\mathbf{I} \quad -\mathbf{I}]\begin{bmatrix}\boldsymbol{\Sigma}_{yy}^{-1} & \mathbf{0} \\ \mathbf{0} & \boldsymbol{\Sigma}_{zz}^{-1}\end{bmatrix}\begin{bmatrix}-\mathbf{y} \\ -\mathbf{z}\end{bmatrix} = \boldsymbol{\Sigma}_{yy}^{-1}\mathbf{y} + \boldsymbol{\Sigma}_{zz}^{-1}\mathbf{z}$$

Hence

$$\mathbf{x} = (\boldsymbol{\Sigma}_{yy}^{-1} + \boldsymbol{\Sigma}_{zz}^{-1})^{-1}(\boldsymbol{\Sigma}_{yy}^{-1}\mathbf{y} + \boldsymbol{\Sigma}_{zz}^{-1}\mathbf{z}) \tag{7.26}$$

and

$$\boldsymbol{\Sigma}_{xx} = (\boldsymbol{\Sigma}_{yy}^{-1} + \boldsymbol{\Sigma}_{zz}^{-1})^{-1} \tag{7.27}$$

This relationship gives the vector mean of two independent vectors in two-dimensional space, each given with its covariance matrix as well. Obviously the same relationship holds for two vectors of more than two components (that is, in multi-dimensional space). Also, either cofactor or weight matrices can be used in place of the covariance matrices.

The case of two determinations only was used as an example. Actually, both equations (7.26) and (7.27) hold for multiple determinations as long as there are no correlations among them. For example, if m determinations $\mathbf{x}_i$, $i = 1, 2, \ldots, m$ are given, each having a covariance matrix $\boldsymbol{\Sigma}_i$, the following relations hold provided $\boldsymbol{\Sigma}_{jk} = \mathbf{0}$ for all $j \neq k$:

$$\mathbf{x} = (\boldsymbol{\Sigma}_1^{-1} + \boldsymbol{\Sigma}_2^{-1} + \cdots + \boldsymbol{\Sigma}_m^{-1})^{-1}(\boldsymbol{\Sigma}_1^{-1}\mathbf{x}_1 + \boldsymbol{\Sigma}_2^{-1}\mathbf{x}_2 + \cdots + \boldsymbol{\Sigma}_m^{-1}\mathbf{x}_m) \tag{7.28}$$

and

$$\boldsymbol{\Sigma}_{xx} = (\boldsymbol{\Sigma}_1^{-1} + \boldsymbol{\Sigma}_2^{-1} + \cdots + \boldsymbol{\Sigma}_m^{-1})^{-1} \tag{7.29}$$

If $\boldsymbol{\Sigma}_{jk} \neq \mathbf{0}$ for some or all $j \neq k$ neither equation (7.28) nor (7.29) will apply, and a more general treatment must be done.

Example 7.11. This example is to rework the problem of Example 7.4, of the five-point level net, using the method of adjustment of indirect observations. Recall that there are $n = 8$ observations, and that the redundancy was $r = 4$. If we introduce the four elevations of points B, C, D, and E as parameters, then $u = 4$. Thus the number of independent condition equations will in this case be $c = r + u = 8$, which is also equal to n. In this case it is likely that we can readily write the condition equations such that each equation includes only one observation. Referring to Figure 7.1, the following *eight* equations can be directly written. (For simplicity in the derivation and computation the elevation of point A will be considered temporarily equal to zero. At the end of the adjustment, the elevation of A, which is

800.00 m will be added to all computed elevations.) In these equations the elevations of points *B*, *C*, *D*, and *E* are designated x_1, x_2, x_3, and x_4, respectively.

$$
\begin{aligned}
&l_1 - x_1 &&= 0\\
&l_2 + x_1 - x_2 &&= 0\\
&l_3 \quad + x_2 &&= 0\\
&l_4 + x_1 \quad - x_3 &&= 0\\
&l_5 \quad + x_3 - x_4 &&= 0\\
&l_6 \quad - x_2 \quad + x_4 &&= 0\\
&l_7 \quad + x_4 &&= 0\\
&l_8 \quad + x_2 - x_3 &&= 0
\end{aligned}
$$

These equations are linear and can therefore be written directly in the form

$$\mathbf{v} + \mathbf{B}\boldsymbol{\Delta} = \mathbf{f}$$

$$
\begin{bmatrix} v_1\\ v_2\\ v_3\\ v_4\\ v_5\\ v_6\\ v_7\\ v_8 \end{bmatrix} +
\begin{bmatrix} -1 & 0 & 0 & 0\\ 1 & -1 & 0 & 0\\ 0 & 1 & 0 & 0\\ 1 & 0 & -1 & 0\\ 0 & 0 & 1 & -1\\ 0 & -1 & 0 & 1\\ 0 & 0 & 0 & 1\\ 0 & 1 & -1 & 0 \end{bmatrix}
\begin{bmatrix} x_1\\ x_2\\ x_3\\ x_4 \end{bmatrix} =
\begin{bmatrix} -l_1\\ -l_2\\ -l_3\\ -l_4\\ -l_5\\ -l_6\\ -l_7\\ -l_8 \end{bmatrix} =
\begin{bmatrix} -25.42 \text{ m}\\ -10.34 \text{ m}\\ 35.20 \text{ m}\\ 15.54 \text{ m}\\ -21.32 \text{ m}\\ -4.82 \text{ m}\\ 31.02 \text{ m}\\ 26.11 \text{ m} \end{bmatrix}
$$

The weight matrix of the observations is

$$\mathbf{W} = \text{diag.}\ [1/1.81,\ 1/0.94,\ 1/1.42,\ 1/1.76,\ 1/1.35,\ 1/0.99,\ 1/1.38,\ 1/1.40]$$

The normal equations coefficient matrix $\mathbf{N}$ is

$$
\mathbf{N} = \begin{bmatrix} 2.1845 & -1.0638 & -0.5682 & 0\\ & 3.4924 & -0.7143 & -1.0101\\ & & 2.0232 & -0.7407\\ \text{symmetric} & & & 2.4755 \end{bmatrix}
$$

and its inverse

$$
\mathbf{N}^{-1} = \begin{bmatrix} 0.806 & 0.445 & 0.505 & 0.333\\ & 0.645 & 0.504 & 0.414\\ & & 0.999 & 0.505\\ \text{symmetric} & & & 0.724 \end{bmatrix}
$$

In a similar fashion the constant term vector may be computed and used to compute the vector $\boldsymbol{\Delta}$,

$$\boldsymbol{\Delta} = [25.220 \quad 35.535 \quad 9.534 \quad 30.846]^t (\text{m})$$

Thus the elevations of points B, C, D, and E may be obtained by adding 800.00 m to each element of the vector $\mathbf{\Delta}$,

$$\text{elevation of point } B = 825.220 \text{ m}$$

$$\text{elevation of point } C = 835.535 \text{ m}$$

$$\text{elevation of point } D = 809.534 \text{ m}$$

$$\text{elevation of point } E = 830.846 \text{ m}$$

which agree exactly with the values computed in Example 7.4 using the method of adjustment of observations only.

The cofactor matrix of these four elevations is equal to $\mathbf{N}^{-1}$, which is the same as $\mathbf{Q}_{xx}$ computed in Example 7.4. If *only* the elevations and their *cofactor* matrix are required, this would conclude the present problem and we would agree that this method of adjustment is shorter than that used in the other example, at least for the current problem. If, however, it is the *covariance* matrix of the elevations that is required, further computation is necessary to determine the a posteriori estimate of the reference variance $\hat{\sigma}_0^2$. First, compute the vector of residuals

$$\mathbf{v} = [-0.199 \quad -0.025 \quad -0.335 \quad -0.147 \quad -0.008 \quad -0.131 \quad 0.174 \quad 0.109]^t \text{ (m)}$$

and

$$\hat{\sigma}_0^2 = 0.0404$$

Multiplying $\mathbf{Q}_{xx}$ by $\hat{\sigma}_0^2$ gives the covariance matrix of the elevations

$$\mathbf{\Sigma}_{xx} = \begin{bmatrix} 0.0326 & 0.0180 & 0.0204 & 0.0135 \\ & 0.0261 & 0.0204 & 0.0167 \\ & & 0.0404 & 0.0204 \\ \text{symmetric} & & & 0.0293 \end{bmatrix} \text{ (m}^2\text{)}$$

which as expected is identical to that computed by the other method. The reader is left the task of determining which technique of adjustment is preferable. He should compare such factors as setting up the condition equations, the size of the normal equations to be inverted, the amount of effort required for precision estimation, and above all the way the computations will be performed (by limited means or by electronic computer). Before leaving this example we perform one last computation to demonstrate the computational check of equation (7.16).

$$\mathbf{B}^t\mathbf{W}\mathbf{v} = \begin{bmatrix} -0.555 & 1.064 & 0 & 0.568 & 0 & 0 & 0 & 0 \\ 0 & -1.064 & 0.704 & 0 & 0 & -1.010 & 0 & 0.714 \\ 0 & 0 & 0 & -0.568 & 0.741 & 0 & 0 & -0.714 \\ 0 & 0 & 0 & 0 & -0.741 & 1.010 & 0.725 & 0 \end{bmatrix} \begin{bmatrix} -0.199 \\ -0.025 \\ -0.335 \\ -0.147 \\ -0.008 \\ -0.131 \\ 0.174 \\ 0.109 \end{bmatrix}$$

$$= \begin{bmatrix} 0.0 \\ 0.0 \\ 0.0 \\ 0.0 \end{bmatrix} = \mathbf{0}$$

7.4. SUMMARY OF EQUATIONS AND CONCLUDING REMARKS

7.4.1. Adjustment of Observations Only

The given information would be a vector of n observations $\boldsymbol{l}$ and its cofactor (or covariance) matrix $\mathbf{Q}$. After determining the redundancy r, a set of r linear condition equations are written

$$\mathbf{A}\mathbf{v} = \mathbf{d} - \mathbf{A}\boldsymbol{l} = \mathbf{f}$$

$\mathbf{v}$ is the $n \times 1$ vector of residuals, and $\mathbf{d}$ and $\mathbf{f}$ are $r \times 1$ vectors of constants

$\mathbf{A}$ is an $r \times n$ coefficient matrix of rank $= r$

$\mathbf{Q}_e = \mathbf{A}\mathbf{Q}\mathbf{A}^t = r \times r$ reduced normal equations coefficient matrix

$\mathbf{k} = \mathbf{Q}_e^{-1}\mathbf{f} = \mathbf{W}_e\,\mathbf{f} =$ vector of r Lagrange multipliers (7.4)

$\mathbf{v} = \mathbf{Q}\mathbf{A}^t\mathbf{k} = \mathbf{Q}\mathbf{A}^t\mathbf{W}_e\,\mathbf{f}$ (7.5)

$\hat{\boldsymbol{l}} = \boldsymbol{l} + \mathbf{v} =$ estimated observations

$\hat{\sigma}_0^2 = \mathbf{v}^t\mathbf{W}\mathbf{v}/r = \mathbf{f}^t\mathbf{k}/r$ (7.6)

$\mathbf{Q}_{vv} = \mathbf{Q}\mathbf{A}^t\mathbf{W}_e\,\mathbf{A}\mathbf{Q}$ rank $(\mathbf{Q}_{vv}) = r$ (7.7), (6.42)

$\mathbf{Q}_{\hat{l}\hat{l}} = \mathbf{Q} - \mathbf{Q}_{vv}$ rank $(\mathbf{Q}_{\hat{l}\hat{l}}) = n - r = n_0$ (6.31), (6.43)

This case of adjustment of observations only is suitable for use when the problem is relatively simple and uninvolved. It is also suitable when either the estimated observations themselves, or rather simple functions of them, are the variables of interest from the adjustment. For good size problems, and for situations in which the values of clearly specified parameters are required, this technique of adjustment is usually not utilized. Either the case covered in Chapter 6 or the simpler technique of indirect observations is often more appropriate.

7.4.2. Adjustment of Indirect Observations As in the preceding section we begin with the observational vector $\boldsymbol{l}$ and its cofactor matrix $\mathbf{Q}$. Since the number of condition equations is specified as equal to the observations n, an $n_0 = n - r$ parameters in $\boldsymbol{\Delta}$, where r is the redundancy, must be carried in the adjustment. The condition equations are of the form

$$\mathbf{v} + \mathbf{B}\boldsymbol{\Delta} = -\boldsymbol{l} + \mathbf{d} = \mathbf{f} \qquad (7.8), (7.9)$$

$\mathbf{d}$ and $\mathbf{f}$ are $n \times 1$ vectors of constants

$\mathbf{B}$ is an $n \times u$ $(= n \times n_0)$ matrix of rank $= u$

$\mathbf{N} = \mathbf{B}^t\mathbf{Q}^{-1}\mathbf{B} = \mathbf{B}^t\mathbf{W}\mathbf{B} =$ reduced normal coefficient matrix

$\mathbf{t} = \mathbf{B}^t\mathbf{W}\mathbf{f} =$ reduced normal equations constant vector

$\boldsymbol{\Delta} = \mathbf{N}^{-1}\mathbf{t}$ (7.14)

(If conditions were originally nonlinear, then $\hat{\mathbf{x}} = \mathbf{x}^0 + \mathbf{\Delta}$, where $\mathbf{x}^0$ is a vector of approximations.)

$$\mathbf{v} = \mathbf{f} - \mathbf{B\Delta} \qquad (7.9)$$

$\hat{\boldsymbol{l}} = \boldsymbol{l} + \mathbf{v}$ = estimated observations

$$\hat{\sigma}_0^2 = \mathbf{v}^t\mathbf{W}\mathbf{v}/r = (\mathbf{f}^t\mathbf{W}\mathbf{f} - \mathbf{\Delta}^t\mathbf{t})/r \qquad (7.17)$$

$$\mathbf{Q}_{\Delta\Delta} = \mathbf{N}^{-1}$$

$$\mathbf{Q}_{vv} = \mathbf{Q} - \mathbf{B}\mathbf{N}^{-1}\mathbf{B}^t$$

$$\mathbf{Q}_{\hat{l}\hat{l}} = \mathbf{B}\mathbf{N}^{-1}\mathbf{B}^t = \mathbf{Q} - \mathbf{Q}_{vv} \qquad (7.19)$$

The technique of adjustment of indirect observations is more closely related to that presented in Chapter 6. The only exception is that the coefficient matrix **A** of the residual vector is in the present case the identity matrix.

8

Examples and General Discussion on Adjustment with Conditions Only

8.1. GENERAL

The purpose of this chapter is to present the reader with a number of worked examples. For each example, we will first present a problem, discuss it in detail, determine the model, analyze the merits of different adjustment techniques, and then choose and apply the one that is considered most appropriate. It may be necessary at times to solve the same problem by more than one technique to stress a particular point.

Example 8.1. Consider mass production of metal cubes of which a particular unit is selected as a sample for check. Let us assume that all face angles in the cube *are* in fact 90-degree angles, and that the volume of the cube is to be checked. Under the assumption above, all that is necessary for computing the volume of a cube is to measure one of its sides. Consider now two possibilities: (a) that the same side of the cube is measured three times; or (b) that three different sides of the cube are measured, each once, as shown in Figure 8.1. In both cases there are two redundant observations and adjustment is necessary. The question is whether there is any difference between the two cases insofar as the adjustment problem is concerned.

For the first case, denote by l_1, l_2, and l_3 the three repetitive measurements of the

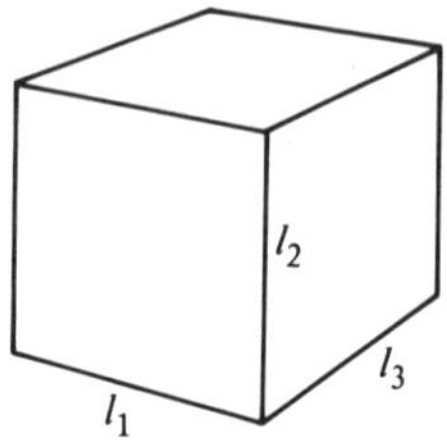

Figure 8-1

one side of the cube. If all three measurements are uncorrelated and of equal precision, the least squares estimate is simply the arithmetic mean of the three observations (see Example 7.9).

$$\hat{l} = \tfrac{1}{3}(l_1 + l_2 + l_3)$$

The volume of the cube is then

$$y = \hat{l}^3$$

For the sake of comparison later on, take the numerical values

$$l_1 = 50.0 \text{ mm} \qquad l_2 = 50.2 \text{ mm} \qquad l_3 = 49.8 \text{ mm}$$

thus

$$\hat{l} = 50.0 \text{ mm} \qquad \text{and} \qquad y = 125 \times 10^3 \text{ mm}^3$$

In the second case, the three observations are of three separate sides of the cube as shown in Figure 8.1. Since it takes only one side to fix a cube, there are 2 degrees of freedom, that is $r = 2$. There are, of course, three possibilities of adjustment techniques based on the three cases presented in the preceding two chapters.

Consider first the case of adjustment with observations only. Thus two condition equations are

$$l_1 - l_2 = 0 \qquad l_2 - l_3 = 0$$

or

$$\begin{bmatrix} 1 & -1 & 0 \\ 0 & 1 & -1 \end{bmatrix} \begin{bmatrix} v_1 \\ v_2 \\ v_3 \end{bmatrix} = \begin{bmatrix} l_2 - l_1 \\ l_3 - l_2 \end{bmatrix} = \begin{bmatrix} 0.2 \text{ mm} \\ -0.4 \text{ mm} \end{bmatrix}$$

With $\mathbf{Q} = \mathbf{I}$ as assumed above, the normal equations are

$$\begin{bmatrix} 2 & -1 \\ -1 & 2 \end{bmatrix} \mathbf{k} = \begin{bmatrix} 0.2 \text{ mm} \\ -0.4 \text{ mm} \end{bmatrix} \qquad \text{or} \qquad \mathbf{k} = \begin{bmatrix} 0 \\ -0.2 \text{ mm} \end{bmatrix}$$

and

$$\mathbf{v} = [0 \quad -0.2 \text{ mm} \quad 0.2 \text{ mm}]^t$$

Thus the least squares estimate of each of the three sides is in fact 50.0 mm with the volume being $y = 125 \times 10^3$.

So it can be seen that the two solutions are identical both as regards to the estimate of the observations and the estimate of the volume. The reason for this possibility is the fact that only one observation is required and also that the type of condition between the observations in both cases is the same—being the equality of all observations—and that the observations were uncorrelated and of equal precision.

For completeness, we work out the above adjustment using the case of explicit observations and parameters. Let x be the parameter representing the best estimate of the cube's side. Then

$$l_1 - x = 0$$

$$l_2 - x = 0$$

$$l_3 - x = 0$$

are three condition equations ($c = r + u = 2 + 1 = 3$), or

$$\begin{bmatrix} v_1 \\ v_2 \\ v_3 \end{bmatrix} + \begin{bmatrix} -1 \\ -1 \\ -1 \end{bmatrix} x = \begin{bmatrix} -l_1 \\ -l_2 \\ -l_3 \end{bmatrix}$$

With $\mathbf{Q} = \mathbf{I}$, the normal equations would simply be

$$3x = (l_1 + l_2 + l_3)$$

Realizing that x is the same as $\hat{l}$, we get

$$\hat{l} = \tfrac{1}{3}(l_1 + l_2 + l_3)$$

which agrees with earlier results.

Example 8.2. In the cases of the preceding example the mathematical equations have all been originally linear and the adjustment proceeded directly. Furthermore the computation of the volume of the cube was a postadjustment operation. In this example we shall deviate from both points and introduce the volume as the parameter required and estimate it directly from the adjustment. If y denotes the volume, the condition equations will be

$$l_1 - y^{1/3} = 0$$

$$l_2 - y^{1/3} = 0$$

$$l_3 - y^{1/3} = 0$$

Linearization yields

$$\begin{bmatrix} v_1 \\ v_2 \\ v_3 \end{bmatrix} + \begin{bmatrix} -\tfrac{1}{3}y^{0-2/3} \\ -\tfrac{1}{3}y^{0-2/3} \\ -\tfrac{1}{3}y^{0-2/3} \end{bmatrix} \boldsymbol{\Delta} = \begin{bmatrix} y^{01/3} - l_1 \\ y^{01/3} - l_2 \\ y^{01/3} - l_3 \end{bmatrix}$$

The normal equations matrix (in this case a scalar) is

$$\mathbf{N} = \tfrac{1}{3}y^{0-4/3}$$

If the approximation is taken as $y^0 = (49.5)^3$, then

$$\underset{3,1}{\mathbf{f}} = \begin{bmatrix} -0.5 \text{ mm} \\ -0.7 \text{ mm} \\ -0.3 \text{ mm} \end{bmatrix} \qquad \text{and} \qquad \mathbf{t} = 0.5y^{0-2/3}$$

Thus

$$\Delta = 3y^{0+4/3}(0.5y^{0-2/3}) = 3675.37 \text{ mm}^3$$

and

$$\hat{y} = 121287.375 + 3675.37 = 124962.745 \text{ mm}^3$$

If this new value of the volume is used as a fresh approximation, another correction Δ may be computed as approximately 37.25 mm^3. This would give the expected answer of 125000.00 mm^3.

The choice of $y^0 = (49.5)^3$ was intentionally done so that both linearization and iteration may be demonstrated. An obvious choice should have been $(50)^3$, but that would have led directly to a $\boldsymbol{\Delta} = \mathbf{0}$, and the adjustment may have appeared trivial.

Example 8.3. For the adjustment case of observations only the linearized condition equations were of the form

$$\underset{r,n}{\mathbf{A}}\;\underset{n,1}{\mathbf{v}} = \underset{r,1}{\mathbf{f}_c} \tag{a}$$

For the case of adjustment of indirect observations, the condition equation were of the form

$$\underset{n,1}{\mathbf{v}} + \underset{n,u}{\mathbf{B}}\;\underset{u,1}{\boldsymbol{\Delta}} = \underset{n,1}{\mathbf{f}_0} \tag{b}$$

Show that for the *same adjustment problem* the following is true: $\mathbf{AB} = \mathbf{0}$.

Solution: Partition the second set of equations (b) into two systems, the first containing u condition equations,

$$\underset{u,1}{\mathbf{v}_1} + \underset{u,u}{\mathbf{B}_1}\;\underset{u,1}{\boldsymbol{\Delta}} = \underset{u,1}{\mathbf{f}_{01}}$$

$$\underset{(n-u),1}{\mathbf{v}_2} + \underset{(n-u),u}{\mathbf{B}_2}\;\underset{u,1}{\boldsymbol{\Delta}} = \underset{u,1}{\mathbf{f}_{02}}$$

Under the assumption that the condition equations are independent, solve the upper set for $\boldsymbol{\Delta}$ (here $\mathbf{B}_1$ will be nonsingular),

$$\boldsymbol{\Delta} = \mathbf{B}_1^{-1}(-\mathbf{v}_1 + \mathbf{f}_{01})$$

and substitute into the lower set

$$\mathbf{v}_2 + \mathbf{B}_2\,\mathbf{B}_1^{-1}(-\mathbf{v}_1 + \mathbf{f}_{01}) = \mathbf{f}_{02}$$

$$[-\mathbf{B}_2\,\mathbf{B}_1^{-1} \quad \mathbf{I}]\begin{bmatrix}\mathbf{v}_1\\ \mathbf{v}_2\end{bmatrix} = [-\mathbf{B}_2\,\mathbf{B}_1^{-1}\mathbf{f}_{01} + \mathbf{f}_{02}]$$

This equation is now of the same form as the first set of condition equations (a), for the adjustment of observations only,

$$\underset{(n-u),n}{\mathbf{A}}\;\underset{n,1}{\mathbf{v}} = \underset{(n-u),1}{\mathbf{f}_c}$$

Thus

$$\mathbf{AB} = [-\mathbf{B}_2\,\mathbf{B}_1^{-1} \quad \mathbf{I}]\begin{bmatrix}\mathbf{B}_1\\ \mathbf{B}_2\end{bmatrix}$$

$$= -\mathbf{B}_2 + \mathbf{B}_2 = \mathbf{0}$$

An Alternative Solution: Using information from precision estimation recall equation (7.7) from adjustment of observations only

$$\underset{n,n}{\mathbf{Q}_{vv}} = \underset{n,n}{\mathbf{Q}}\ \underset{n,r}{\mathbf{A}^t}\ \underset{r,r}{\mathbf{W}_e}\ \underset{r,n}{\mathbf{A}}\ \underset{n,n}{\mathbf{Q}} \tag{7.7}$$

and equation (7.18a) for adjustment of indirect observations

$$\underset{n,n}{\mathbf{Q}_{vv}} = \underset{n,n}{\mathbf{Q}} - \underset{n,u}{\mathbf{B}}\ \underset{u,u}{\mathbf{N}^{-1}}\ \underset{u,n}{\mathbf{B}^t} \tag{7.18a}$$

Since it is an adjustment of the same problem but with two different techniques, first, $\mathbf{W}_e$ is different from $\mathbf{N}^{-1}$ and, second, $\mathbf{Q}_{vv}$ must be the same from both solutions. Thus

$$\mathbf{QA}^t\mathbf{W}_e\,\mathbf{AQ} = \mathbf{Q} - \mathbf{BN}^{-1}\mathbf{B}^t$$

Premultiplying both sides by $\mathbf{A}$, and realizing that $\mathbf{W}_e = (\mathbf{AQA}^t)^{-1}$,

$$(\mathbf{AQA}^t)\mathbf{W}_e\,\mathbf{AQ} = \mathbf{AQ} - \mathbf{ABN}^{-1}\mathbf{B}^t$$

$$\mathbf{AQ} = \mathbf{AQ} - \mathbf{ABN}^{-1}\mathbf{B}^t$$

$$\mathbf{ABN}^{-1}\mathbf{B}^t = \mathbf{0}$$

Postmultiplying by $(\mathbf{WB})$ and recalling that $\mathbf{N} = \mathbf{B}^t\mathbf{WB}$,

$$\mathbf{ABN}^{-1}(\mathbf{B}^t\mathbf{WB}) = \mathbf{0}$$

or

$$\mathbf{AB} = \mathbf{0}$$

As a numerical demonstration refer to Examples 7.4 and 7.11 from the adjustment of a level net given in the preceding chapter:

$$\underset{4,8}{\mathbf{A}} = \begin{bmatrix} 1 & 1 & 1 & 0 & 0 & 0 & 0 & 0\\ 0 & 0 & -1 & 0 & 0 & -1 & 1 & 0\\ 0 & -1 & 0 & 1 & 0 & 0 & 0 & -1\\ 0 & 0 & 0 & 0 & 1 & 1 & 0 & 1 \end{bmatrix}$$

$$\underset{8,4}{\mathbf{B}} = \begin{bmatrix} -1 & 0 & 0 & 0\\ 1 & -1 & 0 & 0\\ 0 & 1 & 0 & 0\\ 1 & 0 & -1 & 0\\ 0 & 0 & 1 & -1\\ 0 & -1 & 0 & 1\\ 0 & 0 & 0 & 1\\ 0 & 1 & -1 & 0 \end{bmatrix} \qquad \underset{4,4}{\mathbf{AB}} = \begin{bmatrix} 0 & 0 & 0 & 0\\ 0 & 0 & 0 & 0\\ 0 & 0 & 0 & 0\\ 0 & 0 & 0 & 0 \end{bmatrix}$$

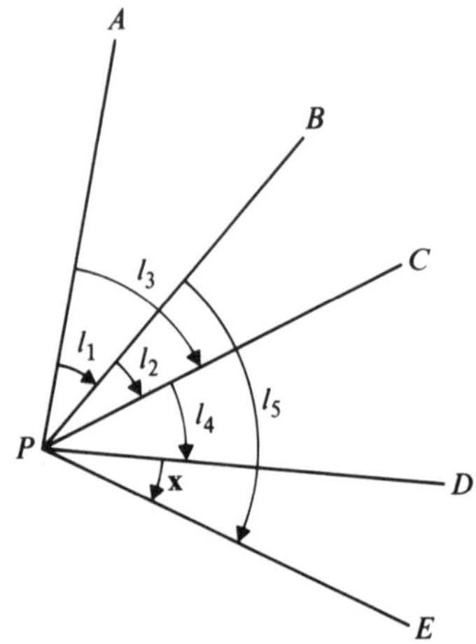

Figure 8-2

Example 8.4. Figure 8.2 shows five plane angles measured at point P with the following values:

l_1 = angle APB = 30°15′15″
l_2 = angle BPC = 20°00′00″
l_3 = angle APC = 50°15′18″
l_4 = angle CPD = 30°00′00″
l_5 = angle BPE = 70°00′01″

All five angles are uncorrelated and of equal precision. Compute the best estimate of the angle DPE, its cofactor, and the estimate of the reference variance from the adjustment.

Solution: The mathematical model of this problem is comprised of five lines in a plane all emanating from one point P. Since there are four angles constructed by those five lines, it would take a minimum of four observed angles to have a unique determination of the problem. Given five observations there is one redundant observation leading to 1 degree of freedom, $r = 1$. If the value of the required angle DPE is considered an unknown parameter x, then with $n = 5$, $r = 1$, and $u = 1$, the number of independent condition equations will be $c = r + u = 2$. Two equations can be readily written from Figure 8.2 as

$$l_1 + l_2 - l_3 = 0$$

$$l_2 + l_4 - l_5 + x = 0$$

or

$$\begin{bmatrix} 1 & 1 & -1 & 0 & 0 \\ 0 & 1 & 0 & 1 & -1 \end{bmatrix} \underset{5,1}{\mathbf{v}} + \begin{bmatrix} 0 \\ 1 \end{bmatrix} \underset{1,1}{\mathbf{x}} = \begin{bmatrix} -l_1 - l_2 + l_3 \\ -l_2 - l_4 + l_5 \end{bmatrix}$$

$$= \begin{bmatrix} 3'' \\ 20°00'01'' \end{bmatrix}$$

Given $\mathbf{Q} = \mathbf{I}$, then

$$\mathbf{Q}_e = \begin{bmatrix} 3 & 1 \\ 1 & 3 \end{bmatrix} \qquad \mathbf{W}_e = \frac{1}{8}\begin{bmatrix} 3 & -1 \\ -1 & 3 \end{bmatrix}$$

and

$$\mathbf{N} = \frac{3}{8} \qquad \mathbf{N}^{-1} = \frac{8}{3} \qquad t = \frac{60^\circ}{8}$$

Therefore

$$x = \mathbf{N}^{-1}\mathbf{t} = 20^\circ 00' 00'' \qquad \text{and} \qquad \mathbf{Q}_{xx} = \mathbf{N}^{-1} = \frac{8}{3}$$

Thus the estimate of the angle DPE is 20 degrees and its cofactor is $\frac{8}{3}$. To compute the a posteriori estimate of the reference variance, first

$$\mathbf{v} = \mathbf{Q}\mathbf{A}^t\mathbf{W}_e(-\mathbf{B}\Delta + \mathbf{f}) = [1'' \quad 1'' \quad -1'' \quad 0 \quad 0]^t$$

and with

$$r = 1 \qquad \hat{\sigma}_0^2 = 3 \text{ sec}^2 \qquad \text{and} \qquad \sigma_x^2 = \hat{\sigma}_0^2\mathbf{Q}_{xx} = 8 \text{ sec}^2$$

Example 8.5. Rework the problem of Example 8.4 using a different technique of least squares.

Solution: There are two other techniques that can be used, one using $\mathbf{A}\mathbf{v} = \mathbf{f}$ and the other using $\mathbf{v} + \mathbf{B}\Delta = \mathbf{f}$. In the first possibility, we would adjust the given observations to obtain $\hat{l}$ and $\mathbf{Q}_{ll}$ and then seek a function relating x and $\hat{l}$ to compute both x and $\mathbf{Q}_{xx}$. In the second possibility, we would attempt to reduce the situation covered in Example 8.4 such that the $\mathbf{A}$ matrix becomes the identity matrix. Let us consider the first case. Here, with $r = 1$ we need to write only one condition equation thus leading to only one normal equation in one unknown. This means that the entire problem would probably be done by hand without need of any computational aid. In the second case, however, we must write *five* condition equations. Furthermore, since $r = 1 = n - u$ and with $n = 5$, it will be necessary to have *four* parameters in the condition equations in order to construct them properly. This means that with only one parameter being of interest we have to carry three more just for the sake of computation but without any real need of them. More serious, however, is the fact that there will be four normal equations, the solution of which will require inverting a square matrix of order 4. This obviously requires much more computation. It is clear then that we should choose the adjustment technique involving only observations as the more economical. From the figure one condition equation is

$$l_1 + l_2 - l_3 = 0$$

This condition equation includes a subset of three observations only from the total of five observations given. The problem can be continued from here and the adjustment answer will be correct *if and only if* there is no correlation between the observations appearing in the condition and those not used. This is true in the present case as

$\mathbf{Q} = \mathbf{I}$. However, we shall change the data of the problem later to demonstrate the point raised here. Continuing now, first,

$$[1 \quad 1 \quad -1]\mathbf{v} = -l_1 - l_2 + l_3 = 3''$$

$$\mathbf{N} = 3, \qquad \mathbf{k} = 1'', \qquad \text{and} \qquad \mathbf{v} = [1'' \quad 1'' \quad -1'']^t$$

Thus

$$\hat{l}_1 = 30°15'16' \qquad \hat{l}_2 = 20°00'01'' \qquad \hat{l}_3 = 50°15'17''$$

$$\hat{l}_4 = l_4 = 30°00'00'' \qquad \hat{l}_5 = l_5 = 70°00'01''$$

$$\mathbf{Q}_{vv} = \tfrac{1}{3}\begin{bmatrix} 1 & 1 & 1 \\ & 1 & 1 \\ \text{symmetric} & & 1 \end{bmatrix} \quad \text{and} \quad \mathbf{Q}_{ll} = \tfrac{1}{3}\begin{bmatrix} 2 & -1 & -1 \\ & 2 & -1 \\ \text{symmetric} & & 2 \end{bmatrix}$$

or, for all five values

$$\mathbf{Q}_{ll} = \tfrac{1}{3}\begin{bmatrix} 2 & -1 & -1 & 0 & 0 \\ & \boxed{2} & -1 & \boxed{0 \quad 0} \\ & & 2 & 0 & 0 \\ & & & 3 & 0 \\ \text{symmetric} & & & & 3 \end{bmatrix}$$

The angle of interest is then

$$\text{angle } DPE = \mathbf{x} = -\hat{l}_2 - \hat{l}_4 + \hat{l}_5 = 20°00°00''$$

$$\mathbf{Q}_{xx} = \tfrac{1}{3}[-1 \quad -1 \quad 1]\begin{bmatrix} 2 & 0 & 0 \\ & 3 & 0 \\ \text{symmetric} & & 3 \end{bmatrix}\begin{bmatrix} -1 \\ -1 \\ 1 \end{bmatrix} = \tfrac{8}{3}$$

Note that the 3×3 matrix in the middle of the above expression is formed from the $\mathbf{Q}_{ll}$ matrix above by deleting the first and third columns and rows and taking the elements delineated by dotted lines. Finally, from the above computed vector of residuals, it is straightforward to compute $\hat{\sigma}_0^2 = 3 \text{ sec}^2$.

Actually the solution according to this case may be considered as simpler than that used in Example 8.4. Both techniques, however, required nothing more than simple computation.

Now we return to the consideration of taking all the observations into the condition equation even though some of them will have zero coefficients. Thus

$$[1 \quad 1 \quad -1 \quad 0 \quad 0]\underset{5,1}{\mathbf{v}} = -l_1 - l_2 + l_3 + 0l_4 + 0l_5 = 3''$$

Since $\mathbf{Q} = \mathbf{I}$, then

$$\mathbf{N} = 3 \quad \mathbf{k} = 1'' \qquad \text{and} \qquad \mathbf{v} = [1'' \quad 1'' \quad -1'' \quad 0 \quad 0]^t$$

which shows that the residuals for l_4 and l_5 are zero. The rest of the solution follows the same steps carried out above.

A very interesting and important case arises when the given observations are correlated a priori. This is treated in the following example.

Example 8.6. Given the same observed angles of Example 8.4 (see Figure 8.2) but with the a priori cofactor matrix

$$\mathbf{Q} = \begin{bmatrix} 2 & 1 & 0 & 0 & 0 \\ 1 & 2 & 1 & 0 & 0 \\ 0 & 1 & 2 & 1 & 0 \\ 0 & 0 & 1 & 2 & 1 \\ 0 & 0 & 0 & 1 & 2 \end{bmatrix}$$

instead of the identity matrix used before. Interest remains in computing the angle x using specifically the technique of adjustment of observations only.

Solution: Suppose that we begin by considering the one condition in terms of the three observations l_1, l_2, l_3 as in the preceding example. Thus

$$[1 \quad 1 \quad -1]\mathbf{v} = -l_1 - l_2 + l_3 = 3''$$

Taking the appropriate $\mathbf{Q}$ for the first three observations, then

$$\mathbf{N} = 6 \qquad \mathbf{k} = 0''.5 \qquad \text{and} \qquad \mathbf{v} = [1''.5 \quad 1'' \quad -0''.5]^t$$

Thus

$$\hat{l}_1 = 30°15'16''.5 \qquad \hat{l}_2 = 20°00'01'' \qquad \hat{l}_3 = 50°15'17''.5$$

Examining the figure, the required angle would be

$$x = l_5 - \hat{l}_2 - l_4 = 20°00'00''$$

Although this answer agrees with the value computed in the preceding two examples, it is in fact the wrong answer. This is because the other two observations, l_4 and l_5, are correlated with the first three and should be carried in the adjustment as newer estimates for them can be obtained. To prove this, we repeat the above adjustment such that all observations are used.

$$[1 \quad 1 \quad -1 \quad 0 \quad 0] \underset{5,1}{\mathbf{v}} = 3''$$

$$\mathbf{N} = [1 \quad 1 \quad -1 \quad 0 \quad 0] \begin{bmatrix} 2 & 1 & 0 & 0 & 0 \\ 1 & 2 & 1 & 0 & 0 \\ 0 & 1 & 2 & 1 & 0 \\ 0 & 0 & 1 & 2 & 1 \\ 0 & 0 & 0 & 1 & 2 \end{bmatrix} \begin{bmatrix} 1 \\ 1 \\ -1 \\ 0 \\ 0 \end{bmatrix} = 6$$

$$\mathbf{k} = 0''.5 \qquad \text{and} \qquad \mathbf{v} = [1''.5 \quad 1'' \quad -0''.5 \quad -0.5'' \quad 0]^t$$
$$\hat{l}_1 = 30°15'16''.5 \qquad \hat{l}_2 = 20°00'01'' \qquad \hat{l}_3 = 50°15'17''.5$$
$$\hat{l}_4 = 29°59'59'' \qquad \hat{l}_5 = 70°00'01''$$
$$x = \hat{l}_5 - \hat{l}_2 - \hat{l}_4 = 20°00'00''.5$$

The value of x is, as expected, different from that computed above, but is the *correct answer*. As a matter of fact it should be made a rule that we must consider first the most general case and then reduce it to a specific one if we can safely do so. In the case when $\mathbf{Q}$ was $\mathbf{I}$ this was possible, but in the presence of correlation this becomes dangerous. The emphasis on using the most general formulation should apply to all techniques of adjustment and not only to the case discussed.

To complete the example, compute the a posteriori cofactor matrices

$$\mathbf{Q}_{vv} = \tfrac{1}{6}\begin{bmatrix} 9 & 6 & -3 & -3 & 0 \\ & 4 & -2 & -2 & 0 \\ & & 1 & 1 & 0 \\ & & & 1 & 0 \\ \text{symmetric} & & & & 0 \end{bmatrix} \qquad \mathbf{Q}_{ll} = \tfrac{1}{6}\begin{bmatrix} 3 & 0 & 3 & 3 & 0 \\ & 8 & 8 & 2 & 0 \\ & & 11 & 5 & 0 \\ & & & 11 & 6 \\ \text{symmetric} & & & & 12 \end{bmatrix}$$

$$\mathbf{Q}_{xx} = \overset{(\hat{l}_2 \quad \hat{l}_4 \quad \hat{l}_5)}{[-1 \quad -1 \quad 1]}\tfrac{1}{6}\begin{bmatrix} 8 & 2 & 0 \\ 2 & 11 & 6 \\ 0 & 6 & 12 \end{bmatrix}\begin{bmatrix} -1 \\ -1 \\ 1 \end{bmatrix} = \frac{23}{6}$$

8.2. COORDINATE TRANSFORMATIONS

Transformation from one system of coordinates to another is a very useful operation that is used frequently in photogrammetry, geodesy, and surveying. There is a whole family of transformations, some of which are applicable to two-dimensional space and others for use in three-dimensional problems. Furthermore the effect of a transformation varies from simple changes of location and direction (without any change in shape and size), to a uniform change in scale (no change in shape), and, finally, to change in shape and size of different degree of nonlinearity.

8.2.1. Transformations in Two-Dimensional Space

1. From polar coordinate system to a cartesian coordinate system

$$\begin{aligned} x_1 &= r \cos \theta \\ x_2 &= r \sin \theta \end{aligned} \tag{8.1}$$

2. From one cartesian coordinate system to another

   a. *Rotation*

$$\begin{aligned} y_1 &= x_1 \cos \alpha - x_2 \sin \alpha \\ y_2 &= x_1 \sin \alpha + x_2 \cos \alpha \end{aligned} \tag{8.2}$$

or

$$\begin{bmatrix} y_1 \\ y_2 \end{bmatrix} = \begin{bmatrix} \cos \alpha & -\sin \alpha \\ \sin \alpha & \cos \alpha \end{bmatrix}\begin{bmatrix} x_1 \\ x_2 \end{bmatrix} \tag{8.3}$$

   b. *Rotation and a Scale Change*

$$\begin{aligned} y_1 &= x_1(s \cos \alpha) - x_2(s \sin \alpha) \\ y_2 &= x_1(s \sin \alpha) + x_2(s \cos \alpha) \end{aligned} \tag{8.4}$$

or

$$\begin{bmatrix} y_1 \\ y_2 \end{bmatrix} = s\begin{bmatrix} \cos \alpha & -\sin \alpha \\ \sin \alpha & \cos \alpha \end{bmatrix}\begin{bmatrix} x_1 \\ x_2 \end{bmatrix} = \begin{bmatrix} a & -b \\ b & a \end{bmatrix}\begin{bmatrix} x_1 \\ x_2 \end{bmatrix} \tag{8.5}$$

c. *Rotation, Shift, and Scale Change*

$$y_1 = ax_1 - bx_2 + c$$
$$y_2 = bx_1 + ax_2 + d \tag{8.6}$$

This transformation is referred to by several names, such as four-parameter transformation, two-dimensional linear conformal transformation, and similarity transformation.

All three types of transformation given above do not allow for any deformation of the space transformed. The maximum change in the shape is a uniform scaling. The following transformations, however, do not possess that property.

a. Six-parameter transformation

$$y_1 = ax_1 + bx_2 + c$$
$$y_2 = dx_1 + ex_2 + f \tag{8.7}$$

b. Eight-parameter projective transformation (fractional linear)

$$y_1 = \frac{a_1 x_1 + a_2 x_2 + a_3}{c_1 x_1 + c_2 x_2 + 1}$$
$$y_2 = \frac{b_1 x_1 + b_2 x_2 + b_3}{c_1 x_1 + c_2 x_2 + 1} \tag{8.8}$$

c. General two-dimensional polynomial transformation

$$y_1 = a_0 + a_1 x_1 + a_2 x_2 + a_3 x_1 x_2 + a_4 x_1^2 + a_5 x_2^2 + \cdots$$
$$y_2 = b_0 + b_1 x_1 + b_2 x_2 + b_3 x_1 x_2 + b_4 x_1^2 + b_5 x_2^2 + \cdots \tag{8.9}$$

d. Conformal two-dimensional polynomial transformation

$$y_1 = A_0 + A_1 x_1 + A_2 x_2 + A_3(x_1^2 - x_2^2) + A_4(2x_1 x_2) + \cdots$$
$$y_2 = B_0 - A_2 x_1 + A_1 x_2 - A_4(x_1^2 - x_2^2) + A_3(2x_1 x_2) + \cdots \tag{8.10}$$

These are only a few examples of such type of transformation and readers who are familiar with computational photogrammetry recognize that there are many more possibilities.

8.2.2. Transformations in Three-Dimensional Space

1. One of the most commonly used transformations is the *seven-parameter transformation* which allows for three rotations, three translations, and one scale change.

$$\underset{3,1}{\mathbf{y}} = s\underset{3,3}{\mathbf{M}}\,\underset{3,1}{\mathbf{x}} + \underset{3,1}{\mathbf{k}} \tag{8.11}$$

where

$\mathbf{y} = [y_1\, y_2\, y_3]^t$ coordinates after transformation
$\mathbf{x} = [x_1\, x_2\, x_3]^t$ coordinates before transformation
$\mathbf{k} = [k_1\, k_2\, k_3]^t$ vector of three shifts
$\mathbf{M}$ = an orthogonal matrix in terms of *only* three independent variables (often taken as three sequential rotation angles)
s = scale change

This transformation is also referred to as a linear three-dimensional conformal transformation.

2. General polynomial in three-dimensions

$$\begin{aligned}
y_1 &= a_0 + a_1x_1 + a_2x_2 + a_3x_3 + a_4x_1^2 + a_5x_2^2 + a_6x_3^2 + a_7x_1x_2 \\
&\quad + a_8x_2x_3 + a_9x_3x_1 + \cdots \\
y_2 &= b_0 + b_1x_1 + b_2x_2 + b_3x_3 + b_4x_1^2 + b_5x_2^2 + b_6x_3^2 + b_7x_1x_2 \\
&\quad + b_8x_2x_3 + b_9x_3x_1 + \cdots \qquad (8.12) \\
y_3 &= c_0 + c_1x_1 + c_2x_2 + c_3x_3 + c_4x_1^2 + c_5x_2^2 + c_6x_3^2 + c_7x_1x_2 \\
&\quad + c_8x_2x_3 + c_9x_3x_1 + \cdots
\end{aligned}$$

3. Polynomials in three-dimensions which are conformal in the three planes†

$$\begin{aligned}
y_1 &= A_0 + A_1x_1 + A_2x_2 - A_3x_3 + A_5(x_1^2 - x_2^2 - x_3^2) + 0 \\
&\quad + 2A_7x_3x_1 + 2A_6x_1x_2 \\
y_2 &= B_0 - A_2x_1 + A_1x_2 + A_4x_3 + A_6(-x_1^2 + x_2^2 - x_3^2) + 2A_7x_2x_3 \\
&\quad + 0 + 2A_5x_1x_2 \\
y_3 &= C_0 + A_3x_1 - A_4x_2 + A_1x_3 + A_7(-x_1^2 - x_2^2 + x_3^2) + 2A_6x_2x_3 \\
&\quad + 2A_5x_3x_1 + 0 \qquad (8.13)
\end{aligned}$$

We have given a few different transformations that perform different functions. Now we discuss aspects of adjustment with regard to these transformations.

Example 8.7. (This is a more extensive example than those given on the two-parameter transformation in previous chapters). A transformation between two coordinate systems $\mathbf{x}$ and $\mathbf{y}$ is given by

$$\begin{aligned}
y_1 &= ax_1 - bx_2 \\
y_2 &= bx_1 + ax_2
\end{aligned}$$

† See Mikhail, 1964, in the Bibliography.

where a and b are the parameters in the transformation, (x_1, x_2) the coordinates before transformation, and (y_1, y_2) the coordinates after transformation. In order to estimate the two parameters, three points of known coordinates in both systems are given. A fourth point is known in the **x**-coordinate system only and its corresponding coordinates in the **y**-coordinate system is required. The data are given in Table 8.1.

TABLE 8-1

POINT NUMBER	x_1	x_2	y_1	y_2
1	0.0	1.0	−2.1	1.1
2	1.0	0.0	1.0	2.0
3	1.0	1.0	−0.9	2.8
4	1.0	2.0	?	?

Every point in the first coordinate system has the covariance matrix

$$\Sigma_{xx} = \begin{bmatrix} 0.01 & 0.0 \\ 0.0 & 0.01 \end{bmatrix}$$

and in the second system the matrix

$$\Sigma_{yy} = \begin{bmatrix} 0.04 & 0.0 \\ 0.0 & 0.04 \end{bmatrix}.$$

The units are all of the same type, for example, metres for coordinates and $(\text{metre})^2$ for covariance matrices' elements.

Required: First compute the estimates of a and b and their covariance matrix, then using these compute the **y** coordinates of point 4 and its covariance matrix. Use point 1 to analyze the question of final coordinates and covariance matrices of the given points in the **y**-coordinate system.

Solution: This problem of two-parameter transformation was given in a little more simplified form in Example 6.2. In that example a detailed analysis of the mathematical model was given and the reader may wish to review that portion before proceeding here. The position of a point in the **y**-coordinate system is obviously the end product since the transformation is given as *from* the **x** *to* the **y** system. The points that are given by both **x** and **y** coordinates present the interesting question of how they should be treated. In Example 6.2 only the **x** coordinates were considered as observations, whereas here *both* sets of coordinates are observations with given a priori covariance matrices. Therefore for the three points there are $n = 12$ observations and the determination of the minimum number of elements, n_0, is an important question. Each one of these points contributes four observations to the amount of known information, and adds two parameters as well. These two parameters are the final **y** coordinates of that point which must eventually be determined. Consequently, for all three points a total of *eight* variables or elements must be considered: two transformation coefficients and six parameters for the y coordinates of the three points. This means that $n_0 = 8$ and then $r = 12 - 8 = 4$ degrees of freedom.

The above analysis for determining n_0 is rather important since it points out important facts. The first point to be observed is that n_0 varies with the amount of given information; that is, it is not independent of n. For example, consider the cases when one, two, or three points are given as in Table 8.2. This fact of different n_0 for different n is often encountered when adjusting transformation problems.

TABLE 8-2

NUMBER OF POINTS	n	n_0	r (redundancy)
1	4	4	0
2	8	6	2
3	12	8	4

Having determined $r = 4$, the question of which technique of least squares to use follows. An attempt could be made to write four conditions among the given 12 observations. However, from the given transformation equations and the treatment of an even simpler case of two-parameter transformation problem in previous examples (for instance, Example 7.2), it is clear that writing such equations is not obvious. Equally unduly complex is to write condition equations for adjustment of indirect observations since this would require the carrying of eight parameters. Since interest is in only two parameters, a and b, and the form of condition equations is already given, then the logical choice would be to have $u = 2$, $c = 6$ and write the linearized conditions for the ith point as

$$\underset{2,4}{\mathbf{A}_i}\ \underset{4,1}{\mathbf{v}_i} + \underset{2,2}{\mathbf{B}_i}\ \underset{2,1}{\Delta} = \underset{2,1}{\mathbf{f}_i}$$

If this equation is written three times for the given three points, then

$$\mathbf{A}_1 \mathbf{v}_1 + \mathbf{B}_1\Delta = \mathbf{f}_1$$

$$\mathbf{A}_2 \mathbf{v}_2 + \mathbf{B}_2\Delta = \mathbf{f}_2$$

$$\mathbf{A}_3 \mathbf{v}_3 + \mathbf{B}_3\Delta = \mathbf{f}_3$$

or

$$\begin{bmatrix} \mathbf{A}_1 & \mathbf{0} & \mathbf{0} \\ \mathbf{0} & \mathbf{A}_2 & \mathbf{0} \\ \mathbf{0} & \mathbf{0} & \mathbf{A}_3 \end{bmatrix} \begin{bmatrix} \mathbf{v}_1 \\ \mathbf{v}_2 \\ \mathbf{v}_3 \end{bmatrix} + \begin{bmatrix} \mathbf{B}_1 \\ \mathbf{B}_2 \\ \mathbf{B}_3 \end{bmatrix} \Delta = \begin{bmatrix} \mathbf{f}_1 \\ \mathbf{f}_2 \\ \mathbf{f}_3 \end{bmatrix}$$

or

$$\underset{6,12}{\mathbf{A}}\ \underset{12.1}{\mathbf{v}} + \underset{6,2}{\mathbf{B}}\ \underset{2,1}{\Delta} = \underset{6,1}{\mathbf{f}}$$

The cofactor matrix $\mathbf{Q}_i$ for each point i is a 4×4 matrix. If the reasonable assumption is made that no correlation exists between the coordinates of different points (either the **x** or the **y** coordinates), then the total cofactor matrix of the three points is

$$\underset{12,12}{\mathbf{Q}} = \begin{bmatrix} \mathbf{Q}_1 & \mathbf{0} & \mathbf{0} \\ \mathbf{0} & \mathbf{Q}_2 & \mathbf{0} \\ \mathbf{0} & \mathbf{0} & \mathbf{Q}_3 \end{bmatrix}$$

Then

$$\mathbf{Q}_e = \mathbf{AQA}^t = \begin{bmatrix} (\mathbf{AQA}^t)_1 & \mathbf{0} & \mathbf{0} \\ \mathbf{0} & (\mathbf{AQA}^t)_2 & \mathbf{0} \\ \mathbf{0} & \mathbf{0} & (\mathbf{AQA}^t)_3 \end{bmatrix}$$

and

$$\mathbf{W}_e = \begin{bmatrix} \mathbf{W}_{e_1} & \mathbf{0} & \mathbf{0} \\ \mathbf{0} & \mathbf{W}_{e_2} & \mathbf{0} \\ \mathbf{0} & \mathbf{0} & \mathbf{W}_{e_3} \end{bmatrix}$$

The normal equations coefficient matrix is

$$\underset{2,2}{\mathbf{N}} = \mathbf{B}^t(\mathbf{AQA}^t)^{-1}\mathbf{B} = \mathbf{B}^t\mathbf{W}_e\,\mathbf{B}$$

$$= [\mathbf{B}_1^t\ \mathbf{B}_2^t\ \mathbf{B}_3^t]\mathbf{W}_e \begin{bmatrix} \mathbf{B}_1 \\ \mathbf{B}_2 \\ \mathbf{B}_3 \end{bmatrix} = (\mathbf{B}^t\mathbf{W}_e\,\mathbf{B})_1 + (\mathbf{B}^t\mathbf{W}_e\,\mathbf{B})_2 + (\mathbf{B}^t\mathbf{W}_e\,\mathbf{B})_3$$

or

$$\underset{2,2}{\mathbf{N}} = \sum_{i=1}^{3} (\mathbf{B}^t\mathbf{W}_e\,\mathbf{B})_i = \sum_{i=1}^{3} \mathbf{N}_i \tag{8.14}$$

Similarly the constant term vector **t** may be developed as

$$\underset{2,1}{\mathbf{t}} = \mathbf{B}^t\mathbf{W}_e\,\mathbf{f} = [\mathbf{B}_1^t\ \mathbf{B}_2^t\ \mathbf{B}_3^t]\mathbf{W}_e \begin{bmatrix} \mathbf{f}_1 \\ \mathbf{f}_2 \\ \mathbf{f}_3 \end{bmatrix}$$

$$= (\mathbf{B}^t\mathbf{W}_e\,\mathbf{f})_1 + (\mathbf{B}^t\mathbf{W}_e\,\mathbf{f})_2 + (\mathbf{B}^t\mathbf{W}_e\,\mathbf{f})_3$$

or

$$\underset{2,1}{\mathbf{t}} = \sum_{i=1}^{3} (\mathbf{B}^t\mathbf{W}_e\,\mathbf{f})_i \tag{8.15}$$

The two results above are interesting because they show that the normal equations may be formed by a summation process. This has the distinct advantage that for programming on a computer, data for each point may be handled at one time, and more important considerable savings in computer space can be effected. Note, for example, that for the **A**, **Q**, and **B** matrices space required is 2×4, 4×4, and 2×2 instead of 6×12, 12×12, and 6×2, respectively. The space for these component matrices can be used over and over for all points instead of having one set for each point. This characteristic of summation is, under the above assumption of block diagonality of **A** and **Q** matrices, common to most transformations. Computer space savings increase as the number of parameters in the transformation increases.

Now we return to our numerical problem and evaluate the matrices for any point i, after writing the equations in the functional form

$$f_{1i} = ax_{1i} - bx_{2i} - y_{1i} = 0$$

$$f_{2i} = bx_{1i} + ax_{2i} - y_{2i} = 0$$

$$\mathbf{A}_i = \begin{bmatrix} \partial f_1/\partial x_1 & \partial f_1/\partial x_2 & \partial f_1/\partial y_1 & \partial f_1/\partial y_2 \\ \partial f_2/\partial x_1 & \partial f_2/\partial x_2 & \partial f_2/\partial y_1 & \partial f_2/\partial y_2 \end{bmatrix}_i = \begin{bmatrix} a^0 & -b^0 & -1 & 0 \\ b^0 & a^0 & 0 & -1 \end{bmatrix}$$

$$= \begin{bmatrix} \underset{2,2}{\mathbf{C}} & \underset{2,2}{-\mathbf{I}} \end{bmatrix} \qquad \text{(independent of "}i\text{")}$$

in which a^0 and b^0 are approximations for the two parameters

$$\mathbf{B}_i = \begin{bmatrix} \partial f_1/\partial a & \partial f_1/\partial b \\ \partial f_2/\partial a & \partial f_2/\partial b \end{bmatrix}_i = \begin{bmatrix} x_1 & -x_2 \\ x_2 & x_1 \end{bmatrix}_i$$

and

$$\underset{4,4}{\boldsymbol{\Sigma}_i} = \begin{bmatrix} \boldsymbol{\Sigma}_{xx} & \mathbf{0} \\ \mathbf{0} & \boldsymbol{\Sigma}_{yy} \end{bmatrix}_i = 10^{-2} \begin{bmatrix} 1 & 0 & 0 & 0 \\ & 1 & 0 & 0 \\ & & 4 & 0 \\ \text{symmetric} & & & 4 \end{bmatrix} = \sigma_0^2 \mathbf{Q}_i$$

With the a priori $\sigma_0^2 = 0.01$, the cofactor matrix of the observations will be taken as

$$\mathbf{Q}_i = \begin{bmatrix} 1 & 0 & 0 & 0 \\ & 1 & 0 & 0 \\ & & 4 & 0 \\ \text{symmetric} & & & 4 \end{bmatrix} = \begin{bmatrix} \mathbf{I} & \mathbf{0} \\ \mathbf{0} & 4\mathbf{I} \end{bmatrix}$$

then

$$\mathbf{Q}_{e_i} = \mathbf{A}_i \mathbf{Q}_i \mathbf{A}_i^t = \mathbf{C}\mathbf{C}^t + 4\mathbf{I} = (a^{02} + b^{02})\mathbf{I} + 4\mathbf{I}$$

$$= (a^{02} + b^{02} + 4)\mathbf{I}$$

$$\mathbf{W}_{e_i} = (a^{02} + b^{02} + 4)^{-1}\mathbf{I} = \alpha\mathbf{I}$$

where $\alpha = 1/(a^{02} + b^{02} + 4)$, and

$$\mathbf{N}_i = \mathbf{B}_i^t \mathbf{W}_{e_i} \mathbf{B}_i = \alpha(x_1^2 + x_2^2)_i \mathbf{I}$$

Thus the total normal equation matrix will then be

$$\mathbf{N} = \alpha \sum_{i=1}^{3} (x_1^2 + x_2^2)_i \mathbf{I} = \alpha(1 + 1 + 2)\mathbf{I} = 4\alpha\mathbf{I}$$

noting that α is the only value that varies from one iteration to the next because of updating the approximations a^0 and b^0.

In a similar manner the constant term vector is

$$\mathbf{t} = \sum_{i=1}^{3} \mathbf{t}_i = \sum_{i=1}^{3} \alpha \mathbf{B}_i^t \mathbf{f}_i = \alpha \sum_{i=1}^{3} \mathbf{B}_i^t \mathbf{f}_i = \alpha \begin{bmatrix} \sum (x_1 f_1 + x_2 f_2)_i \\ \sum (-x_2 f_1 + x_1 f_2)_i \end{bmatrix}$$

It is important to note that the vector **t** is more dependent on the approximations a^0 and b^0 because it is directly related to the vector of so-called discrepancies **f**. The latter vector will approach a minimum when the solution is iterated until practically no improvement takes place.

Using the matrix **N** and vector **t** the correction vector **Δ** may be computed as

$$\underset{2,2}{\boldsymbol{\Delta}} = \mathbf{N}^{-1}\mathbf{t} = \frac{1}{4} \begin{bmatrix} \sum_{i=1}^{3} (x_1 f_1 + x_2 f_2)_i \\ \sum_{i=1}^{3} (-x_2 f_1 + x_1 f_2)_i \end{bmatrix}$$

Now, consider the question of choosing approximations for the parameters. Taking the rather easy choice first begin by $a^0 = b^0 = 0$. Thus

Case 1. $a^0 = 0$ *and* $b^0 = 0$

$$\mathbf{f}_1 = \begin{bmatrix} -2.1 \\ 1.1 \end{bmatrix} \qquad \mathbf{f}_2 = \begin{bmatrix} 1.0 \\ 2.0 \end{bmatrix} \qquad \mathbf{f}_3 = \begin{bmatrix} -0.9 \\ 2.8 \end{bmatrix}$$

$$\sum (x_1 f_1 + x_2 f_2)_i = 1.1 + 1.0 + 1.9 = 4.0$$

$$\sum (-x_2 f_1 + x_1 f_2)_i = 2.1 + 2.0 + 3.7 = 7.8$$

Then

$$\boldsymbol{\Delta} = \frac{1}{4} \begin{bmatrix} 4.0 \\ 7.8 \end{bmatrix} = \begin{bmatrix} 1.0 \\ 1.95 \end{bmatrix}$$

and

$$\mathbf{Q}_{\Delta\Delta} = \mathbf{N}^{-1} = \frac{1}{4\alpha} \mathbf{I} = \mathbf{I}$$

since $\alpha = \frac{1}{4}$ when $a^0 = b^0 = 0$ which with a priori $\sigma_0^2 = 0.01$ leads to the covariance matrix of the parameters

$$\boldsymbol{\Sigma}_{\Delta\Delta} = \begin{bmatrix} 0.01 & 0 \\ 0 & 0.01 \end{bmatrix}$$

This, in fact, ends only the first iteration, and since the corrections are obviously quite large, at least another iteration is necessary. To start the second iteration, therefore,

$$a^0 = 1.0 \qquad b^0 = 1.95$$

and

$$\mathbf{f}_1 = \begin{bmatrix} -0.15 \\ 0.10 \end{bmatrix} \qquad \mathbf{f}_2 = \begin{bmatrix} 0.0 \\ 0.05 \end{bmatrix} \qquad \mathbf{f}_3 = \begin{bmatrix} +0.05 \\ -0.15 \end{bmatrix}$$

$$\sum (x_1 f_1 + x_2 f_2)_i = 0.10 + 0.0 - 0.10 = 0$$

$$\sum (-x_2 f_1 + x_1 f_2)_i = 0.15 + 0.05 - 0.20 = 0$$

With a zero $\mathbf{t}$ vector then $\boldsymbol{\Delta} = [0 \quad 0]^t$ and the final parameters are $\hat{a} = 1.0$ and $\hat{b} = 1.95$

It is very important to realize that in order to get the correct cofactor matrix of the parameters, the last $\mathbf{N}^{-1}$ must be computed, otherwise the one already available may be grossly in error. In this case the new $\mathbf{N}^{-1}$ is

$$\mathbf{N}^{-1} = \frac{1}{4\alpha}\mathbf{I} = \frac{a^{02} + b^{02} + 4}{4}\mathbf{I} = \frac{8.8025}{4}\mathbf{I} = 2.2006\mathbf{I}$$

Thus the covariance matrix of the final estimates of the parameters is

$$\boldsymbol{\Sigma}_{xx} = \begin{bmatrix} 0.022 & 0.0 \\ 0.0 & 0.022 \end{bmatrix}$$

Comparing this matrix with the one computed at the end of the first iteration reveals a variation of over 200%, indicating how serious an error may be committed if proper attention is not paid when solving a problem by iteration.

Case 2: In this case consider an approach of computing approximations for the parameters with reasonable effort. For example, we might apply a minimum solution by choosing, say, the first point to solve for a and b. Thus

$$-2.1 = -b \qquad \text{or} \qquad b^0 = 2.0 \qquad \text{(rounded off to whole number)}$$

$$1.1 = a \qquad \text{or} \qquad a^0 = 1.0 \qquad \text{(rounded off to whole number)}$$

(Note that if the second point was used, the above approximations would have been obtained without rounding off. If the third point is used, a little more computation will be involved but with about the same resulting approximations.)

Another possibility, which in principle would yield slightly better approximations, is to compute the means of all coordinates and use them to solve a minimum case. Although this is not a demanding effort in the present example, it could be a rather unwarranted computation in more involved problems.

With $a^0 = 1.0$ and $b^0 = 2.0$, then

$$\mathbf{f}_1 = \begin{bmatrix} -0.10 \\ 0.10 \end{bmatrix} \qquad \mathbf{f}_2 = \begin{bmatrix} 0.0 \\ 0.0 \end{bmatrix} \qquad \mathbf{f}_3 = \begin{bmatrix} 0.10 \\ -0.20 \end{bmatrix}$$

$$\sum (x_1 f_1 + x_2 f_2)_i = 0.10 + 0.0 - 0.10 = 0$$

$$\sum (-x_2 f_1 + x_1 f_2)_i = 0.10 + 0.0 - 0.3 = -0.2$$

$$\boldsymbol{\Delta} = \frac{1}{4}\begin{bmatrix} 0 \\ -0.2 \end{bmatrix} = \begin{bmatrix} 0 \\ -0.05 \end{bmatrix}$$

leading directly to the final result of $\hat{a} = 1.0$ and $\hat{b} = 1.95$. Now, to compute $\mathbf{Q}_{\Delta\Delta}$ we should advisably use $\hat{a}$ and $\hat{b}$ directly to compute the final $\mathbf{N}^{-1}$. If we were to use the original approximations, we would get

$$\mathbf{Q}_{\Delta\Delta} = \mathbf{N}^{-1} = \frac{a^{02} + b^{02} + 4}{4}\mathbf{I} = 2.25\mathbf{I}$$

which would lead to

$$\boldsymbol{\Sigma}_{xx} = \begin{bmatrix} 0.0225 & 0 \\ 0 & 0.0225 \end{bmatrix}$$

The variation in this case is only slightly over 2% and this is because the computed Δ is about the same percentage different from the final answer. Although such difference may be sufficiently small to be negligible, it is advisable always to update the cofactor matrix at the last iteration.

An important final remark is worth mentioning here, which is that we should always obtain the best possible approximation vector that requires a reasonable effort of computation. As has been shown above, with good approximations it was not even necessary to perform another iteration. The reader must be cautioned, however, that this is true here only because the condition equations are almost linear (of course, if they were in fact linear, no iteration would be necessary). Nevertheless, we would expect that, in general, choosing better approximations would lead to fewer iterations.

Coordinates of Point 4 in the **y** *System*

$$\hat{y}_{14} = \hat{a}x_{14} - \hat{b}x_{24} = -2.90$$

$$\hat{y}_{24} = \hat{b}x_{14} + \hat{a}x_{24} = 3.95$$

To compute the covariance matrix of the vector $\hat{\mathbf{y}}_4$ we should realize that all variables $(x_1, x_2)_4$ and $(\hat{a}, \hat{b})$ are random variables with known covariance matrices. Since point 4 was not one of the points used in estimating $\hat{a}$, $\hat{b}$, there would be no correlation between the covariance matrix of the coordinates and that of the parameters. To facilitate the computation introduce the auxiliary vector

$$\mathbf{r} = [\hat{a} \quad \hat{b} \quad x_{14} \quad x_{24}]^t$$

and thus

$$\underset{4,4}{\boldsymbol{\Sigma}_{rr}} = \begin{bmatrix} 0.022\mathbf{I}_2 & \mathbf{0} \\ \mathbf{0} & 0.01\mathbf{I}_2 \end{bmatrix}$$

Hence

$$\boldsymbol{\Sigma}_{\hat{y}_4\hat{y}_4} = \mathbf{J}\boldsymbol{\Sigma}_{rr}\,\mathbf{J}^t$$

where

$$\mathbf{J} = \frac{\partial \mathbf{y}_4}{\partial \mathbf{r}} = \begin{bmatrix} x_{14} & -x_{24} & \hat{a} & -\hat{b} \\ x_{24} & x_{14} & \hat{b} & \hat{a} \end{bmatrix} = \begin{bmatrix} 1 & -2 & 1 & -1.95 \\ 2 & 1 & 1.95 & 1 \end{bmatrix}$$

Thus

$$\boldsymbol{\Sigma}_{\hat{y}_4\hat{y}_4} = \begin{bmatrix} 0.158 & 0.0 \\ 0.0 & 0.158 \end{bmatrix}$$

The Coordinates of Point 1: Point 1 was used in the transformation to estimate the two parameters a and b. It was therefore given with coordinates in both systems of coordinates **x** and **y**. However, using the same procedure that was applied to point 4 above from the known **x** coordinates and estimated parameters $\hat{a}$ and $\hat{b}$, another pair of "transformed" **y** coordinates may be computed:

$$y'_{11} = \hat{a}x_{11} - \hat{b}x_{21} = -1.95$$

$$y'_{12} = bx_{11} + ax_{21} = 1.00$$

These coordinates are obviously different from those given a priori, and it is not the final position of the point. The final coordinates in the y system will be computed shortly, but first we evaluate the covariance matrix of these coordinates. Again, let an auxiliary vector be defined as

$$\mathbf{t} = [a \quad b \quad x_{11} \quad x_{21}]^t$$

The covariance matrix $\underset{4,\,4}{\boldsymbol{\Sigma}_{tt}}$ may be constructed in the form

$$\boldsymbol{\Sigma}_{tt} = \begin{bmatrix} 0.022\mathbf{I} & \boldsymbol{\Sigma}_{\Delta x1} \\ \boldsymbol{\Sigma}_{x1\Delta} & 0.01\mathbf{I}_2 \end{bmatrix}$$

Unlike the case of point 4, the matrix $\boldsymbol{\Sigma}_{\Delta x_1}$ is not necessarily zero because point 1 was used in the transformation and therefore $(\hat{a}, \hat{b})$ and $(x_1, x_2)_1$ are expected to be correlated. To compute this covariance matrix it should be realized that $(x_1, x_2)_1$ is a part of the total observation vector $l_{12,\,1}$. Referring to equation (6.32) the total cofactor matrix would be

$$\underset{2,\,12}{\mathbf{Q}_{\Delta l}} = -\underset{2,\,2}{\mathbf{N}^{-1}}\ \underset{2,\,6}{\mathbf{B}^t}\ \Big(\underset{\leftarrow 6,\,6\rightarrow}{\mathbf{AQA}^t}\Big)^{-1}\ \underset{6,\,12}{\mathbf{A}}\ \underset{12,\,12}{\mathbf{Q}}$$

$$= -\mathbf{N}^{-1}[\mathbf{B}_1^t\ \mathbf{B}_2^t\ \mathbf{B}_3^t]\begin{bmatrix} \mathbf{W}_{e_1} & & \\ & \mathbf{W}_{e_2} & \\ & & \mathbf{W}_{e_3} \end{bmatrix}\begin{bmatrix} \mathbf{A}_1 & & \\ & \mathbf{A}_2 & \\ & & \mathbf{A}_3 \end{bmatrix}\begin{bmatrix} \mathbf{Q}_1 & & \\ & \mathbf{Q}_2 & \\ & & \mathbf{Q}_3 \end{bmatrix}$$

$$= -\underset{2,\,2}{\mathbf{N}^{-1}}\left[\Big(\underset{\leftarrow 2,\,4\rightarrow}{\mathbf{B}^t\mathbf{W}_e\,\mathbf{AQ}}\Big)_1 \Big(\underset{\leftarrow 2,\,4\rightarrow}{\mathbf{B}^t\mathbf{W}_e\,\mathbf{AQ}}\Big)_2 \Big(\underset{\leftarrow 2,\,4\rightarrow}{\mathbf{B}^t\mathbf{W}_e\,\mathbf{AQ}}\Big)_3\right]$$

Since we are interested in point 1 only, then

$$\underset{2,\,4}{\mathbf{Q}_{\Delta l_1}} = -\mathbf{N}^{-1}\mathbf{B}_1^t\,\mathbf{W}_{e_1}\mathbf{A}_1\,\mathbf{Q}_1 = -\mathbf{N}^{-1}\mathbf{B}_1^t\,\mathbf{W}_{e_1}[\mathbf{C}\mathbf{Q}_{x_1} - \mathbf{Q}_{y_1}] \tag{8.16}$$

where

$$\mathbf{C} = \begin{bmatrix} \hat{a} & -\hat{b} \\ \hat{b} & \hat{a} \end{bmatrix}, \qquad \mathbf{Q}_{x_1} = \mathbf{I}_2\,, \qquad \text{and} \qquad \mathbf{Q}_{y_1} = 4\mathbf{I}_2$$

The cofactor matrix required is that which relates the estimated parameters and the **x** coordinates only. Thus

$$\mathbf{Q}_{\Delta x_1} = -\mathbf{N}^{-1}\mathbf{B}_1^t\,\mathbf{W}_{e1}\mathbf{C}\mathbf{Q}_{x1}$$

or

$$\mathbf{Q}_{\Delta x_1} = \frac{-1}{4\alpha}\mathbf{I}_2\begin{bmatrix} 0 & +1 \\ -1 & 0 \end{bmatrix}\alpha\mathbf{I}_2\begin{bmatrix} 1 & -1.95 \\ 1.95 & 1 \end{bmatrix}\mathbf{I}_2$$

$$= \frac{1}{4}\begin{bmatrix} -1.95 & -1 \\ 1 & -1.95 \end{bmatrix} = \begin{bmatrix} -0.4875 & -0.25 \\ 0.25 & -0.4875 \end{bmatrix}$$

or, the covariance matrix (remembering that $\sigma_0^2 = 0.01$)

$$\boldsymbol{\Sigma}_{\Delta x_1} = 0.01\begin{bmatrix} -0.4875 & -0.25 \\ 0.25 & -0.4875 \end{bmatrix}$$

The covariance matrix of the vector $\mathbf{t}$ is therefore

$$\underset{4,4}{\boldsymbol{\Sigma}_{tt}} = 10^{-2}\begin{bmatrix} 2.2 & 0 & -0.4875 & -0.25 \\ & 2.2 & 0.25 & -0.4875 \\ & & 1.0 & 0 \\ \text{symmetric} & & & 1.0 \end{bmatrix}$$

Finally, the covariance matrix of the transformed coordinates $\boldsymbol{\Sigma}_{y_1'y_1'}$ is computed as

$$\boldsymbol{\Sigma}_{y_1'y_1'} = \mathbf{J}'\boldsymbol{\Sigma}_{tt}\,\mathbf{J}'^t$$

where

$$\mathbf{J}' = \begin{bmatrix} x_{11} & -x_{21} & \hat{a} & -\hat{b} \\ x_{21} & x_{11} & \hat{b} & \hat{a} \end{bmatrix} = \begin{bmatrix} 0 & -1 & 1 & -1.95 \\ 1 & 0 & 1.95 & 1 \end{bmatrix}$$

or

$$\boldsymbol{\Sigma}_{y_1'y_1'} = 10^{-2}\begin{bmatrix} 4.6019 & 0 \\ 0 & 4.6019 \end{bmatrix}$$

Thus the "transformed" coordinates of point 1 are

$$\mathbf{y}_1' = [-1.95 \quad 1.00]^t$$

and the covariance matrix is

$$\boldsymbol{\Sigma}_{y_1'y_1'} = 10^{-2}\begin{bmatrix} 4.6019 & 0 \\ 0 & 4.6019 \end{bmatrix}$$

A more interesting question still remains, however; and that is what is the best estimate $\hat{\mathbf{y}}_1$ for the coordinates of this point in the $\mathbf{y}$ system and its covariance matrix? To answer this question, recall the available information.

First, the given observational vector $\mathbf{y} = [-2.1 \quad 1.1]^t$ and its covariance matrix $\boldsymbol{\Sigma}_{yy} = 0.04\mathbf{I}_2$; and another estimate $\mathbf{y}' = [-1.95 \quad 1.0]^t$ and its covariance matrix $\boldsymbol{\Sigma}_{y'y'} = 0.046019\mathbf{I}_2$. (We have dropped the subscript 1 for simplicity). These two vectors together may be constructed as an observational vector

$$\underset{4,1}{\boldsymbol{l}} = \begin{bmatrix} \mathbf{y}' \\ \mathbf{y} \end{bmatrix} = [-1.95 \quad 1 \quad -2.1 \quad 1.1]^t$$

To construct the covariance matrix $\boldsymbol{\Sigma}_{ll}$ of these observations, care must be exercised in establishing whether the two vectors $\mathbf{y}$ and $\mathbf{y}'$ are correlated or not. The vector $\mathbf{y}'$ was computed in the adjustment and we must find out if $\mathbf{y}$ was involved in its computation. Recalling those equations from which $\mathbf{y}'$ was computed

$$\mathbf{y}' = \begin{bmatrix} x_1 & -x_2 \\ x_2 & x_1 \end{bmatrix}\begin{bmatrix} \hat{a} \\ \hat{b} \end{bmatrix} = \mathbf{B}\boldsymbol{\Delta}$$

In the process of estimating $\boldsymbol{\Delta} = [\hat{a} \quad \hat{b}]^t$, the vector $\mathbf{y}$ was in fact used in the least squares solution. Therefore, working with cofactors for the moment, $\mathbf{Q}_{y'y}$ may be computed as follows: Given

$$\mathbf{y}' = \mathbf{B}\boldsymbol{\Delta} \qquad \text{and} \qquad \mathbf{y} = \mathbf{y}$$

and

$$\mathbf{Q}_{\Delta y_1} = -\mathbf{N}^{-1}\mathbf{B}_1^t \mathbf{W}_{e_1}(-\mathbf{Q}_{y_1}) \qquad \text{[from equation (8.16)]}$$

Again, dropping the subscript 1, then

$$\mathbf{Q}_{\Delta y} = \mathbf{N}^{-1}\mathbf{B}^t\mathbf{W}_e \mathbf{Q}_y$$

Then

$$\mathbf{Q}_{y'y} = \frac{\partial \mathbf{y}'}{\partial \Delta} \mathbf{Q}_{\Delta y} \frac{\partial \mathbf{y}}{\partial \mathbf{y}}$$

or

$$\mathbf{Q}_{y'y} = \mathbf{B}\mathbf{N}^{-1}\mathbf{B}^t\mathbf{W}_e \mathbf{Q}_y$$

$$= \begin{bmatrix} 0 & -1 \\ 1 & 0 \end{bmatrix} \frac{1}{4\alpha}\mathbf{I}_2 \begin{bmatrix} 0 & 1 \\ -1 & 0 \end{bmatrix} \alpha\mathbf{I}_2 \cdot 4\mathbf{I}_2 = \begin{bmatrix} 1 & 0 \\ 0 & 1 \end{bmatrix} = \mathbf{I}_2$$

Therefore

$$\mathbf{Q}_{ll} = \begin{bmatrix} 4.6019\mathbf{I}_2 & \mathbf{I}_2 \\ \mathbf{I}_2 & 4\mathbf{I}_2 \end{bmatrix} \qquad \underset{4,\,4}{\mathbf{W}_{ll}} = \frac{1}{17.4076}\begin{bmatrix} 4\mathbf{I} & -\mathbf{I} \\ -\mathbf{I} & 4.6019\mathbf{I} \end{bmatrix}$$

To obtain the best estimate $\hat{\mathbf{y}}$ we apply the least squares method of adjustment of indirect observations. Also, to further the reader's familiarity with shorter notations, we perform the operations directly in vectors. Thus

$$\mathbf{v}_1 - \hat{\mathbf{y}} = -\mathbf{y}'$$

$$\mathbf{v}_2 - \hat{\mathbf{y}} = -\mathbf{y}$$

or

$$\underset{4,\,1}{\mathbf{v}} + \underset{4,\,2}{\begin{bmatrix} -\mathbf{I} \\ \\ -\mathbf{I} \end{bmatrix}} \underset{2,\,1}{\hat{\mathbf{y}}} = \underset{4,\,1}{\begin{bmatrix} -\mathbf{y}' \\ \\ -\mathbf{y} \end{bmatrix}}$$

which is of the form $\mathbf{v} + \mathbf{B}\Delta = \mathbf{f}$. Next,

$$\mathbf{N} = \frac{6.6019}{17.4076}\mathbf{I}, \qquad \mathbf{N}^{-1} = 2.6368\mathbf{I}, \qquad \text{and} \qquad \mathbf{t} = \frac{1}{17.4076}(3\mathbf{y}' + 3.6019\mathbf{y})$$

$$\mathbf{y} = \mathbf{N}^{-1}\mathbf{t} = \frac{1}{6.6019}(3\mathbf{y}' + 3.6019\mathbf{y}) = \begin{bmatrix} -2.03184 \\ 1.05456 \end{bmatrix}$$

$$\mathbf{Q}_{\hat{y}\hat{y}} = \mathbf{N}^{-1} = 2.6368\mathbf{I} \qquad \mathbf{\Sigma}_{\hat{y}\hat{y}} = (0.01)(2.6368)\mathbf{I} = 0.026368\mathbf{I}$$

Thus the final estimate of the $\mathbf{y}$ coordinates of point 1 are $(-2.0384, 1.05456)$ and its covariance matrix is $0.026368\mathbf{I}$.

In the preceding analysis and computation many rather important steps were involved. Such steps demonstrate several features of the concepts of the functional and, particularly, the stochastic models. We shall give the discussion of these features at the end of the example. We shall now continue the analysis of obtaining the final $\mathbf{y}$-coordinate estimates for the reference points and the covariance matrices by another process.

It should be remembered that both the **x** and **y** coordinates were considered as observations. This means that both sets of coordinates, for each of the reference points, receive residuals after the adjustment in order to obtain the estimated $\hat{\mathbf{x}}$ and $\hat{\mathbf{y}}$ coordinates. These estimated coordinates must satisfy the transformation equations since this is one of the basic premises of the least squares adjustment. Therefore when $\hat{\mathbf{x}}$ coordinates are transformed by $\hat{a}$ and $\hat{b}$, the resulting transformed coordinates must be equal to the $\hat{\mathbf{y}}$ coordinates, and that these coordinates are the final best estimate of the position of the point in the **y** system. The total vector of residuals is (recalling that the final $\mathbf{\Delta}$ is zero)

$$\begin{bmatrix} \mathbf{v}_1 \\ \mathbf{v}_2 \\ \mathbf{v}_3 \end{bmatrix} = \begin{bmatrix} \mathbf{Q}_1 & & \\ & \mathbf{Q}_2 & \\ & & \mathbf{Q}_3 \end{bmatrix} \begin{bmatrix} \mathbf{A}_1^t & & \\ & \mathbf{A}_2^t & \\ & & \mathbf{A}_3^t \end{bmatrix} \begin{bmatrix} \mathbf{W}_{e_1} & & \\ & \mathbf{W}_{e_2} & \\ & & \mathbf{W}_{e_3} \end{bmatrix} \begin{bmatrix} \mathbf{f}_1 \\ \mathbf{f}_2 \\ \mathbf{f}_3 \end{bmatrix}$$

from which we can write for the residual vector of point 1

$$\mathbf{v}_1 = \mathbf{Q}_1 \mathbf{A}_1^t \mathbf{W}_{e_1} \mathbf{f}_1$$

Continuing further with the partitioning

$$\begin{bmatrix} \mathbf{v}_x \\ \mathbf{v}_y \end{bmatrix} = \begin{bmatrix} \mathbf{Q}_{xx} & \mathbf{0} \\ \mathbf{0} & \mathbf{Q}_{yy} \end{bmatrix}_1 \begin{bmatrix} \mathbf{C}^t \\ -\mathbf{I} \end{bmatrix} \alpha \mathbf{I} \mathbf{f}_1$$

$$\mathbf{v}_{x_1} = \alpha \mathbf{Q}_{xx} \mathbf{C}^t \mathbf{f}_1 = \alpha \mathbf{C}^t \mathbf{f}_1$$

$$\mathbf{v}_{y_1} = -\alpha \mathbf{Q}_{yy} \mathbf{f}_1 = -4\alpha \mathbf{f}_1$$

with

$$\alpha = (\hat{a}^2 + \hat{b}^2 + 4)^{-1} = \frac{1}{8.8025} = 0.1136$$

$$\mathbf{C} = \begin{bmatrix} \hat{a} & -\hat{b} \\ \hat{b} & \hat{a} \end{bmatrix} = \begin{bmatrix} 1.0 & -1.95 \\ 1.95 & 1.0 \end{bmatrix}, \quad \mathbf{f} = \begin{bmatrix} -0.15 \\ 0.10 \end{bmatrix}$$

then

$$\mathbf{v}_x = \begin{bmatrix} 0.00511 \\ 0.04459 \end{bmatrix} \quad \text{and} \quad \mathbf{v}_y = \begin{bmatrix} 0.0682 \\ -0.0454 \end{bmatrix}$$

The estimated "observations" are therefore

$$\hat{\mathbf{x}} = \mathbf{x} + \mathbf{v}_x = \begin{bmatrix} 0.00511 \\ 1.04459 \end{bmatrix} \qquad \hat{\mathbf{y}} = \mathbf{y} + \mathbf{v}_y = \begin{bmatrix} -2.0318 \\ 1.0546 \end{bmatrix}$$

Using $\hat{a}$, $\hat{b}$, and $\hat{\mathbf{x}}$, an estimate for the **y** coordinates may be computed again

$$\hat{\mathbf{y}} = \begin{bmatrix} \hat{a} & -\hat{b} \\ \hat{b} & \hat{a} \end{bmatrix} \begin{bmatrix} \hat{x}_1 \\ \hat{x}_2 \end{bmatrix} = \begin{bmatrix} 1.0 & -1.95 \\ 1.95 & 1.0 \end{bmatrix} \begin{bmatrix} 0.00511 \\ 1.04459 \end{bmatrix} = \begin{bmatrix} -2.03183 \\ 1.05455 \end{bmatrix}$$

Thus all three procedures of computing $\hat{\mathbf{y}}$ for reference point number 1 resulted in exactly the same values. To complete the analysis, evaluate the corresponding cofactor matrix $\mathbf{Q}_{\hat{y}\hat{y}}$. First

$$\mathbf{Q}_{vv} = \mathbf{Q}\mathbf{A}^t\mathbf{W}_e(\mathbf{I} - \mathbf{B}\mathbf{N}^{-1}\mathbf{B}^t\mathbf{W}_e)\mathbf{A}\mathbf{Q}$$

Remembering that $\mathbf{Q}$, $\mathbf{A}$, and $\mathbf{W}_e$ are block diagonal matrices, examine the expression

$$\underset{6,6}{\mathbf{K}} = \underset{6,2}{\mathbf{B}}\ \underset{2,2}{\mathbf{N}^{-1}}\ \underset{2,6}{\mathbf{B}^t}\ \underset{6,6}{\mathbf{W}_e} = \begin{bmatrix} \mathbf{B}_1 \\ \mathbf{B}_2 \\ \mathbf{B}_3 \end{bmatrix} \mathbf{N}^{-1}[\mathbf{B}_1^t\ \mathbf{B}_2^t\ \mathbf{B}_3^t] \begin{bmatrix} \mathbf{W}_{e_1} & & \\ & \mathbf{W}_{e_2} & \\ & & \mathbf{W}_{e_3} \end{bmatrix}$$

$$\begin{bmatrix} \mathbf{B}_1\mathbf{N}^{-1}\mathbf{B}_1^t\mathbf{W}_{e_1} & \mathbf{B}_1\mathbf{N}^{-1}\mathbf{B}_2^t\mathbf{W}_{e_2} & \mathbf{B}_1\mathbf{N}^{-1}\mathbf{B}_3^t\mathbf{W}_{e_3} \\ \mathbf{B}_2\mathbf{N}^{-1}\mathbf{B}_1^t\mathbf{W}_{e_1} & \mathbf{B}_2\mathbf{N}^{-1}\mathbf{B}_2^t\mathbf{W}_{e_2} & \mathbf{B}_2\mathbf{N}^{-1}\mathbf{B}_3^t\mathbf{W}_{e_3} \\ \mathbf{B}_3\mathbf{N}^{-1}\mathbf{B}_1^t\mathbf{W}_{e_1} & \mathbf{B}_3\mathbf{N}^{-1}\mathbf{B}_2^t\mathbf{W}_{e_2} & \mathbf{B}_3\mathbf{N}^{-1}\mathbf{B}_3^t\mathbf{W}_{e_3} \end{bmatrix}$$

$$= \begin{bmatrix} \mathbf{K}_{11} & \mathbf{K}_{12} & \mathbf{K}_{13} \\ \mathbf{K}_{21} & \mathbf{K}_{22} & \mathbf{K}_{23} \\ \mathbf{K}_{31} & \mathbf{K}_{32} & \mathbf{K}_{33} \end{bmatrix}$$

and then

$$\underset{12,12}{\mathbf{Q}_{vv}} =$$

$$\begin{bmatrix} \mathbf{Q}_1\mathbf{A}_1^t\mathbf{W}_{e1} & & \\ & \mathbf{Q}_2\mathbf{A}_2^t\mathbf{W}_{e2} & \\ & & \mathbf{Q}_3\mathbf{A}_3^t\mathbf{W}_{e3} \end{bmatrix} (\mathbf{I}-\mathbf{K}) \begin{bmatrix} \mathbf{A}_1\mathbf{Q}_1 & & \\ & \mathbf{A}_2\mathbf{Q}_2 & \\ & & \mathbf{A}_3\mathbf{Q}_3 \end{bmatrix}$$

From the above structure of $\mathbf{Q}_{vv}$, any submatrix of interest can be directly extracted. For example, the 4×4 submatrix of cofactors for the first point (with respect to itself) is

$$\mathbf{Q}_{v_1v_1} = \mathbf{Q}_1\mathbf{A}_1^t\mathbf{W}_{e1}(\mathbf{I} - \mathbf{B}_1\mathbf{N}^{-1}\mathbf{B}_1^t\mathbf{W}_{e1})\mathbf{A}_1\mathbf{Q}_1$$

for which

$$(\mathbf{I} - \mathbf{B}_1\mathbf{N}^{-1}\mathbf{B}_1^t\mathbf{W}_{e1}) = \mathbf{I} - \mathbf{B}_1\frac{1}{4\alpha}\mathbf{I}\mathbf{B}_1^t\alpha\mathbf{I} = \mathbf{I} - \frac{1}{4}\mathbf{B}_1\mathbf{B}_1^t$$

$$= \mathbf{I} - \frac{(x_1^2 + x_2^2)}{4}\mathbf{I} = 0.75\mathbf{I}$$

and

$$\mathbf{Q}_{v_1v_1} = \begin{bmatrix} \mathbf{I} & \mathbf{0} \\ \mathbf{0} & 4\mathbf{I} \end{bmatrix} \begin{bmatrix} \mathbf{C}^t \\ -\mathbf{I} \end{bmatrix} 0.75\alpha\mathbf{I}[\mathbf{C} \quad -\mathbf{I}] \begin{bmatrix} \mathbf{I} & \mathbf{0} \\ \mathbf{0} & 4\mathbf{I} \end{bmatrix}$$

$$= 0.0852\begin{bmatrix} \mathbf{C}^t\mathbf{C} & 4\mathbf{C}^t \\ 4\mathbf{C} & 16\mathbf{I} \end{bmatrix}$$

$$\mathbf{Q}_{v_1v_1} = \begin{bmatrix} 0.4092 & 0 & -0.6632 & -0.3408 \\ & 0.4092 & 0.3408 & -0.6632 \\ & & 1.3632 & 0 \\ \text{symmetric} & & & 1.3632 \end{bmatrix}$$

$$\mathbf{Q}_{l_1l_1} = \mathbf{Q}_1 - \mathbf{Q}_{v_1v_1} = \left[\begin{array}{cc|cc} 0.5908 & 0 & 0.6632 & 0.3408 \\ & 0.5908 & -0.3403 & 0.6632 \\ \hline & & 2.6368 & 0 \\ \text{symmetric} & & & 2.6368 \end{array}\right]$$

$$= \begin{bmatrix} \mathbf{Q}_{\hat{x}\hat{x}} & \mathbf{Q}_{\hat{x}\hat{y}} \\ \mathbf{Q}_{\hat{y}\hat{x}} & \mathbf{Q}_{\hat{y}\hat{y}} \end{bmatrix}$$

with the numerical values of the submatrices easily obtained from the partitioning lines.

At this point, it is clear that $\mathbf{Q}_{\hat{y}\hat{y}} = 2.6368\mathbf{I}$ is exactly the same as was computed before. To complete the computation, derive $\mathbf{Q}_{\hat{y}\hat{y}}$ by propagation applied to $\hat{\mathbf{y}} = \mathbf{C}\hat{\mathbf{x}}$. As before, both $\hat{a}$, $\hat{b}$ (elements of $\mathbf{C}$) and $\hat{x}_1$, $\hat{x}_2$ (elements of $\hat{\mathbf{x}}$) are random variables. Placing these in a vector

$$\mathbf{t}' = [\hat{a} \quad \hat{b} \quad \hat{x}_1 \quad \hat{x}_2]^t$$

the corresponding cofactor matrix would be

$$\mathbf{Q}_{t't'} = \begin{bmatrix} Q_{\Delta\Delta} & Q_{\Delta\hat{x}} \\ Q_{\hat{x}\Delta} & Q_{\hat{x}\hat{x}} \end{bmatrix}$$

From paragraph 6.2.3 the cross cofactor matrix between parameters and original observations is the same as between parameters and estimated observations, thus

$$\mathbf{Q}_{\Delta\hat{x}} = \mathbf{Q}_{\Delta x} = \begin{bmatrix} -0.4875 & -0.25 \\ 0.25 & -0.4875 \end{bmatrix}$$

Having

$$\mathbf{Q}_{\Delta\Delta} = 2.2006\mathbf{I} \qquad \text{and} \qquad \mathbf{Q}_{\hat{x}\hat{x}} = 0.5908\mathbf{I}$$

then

$$\mathbf{Q}_{t't'} = \begin{bmatrix} 2.2006 & 0 & -0.4875 & -0.25 \\ & 2.2006 & 0.25 & -0.4875 \\ & & 0.5908 & 0 \\ \text{symmetric} & & & 0.5908 \end{bmatrix}$$

Cofactor propagation yields

$$\mathbf{Q}_{\hat{y}\hat{y}} = \mathbf{J}\mathbf{Q}_{\hat{x}}\,\mathbf{J}^t$$

where

$$\mathbf{J} = \begin{bmatrix} \hat{x}_1 & -\hat{x}_2 & \hat{a} & -\hat{b} \\ \hat{x}_2 & \hat{x}_1 & \hat{b} & \hat{a} \end{bmatrix} = \begin{bmatrix} 0 & -1.0 & 1.0 & -1.95 \\ 1.0 & 0 & 1.95 & 1.0 \end{bmatrix}$$

or

$$\mathbf{Q}_{\hat{y}\hat{y}} = \begin{bmatrix} 2.6368 & 0 \\ 0 & 2.6368 \end{bmatrix} = 2.6368\mathbf{I}$$

which is identical to the two answers already obtained.

General Discussion

This example has been treated extensively because several important aspects were demonstrated in the process of solving what appeared to be a relatively simple problem. In the course of this example a number of points should be recognized.

1. Either or both sets of coordinates may be taken as observations. In Example 6.2, only the **x** coordinates were considered as observations, but in

this example both sets were considered. A comparison of both examples would be useful for recognizing differences in procedure.

2. The technique of adjusting observations and parameters is more suitable for transformation problems than other techniques, particularly when all coordinates are taken as observations with cofactor matrices different from unit matrices.

3. If the a priori cofactor matrix of the observations is block diagonal (by point), then the normal equations can be formed in a summation process in which only the contribution of one point is computed at one time. This technique may lead to considerable savings in computer space and possibly computing time.

4. We should try to obtain the best approximations for the parameters that are practicable, without excessive computational effort. One practical way with transformation problems, and for that matter most other problems, is to perform a unique solution using a minimum data set. Having good approximations may result in having to iterate the solution only once or twice.

5. It is possible that the first set of corrections $\mathbf{\Delta}$ will be all that is needed to yield the required answer. In such cases it might happen that the magnitude of $\mathbf{\Delta}$ is large relative to the vector of approximations $\mathbf{x}^0$. Here we must be sure to update the inverse of the normal equations, $\mathbf{N}^{-1}$, using $\mathbf{x} = \mathbf{x}^0 + \mathbf{\Delta}$ in order to obtain the correct estimate of $\mathbf{Q}_{\hat{x}\hat{x}}$. If this is not done, the error could be considerable (200% in this example).

6. The computation of transformed coordinates brought to light several quite important points. For example, it is very important when propagating cofactors (or covariances) to determine *all* the random variables in the equations *and* to construct the correct cofactor (or covariance) matrix of the independent variables (that is, the variables on the right side of the equal sign). Of utmost importance here is the determination of correlation whenever it exists.

7. A fact, which is probably obvious but does merit mentioning, is that the estimated observations $\hat{l}$ satisfy the condition equations. In the case of transformation, realizing this fact facilitates considerably the computation of the final estimates of the coordinates of the reference points used in the transformation.

8. This last remark is perhaps the most important as it relates to the fundamental concepts of observations and the stochastic model. At the beginning of the example, the given coordinates were considered the observations with their given cofactor matrices. In the computations subsequent to estimating the transformation parameters, these parameter estimates ($\hat{a}$, $\hat{b}$) were then treated as observations, or stochastic variables. The same applies to the estimated observations as well. In these computations several quite different procedures were used to compute the same quantity of interest. The reason all the answers turned out to be identical is that in each case the

proper stochastic model was constructed and the resulting correct cofactor matrix used. Any deviation from that would have resulted in inconsistent answers. In brief, then, observations may be the actual numeric outputs of the operations we call "measurements," but may also be derived quantities. Whichever quantities we choose to term observations, we must set up the proper stochastic model that fits our chosen values.

Example 8.8. *Four-Parameter Transformation.* As before, the four-parameter transformation is given by

$$\begin{aligned} y_1 &= ax_1 - bx_2 + c \\ y_2 &= bx_1 + ax_2 + d \end{aligned} \tag{8.6}$$

This transformation is one of the most commonly used types in photogrammetry and allied areas. Under certain assumptions, we can apply a simplified approach to least squares adjustment by writing the conditions as

$$\begin{aligned} v_1 + ax_1 - bx_2 + c - y_1 &= 0 \\ v_2 + bx_1 + ax_2 + d - y_2 &= 0 \end{aligned}$$

or

$$\begin{bmatrix} v_1 \\ v_2 \end{bmatrix} + \begin{bmatrix} x_1 & -x_2 & 1 & 0 \\ x_2 & x_1 & 0 & 1 \end{bmatrix} \begin{bmatrix} a \\ b \\ c \\ d \end{bmatrix} - \begin{bmatrix} y_1 \\ y_2 \end{bmatrix} = 0$$

This is a pair of condition equations that arises for each point of known coordinates in both the **x** and **y** systems. Obviously two such points are necessary and sufficient to compute the four parameters of the transformation (a, b, c, and d) without adjustment. Whenever more than two points are given, a least squares adjustment becomes necessary.

Suppose that we have m points that are known in both coordinate systems. It can easily be shown that (see the pre Example, 8.7), in case $\mathbf{W} = \mathbf{I}$,

$$\begin{bmatrix} \sum_1^m (x_1^2 + x_2^2) & 0 & \sum_1^m x_1 & \sum_1^m x_2 \\ & \sum_1^m (x_1^2 + x_2^2) & -\sum_1^m x_2 & \sum_1^m x_1 \\ & & m & 0 \\ \text{symmetric} & & & m \end{bmatrix} \begin{bmatrix} a \\ b \\ c \\ d \end{bmatrix} = \begin{bmatrix} \sum_1^m (x_1 y_1 + x_2 y_2) \\ \sum_1^m (x_1 y_2 - x_2 y_1) \\ \sum_1^m y_1 \\ \sum_1^m y_2 \end{bmatrix} \tag{8.17}$$

From these four normal equations the four parameters may be obtained. In the preceding treatment no attempt was made to specify the observations. As a matter of fact, we simply attached a residual to each equation in a manner similar to that used in statistical regression problems. On analysis, however, we can see that each residual, v, is composed of two parts, one associated with a **y** coordinate and the other with the correspondingly "transformed" coordinate from the **x** system. We shall see subsequently how this type of adjustment compares to what we would normally do. But first let us note a computational simplification.

If in the **x** system, we would use the m points of the transformation and compute

$$\bar{x}_1 = \frac{\sum x_1}{m} \quad \bar{x}_2 = \frac{\sum x_2}{m}$$

and subtract $\bar{x}_1$ from x_1 coordinates, and $\bar{x}_2$ from x_2 coordinates, for *all* points in the **x** system, then the normal equations matrix above would reduce to a diagonal form. This is because in such a case

$$\sum_1^m x_1 = 0 \qquad \text{and} \qquad \sum_1^m x_2 = 0$$

Thus

$$\begin{aligned}
\hat{a} &= \frac{\sum_1^m (x_1 y_1 + x_2 y_2)}{\sum_1^m (x_1^2 + x_2^2)} \\
\hat{b} &= \frac{\sum_1^m (x_1 y_2 - x_2 y_1)}{\sum_1^m (x_1^2 + x_2^2)} \\
\hat{c} &= \frac{\sum_1^m y_1}{m} \\
\hat{d} &= \frac{\sum_1^m y_2}{m}
\end{aligned} \tag{8.18}$$

noting that x_1 and x_2 are referred to the *centroid* as an origin. In this case also, the parameters would be uncorrelated and have the diagonal cofactor matrix as follows:

$$\underset{4,4}{\mathbf{Q}_{\Delta\Delta}} = \mathbf{N}^{-1} = \begin{bmatrix} [\sum_1^m (x_1^2 + x_2^2)]^{-1} & & & \\ & [\sum_1^m (x_1^2 + x_2^2)]^{-1} & & \\ & & 1/m & \\ & & & 1/m \end{bmatrix} \tag{8.19}$$

The covariance matrix of the parameters can be computed if the a posteriori estimate of the reference variance is computed. To do this first compute the residuals

$$v_1 = y_1 - \hat{a}x_1 + \hat{b}x_2 - \hat{c}$$
$$v_2 = y_2 - \hat{b}x_1 - \hat{a}x_2 - \hat{d}$$

Then

$$\mathbf{v}^t\mathbf{W}\mathbf{v} = \mathbf{v}^t\mathbf{v} = \sum_1^m [(y_1 - \hat{a}x_1 + \hat{b}x_2 - \hat{c})^2 + (y_2 - \hat{b}x_1 - \hat{a}x_2 - \hat{d})^2]$$

and the reference variance $\hat{\sigma}_0^2 = \mathbf{v}^t\mathbf{v}/r$ where r = number of degrees of freedom = $2m - 4$.

This gives the following covariance matrix for the parameters:

$$\underset{4,4}{\Sigma_{\Delta\Delta}} = \frac{1}{r}\sum_{1}^{m}[(y_1 - ax_1 + bx_2 - c)^2 + (y_2 - bx_1 - ax_2 - d)^2] \times \begin{bmatrix} [\sum_1^m (x_1^2 + x_2^2)]^{-1} & & & \\ & [\sum_1^m (x_1^2 + x_2^2)]^{-1} & & \\ & & 1/m & \\ & & & 1/m \end{bmatrix} \tag{8.20}$$

The $\mathbf{y}$ coordinates of any point of given $\mathbf{x}$ coordinates may be computed from the transformation equations using the estimated parameters ($\hat{a}$, $\hat{b}$, $\hat{c}$, and $\hat{d}$). The corresponding covariance matrix of these coordinates may be computed by propagation techniques.

Returning now to the original transformation equations, rewrite them in functional form

$$f_1 = ax_1 - bx_2 + c - y_1 = 0$$
$$f_2 = bx_1 + ax_2 + d - y_2 = 0$$

As would normally be done we consider the general situation in which both the $\mathbf{x}$ and $\mathbf{y}$ coordinates are observations. Thus

$$\begin{bmatrix} a^0 & -b^0 & -1 & 0 \\ b^0 & a^0 & 0 & -1 \end{bmatrix}\begin{bmatrix} v_{x_1} \\ v_{x_2} \\ v_{y_1} \\ v_{y_2} \end{bmatrix} + \begin{bmatrix} x_1 & -x_2 & 1 & 0 \\ x_2 & x_1 & 0 & 1 \end{bmatrix}\begin{bmatrix} \delta_a \\ \delta_b \\ \delta_c \\ \delta_d \end{bmatrix} = \begin{bmatrix} y_1 - a^0x_1 + b^0x_2 - c^0 \\ y_2 - b^0x_1 - a^0x_2 - d^0 \end{bmatrix}$$

is the linearized form of the condition equations for one point. It should be noted that in the simplified case above the equations were directly linear, but when the regular adjustment treatment is pursued, the equations become nonlinear.

The first and simplest case to consider is when all coordinates (both $\mathbf{x}$ and $\mathbf{y}$) are uncorrelated and of equal precision, or when $\mathbf{W} = \mathbf{Q} = \mathbf{I}$. If we choose the first approximations to be $a^0 = b^0 = c^0 = d^0 = 0$, the matrix $(\mathbf{AQA}^t)^{-1}$ reduces to the identity matrix and $\mathbf{f} = [y_1\, y_2]^t$. In this situation the normal equations become identical to those given by equation (8.17). Taking, for the sake of simplicity of derivation, the case of referring the $\mathbf{x}$ coordinates to their centroid, the first solution becomes that given by equation (8.18); that is,

$$\begin{aligned} \delta a_1 &= \frac{\sum (x_1y_1 + x_2y_2)}{\sum (x_1^2 + x_2^2)} = a_1^0 \\ \delta b_1 &= \frac{\sum (x_1y_2 - x_2y_1)}{\sum (x_1^2 + x_2^2)} = b_1^0 \\ \delta c_1 &= \frac{\sum y_1}{m} = c_1^0 \\ \delta d_1 &= \frac{\sum y_2}{m} = d_1^0 \end{aligned} \tag{8.21}$$

because the first approximations were all zero. The second iteration can be performed in the following manner:

$$(\mathbf{AQA}')^{-1} = (1 + a_1^{02} + b_1^{02})^{-1}\mathbf{I} = \alpha\mathbf{I}$$

The scalar factor α is the *only change that will occur to the normal equations.* Next

$$\mathbf{t} = \mathbf{B}'(\mathbf{AQA}')^{-1}\mathbf{f} = \alpha\mathbf{B}'\mathbf{f}$$

Each element of **t** may be derived as follows:

$$\begin{aligned} t_1 &= \alpha \sum [x_1(y_1 - a_1^0 x_1 + b_1^0 x_2 - c_1^0)] \\ &\quad + \alpha \sum [x_2(y_2 - b_1^0 x_1 - a_1^0 x_2 - d_1^0)] \\ &= \alpha[\sum x_1 y_1 - a_1^0 \sum x_1^2 + b_1^0 \sum x_1 x_2 - c_1^0 \sum x_1] \\ &\quad + \alpha[\sum x_2 y_2 - b_1^0 \sum x_1 x_2 - a_1^0 \sum x_2^2 - d_1^0 \sum x_2] \end{aligned}$$

Since $\sum x_1 = 0$ and $\sum x_2 = 0$, and referring to equation (8.21), then

$$\begin{aligned} t_1 &= \alpha[\sum (x_1 y_1 + x_2 y_2) - a_1^0 \sum (x_1^2 + x_2^2)] = 0 \\ t_2 &= \alpha \sum [-x_2(y_1 - a_1^0 x_1 + b_1^0 x_2 - c_1^0)] \\ &\quad + \alpha \sum [x_1(y_2 - b_1^0 x_1 - a_1^0 x_2 - d_1^0)] \\ &= \alpha[-\sum x_2 y_1 + a_1^0 \sum x_1 x_2 - b_1^0 \sum x_2^2 + c_1^0 \sum x_2] \\ &\quad + \alpha[\sum x_1 y_2 - b_1^0 \sum x_1^2 - a_1^0 \sum x_1 x_2 - d_1^0 \sum x_1] \\ &= \alpha[\sum (x_1 y_2 - x_2 y_1) - b_1^0 \sum (x_1^2 + x_2^2)] = 0 \\ t_3 &= \alpha \sum (y_1 - a_1^0 x_1 + b_1^0 x_2 - c_1^0) \\ &= \alpha(\sum y_1 - a_1^0 \sum x_1 + b_1^0 \sum x_2 - mc_1^0) = \alpha(\sum y_1 - \sum y_1) = 0 \\ t_4 &= \alpha \sum [y_2 - b_1^0 x_1 - a_1^0 x_2 - d_1^0] = \alpha(\sum y_2 - md_1^0) = 0 \end{aligned}$$

Thus $\mathbf{t} = \mathbf{0}$ and $\mathbf{\Delta}_2 = \mathbf{0}$, which means that the final estimates of the parameters are those given by equation (8.21). To complete this section, we must evaluate the updated normal equations in order to have the proper cofactor matrix, or

$$\mathbf{Q}_{\Delta\Delta} = \frac{1}{\alpha}\mathbf{N}^{-1} \tag{8.22}$$

where $\mathbf{N}^{-1}$ is that given by equation (8.19).

At first glance this may lead us to believe that the simplified approach, although giving the exact parameter estimates as the regular approach, does not yield the *same cofactor matrix* for the parameters as exhibited by the factor $1/\alpha$ in equation (8.22). However, since cofactor matrices are related to the covariance matrices by scale factors and since the only difference between equations (8.22) and (8.19) is a scale factor ($1/\alpha$), it would be advisable to continue the derivation and obtain the covariance matrix of the parameters. Therefore the vector of residuals is

$$\mathbf{v} = \mathbf{QA}'(\mathbf{AQA}')^{-1}(-\mathbf{B\Delta} + \mathbf{f})$$

but $\mathbf{Q} = \mathbf{I}$, $(\mathbf{AQA}')^{-1} = \alpha\mathbf{I}$, and $\mathbf{\Delta} = \mathbf{0}$; then

$$\mathbf{v} = \alpha\mathbf{A}'\mathbf{f}$$

The quadratic form is therefore, $\mathbf{v}^t\mathbf{W}\mathbf{v} = \mathbf{v}^t\mathbf{v} = \alpha^2\mathbf{f}^t\mathbf{A}\mathbf{A}^t\mathbf{f}$ but, again $\mathbf{A}\mathbf{A}^t = 1/\alpha\mathbf{I}$ so

$$\mathbf{v}^t\mathbf{v} = \alpha\mathbf{f}^t\mathbf{f}$$

or

$$\mathbf{v}^t\mathbf{W}\mathbf{v} = \alpha \sum_1^m [(y_1 - \hat{a}x_1 + \hat{b}x_2 - \hat{c})^2 + (y_2 - \hat{b}x_1 - \hat{a}x_2 - \hat{d})^2]$$

With r as the number of degrees of freedom, the covariance matrix of the parameters becomes identical to that in equation (8.20) indicating that the *estimate of the covariance matrix of the parameters* obtained from the simplified treatment given at the beginning of this example is also the correct one. Consequently, such a simplified use of adjustment is totally practical and rigorous as long as the *observations, all of them, have a unit cofactor matrix.* The question then arises as to whether it also applies in other situations of cofactor matrices that are different from the identity matrix. The first of these would be a diagonal matrix, with the x's having the same precision but which is different from that for the y's, or

$$\underset{4,4}{\mathbf{Q}} = \begin{bmatrix} q_x & & & \\ & q_x & & \\ & & q_y & \\ & & & q_y \end{bmatrix} \quad \text{for each point}$$

Thus again if $a^0 = b^0 = c^0 = d^0 = 0$, then

$$(\mathbf{A}\mathbf{Q}\mathbf{A}^t)^{-1} = \begin{bmatrix} q_y & 0 \\ 0 & q_y \end{bmatrix} = \underset{2,2}{\mathbf{W}_{yy}} = q_y^{-1}\mathbf{I}$$

The normal equations matrix

$$\mathbf{N} = \mathbf{B}^t\mathbf{W}_{yy}\,\mathbf{B} = q_y^{-1}\begin{bmatrix} \sum(x_1^2 + x_2^2) & & & \\ & \sum(x_1^2 + x_2^2) & & \\ & & m & \\ & & & m \end{bmatrix} \tag{8.23}$$

and

$$\mathbf{t} = \mathbf{B}^t\mathbf{W}_{yy}\,\mathbf{f} = q_y^{-1}\begin{bmatrix} \sum(x_1y_1 + x_2y_2) \\ \sum(x_1y_2 - x_2y_1) \\ \sum y_1 \\ \sum y_2 \end{bmatrix}$$

The solution of this system is identical to that given by equation (8.21). The second iteration must then be performed.

$$(\mathbf{A}\mathbf{Q}\mathbf{A}^t)^{-1} = \left\{ \begin{bmatrix} a^0 & -b^0 & -1 & 0 \\ b^0 & a^0 & 0 & -1 \end{bmatrix} \begin{bmatrix} q_x & & & \\ & q_x & & \\ & & q_y & \\ & & & q_y \end{bmatrix} \begin{bmatrix} a^0 & b^0 \\ -b^0 & a^0 \\ -1 & 0 \\ 0 & -1 \end{bmatrix} \right\}^{-1}$$

$$= \begin{bmatrix} (a^{02} + b^{02})q_x + q_y & 0 \\ 0 & (a^{02} + b^{02})q_x + q_y \end{bmatrix}^{-1} = \beta\mathbf{I}$$

where

$$\beta = [(a^{02} + b^{02})q_x + q_y]^{-1}$$

The normal equation matrix will change only by being multiplied by the scalar β. The constant term vector will however be

$$\mathbf{t} = \mathbf{B}'\mathbf{W}\mathbf{f} = \beta\mathbf{B}'\mathbf{f}$$

Since this is, to a scalar variation, the same as that computed previously, it follows that $\mathbf{t} = \mathbf{0}$ and the second correction vector is also zero. Thus the estimates of the parameters are identical to those given by equation (8.21). The cofactor matrix of the parameters can be shown to be

$$\mathbf{Q}_{\Delta\Delta} = q_y\beta^{-1}\mathbf{N}^{-1}$$

where $\mathbf{N}$ is computed from equation (8.23). Finally, in order to compute the covariance matrix $\mathbf{\Sigma}_{\Delta\Delta}$ then

$$\mathbf{v} = \mathbf{QA}'(\mathbf{AQA}')^{-1}(-\mathbf{B\Delta} + \mathbf{f}) = \beta\mathbf{QA}'\mathbf{f}$$

and

$$\begin{aligned}\mathbf{v}'\mathbf{W}\mathbf{v} &= \beta^2\mathbf{f}'\mathbf{AQWQA}'\mathbf{f} = \beta^2\mathbf{f}'(\mathbf{AQA}')\mathbf{f} = \beta\mathbf{f}'\mathbf{f} \\ &= \beta\sum[(y_1 - \hat{a}x_1 + \hat{b}x_2 - \hat{c})^2 + (y_2 - \hat{b}x_1 - \hat{a}x_2 - \hat{d})^2]\end{aligned}$$

The covariance matrix of the parameters $\mathbf{\Sigma}_{\Delta\Delta}$ is thus identical to that of equation (8.20).

Clearly the simplified adjustment is still rigorous and practical in the case in which $\mathbf{Q}_{xx}$ is different from $\mathbf{Q}_{yy}$ and each is a scalar matrix.

The next case would be no correlation, but each coordinate in each coordinate system has a different cofactor, and that each point has the same cofactor matrix, or,

$$\mathbf{Q} = \text{diag.}\,\{q_{x1}, q_{x2}, q_{y1}, q_{y2}\} \qquad \textit{for each point}$$

If the first set of approximations are again taken to be $a^0 = b^0 = c^0 = d^0 = 0$,

$$(\mathbf{AQA}')^{-1} = \begin{bmatrix} q_{y1} & 0 \\ 0 & q_{y2} \end{bmatrix}^{-1} = \begin{bmatrix} w_{y1} & 0 \\ 0 & w_{y2} \end{bmatrix} = \mathbf{W}_{yy}$$

$$\mathbf{N} = \mathbf{B}'\mathbf{W}_{yy}\,\mathbf{B} = \begin{bmatrix} \sum(w_{y_1}x_1^2 + w_{y_2}x_2^2) & & & \\ & \sum(w_{y_1}x_1^2 + w_{y_2}x_2^2) & & \\ & & mw_{y_1} & \\ & & & mw_{y_2} \end{bmatrix}$$

and

$$\mathbf{t} = \mathbf{B}'\mathbf{W}_y\,\mathbf{f} = \begin{bmatrix} \sum(w_{y_1}x_1y_1 + w_{y_2}x_2y_2) \\ \sum(-w_{y_1}x_2y_1 + w_{y_2}x_1y_2) \\ \sum w_{y_1}y_1 \\ \sum w_{y_2}y_2 \end{bmatrix}$$

Then

$$a_1^0 = \frac{\sum (w_{y1}x_1y_1 + w_{y2}x_2y_2)}{\sum (w_{y1}x_1^2 + w_{y2}x_2^2)}$$

$$b_1^0 = \frac{\sum (-w_{y1}x_2y_1 + w_{y2}x_1y_2)}{\sum (w_{y1}x_1^2 + w_{y2}x_2^2)}$$

$$c_1^0 = \frac{\sum w_{y1}y_1}{mw_{y1}} = \frac{\sum y_1}{m}$$

$$d_1^0 = \frac{\sum w_{y2}y_2}{mw_{y2}} = \frac{\sum y_2}{m}$$

Second iteration

$$(\mathbf{AQA}^t)^{-1} = \left\{ \begin{bmatrix} a_1^0 & -b_1^0 & -1 & 0 \\ b_1^0 & a_1^0 & 0 & -1 \end{bmatrix} \begin{bmatrix} q_{x1} & & & \\ & q_{x2} & & \\ & & q_{y1} & \\ & & & q_{y2} \end{bmatrix} \begin{bmatrix} a_1^0 & b_1^0 \\ -b_1^0 & a_1^0 \\ -1 & 0 \\ 0 & -1 \end{bmatrix} \right\}^{-1}$$

$$= \begin{bmatrix} \gamma_{11} & \gamma_{12} \\ \gamma_{12} & \gamma_{22} \end{bmatrix} \tag{8.24}$$

From equation (8.24) it becomes obvious that the situation is no longer simple and that iterating the solution is necessary. Consequently, we can no longer use the simplified treatment and the solution must be carried out rigorously and iterated sufficiently. Of course, if we do not already have a program to handle the general case ($\mathbf{Av} + \mathbf{B\Delta} = \mathbf{f}$), or if the transformation is a part of a more elaborate program, a simpler computational approach may be effected using the relations above. After the iterations are terminated, the cofactor matrix of the parameters will be the inverse of the last normal equations' coefficient matrix.

Other cases that may be considered for four-parameter transformation would be those involving correlation. First, there would be cases in which $\mathbf{Q}_{xx}$ and $\mathbf{Q}_{yy}$ are both full matrices, but $\mathbf{Q}_{xy} = \mathbf{0}$, meaning no correlation between the **x** coordinates and **y** coordinates. Finally, there would be the case in which **Q** is, in principle, a full matrix. All such cases become rather too involved to analyze by algebraic derivations as was done above. Instead, it would be by far better to perform the adjustment by rigorous and direct means using conditions of the form $\mathbf{Av} + \mathbf{B\Delta} = \mathbf{f}$.

Before leaving the subject of four-parameter transformation, we should point out one of the common cases that often arises in practice. It is the case when the **y** coordinates are of such a superior quality that they may be considered as constants in the adjustment. Although we can derive the appropriate relations from the start by applying the above assumption (something that is left to the reader as an exercise), the case can be easily accommodated with what we have presented so far. Instead of strictly enforcing the **y** coordinates to be constants, we introduce them as observations, but with *very small cofactors* compared to the **x** coordinates. This has practically the same effect, and no individual treatment is therefore necessary. More elaboration on this aspect will be given under the unified approach to adjustment in Chapter 12, Part III.

Example 8.9. *Six-Parameter Transformation.* The basic equations are

$$y_1 = ax_1 + bx_2 + c$$

$$y_2 = dx_1 + ex_2 + f$$

One of the simplest cases is that in which *only* the **y** coordinates are considered as the observations, with unit cofactor matrices. The adjustment problem is then linear and the condition equations for one point are

$$\begin{bmatrix} v_1 \\ v_2 \end{bmatrix} + \begin{bmatrix} -x_1 & -x_2 & -1 & 0 & 0 & 0 \\ 0 & 0 & 0 & -x_1 & -x_2 & -1 \end{bmatrix} \begin{bmatrix} a \\ b \\ c \\ d \\ e \\ f \end{bmatrix} = \begin{bmatrix} -y_1 \\ -y_2 \end{bmatrix} \tag{8.25}$$

Assuming that the **x** coordinates are shifted to their centroid as origin the normal equations take the form

$$\left[\begin{array}{ccc|ccc} \sum x_1^2 & \sum x_1 x_2 & 0 & 0 & 0 & 0 \\ & \sum x_2^2 & 0 & 0 & 0 & 0 \\ & & m & 0 & 0 & 0 \\ \hline & & & \sum x_1^2 & \sum x_1 x_2 & 0 \\ & & & & \sum x_2^2 & 0 \\ \text{symmetric} & & & & & m \end{array}\right] \begin{bmatrix} a \\ b \\ c \\ d \\ e \\ f \end{bmatrix} = \left[\begin{array}{c} \sum x_1 y_1 \\ \sum x_2 y_1 \\ \sum y_1 \\ \hline \sum x_1 y_2 \\ \sum x_2 y_2 \\ \sum y_2 \end{array}\right] \tag{8.26}$$

An interesting observation can be made regarding equation (8.26), which is that the normal equations' coefficient matrix can be divided into two *distinct and identical* 3 × 3 matrices. The reason is that each of the two condition equations contains three parameters that are completely different from the three appearing in the other equation. This leads to an economical way of solving the normal equations, as equation (8.26) may be rewritten as

$$\begin{bmatrix} \sum x_1^2 & \sum x_1 x_2 & 0 \\ & \sum x_2^2 & 0 \\ \text{symmetric} & & m \end{bmatrix} \left[\begin{array}{c|c} a & d \\ b & e \\ c & f \end{array}\right] = \left[\begin{array}{c|c} \sum x_1 y_1 & \sum x_1 y_2 \\ \sum x_2 y_1 & \sum x_2 y_2 \\ \sum y_1 & \sum y_2 \end{array}\right] \tag{8.27}$$

The form of equation (8.27) requires the inversion of only a 3 × 3 matrix, and the inverse when multiplied by the right-hand side yields all six parameters in the form of two columns of three elements each. Such a scheme can be used under the same assumptions of observations and cofactor matrices to treat transformation equations of many parameters as long as such parameters are distinctly divided between the two condition equations. A good example of this is given by equation (8.9) where 12 or more parameters may be involved. Another example, which requires the extension of the technique to three equal groups of parameters, is found in the adjustment of transformation equations given by equation (8.12).

Returning to a more general case in which the **x** coordinates are also considered as observations, the problem becomes nonlinear. The linearized condition equations become

$$\begin{bmatrix} a^0 & b^0 & -1 & 0 \\ d^0 & e^0 & 0 & -1 \end{bmatrix} \begin{bmatrix} v_{x1} \\ v_{x2} \\ v_{y1} \\ v_{y2} \end{bmatrix} + \begin{bmatrix} x_1 & x_2 & 1 & 0 & 0 & 0 \\ 0 & 0 & 0 & x_1 & x_2 & 1 \end{bmatrix} \begin{bmatrix} \delta_a \\ \delta_b \\ \delta_c \\ \delta_d \\ \delta_e \\ \delta_f \end{bmatrix}$$

$$= \begin{bmatrix} y_1 - a^0 x_1 - b^0 x_2 - c^0 \\ y_2 - d^0 x_1 - e^0 x_2 - f^0 \end{bmatrix} \tag{8.28}$$

Even under the assumption of unit cofactor matrices, the matrix

$$(\mathbf{AQA}^t)^{-1} = \begin{bmatrix} (1 + a^{02} + b^{02}) & (a^0 d^0 + b^0 e^0) \\ (a^0 d^0 + b^0 e^0) & (1 + d^{02} + e^{02}) \end{bmatrix}^{-1}$$

does not reduce to a simple enough form. Consequently, even though the first iteration solution is identical to that obtained above when $a^0 = b^0 = c^0 = d^0 = e^0 = f^0 = 0$, the second and subsequent iterations must be performed until the solution converges. A less than rigorous solution would be obtained if equation (8.25) were used instead of equation (8.28). Of course this deviation from rigor increases as the cofactor matrix of the observations increases in complexity from an identity matrix to a full matrix.

Example 8.10. *Seven-Parameter Transformation.* A seven-parameter transformation is a conformal transformation between two 3-dimensional coordinate systems allowing for a uniform scale change, an orthogonal rotation matrix that rotates the space solidly, and a vector of three mutually orthogonal shifts. Therefore it preserves shape (particularly angles) and size to the extent of the uniform scale change. It is normally written as

$$\underset{3,1}{\mathbf{y}} = s\underset{3,3}{\mathbf{M}}\ \underset{3,1}{\mathbf{x}} + \underset{3,1}{\mathbf{k}}$$

which was given by equation (8.11) and explained thereafter. It is important to recall that **M** is an orthogonal matrix or that $\mathbf{MM}^t = \mathbf{I}$. Rewriting that equation for one point in a functional form

$$s\mathbf{Mx} + \mathbf{k} - \mathbf{y} = \mathbf{0} \tag{8.29}$$

and considering, for the sake of generality, both **x** and **y** as observations, we linearize equation (8.29) as $\mathbf{Av} + \mathbf{B\Delta} = \mathbf{f}$ where

$$\underset{3,6}{\mathbf{A}} = \begin{bmatrix} \underset{3,3}{s^0\mathbf{M}^0} & \underset{3,3}{-\mathbf{I}} \end{bmatrix}$$

$$\underset{6,1}{\mathbf{v}} = \begin{bmatrix} \underset{3,1}{\mathbf{v}_x} \\ \underset{3,1}{\mathbf{v}_y} \end{bmatrix} \qquad \underset{3,7}{\mathbf{B}} = \begin{bmatrix} \underset{3,1}{\mathbf{M}^0\mathbf{x}} & \underset{3,1}{s^0\mathbf{M}_1^0\mathbf{x}} & \underset{3,1}{s^0\mathbf{M}_2^0\mathbf{x}} & \underset{3,1}{s^0\mathbf{M}_3^0\mathbf{x}} & \underset{3,3}{\mathbf{I}} \end{bmatrix}$$

in which $\mathbf{M}_1$, $\mathbf{M}_2$, and $\mathbf{M}_3$ are the partial derivatives of **M** with respect to each of the three independent parameters, respectively.

$$\underset{7,1}{\mathbf{\Delta}} = [\delta s \quad \delta_1 \quad \delta_2 \quad \delta_3 \quad \delta k_1 \quad \delta k_2 \quad \delta k_3]^t$$

in which δ_1, δ_2, and δ_3 designate corrections to approximations of the three parameters of **M**, respectively. Consider first the simplest case when $\mathbf{Q} = \mathbf{W} = \mathbf{I}$.

Examining the term $\mathbf{Q}_e = \mathbf{AQA}^t$ gives

$$\underset{3,3}{\mathbf{Q}_e} = \mathbf{AQA}^t = \mathbf{AA}^t = [s^0\mathbf{M}^{0t} \; -\mathbf{I}] \begin{bmatrix} s^0\mathbf{M}^{0t} \\ -\mathbf{I} \end{bmatrix} = s^{02}\mathbf{I} + \mathbf{I}$$

or

$$\mathbf{Q}_e = (1 + s^{02})\mathbf{I} \qquad \text{and} \qquad \mathbf{W}_e = (1 + s^{02})^{-1}\mathbf{I} = \alpha\mathbf{I}$$

The contribution from one point to the normal equations' matrix and constant vector becomes

$$\mathbf{N}_i = \mathbf{B}_i^t \mathbf{W}_{e_i} \mathbf{B}_i = \alpha \mathbf{B}_i^t \mathbf{B}_i \qquad \text{and} \qquad \mathbf{t}_i = \alpha \mathbf{B}_i^t \mathbf{f}_i$$

from which

$$\mathbf{\Delta} = \left[\sum_i (\mathbf{B}^t\mathbf{B})_i\right]^{-1} \left[\sum_i (\mathbf{B}^t\mathbf{f}_i)\right]$$

which is the solution that would be obtained if we simply wrote the equations in the form $\mathbf{v} + \mathbf{B\Delta} = \mathbf{f}$. In both cases, however, iteration is a necessary process because the equations are nonlinear in the parameters. The scalar α, although having no effect on the computation of $\mathbf{\Delta}$, should be evaluated in order to compute the cofactor matrix of the parameters. It depends only on the scale factor s and can therefore be computed only once after the iterative procedure is ended. Thus

$$\mathbf{Q}_{\Delta\Delta} = \alpha^{-1}\left[\sum_i (\mathbf{B}_i^t \mathbf{B}_i)\right]^{-1} = (1 + s^{02})\left[\sum_i (\mathbf{B}_i^t \mathbf{B}_i)\right]^{-1}$$

In order to compute the corresponding covariance matrix, we need to evaluate the a posteriori value of the reference variance.

$$\mathbf{v} = \mathbf{QA}^t(\mathbf{AQA}^t)^{-1}(-\mathbf{B\Delta} + \mathbf{f}) = \alpha\mathbf{A}^t(-\mathbf{B\Delta} + \mathbf{f})$$

If $\mathbf{\Delta}$ is practically zero at the end of iterations, then $\mathbf{v} = \alpha\mathbf{A}^t\mathbf{f}$ and

$$\mathbf{v}^t\mathbf{W}\mathbf{v} = \alpha^2\mathbf{f}^t\mathbf{AA}^t\mathbf{f} = \alpha\mathbf{f}^t\mathbf{f}, \qquad \hat{\sigma}_0^2 = \alpha\mathbf{f}^t\mathbf{f}/r$$

and

$$\Sigma_{\Delta\Delta} = \hat{\sigma}_0^2 \mathbf{Q}_{\Delta\Delta} = \alpha\alpha^{-1} \frac{\mathbf{f}^t\mathbf{f}}{r} \left[\sum_i (\mathbf{B}^t\mathbf{B})_i\right]^{-1}$$

$$\Sigma_{\Delta\Delta} = \frac{1}{r} \mathbf{f}^t\mathbf{f} \left[\sum_i (\mathbf{B}^t\mathbf{B})_i\right]^{-1}$$

which is an interesting result because it is independent of α and would be directly obtainable from the case of $\mathbf{v} + \mathbf{B}\Delta = \mathbf{f}$.

The second case to discuss is when $\mathbf{Q}_{xx} = a\mathbf{I}$ and $\mathbf{Q}_{yy} = b\mathbf{I}$ for each point used in the transformation. Again, first evaluate the term $\mathbf{Q}_e$ for each point

$$\mathbf{Q}_e = \mathbf{A}\mathbf{Q}\mathbf{A}^t = [s^0\mathbf{M}^0 \; -\mathbf{I}] \begin{bmatrix} a\mathbf{I} & \mathbf{0} \\ 0 & b\mathbf{I} \end{bmatrix} \begin{bmatrix} s^0\mathbf{M}^{0t} \\ -\mathbf{I} \end{bmatrix}$$

$$= as^{02}\mathbf{I} + b\mathbf{I} = (as^{02} + b)\mathbf{I}$$

$$\mathbf{W}_e = (as^{02} + b)^{-1}\mathbf{I} = \beta\mathbf{I}$$

The contribution of one point to the normal equation is

$$\mathbf{N}_i = \beta(\mathbf{B}^t\mathbf{B})_i \qquad \text{and} \qquad \mathbf{t}_i = \beta(\mathbf{B}^t\mathbf{f})_i$$

and thus

$$\Delta = \left[\sum_i (\mathbf{B}^t\mathbf{B})_i\right]^{-1} \left[\sum_i (\mathbf{B}^t\mathbf{f})_i\right]$$

which is the same as the simplified solution from $\mathbf{v} + \mathbf{B}\Delta = \mathbf{f}$. Next

$$\mathbf{Q}_{\Delta\Delta} = \mathbf{N}^{-1} = \beta^{-1} \left[\sum_i (\mathbf{B}^t\mathbf{B})_i\right]^{-1}$$

and

$$\mathbf{v} = \mathbf{Q}\mathbf{A}^t\mathbf{W}_e(-\mathbf{B}\Delta + \mathbf{f}) = \beta\mathbf{Q}\mathbf{A}^t\mathbf{f}$$

$$\mathbf{v}^t\mathbf{W}\mathbf{v} = \beta^2\mathbf{f}^t\mathbf{A}\mathbf{Q}\mathbf{W}\mathbf{Q}\mathbf{A}^t\mathbf{f} = \beta\mathbf{f}^t\mathbf{f}$$

$$\hat{\sigma}_0^2 = \frac{1}{r}\beta\mathbf{f}^t\mathbf{f}$$

$$\Sigma_{\Delta\Delta} = \hat{\sigma}_0^2 \, \mathbf{Q}_{\Delta\Delta} = \frac{1}{r} \mathbf{f}^t\mathbf{f} \left[\sum_i (\mathbf{B}^t\mathbf{B})_i\right]^{-1}$$

Another case would be when

$$\mathbf{Q} = \text{diag.}\, \{q_{x1}, q_{x2}, q_{x3}, q_{y1}, q_{y2}, q_{y3}\} = \begin{bmatrix} \mathbf{Q}_{xx} & \mathbf{0} \\ \mathbf{0} & \mathbf{Q}_{yy} \end{bmatrix}$$

$$\mathbf{Q}_e = \mathbf{A}\mathbf{Q}\mathbf{A}^t = [s^0\mathbf{M}^0 \quad \mathbf{I}] \begin{bmatrix} \mathbf{Q}_{xx} & \mathbf{0} \\ \mathbf{0} & \mathbf{Q}_{yy} \end{bmatrix} \begin{bmatrix} s^0\mathbf{M}^{0t} \\ -\mathbf{I} \end{bmatrix}$$

$$= s^{02}\mathbf{M}^0\mathbf{Q}_{xx}\,\mathbf{M}^{0t} + \mathbf{Q}_{yy}$$

which is a matrix that is not necessarily diagonal.

Consequently, from this case on (that is, with more general cofactor matrices), we should not bother with seeking simpler approaches and it would be advisable to apply directly the general case of $\mathbf{Av} + \mathbf{B\Delta} = \mathbf{f}$.

For those readers who might want to program this transformation, a little more detail is given about the orthogonal matrix $\mathbf{M}$. One way of constructing it is in the form of three sequential rotation matrices, or $\mathbf{M} = \mathbf{M}_\kappa \mathbf{M}_\phi \mathbf{M}_\omega$ where

$$\mathbf{M}_\omega = \begin{bmatrix} 1 & 0 & 0 \\ 0 & \cos\omega & \sin\omega \\ 0 & -\sin\omega & \cos\omega \end{bmatrix} \quad \text{rotation about } x_1 \text{ axis}$$

$$\mathbf{M}_\phi = \begin{bmatrix} \cos\phi & 0 & -\sin\phi \\ 0 & 1 & 0 \\ +\sin\phi & 0 & \cos\phi \end{bmatrix} \quad \text{rotation about once-rotated } x_2 \text{ axis}$$

$$\mathbf{M}_\kappa = \begin{bmatrix} \cos\kappa & +\sin\kappa & 0 \\ -\sin\kappa & \cos\kappa & 0 \\ 0 & 0 & 1 \end{bmatrix} \quad \text{rotation about twice-rotated } x_3 \text{ axis}$$

The partial derivatives of $\mathbf{M}$ with respect to ω, ϕ, and κ, respectively, can be shown to be

$$\mathbf{M}_1 = \frac{\partial \mathbf{M}}{\partial \omega} = \mathbf{M}_\kappa \mathbf{M}_\phi \frac{\partial \mathbf{M}_\omega}{\partial \omega} = \mathbf{M} \begin{bmatrix} 0 & 0 & 0 \\ 0 & 0 & +1 \\ 0 & -1 & 0 \end{bmatrix}$$

$$\mathbf{M}_2 = \frac{\partial \mathbf{M}}{\partial \phi} = \mathbf{M}_\kappa \frac{\partial \mathbf{M}_\phi}{\partial \phi} \mathbf{M}_\omega = \mathbf{M} \begin{bmatrix} 0 & \sin\omega & \cos\omega \\ -\sin\omega & 0 & 0 \\ \cos\omega & 0 & 0 \end{bmatrix}$$

$$= \begin{bmatrix} 0 & 0 & -\cos\kappa \\ 0 & 0 & \sin\kappa \\ +\cos\kappa & -\sin\kappa & 0 \end{bmatrix} \mathbf{M}$$

$$\mathbf{M}_3 = \frac{\partial \mathbf{M}}{\partial \kappa} = \frac{\partial \mathbf{M}_\kappa}{\partial \kappa} \mathbf{M}_\phi \mathbf{M}_\omega = \begin{bmatrix} 0 & +1 & 0 \\ -1 & 0 & 0 \\ 0 & 0 & 0 \end{bmatrix} \mathbf{M}$$

9

Least Squares Adjustment with Conditions and Constraints

(Adjustment with Functionally Dependent Parameters)

9.1. INTRODUCTION

A constraint equation has been defined as that which relates only parameters to each other. Its presence in the functional model implies that the parameters are functionally dependent.

Parameter constraints occur in practice when some or all of the parameters in the adjustment must conform to some relationships arising from either geometric or physical characteristics of the model. For example, in photogrammetry, points on a lake shore can be constrained to have the same elevation, which may or may not be known; points chosen on a straight portion of a railroad are constrained to lie on a straight line; two points in the photogrammetric model may be constrained to known distance; camera stations from an orbital vehicle may be constrained to an orbit; and so on. The functional dependence of the parameters leads to having as many dependent parameters as there are constraint equations. This can be explained in the following manner.

First, from n observations and n_0 minimum number of model variables, the redundancy r is determined for which r conditions may be formulated among the observations. If we now wish to include u' *functionally independent* parameters, the number of conditions increases accordingly to $(r + u')$ with a maximum number of n. Progressing further we would like to accept the possibility of *functionally dependent* parameters in the model. Thus if $(u' + 1)$ parameters are considered, then one of them will be dependent on the u' independent parameters. For the parameter, one *constraint equation* must be written to reflect that dependency. In general, if as usual there are u parameters $(u > u')$, some of which are functionally dependent, then s constraints must be written to account for (and be equal to) the number of dependent parameters, or $s = u - u'$. Evidently s must always be less than u, or the constraints become an inconsistent set of equations,

$$s < u \tag{9.1}$$

Since the s constraints are equal to the number of dependent parameters, it follows from equation (6.1) that

$$c + s = r + u \tag{9.2}$$

or the number of conditions plus constraints must be equal to the redundancy plus the total number of (dependent) parameters.

The number of conditions still must not be larger than the number of observations as given by equation (6.2). However, the limitations placed on the number of parameters in equation (6.3) when the parameters are functionally independent do not hold here. In other words, there is essentially no upper limit on the number of parameters in this case, as u can readily exceed n_0, *provided* the proper number of constraints is formulated according to equation (9.2).

9.2. GENERAL CASE FOR ADJUSTMENT WITH CONDITIONS AND CONSTRAINTS

The two sets of equations, conditions and constraints, are

$$\underset{c,n}{\mathbf{A}}\ \underset{n,1}{\mathbf{v}} + \underset{c,u}{\mathbf{B}}\ \underset{u,1}{\mathbf{\Delta}} = \underset{c,1}{\mathbf{f}} \tag{9.3a}$$

$$\underset{s,u}{\mathbf{C}}\ \underset{u,1}{\mathbf{\Delta}} = \underset{s,1}{\mathbf{g}} \tag{9.3b}$$

with

$$\text{rank}\ (\mathbf{A}) = c \qquad \text{rank}\ (\mathbf{B}) = u \qquad \text{rank}\ (\mathbf{C}) = s \tag{9.3c}$$

In the presence of constraints it is, in general, possible that $u > c$ leading to rank of $\mathbf{B}$ being c. In such a situation, part of the development to follow (solution by partitioning) would not work. Here $u < c$ is selected and the other possibility (for $u > c$) is addressed later as a special case. Two different procedures are possible and each is presented in a separate section.

9.2.1. Direct Treatment Since there are two separate sets of equations, then two sets or vectors of Lagrange multipliers

$$\underset{n,1}{\mathbf{k}} \quad \text{and} \quad \underset{s,1}{\mathbf{k}_c}$$

are needed in order to use the "constrained minima" technique. The symbol **k** without subscript is still retained to mean that for the conditions. Consequently, the quadratic form to be minimized is

$$\begin{aligned} \phi &= \mathbf{v}^t\mathbf{W}\mathbf{v} - 2\mathbf{k}^t(\mathbf{A}\mathbf{v} + \mathbf{B}\boldsymbol{\Delta} - \mathbf{f}) - 2\mathbf{k}_c^t(\mathbf{C}\boldsymbol{\Delta} - \mathbf{g}) \\ &= \text{minimum} \end{aligned} \tag{9.4}$$

Equation (9.4) contains two free variable vectors, **v** and **Δ**, thus

$$\frac{\partial \phi}{\partial \mathbf{v}} = 2\mathbf{v}^t\mathbf{W} - 2\mathbf{k}^t\mathbf{A} = \mathbf{0}^t$$

$$\frac{\partial \phi}{\partial \boldsymbol{\Delta}} = -2\mathbf{k}^t\mathbf{B} - 2\mathbf{k}_c^t\,\mathbf{C} = \mathbf{0}^t$$

or

$$-\mathbf{W}\mathbf{v} + \mathbf{A}^t\mathbf{k} = \mathbf{0} \tag{9.5}$$

$$\mathbf{B}^t\mathbf{k} + \mathbf{C}^t\mathbf{k}_c = \mathbf{0} \tag{9.6}$$

The relations (9.5) and (9.6) represent a set of $(n + u)$ linear equations which when combined with equations (9.3a) and (9.3b) give a total of $(n + c + u + s)$ equations in as many unknowns. The unknown vectors are

$$\underset{n,1}{\mathbf{v}}, \quad \underset{c,1}{\mathbf{k}}, \quad \underset{u,1}{\boldsymbol{\Delta}}, \quad \text{and} \quad \underset{s,1}{\mathbf{k}_c}$$

The four sets of equations combined constitute the total system of normal equations. The least squares problem is resolved by solving that system, or

$$\begin{bmatrix} \mathbf{v} \\ \mathbf{k} \\ \boldsymbol{\Delta} \\ \mathbf{k}_c \end{bmatrix} = \begin{bmatrix} -\mathbf{W} & \mathbf{A}^t & \mathbf{0} & \mathbf{0} \\ \mathbf{A} & \mathbf{0} & \mathbf{B} & \mathbf{0} \\ \mathbf{0} & \mathbf{B}^t & \mathbf{0} & \mathbf{C}^t \\ \mathbf{0} & \mathbf{0} & \mathbf{C} & \mathbf{0} \end{bmatrix}^{-1} \begin{bmatrix} \mathbf{0} \\ \mathbf{f} \\ \mathbf{0} \\ \mathbf{g} \end{bmatrix} \tag{9.7}$$

This matrix has a rank of $(n + c + u + s)$ and is therefore nonsingular (assuming of course that the model is properly constructed). Difficulties in inverting the coefficient matrix would be experienced only if the mathematical model is improperly constructed. Therefore the functional model must be correctly established and the proper conditional and constraint equations written. (Of course unusual cases may exist in which the coefficient matrix is ill conditioned.)

In general, the inversion in equation (9.7) could involve quite a large matrix. Of the four vectors computed from that equation only $\mathbf{\Delta}$, and perhaps $\mathbf{v}$, may be of direct interest. Therefore, for many problems a solution by partitioning is more appropriate.

Thus from equations (9.5), (9.3a), and (9.6), we get, respectively,

$$\mathbf{v} = \mathbf{W}^{-1}\mathbf{A}^t\mathbf{k} = \mathbf{Q}\mathbf{A}^t\mathbf{k} \tag{9.8}$$

$$\begin{aligned} \mathbf{k} &= (\mathbf{A}\mathbf{Q}\mathbf{A}^t)^{-1}(-\mathbf{B}\mathbf{\Delta} + \mathbf{f}) \\ &= \mathbf{W}_e(-\mathbf{B}\mathbf{\Delta} + \mathbf{f}) \end{aligned} \tag{9.9}$$

$$-(\mathbf{B}^t\mathbf{W}_e\,\mathbf{B})\mathbf{\Delta} + \mathbf{B}^t\mathbf{W}_e\,\mathbf{f} + \mathbf{C}^t\mathbf{k}_c = \mathbf{0}$$

or

$$-\mathbf{N}\mathbf{\Delta} + \mathbf{C}^t\mathbf{k}_c = -\mathbf{t} \tag{9.10}$$

Solving for $\mathbf{\Delta}$ and $\mathbf{k}_c$ from equations (9.10) and (9.3b) gives

$$\mathbf{\Delta} = \mathbf{N}^{-1}(\mathbf{t} + \mathbf{C}^t\mathbf{k}_c) = \mathbf{N}^{-1}\mathbf{t} + \mathbf{N}^{-1}\mathbf{C}^t\mathbf{k}_c = \mathbf{\Delta}^0 + \boldsymbol{\delta}\mathbf{\Delta} \tag{9.11}$$

$$\mathbf{k}_c = (\mathbf{C}\mathbf{N}^{-1}\mathbf{C}^t)^{-1}(\mathbf{g} - \mathbf{C}\mathbf{N}^{-1}\mathbf{t}) = \mathbf{M}^{-1}(\mathbf{g} - \mathbf{C}\mathbf{\Delta}^0) \tag{9.12}$$

where

$$\mathbf{M} = (\mathbf{C}\mathbf{N}^{-1}\mathbf{C}^t), \qquad \mathbf{\Delta}^0 = \mathbf{N}^{-1}\mathbf{t}, \qquad \boldsymbol{\delta}\mathbf{\Delta} = \mathbf{N}^{-1}\mathbf{C}^t\mathbf{k}_c \tag{9.13}$$

The inverse of $\mathbf{M}$ exists because its rank is equal to its order s.

The vector $\mathbf{\Delta}^0$ represents the solution for the parameter vector in the absence of constraints. The final parameter vector is therefore

$$\begin{aligned} \mathbf{\Delta} &= \mathbf{\Delta}^0 + \mathbf{N}^{-1}\mathbf{C}^t(\mathbf{C}\mathbf{N}^{-1}\mathbf{C}^t)^{-1}(\mathbf{g} - \mathbf{C}\mathbf{\Delta}^0) \\ &= \mathbf{\Delta}^0 + \mathbf{N}^{-1}\mathbf{C}^t\mathbf{M}^{-1}(\mathbf{g} - \mathbf{C}\mathbf{\Delta}^0) \end{aligned} \tag{9.14}$$

which is the u partially *reduced normal equations*, assuming that $\mathbf{\Delta}$ is the unknown vector of primary importance.

Several observations can be made regarding the treatment of constraint equations. First, equation (9.14) shows that the parameter vector $\mathbf{\Delta}$ can be computed in two parts: $\mathbf{\Delta}^0$ as the least squares solution of the problem as if the constraints did not exist; then, with matrices $\mathbf{C}$ and $\mathbf{g}$ from the constraint equations $\boldsymbol{\delta}\mathbf{\Delta}$ can be computed. Once $\mathbf{\Delta}$ is computed, the rest of the problem proceeds in a manner identical to that covered previously. For example, using $\mathbf{\Delta}$, $\mathbf{k}$ may be computed from equation (9.9) and substituted into equation (9.8) to compute $\mathbf{v}$. Adding $\mathbf{v}$ to $\boldsymbol{l}$ gives $\hat{\boldsymbol{l}}$. The a posteriori estimate for the reference variance, $\hat{\sigma}_0^2$ may be computed from equation (6.25).

9.2.2. Precision Estimation for the Direct Treatment Applying the propagation rule, it can be shown that

$$\mathbf{Q}_{\Delta\Delta} = \mathbf{N}^{-1}(\mathbf{I} - \mathbf{C}^t\mathbf{M}^{-1}\mathbf{C}\mathbf{N}^{-1}) \tag{9.15a}$$

or

$$\mathbf{Q}_{\Delta\Delta} = \mathbf{N}^{-1}[\mathbf{I} - \mathbf{C}^t(\mathbf{C}\mathbf{N}^{-1}\mathbf{C}^t)^{-1}\mathbf{C}\mathbf{N}^{-1}] \tag{9.15b}$$

An interesting equality exists on the basis of equation (9.15b), if we post-multiply both its sides by $\mathbf{C}^t$, thus

$$\begin{aligned}\mathbf{Q}_{\Delta\Delta}\mathbf{C}^t &= \mathbf{N}^{-1}\mathbf{C}^t - \mathbf{N}^{-1}\mathbf{C}^t(\mathbf{C}\mathbf{N}^{-1}\mathbf{C}^t)^{-1}\mathbf{C}\mathbf{N}^{-1}\mathbf{C}^t \\ &= \mathbf{N}^{-1}\mathbf{C}^t - \mathbf{N}^{-1}\mathbf{C}^t\end{aligned}$$

or

$$\mathbf{Q}_{\Delta\Delta}\mathbf{C}^t = \mathbf{0} \tag{9.16}$$

which may be used for checking purposes. Similarly,

$$\mathbf{Q}_{vv} = \mathbf{Q}\mathbf{A}^t\mathbf{W}_e\mathbf{A}\mathbf{Q} - \mathbf{Q}\mathbf{A}^t\mathbf{W}_e\mathbf{B}(\mathbf{N}^{-1} - \mathbf{N}^{-1}\mathbf{C}^t\mathbf{M}^{-1}\mathbf{C}\mathbf{N}^{-1})\mathbf{B}^t\mathbf{W}_e\mathbf{A}\mathbf{Q} \tag{9.17a}$$

which, from equation (9.15a) becomes

$$\mathbf{Q}_{vv} = \mathbf{Q}\mathbf{A}^t\mathbf{W}_e\mathbf{A}\mathbf{Q} - \mathbf{Q}\mathbf{A}^t\mathbf{W}_e\mathbf{B}\mathbf{Q}_{\Delta\Delta}\mathbf{B}^t\mathbf{W}_e\mathbf{A}\mathbf{Q} \tag{9.17b}$$

The reader can ascertain that equation (9.17b) is a general relationship since it is identical to equation (6.30). From equation (9.17b) an interesting equality is possible if both sides are premultiplied by $\mathbf{A}$ and postmultiplied by $\mathbf{A}^t$, thus

$$\mathbf{A}\mathbf{Q}_{vv}\mathbf{A}^t = (\mathbf{A}\mathbf{Q}\mathbf{A}^t)\mathbf{W}_e(\mathbf{A}\mathbf{Q}\mathbf{A}^t) - (\mathbf{A}\mathbf{Q}\mathbf{A}^t)\mathbf{W}_e\mathbf{B}\mathbf{Q}_{\Delta\Delta}\mathbf{B}^t\mathbf{W}_e(\mathbf{A}\mathbf{Q}\mathbf{A}^t)$$

or

$$\mathbf{A}\mathbf{Q}_{vv}\mathbf{A}^t = \mathbf{Q}_e - \mathbf{B}\mathbf{Q}_{\Delta\Delta}\mathbf{B}^t \tag{9.18}$$

Finally, the cofactor matrix for the estimated observations is given by the same standard relation or

$$\mathbf{Q}_{ll} = \mathbf{Q} - \mathbf{Q}_{vv} \tag{6.31}$$

9.2.3. Solution by Elimination of Constraints The constraint equations (9.3b) may be used to solve, functionally, for as many parameters as there are constraints. These parameters are then substituted in turn into the first set of condition equations, thus reducing the number of parameters. The resulting condition equations can be utilized in a direct least squares solution for the remaining parameters. Such an approach has two advantages: First, it avoids the direct use of the constraints into the least squares estimation, and, second, it leads to a reduction of the number of parameters to be simultaneously estimated. However, the process of eliminating part of the parameters using the constraints needs care since the constraint equations may not, and often do not, include all the parameters. This may make it difficult to standardize the elimination procedure in a computer program. In rather complex cases, prior evaluation of the equations may be necessary to effect a rearrangement of their order so that they may be used in a prewritten program.

Partitioning the parameter vector $\mathbf{\Delta}$ into two subvectors

$$\underset{s,1}{\mathbf{\Delta}_1} \quad \text{and} \quad \underset{a,1}{\mathbf{\Delta}_2}$$

where $a = u - s$, gives

$$\underset{c,n}{\mathbf{A}}\,\underset{n,1}{\mathbf{v}} + \underset{c,s}{\mathbf{B}_1}\,\underset{s,1}{\mathbf{\Delta}_1} + \underset{c,a}{\mathbf{B}_2}\,\underset{a,1}{\mathbf{\Delta}_2} = \underset{c,1}{\mathbf{f}} \tag{9.19a}$$

$$\underset{s,s}{\mathbf{C}_{11}}\,\underset{s,1}{\mathbf{\Delta}_1} + \underset{s,a}{\mathbf{C}_{12}}\,\underset{a,1}{\mathbf{\Delta}_2} = \underset{s,1}{\mathbf{g}} \tag{9.19b}$$

Using equation (9.19b), we can solve for $\mathbf{\Delta}_1$, in terms of $\mathbf{\Delta}_2$,

$$\mathbf{\Delta}_1 = \mathbf{C}_{11}^{-1}(\mathbf{g} - \mathbf{C}_{12}\mathbf{\Delta}_2) \tag{9.20}$$

Care must be exercised in the partitioning operation to render $\mathbf{C}_{11}$ nonsingular. If the constraint equations as they are generated lead to a singular matrix, they must be rearranged to avoid that difficulty. Thus

$$\mathbf{Av} + (-\mathbf{B}_1\mathbf{C}_{11}^{-1}\mathbf{C}_{12} + \mathbf{B}_2)\mathbf{\Delta}_2 = (\mathbf{f} - \mathbf{B}_1\mathbf{C}_{11}^{-1}\mathbf{g})$$

and with the auxiliaries

$$\bar{\mathbf{B}} = (-\mathbf{B}_1\mathbf{C}_{11}^{-1}\mathbf{C}_{12} + \mathbf{B}_2) \tag{9.21a}$$

$$\bar{\mathbf{f}} = (\mathbf{f} - \mathbf{B}_1\mathbf{C}_{11}^{-1}\mathbf{g}) \tag{9.21b}$$

becomes

$$\mathbf{Av} + \bar{\mathbf{B}}\mathbf{\Delta}_2 = \bar{\mathbf{f}} \tag{9.22}$$

Equation (9.22) represents a system of c condition equations including only $a = u - s$ parameters. In a straightforward manner, the solution for $\mathbf{\Delta}_2$ by least squares would be

$$\mathbf{\Delta}_2 = \bar{\mathbf{N}}^{-1}\bar{\mathbf{t}} \tag{9.23}$$

where

$$\underset{a,a}{\bar{\mathbf{N}}} = \underset{a,c}{\bar{\mathbf{B}}^t}\left(\underset{c,c}{\mathbf{AQA}^t}\right)^{-1}\bar{\mathbf{B}} = \bar{\mathbf{B}}^t\mathbf{W}_e\,\bar{\mathbf{B}} \tag{9.24a}$$

$$\underset{a,1}{\bar{\mathbf{t}}} = \bar{\mathbf{B}}^t(\mathbf{AQA}^t)^{-1}\bar{\mathbf{f}} = \bar{\mathbf{B}}^t\mathbf{W}_e\,\bar{\mathbf{f}} \tag{9.24b}$$

Once $\mathbf{\Delta}_2$ is computed, $\mathbf{\Delta}_1$ may be evaluated from equation (9.20) and the total vector of parameters would simply be

$$\mathbf{\Delta} = \begin{bmatrix} \mathbf{\Delta}_1 \\ \mathbf{\Delta}_2 \end{bmatrix} \tag{9.25}$$

With the value of $\boldsymbol{\Delta}$ thus computed, the rest of the variables can be evaluated exactly as before.

9.2.4. Precision Estimation for the Solution by Constraint Elimination

From the propagation rule,

$$\mathbf{Q}_{\Delta_2\Delta_2} = \bar{\mathbf{N}}^{-1} \tag{9.26}$$

$$\mathbf{Q}_{\Delta_1\Delta_1} = \mathbf{C}_{11}^{-1}\mathbf{C}_{12}\bar{\mathbf{N}}^{-1}\mathbf{C}_{12}^{t}(\mathbf{C}_{11}^{-1})^{t} \tag{9.27}$$

$$\mathbf{Q}_{\Delta_1\Delta_2} = -\mathbf{C}_{11}^{-1}\mathbf{C}_{12}\,\bar{\mathbf{N}}_{1}^{-1} \tag{9.28}$$

Thus the total cofactor matrix for $\boldsymbol{\Delta}$ is

$$\mathbf{Q}_{\Delta\Delta} = \begin{bmatrix} \mathbf{C}_{11}^{-1}\mathbf{C}_{12}\,\bar{\mathbf{N}}^{-1}\mathbf{C}_{12}^{t}(\mathbf{C}_{11}^{-1})^{t} & -\mathbf{C}_{11}^{-1}\mathbf{C}_{12}\,\bar{\mathbf{N}}^{-1} \\ -\bar{\mathbf{N}}^{-1}\mathbf{C}_{12}^{t}(\mathbf{C}_{11}^{-1})^{t} & \bar{\mathbf{N}}^{-1} \end{bmatrix} \tag{9.29}$$

Then

$$\mathbf{Q}_{vv} = \mathbf{Q}\mathbf{A}^{t}[\mathbf{W}_e - \mathbf{W}_e\,\mathbf{B}\mathbf{Q}_{\Delta\Delta}\,\mathbf{B}^{t}\mathbf{W}_e]\mathbf{A}\mathbf{Q} \tag{9.30}$$

in which $\mathbf{Q}_{\Delta\Delta}$ would be evaluated from equation (9.29). As before, the cofactor matrix of estimated observations $\mathbf{Q}_{\hat{l}\hat{l}}$ can be computed from equation (6.31).

Example 9.1. Figure 9.1 shows a simple problem of trilateration to determine the two coordinates of point R. Three ranges to three perfectly known points, P, O, and Q, are measured with the following results:

$$l_1 = \overline{RP} = 5.10 \text{ m}$$

$$l_2 = \overline{RO} = 7.07 \text{ m}$$

$$l_3 = \overline{RQ} = 4.90 \text{ m}$$

All three observations are of equal precision and uncorrelated. The coordinates of the fixed points are $P = (0, 5)$, $O = (0, 0)$, and $Q = (5, 0)$.

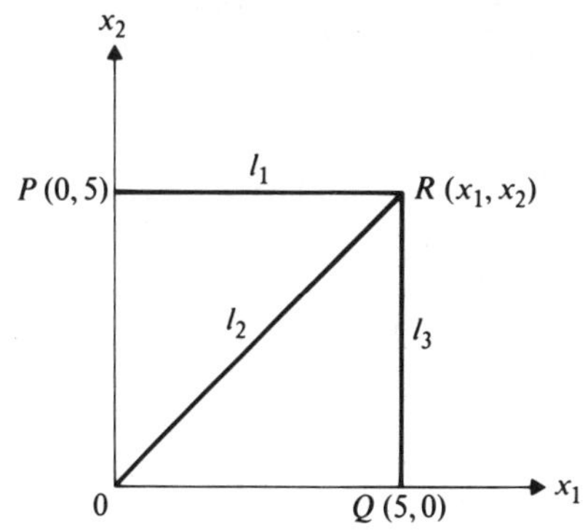

Figure 9.1

a. Assuming the approximate values for the unknown coordinates to be $x_1^0 = 5.0$ m and $x_2^0 = 5.0$ m, compute the correction vector from the first iteration.

b. If x_2 is to be constrained to the value 5.0 m, compute the corresponding value of x_1.

Solution: The model here is relatively simple, being composed of a plane with a two-dimensional cartesian coordinate system. There are four points in that plane involved in the problem, three of which are given as perfectly known. In order to determine the coordinates of the fourth point, two ranges will be the minimum required. (The point of intersection in the first quadrant will be selected as implied from Figure 9.1.) Since three ranges are given, there exists 1 degree of freedom, that is, $r = 1$. Given $u = 2$, the number of condition equations will be $c = 3$. Obviously a condition equation can be written for each given range, thus

$$\begin{aligned}
l_1^2 - (x_1 - 0)^2 - (x_2 - 5)^2 &= 0 \\
l_2^2 - (x_1 - 0)^2 - (x_2 - 0)^2 &= 0 \\
l_3^2 - (x_1 - 5)^2 - (x_2 - 0)^2 &= 0
\end{aligned}$$

If these equations are left as they are, an **A** matrix would have to be evaluated since they are nonlinear in the observations. However, with simple manipulation we can reduce **A** to the identity matrix and work with the simpler case of adjustment of indirect observations. (Of course, as a price for this, the condition equations may become slightly more complicated, thus requiring a little more work in linearization.)

$$\begin{aligned}
l_1 - [x_1^2 + (x_2 - 5)^2]^{1/2} &= 0 \\
l_2 - (x_1^2 + x_2^2)^{1/2} &= 0 \\
l_3 - [(x_1 - 5)^2 + x_2^2]^{1/2} &= 0
\end{aligned}$$

Linearization to the form $\mathbf{v} + \mathbf{B}\boldsymbol{\Delta} = \mathbf{f}$ leads to

$$\underset{3,2}{\mathbf{B}} = \begin{bmatrix} \dfrac{-x_1^0}{[x_1^{02} + (x_2^0 - 5)^2]^{1/2}} & \dfrac{-(x_2^0 - 5)}{[x_1^{02} + (x_2^0 - 5)^2]^{1/2}} \\[2ex] \dfrac{-x_1^0}{(x_1^{02} + x_2^{02})^{1/2}} & \dfrac{-x_2^0}{(x_1^{02} + x_2^{02})^{1/2}} \\[2ex] \dfrac{-(x_1^0 - 5)}{[(x_1^0 - 5)^2 + x_2^{02})]^{1/2}} & \dfrac{-x_2^0}{[(x_1^0 - 5)^2 + x_2^{02})]^{1/2}} \end{bmatrix} = \begin{bmatrix} -1 & 0 \\[2ex] \dfrac{-1}{\sqrt{2}} & \dfrac{-1}{\sqrt{2}} \\[2ex] 0 & -1 \end{bmatrix}$$

$$\underset{3,1}{\mathbf{f}} = \begin{bmatrix} [x_1^{02} + (x_2^0 - 5)^2]^{1/2} - l_1 \\ (x_1^{02} + x_2^{02})^{1/2} - l_2 \\ [(x_1^0 - 5)^2 + x_2^{02}]^{1/2} - l_3 \end{bmatrix} = \begin{bmatrix} -0.1 \\ 0 \\ 0.1 \end{bmatrix}$$

Given $\mathbf{W} = \mathbf{I}$ and the above values, then

$$\underset{2,2}{\mathbf{N}} = \begin{bmatrix} \frac{3}{2} & \frac{1}{2} \\ \frac{1}{2} & \frac{3}{2} \end{bmatrix}$$

$$\mathbf{t} = \begin{bmatrix} 0.1 \\ -0.1 \end{bmatrix} \qquad \boldsymbol{\Delta} = \begin{bmatrix} 0.1 \\ -0.1 \end{bmatrix}$$

which is the first answer required.

Given the constraint, n_0 will reduce to 1 and the redundancy will increase to $r = 2$; thus $c = 3$, $u = 2$, and $s = 1$. First, using the *direct treatment* the constraint equation is simply $x_2 - 5 = 0$. This equation, although linear, must, however, be linearized. This is because the condition equations are nonlinear and have therefore to be linearized. Such linearization leads to estimating a correction vector $\boldsymbol{\Delta}$ instead of the coordinates themselves. Consequently, x_2 in the constraint equation is replaced by $(x_2^0 + \delta x_2)$, where x_2^0 *must be the same approximate* value used in the condition equations. Thus with the given $x_2^0 = 5$ m, the constraint equation becomes

$$x_2^0 + \delta x_2 - 5 = 0 \qquad \text{or} \qquad \delta x_2 = 0$$

which, in the usual matrix from of $\mathbf{C\Delta} = \mathbf{g}$, becomes

$$[0 \quad 1]\begin{bmatrix}\delta x_1 \\ \delta x_2\end{bmatrix} = 0$$

Using the $\mathbf{C}$ matrix and $\mathbf{N}^{-1}$ above leads to

$$\mathbf{M} = \tfrac{3}{4}$$

and

$$\boldsymbol{\delta\Delta} = \begin{bmatrix}-0.033 \\ 0.1\end{bmatrix}$$

Thus

$$\boldsymbol{\Delta} = \boldsymbol{\Delta}^0 + \delta\boldsymbol{\Delta} = \begin{bmatrix}0.1 \\ -0.1\end{bmatrix} + \begin{bmatrix}-0.033 \\ 0.1\end{bmatrix} = \begin{bmatrix}0.067 \\ 0.0\end{bmatrix}$$

which implies that there will be no correction to $x_2^0 = 5.00$ m, thus satisfying the constraint equation of fixing the value of x_2 at 5.00 m. In the second process, we eliminate x_2 from the original condition equations using the constraint equations. Thus with $x_2 = 5$, δx_2 must be zero (since $x_2^0 = 5$) and the original condition equations become

$$\mathbf{v} + \begin{bmatrix}-1 \\ \dfrac{-1}{\sqrt{2}} \\ 0\end{bmatrix}(\delta x_1) = \begin{bmatrix}-0.1 \\ 0 \\ 0.1\end{bmatrix} - \begin{bmatrix}0 \\ \dfrac{-1}{\sqrt{2}} \\ -1\end{bmatrix}(0) = \begin{bmatrix}-0.1 \\ 0 \\ 0.1\end{bmatrix}$$

and

$$\underset{1,1}{\bar{\mathbf{N}}} = \tfrac{3}{2}, \qquad \bar{\mathbf{t}} = 0.1, \qquad \mathbf{x}_1 = 0.067$$

which is exactly the same value we obtained above.

Example 9.2

Given: The problem of Example 6.1 is expanded to allow for the use of constraints. Three terrestrial camera stations S_1, S_2, and S_3 photograph two points A and B (see Figure 9.2). All three camera axes are parallel and normal to the base which is taken as the X_1 axis. The X_2 axis coincides with the optical axis of camera

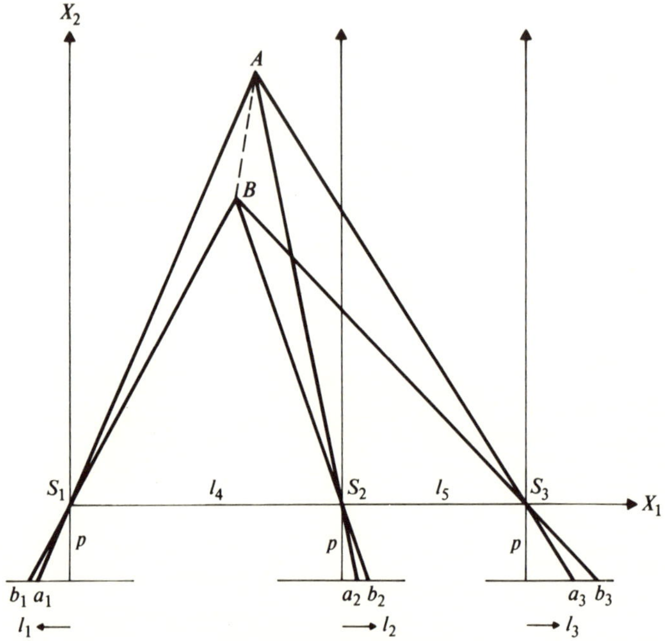

Figure 9.2. (Not to Scale)

S_1. The measured values of the two bases are $l_4 = 10$ m and $l_5 = 8$ m, each with a standard deviation, $\sigma = 0.05$ m. Denoting by l_1, l_2, and l_3 the observations on the three photographs S_1, S_2, and S_3, respectively, the observation are

POINT	l_1 mm	l_2 mm	l_3 mm
a	14.1	6.1	22.1
b	16.6	7.1	26.3

The standard deviation for all photo observations is 0.10 mm, and the principal distance is $p = 100$ mm (assumed without error, that is, a constant).

Required: Using the data given above compute the coordinates of both points A and B in an x_1 and x_2 coordinate system. Then constraining the distance between the two points to 7.8 m, compute the new set of coordinates, using both the direct treatment and the method of elimination.

Solution: To demonstrate first the use of the direct treatment, solve the first part by the method of combined observations and parameters. The number of condition equations is computed as follows:

+ Minimum number of variables to determine the geometry of the problem is six (x_1 for S_2 and S_3, and x_1 and x_2 for both points A and B), or $n_0 = 6$.
+ The total number of observed quantities is $n = 8$.
+ The redundancy is $r = 2$.
+ The number of parameters to be solved for is $u = 4$ (coordinates of A and B).
+ The total number of independent condition equations is $c = r + u = 6$.

Referring to Figure 9.2, write one condition equation for each photogrammetric ray:

$$
\begin{aligned}
l_{1a}x_{2a} - px_{1a} &= 0 \\
l_{1b}x_{2b} - px_{1b} &= 0 \\
l_{2a}x_{2a} - pl_4 + px_{1a} &= 0 \\
l_{2b}x_{2b} - pl_4 + px_{1b} &= 0 \\
l_{3a}x_{2a} - pl_4 - pl_5 + px_{1a} &= 0 \\
l_{3b}x_{2b} - pl_4 - pl_5 + px_{1b} &= 0
\end{aligned}
$$

Linearization:

$$
\mathbf{Av} = \begin{bmatrix} x_{2a}^0 & 0 & 0 & & & & & 0 \\ 0 & x_{2b}^0 & 0 & & & & & 0 \\ & 0 & x_{2a}^0 & 0 & 0 & 0 & -p & 0 \\ \vdots & \vdots & 0 & x_{2b}^0 & 0 & 0 & -p & 0 \\ & & \vdots & 0 & x_{2a}^0 & 0 & -p & -p \\ 0 & 0 & 0 & 0 & 0 & x_{2b}^0 & -p & -p \end{bmatrix} \begin{bmatrix} v_{1a} \\ v_{1b} \\ v_{2a} \\ v_{2b} \\ v_{3a} \\ v_{3b} \\ v_4 \\ v_5 \end{bmatrix}
$$

$$
\mathbf{B\Delta} = \begin{bmatrix} -p & l_{1a} & 0 & 0 \\ 0 & 0 & -p & l_{1b} \\ p & l_{2a} & 0 & 0 \\ 0 & 0 & p & l_{2b} \\ p & l_{3a} & 0 & 0 \\ 0 & 0 & p & l_{3b} \end{bmatrix} \begin{bmatrix} \delta x_{1a} \\ \delta x_{2a} \\ \delta x_{1b} \\ \delta x_{2b} \end{bmatrix}
$$

$$
\underset{6,1}{\mathbf{f}} = \begin{bmatrix} px_{1a}^0 - l_{1a}x_{2a}^0 \\ px_{1b}^0 - l_{1b}x_{2b}^0 \\ pl_4 - px_{1a}^0 - l_{2a}x_{2a}^0 \\ pl_4 - px_{1b}^0 - l_{2b}x_{2b}^0 \\ pl_4 + pl_5 - px_{1a}^0 - l_{3a}x_{2a}^0 \\ pl_4 + pl_5 - px_{1b}^0 - l_{3b}x_{2b}^0 \end{bmatrix}
$$

Using simple geometric relationships compute the following approximations:

$x_{1a}^0 = 8.00$ m, $x_{2a}^0 = 51.00$ m, $x_{1b}^0 = 7.00$ m, and $x_{2b}^0 = 41.00$ m from which, with the observations, evaluate the numerical values of the matrices as

$$\underset{6,8}{\mathbf{A}} = \begin{bmatrix} 51.0 & 0 & 0 & 0 & 0 & 0 & 0 & 0 \\ 0 & 41.0 & 0 & 0 & 0 & 0 & 0 & 0 \\ 0 & 0 & 51.0 & 0 & 0 & 0 & -100.0 & 0 \\ 0 & 0 & 0 & 41.0 & 0 & 0 & -100.0 & 0 \\ 0 & 0 & 0 & 0 & 51.0 & 0 & -100.0 & -100.0 \\ 0 & 0 & 0 & 0 & 0 & 41.0 & -100.0 & -100.0 \end{bmatrix}$$

This matrix includes parameter approximations and therefore would have to be recomputed each iteration.

$$\underset{6,4}{\mathbf{B}} = \begin{bmatrix} -100.0 & 14.1 & 0 & 0 \\ 0 & 0 & -100.0 & 16.6 \\ 100.0 & 6.1 & 0 & 0 \\ 0 & 0 & 100.0 & 7.1 \\ 100.0 & 22.1 & 0 & 0 \\ 0 & 0 & 100.0 & 26.3 \end{bmatrix}$$

This matrix contains no parameter approximations. It remains unaltered through the iterative process.

$$\underset{6,1}{\mathbf{f}} = [80.90 \quad 19.40 \quad -111.10 \quad 8.90 \quad -127.10 \quad 21.70]^t$$

This vector needs to be recomputed every iteration.

From the given standard deviations the cofactor matrix of the given observation (assuming a priori $\sigma_0^2 = 1$) is

$$\underset{8,8}{\mathbf{Q}} = \text{diag.}\left\{\underbrace{0.01, 0.01, 0.01, 0.01, 0.01, 0.01}_{\text{mm}^2}, \underbrace{0.0025, 0.0025}_{\text{m}^2}\right\}$$

With these numerical values, the first correction vector to the coordinate approximations may be computed as (by a computer program)

$$\mathbf{\Delta}_{(1)}^0 = \begin{bmatrix} \delta x_{1a} \\ \delta x_{2a} \\ \delta x_{1b} \\ \delta x_{2b} \end{bmatrix} = \begin{bmatrix} -1.005 \text{ m} \\ -1.284 \text{ m} \\ -0.019 \text{ m} \\ +0.969 \text{ m} \end{bmatrix}$$

The updated values of the coordinates of the two points are therefore

$$x_{1a} = 6.995 \text{ m} \qquad x_{1b} = 6.981 \text{ m}$$

$$x_{2a} = 49.716 \text{ m} \qquad x_{2b} = 41.969 \text{ m}$$

With these values we recompute the matrices **A** and **f**, form normal equations, and solve for a new correction vector.

$$\mathbf{\Delta}_{(2)}^0 = [0.00026 \quad -0.00042 \quad 0.00026 \quad -0.00038]^t(\text{m})$$

which is zero to three decimal places, thus the iteration procedure may be terminated. The final coordinates required in the problem are then

$$\hat{x}_{1a} = 6.995 \text{ m} \qquad \hat{x}_{2a} = 49.716 \text{ m} \qquad \hat{x}_{1b} = 6.982 \text{ m} \qquad \hat{x}_{2b} = 41.968 \text{ m}$$

Introduction of Distance Constraint: The second part of the required information for this problem is to compute the coordinates of points A and B when the distance between them is constrained to be exactly 7.8 m. From the coordinates computed above, the said distance is less than 7.8 m and therefore when the constraint is introduced, these coordinates will change.

Without the constraint, n_0 is equal to six; that is, six independent variables are necessary for a unique determination of the geometry of the problem. This is true because all the parameters are functionally independent. With *one* geometric constraint, the functional model must be re-examined and a new n_0 computed. It can be easily shown that in this case $n_0 = 5$. Thus $n = 8$, $u = 4$, $c = 6$, $s = 1$, and all pertinent relationships are satisfied. Therefore

$$0 < u = 4 < n_0 = 5$$

$$c + s = 6 + 1 = (n - n_0) + u = 3 + 4$$

$$(n - n_0) = 3 < c = 6 < n = 8$$

$$s = 1 < u = 4$$

The constraint equation is

$$(x_{1b} - x_{1a})^2 + (x_{2b} - x_{2a})^2 - (7.8)^2 = 0$$

which when linearized yields

$$\mathbf{C\Delta} = [-2(x_{1b}^0 - x_{1a}^0) \quad -2(x_{2b}^0 - x_{2a}^0) \quad 2(x_{1b}^0 - x_{1a}^0) \quad 2(x_{2b}^0 - x_{2a}^0)] \times \begin{bmatrix} \delta x_{1a} \\ \delta x_{2a} \\ \delta x_{1b} \\ \delta x_{2b} \end{bmatrix}$$

$$g = [(7.8)^2 - (x_{1b}^0 - x_{1a}^0)^2 - (x_{2b}^0 - x_{2a}^0)^2]$$

Using the *same starting approximations*, the $\mathbf{C}$ and $\mathbf{g}$ matrices become

$$C = [2.0 \quad 20.0 \quad -2.0 \quad -20.0] \qquad \text{and} \qquad \mathbf{g} = -40.16$$

With these matrices, the addition to $\mathbf{\Delta}_{(1)}^0$ due to the constraint for the first iteration may be computed as the vector

$$\delta\mathbf{\Delta}_{(1)} = [-0.006 \quad 0.267 \quad 0.011 \quad -0.078]^t(\text{m})$$

Thus the total correction vector for the first iteration is

$$\mathbf{\Delta}_{(1)} = [-1.011 \quad -1.017 \quad -0.008 \quad 0.891]^t(\text{m})$$

The new approximation vector is therefore

$$\mathbf{x}_{(1)}^0 = [6.989 \quad 49.983 \quad 6.992 \quad 41.891]^t(\text{m})$$

Since this vector is different from that computed after the first iteration of the solution without constraint, it is very important to note that the value $\mathbf{\Delta}_{(2)}^0$, computed there, *cannot be used here.* This is because the second iteration is now starting with a different approximation vector than that used for the linearization of condition equations used to compute $\mathbf{\Delta}_{(2)}^0$ for the case of no constraint.

With $\mathbf{x}^0_{(1)}$ computed above, a new set of matrices **A**, **f**, **C**, and **g** must be evaluated. From these a new total correction vector $\boldsymbol{\Delta}_{(2)}$ can be obtained as

$$\boldsymbol{\Delta}_{(2)} = [0.005 \quad -0.224 \quad -0.009 \quad 0.063]^t(\text{m})$$

Some of the elements of that vector are still significant and a third iteration should be performed. The new approximation vector is

$$\mathbf{x}^0_{(2)} = [6.994 \quad 49.761 \quad 6.983 \quad 41.954]^t(\text{m})$$

and the total correction vector from the third iteration is

$$\boldsymbol{\Delta}_{(3)} = [0.00013 \quad -0.00417 \quad -0.00013 \quad 0.00112]^t(\text{m})$$

which is sufficiently small to terminate the relinearization process. The final answer for the coordinates of points A and B with distance between them constrained to 7.8 m is

$$\hat{x}_{1a} = 6.994 \text{ m} \qquad \hat{x}_{2a} = 49.757 \text{ m} \qquad \hat{x}_{1b} = 6.983 \text{ m} \qquad \hat{x}_{2b} = 41.955 \text{ m}$$

A check on the distance between the points computed from the above coordinates would show that it is equal to 7.8 m to within the fifth decimal place.

The a posteriori cofactor matrices are as follows.

a. *Without Constraint.* After terminating the iterations, the final results are

$$\mathbf{Q}_{\Delta\Delta(f)} = 10^{-3}\begin{bmatrix} 1.693 & 0.784 & 3.470 & 4.419 \\ & 1.430 & 5.225 & 3.342 \\ & & 75.348 & 31.902 \\ \text{symmetric} & & & 46.003 \end{bmatrix}(\text{m}^2) \qquad \text{rank} = 4 = u$$

$$\mathbf{Q}_{ll(f)} = 10^{-3}$$

$$\times \begin{bmatrix} 8.932 & 0.291 & -2.401 & 0.654 & 1.330 & -0.363 & 0.363 & -0.454 \\ & 9.031 & 0.654 & -2.178 & -0.363 & 1.209 & 0.431 & -0.538 \\ & & 4.602 & 1.471 & 2.997 & -0.817 & 0.817 & -1.020 \\ & & & 5.103 & 0.817 & 2.719 & 0.968 & -1.209 \\ & & & & 8.336 & 0.454 & -0.454 & 0.560 \\ & & & & & 8.491 & -0.538 & 0.671 \\ & & & & & & 2.061 & 0.548 \\ \text{symmetric} & & & & & & & 1.815 \end{bmatrix}$$

$$\text{rank }(\mathbf{Q}_{ll})_{(f)} = 6$$

$$\hat{\sigma}^2_{0(f)} = 0.8228$$

$$\hat{\sigma}_{0(f)} = 0.9071$$

b. *With Distance Constraint.* After the last (sixth iteration) the results are

$$\mathbf{Q}_{\Delta\Delta(f)} = 10^{-3}\begin{bmatrix} 1.678 & 0.815 & 4.184 & 4.185 \\ & 1.368 & 3.803 & 3.802 \\ & & 42.532 & 42.533 \\ \text{symmetric} & & & 42.533 \end{bmatrix} \qquad \text{rank} = 3 = u - s$$

$$\mathbf{Q}_{ll(f)} = 10^{-3}$$

$$\times \begin{bmatrix} 5.422 & 2.774 & -2.665 & 0.815 & -1.914 & 1.961 & 0.647 & -0.207 \\ & 7.274 & 0.841 & -2.291 & 1.935 & -0.436 & 0.230 & -0.712 \\ & & 4.581 & 1.483 & 2.754 & -0.642 & 0.838 & -1.001 \\ & & & 5.097 & -0.668 & 2.612 & 0.056 & -1.220 \\ & & & & 5.330 & 2.604 & -0.191 & 0.794 \\ & & & & & 6.951 & -0.726 & 0.508 \\ & & & & & & 2.038 & 0.528 \\ \text{symmetric} & & & & & & & 1.798 \end{bmatrix}$$

$$\text{rank } \mathbf{Q}_{ll(f)} = 5$$

$$\hat{\sigma}^2_{0(f)} = 0.5644$$

$$\hat{\sigma}_{0(f)} = 0.7513$$

Example 9.3. In this example the problem of the three photographs with distance constraint in the preceding example is reworked using the technique of eliminating the constraint. Recall that

$$n = 8, \quad n_0 = 5, \quad r = 3, \quad u = 4, \quad c = 6, \quad \text{and} \quad s = 1$$

With the approximations

$$\underset{4,1}{\mathbf{x}^0} = [x^0_{1a} \quad x^0_{2a} \quad x^0_{1b} \quad x^0_{2b}]^t = [8.0 \quad 51.0 \quad 7.0 \quad 41.0]^t$$

the constraint equation was

$$[2 \quad 20 \quad -2 \quad -20][\delta x_{1a} \quad \delta x_{2a} \quad \delta x_{1b} \quad \delta x_{2b}]^t = -40.16$$

Since $s = 1$, the partitioning becomes (note that $\mathbf{\Delta}_1$ is a 1×1 only and corresponds to the correction δx_{1a}).

$$[2]\mathbf{\Delta}_1 + [20 \quad -2 \quad -20]\mathbf{\Delta}_2 = -40.16$$

or

$$\mathbf{C}_{11} = [2] \qquad \mathbf{C}_{12} = [20 \quad -2 \quad -20] \qquad \mathbf{g} = [-40.16]$$

Further, partition the **B** matrix (evaluated at the same approximation vector $\mathbf{x}^0$) into

$$\underset{c,s}{\mathbf{B}_1} \quad \text{and} \quad \underset{c,a}{\mathbf{B}_2}$$

or

$$\underset{6,1}{\mathbf{B}_1} = \begin{bmatrix} -100 \\ 0 \\ 100 \\ 0 \\ 100 \\ 0 \end{bmatrix} \quad \text{and} \quad \underset{6,3}{\mathbf{B}_2} = \begin{bmatrix} 14.1 & 0 & 0 \\ 0 & -100 & 16.6 \\ 6.1 & 0 & 0 \\ 0 & 100 & 7.1 \\ 22.1 & 0 & 0 \\ 0 & 100 & 26.3 \end{bmatrix}$$

With $\mathbf{B}_1$, $\mathbf{B}_2$, $\mathbf{f}$, $\mathbf{C}_{11}$, $\mathbf{C}_{12}$, and $\mathbf{g}$ the matrices $\bar{\mathbf{B}}$ and $\bar{\mathbf{f}}$ are

$$\underset{6,\,3}{\bar{\mathbf{B}}} = \begin{bmatrix} 1014.1 & -100.0 & -1000.0 \\ 0 & -100.0 & 16.6 \\ -993.9 & 100.0 & 1000.0 \\ 0 & 100.0 & 7.1 \\ -977.9 & 100.0 & 1000.0 \end{bmatrix} \qquad \bar{\mathbf{f}} = \begin{bmatrix} -1927.1 \\ 19.4 \\ 1896.9 \\ 8.9 \\ 1880.9 \end{bmatrix}$$

In addition to $\bar{\mathbf{B}}$ and $\bar{\mathbf{f}}$, the matrices $\mathbf{A}$ and $\mathbf{Q}$ from the preceding example are used in equations (9.23) and (9.24) to compute $\boldsymbol{\Delta}_2$ as

$$\boldsymbol{\Delta}_2 = \begin{bmatrix} \delta x_{2a} \\ \delta x_{1b} \\ \delta x_{2b} \end{bmatrix} = \begin{bmatrix} -1.017 \\ -0.008 \\ 0.891 \end{bmatrix} \text{(m)}$$

and using it in equation (9.20) gives

$$\boldsymbol{\Delta}_1 = \delta x_{1a} = 1.011\text{(m)}$$

These four values are the same as those computed after the first iteration of the second part of Example 9.2. After the proper iteration, the final estimates for the coordinates are

$$\hat{x}_{1a} = 6.994 \text{ m} \qquad \hat{x}_{2a} = 49.756 \text{ m} \qquad \hat{x}_{1b} = 6.983 \text{ m} \qquad \hat{x}_{2b} = 41.957 \text{ m}$$

As regards a posteriori estimates, at the end of the iterations the results obtained are precisely the same as those given in Example 9.2.

9.3. SPECIAL CASES

From the general case covered in Section 9.2, several special cases can be obtained and solved using the relations derived so far. A more unusual case which requires a slightly different treatment is given in the following section.

9.3.1. Adjustment with Constraints When the Parameters Exceed the Conditions It may be recalled that at the early discussion on adjustment with constraints it was indicated that the number of parameters u is rather arbitrary, particularly in regard to its upper limit. It was also said that although u has no upper limit, the number of condition equations *must* not exceed the number of observations n. Consequently, a rather interesting situation arises when there are more parameters than condition equations. This special case is therefore characterized by the inequality

$$c < u \tag{9.31}$$

while all other previously given relationships hold. In light of equation (9.31), the ranks of the matrices are

$$\text{rank}\,(\mathbf{A}) = c \qquad \text{rank}\,(\mathbf{B}) = c \qquad \text{rank}\,(\mathbf{C}) = s \tag{9.32}$$

Because of equation (9.32) both treatments given in Section 9.2 must be modified to fit the present case.

Direct Treatment Following the development in Section 9.2.1, the partitioning must stop with **N**, because in the present case it is of order u, but of rank $c < u$. Combining equations (9.10) and (9.3b) gives

$$\begin{bmatrix} -\mathbf{N} & \mathbf{C}^t \\ \mathbf{C} & \mathbf{0} \end{bmatrix} \begin{bmatrix} \mathbf{\Delta} \\ \mathbf{k}_c \end{bmatrix} = \begin{bmatrix} -\mathbf{t} \\ \mathbf{g} \end{bmatrix} \tag{9.33a}$$

which may be solved directly

$$\begin{bmatrix} \mathbf{\Delta} \\ \mathbf{k}_c \end{bmatrix} = \begin{bmatrix} \underset{u,u}{-\mathbf{N}} & \underset{u,s}{\mathbf{C}^t} \\ \underset{s,u}{\mathbf{C}} & \underset{s,s}{\mathbf{0}} \end{bmatrix}^{-1} \begin{bmatrix} -\mathbf{t} \\ \mathbf{g} \end{bmatrix} \tag{9.33b}$$

and from which $\mathbf{\Delta}$ can be obtained. Note that although **N** is singular, the coefficient matrix in equation (9.33b) is nonsingular. With respect to precision estimation, special consideration is needed. Denote the inverse in equation (9.33b) by

$$\mathbf{\Omega} = \begin{bmatrix} -\mathbf{N} & \mathbf{C}^t \\ \mathbf{C} & \mathbf{0} \end{bmatrix}^{-1} = \begin{bmatrix} \boldsymbol{\alpha} & \boldsymbol{\beta}^t \\ \boldsymbol{\beta} & \boldsymbol{\gamma} \end{bmatrix} \tag{9.34}$$

and expand the constant term vector to

$$\mathbf{a} = \begin{bmatrix} -\mathbf{t} \\ \mathbf{g} \end{bmatrix} = \begin{bmatrix} -\mathbf{B}^t\mathbf{W}_e\mathbf{f} \\ \mathbf{g} \end{bmatrix} = \begin{bmatrix} \mathbf{B}^t\mathbf{W}_e\mathbf{A}\mathbf{l} \\ \mathbf{0} \end{bmatrix} + \begin{bmatrix} -\mathbf{B}^t\mathbf{W}_e\mathbf{d} \\ \mathbf{g} \end{bmatrix} \tag{9.35}$$

The last vector in equation (9.35) is a constant, and thus from cofactor propagation

$$\mathbf{Q}_{aa} = \begin{bmatrix} \mathbf{B}^t\mathbf{W}_e\mathbf{A} \\ \mathbf{0} \end{bmatrix} \mathbf{Q}[\mathbf{A}^t\mathbf{W}_e\mathbf{B} \quad \mathbf{0}^t] = \begin{bmatrix} \mathbf{B}^t\mathbf{W}_e\mathbf{B} & \mathbf{0} \\ \mathbf{0} & \mathbf{0} \end{bmatrix} = \begin{bmatrix} \mathbf{N} & \mathbf{0} \\ \mathbf{0} & \mathbf{0} \end{bmatrix} \tag{9.36}$$

Applying propagation principle to equation (9.33b) also (referring to $[\mathbf{\Delta} \quad \mathbf{k}_c]^t$ by **u**) we get

$$\mathbf{Q}_{uu} = \begin{bmatrix} \mathbf{Q}_{\Delta\Delta} & \mathbf{Q}_{\Delta k_c} \\ \mathbf{Q}_{k_c\Delta} & \mathbf{Q}_{k_ck_c} \end{bmatrix} = \mathbf{\Omega}\mathbf{Q}_{aa}\mathbf{\Omega} \qquad (\mathbf{\Omega} \text{ is symmetric})$$

$$= \begin{bmatrix} \boldsymbol{\alpha} & \boldsymbol{\beta}^t \\ \boldsymbol{\beta} & \boldsymbol{\gamma} \end{bmatrix} \begin{bmatrix} \mathbf{N} & \mathbf{0} \\ \mathbf{0} & \mathbf{0} \end{bmatrix} \begin{bmatrix} \boldsymbol{\alpha} & \boldsymbol{\beta}^t \\ \boldsymbol{\beta} & \boldsymbol{\gamma} \end{bmatrix} = \begin{bmatrix} \boldsymbol{\alpha}\mathbf{N}\boldsymbol{\alpha} & \boldsymbol{\alpha}\mathbf{N}\boldsymbol{\beta}^t \\ \boldsymbol{\beta}\mathbf{N}\boldsymbol{\alpha} & \boldsymbol{\beta}\mathbf{N}\boldsymbol{\beta}^t \end{bmatrix}. \tag{9.37}$$

from which

$$\mathbf{Q}_{\Delta\Delta} = \boldsymbol{\alpha}\mathbf{N}\boldsymbol{\alpha} \tag{9.38}$$

but from equation (9.34) we may write

$$\begin{bmatrix} -\mathbf{N} & \mathbf{C}^t \\ \mathbf{C} & \mathbf{0} \end{bmatrix} \begin{bmatrix} \boldsymbol{\alpha} & \boldsymbol{\beta}^t \\ \boldsymbol{\beta} & \boldsymbol{\gamma} \end{bmatrix} = \begin{bmatrix} \mathbf{I} & \mathbf{0} \\ \mathbf{0} & \mathbf{I} \end{bmatrix} \tag{9.39}$$

then

$$-\mathbf{N}\boldsymbol{\alpha} + \mathbf{C}^t\boldsymbol{\beta} = \mathbf{I}$$

$$\mathbf{N}\boldsymbol{\alpha} = -\mathbf{I} + \mathbf{C}^t\boldsymbol{\beta} \tag{9.40}$$

$$\mathbf{C}\boldsymbol{\alpha} = \mathbf{0}$$

$$\boldsymbol{\alpha}\mathbf{C}^t = \mathbf{0} \qquad (\boldsymbol{\alpha} \text{ is symmetric}) \tag{9.41}$$

Substituting equation (9.40) into (9.38)

$$\mathbf{Q}_{\Delta\Delta} = \boldsymbol{\alpha}(-\mathbf{I} + \mathbf{C}^t\boldsymbol{\beta})$$

$$= -\boldsymbol{\alpha} + \boldsymbol{\alpha}\mathbf{C}^t\boldsymbol{\beta}$$

or since $\boldsymbol{\alpha}\mathbf{C}^t = \mathbf{0}$ from equation (9.41)

$$\mathbf{Q}_{\Delta\Delta} = -\boldsymbol{\alpha} \tag{9.42}$$

Next

$$\mathbf{Q}_{\Delta k_c} = \boldsymbol{\alpha}\mathbf{N}\boldsymbol{\beta}^t = (-\mathbf{I} + \boldsymbol{\beta}^t\mathbf{C})\boldsymbol{\beta}^t$$

$$= -\boldsymbol{\beta}^t + \boldsymbol{\beta}^t\mathbf{C}\boldsymbol{\beta}^t$$

$$\mathbf{C}\boldsymbol{\beta}^t = \mathbf{I}$$

$$\mathbf{Q}_{\Delta k_c} = \boldsymbol{\beta}^t - \boldsymbol{\beta}^t = \mathbf{0} \tag{9.43}$$

$$\mathbf{Q}_{k_c k_c} = \boldsymbol{\beta}\mathbf{N}\boldsymbol{\beta}^t$$

$$-\mathbf{N}\boldsymbol{\beta}^t + \mathbf{C}^t\boldsymbol{\gamma} = \mathbf{0}$$

$$\mathbf{N}\boldsymbol{\beta}^t = \mathbf{C}^t\boldsymbol{\gamma}$$

$$\mathbf{Q}_{k_c k_c} = \boldsymbol{\beta}\mathbf{C}^t\boldsymbol{\gamma}$$

$$\mathbf{Q}_{k_c k_c} = \boldsymbol{\gamma} \tag{9.44}$$

Collecting all the cofactor matrices then

$$\begin{bmatrix} \mathbf{Q}_{\Delta\Delta} & \mathbf{Q}_{\Delta k_c} \\ \mathbf{Q}_{k_c\Delta} & \mathbf{Q}_{k_c k_c} \end{bmatrix} = \begin{bmatrix} -\boldsymbol{\alpha} & \mathbf{0} \\ \mathbf{0} & \boldsymbol{\gamma} \end{bmatrix} \neq \begin{bmatrix} -\mathbf{N} & \mathbf{C}^t \\ \mathbf{C} & \mathbf{0} \end{bmatrix}^{-1} \tag{9.45}$$

These results when checked with those developed in equations (6.35) through (6.41) show that the two problems are essentially the same. The only difference is that here the upper left submatrix is singular, whereas in the other case it was not and there the solution was possible using inversion by partitioning.

CONSTRAINT ELIMINATION From the general relationship (9.2)

$$c + s = r + u$$

the number of elements in the subvector $\underset{a,1}{\mathbf{\Delta}_2}$ remaining after eliminating $\underset{s,1}{\mathbf{\Delta}_1}$ from $\underset{u,1}{\mathbf{\Delta}}$ (see Section 9.2.3) is given by

$$a = u - s = c - r \tag{9.46}$$

Equation (9.46) indicates that the number of parameters a remaining after eliminating the constraints is less than the number of conditions c. Henceforth the matrix $\bar{\mathbf{B}}$ formed by equation (9.21a) will have a rank equal to a, leading to a nonsingular coefficient matrix $\bar{\mathbf{N}}$ in equation (9.23). Consequently, the solutions for $\mathbf{\Delta}_2$ in equation (9.23) and $\mathbf{\Delta}_1$ in equation (9.20) are both possible in the present case. This means that all equations and steps given in Section 9.2.3 apply directly to this case without any modification.

Example 9.4.† Suppose that there are $n = 2$ observations, $c = 2$ condition equations (with an $\mathbf{A} = \mathbf{I}$ for simplicity), $u = 3$ parameters, and $s = 2$ constraint equations.

$$\begin{bmatrix} v_1 \\ v_2 \end{bmatrix} + \begin{bmatrix} -3 & 1 & -1 \\ 1 & -4 & -2 \end{bmatrix} \begin{bmatrix} x_1 \\ x_2 \\ x_3 \end{bmatrix} = \begin{bmatrix} 2 \\ 18 \end{bmatrix} \quad \text{or} \quad \underset{2,1}{\mathbf{v}} + \underset{2,3}{\mathbf{B}}\ \underset{3,1}{\mathbf{\Delta}} = \underset{2,1}{\mathbf{f}}$$

$$\begin{bmatrix} -1 & 1 & 1 \\ 1 & -3 & 1 \end{bmatrix} \begin{bmatrix} x_1 \\ x_2 \\ x_3 \end{bmatrix} = \begin{bmatrix} 2 \\ 26 \end{bmatrix} \quad \text{or} \quad \underset{2,3}{\mathbf{C}}\ \underset{3,1}{\mathbf{\Delta}} = \underset{2,1}{\mathbf{g}}$$

noting that $c + s = 2 + 2 = 4$ and $u + r = 3 + 1 = 4$ (one redundancy). It is required to compute, by two methods, the estimates of x_1, x_2, x_3, and their cofactor matrix. Take the a priori weight matrix

$$\mathbf{W} = \begin{bmatrix} 2 & 0 \\ 0 & 0.5 \end{bmatrix}$$

Direct Solution:

$$\mathbf{N} = \begin{bmatrix} 18.5 & -8 & 5 \\ -8 & 10 & 2 \\ 5 & 2 & 4 \end{bmatrix} \qquad \mathbf{t} = \begin{bmatrix} -3 \\ -32 \\ -22 \end{bmatrix}$$

† The numbers for this example are adopted from an example in reference Linkwitz, 1971, in the Bibliography.

According to equation (9.33b)

$$\begin{bmatrix} \underset{3,1}{\boldsymbol{\Delta}} \\ \underset{2,1}{\mathbf{k}_c} \end{bmatrix} = \begin{bmatrix} -\mathbf{N} & \mathbf{C}^t \\ \mathbf{C} & \mathbf{0} \end{bmatrix}^{-1} \begin{bmatrix} -\mathbf{t} \\ \mathbf{g} \end{bmatrix}$$

$$= \left[\begin{array}{ccc:cc} -18.5 & 8 & -5 & -1 & 1 \\ & -10 & -2 & 1 & -3 \\ & & -4 & 1 & 1 \\ \hdashline & & & 0 & 0 \\ \text{symmetric} & & & & 0 \end{array}\right]^{-1} \left[\begin{array}{c} 3 \\ 32 \\ 22 \\ \hdashline 2 \\ 26 \end{array}\right]$$

$$= \left[\begin{array}{ccc:cc} -0.05 & -0.025 & -0.025 & -0.275 & -0.125 \\ & -0.0125 & -0.0125 & 0.1125 & -0.3125 \\ \boxed{1} & & -0.0125 & 0.6125 & 0.1875 \\ \hdashline & & & 2.1125 & -0.8125 \\ \text{symmetric} & & & & 0.3125 \end{array}\right] \left[\begin{array}{c} 3 \\ 32 \\ 22 \\ \hdashline 2 \\ 26 \end{array}\right]$$

From which

$$\boldsymbol{\Delta} = \begin{bmatrix} x_1 \\ x_2 \\ x_3 \end{bmatrix} = \begin{bmatrix} -5.30 \\ -8.65 \\ +5.35 \end{bmatrix}$$

The cofactor matrix of the parameters $\mathbf{Q}_{\Delta\Delta}$ is the negative of submatrix 1 according to equation (9.42).

Solution by Elimination: First partition $\boldsymbol{\Delta}$ to

$$\begin{bmatrix} \boldsymbol{\Delta}_1 \\ \\ \boldsymbol{\Delta}_2 \end{bmatrix} = \left[\begin{array}{c} x_1 \\ x_2 \\ \hdashline x_3 \end{array}\right] \qquad \text{and} \qquad \mathbf{v} + \mathbf{B}_1 \boldsymbol{\Delta}_1 + \mathbf{B}_2 \boldsymbol{\Delta}_2 = \mathbf{f}$$

and correspondingly the constraints into $\mathbf{C}_{11}\boldsymbol{\Delta}_1 + \mathbf{C}_{12}\boldsymbol{\Delta}_2 = \mathbf{g}$

$$\mathbf{C}_{11} = \begin{bmatrix} -1 & 1 \\ 1 & -3 \end{bmatrix} \qquad \mathbf{C}_{12} = \begin{bmatrix} 1 \\ 1 \end{bmatrix} \qquad \underset{2,2}{\mathbf{B}_1} = \begin{bmatrix} -3 & 1 \\ 1 & -4 \end{bmatrix} \qquad \mathbf{B}_2 = \begin{bmatrix} -1 \\ -2 \end{bmatrix}$$

and compute

$$\bar{\mathbf{N}} = 80 \qquad \bar{\mathbf{t}} = 428 \qquad \boldsymbol{\Delta}_2 = x_3 = 5.35$$

$$\boldsymbol{\Delta}_1 = \begin{bmatrix} x_1 \\ x_2 \end{bmatrix} = \begin{bmatrix} -5.30 \\ -8.65 \end{bmatrix}$$

Thus the total answer vector is identical to that computed above. To compute $\mathbf{Q}_{\Delta\Delta}$, equation (9.29) is applied and the result is identical to that obtained in the direct method.

9.4. CONSTRAINTS WITH ADDED PARAMETERS

In all the cases covered in the previous sections of this chapter a tacit assumption was made that the constraint equations involve only the parameters used in the condition equations. In practice, however, there are situations in which the writing of constraints may introduce a new set of parameters, which may not be of interest themselves. Take the simple example of constraining a set of points in a plane to lie on a circle. Writing an equation of a circle may require, in addition to the coordinates of the points in question (which would belong to the parameters of the problem), a set of three added parameters. These may be, for example, the two coordinates of its center and its radius. It is, of course, possible to combine the added parameters with the original parameters and treat the problem with one of the cases already discussed. Although this is in principle possible (although singular matrices could be encountered), there may be situations in which keeping the two sets of parameters separate is easier to handle. Consequently, for the sake of completeness, this will be briefly treated hereafter.

As introduced in Section 5.3 (see Table 5.1) the symbol s' denotes the number of constraint equations when a set of q added parameters exists. In this case the total number of equations will be $(c + s')$ and the total number of parameters $(u + q)$. Therefore equations (9.1) and (9.2) become

$$s' < u + q \tag{9.47}$$

$$(c + s') = r + (u + q) \tag{9.48}$$

and furthermore

$$s' > q \tag{9.49}$$

or the number of constraints must exceed the number of added parameters, otherwise the constraints would not contribute to the adjustment problem.

The linear, or linearized, equations for the case of added parameters take the form

$$\underset{c,n}{\mathbf{A}}\ \underset{n,1}{\mathbf{v}} + \underset{c,u}{\mathbf{B}}\ \underset{u,1}{\mathbf{\Delta}} = \underset{c,1}{\mathbf{f}} \tag{9.50a}$$

$$\underset{s',u}{\mathbf{D}_1}\ \underset{u,1}{\mathbf{\Delta}} + \underset{s',q}{\mathbf{D}_2}\ \underset{q,1}{\mathbf{\Delta}'} = \underset{s',1}{\mathbf{h}} \tag{9.50b}$$

with rank $(\mathbf{A}) = c$, rank $(\mathbf{B}) =$ the smaller of c or u, rank $(\mathbf{D}_1) = s'$, and rank $(\mathbf{D}_2) = q$. The vector $\mathbf{\Delta}$ is as usual the original parameter vector of interest, whereas $\mathbf{\Delta}'$ is the vector of added parameters that appears only in the constraints.

9.4.1. The General Case In this case the selection of $u \leq c$, which makes rank $(\mathbf{B}) = u$, is specified.

Direct Treatment The minimum principle is in this case

$$\phi = \mathbf{v}^t\mathbf{W}\mathbf{v} - 2\mathbf{k}^t(\mathbf{A}\mathbf{v} + \mathbf{B}\mathbf{\Delta} - \mathbf{f}) - 2\mathbf{k}_c^t(\mathbf{D}_1\mathbf{\Delta} + \mathbf{D}_2\mathbf{\Delta}' - \mathbf{h}) \tag{9.51}$$

The partial derivatives of ϕ with respect to the three free variables $\mathbf{v}$, $\mathbf{\Delta}$, and $\mathbf{\Delta}'$ may be evaluated and equated to zero, to obtain a minimum for ϕ. This leads to three equations that may be combined with equations (9.50a) and (9.50b) to give the following *total system of normal equations:*

$$\begin{array}{c} \\ n \\ c \\ u \\ s' \\ q \end{array}
\begin{array}{c} \begin{array}{ccccc} n & c & u & s' & q \end{array} \\
\begin{bmatrix} -\mathbf{W} & \mathbf{A}^t & \mathbf{0} & \mathbf{0} & \mathbf{0} \\ \mathbf{A} & \mathbf{0} & \mathbf{B} & \mathbf{0} & \mathbf{0} \\ \mathbf{0} & \mathbf{B}^t & \mathbf{0} & \mathbf{D}_1^t & \mathbf{0} \\ \mathbf{0} & \mathbf{0} & \mathbf{D}_1 & \mathbf{0} & \mathbf{D}_2 \\ \mathbf{0} & \mathbf{0} & \mathbf{0} & \mathbf{D}_2^t & \mathbf{0} \end{bmatrix} \end{array}
\begin{bmatrix} \mathbf{v} \\ \mathbf{k} \\ \mathbf{\Delta} \\ \mathbf{k}_c \\ \mathbf{\Delta}' \end{bmatrix} = \begin{bmatrix} \mathbf{0} \\ \mathbf{f} \\ \mathbf{0} \\ \mathbf{h} \\ \mathbf{0} \end{bmatrix} \tag{9.52}$$

Obviously this equation is an extension of equation (9.7) by the addition of the $\mathbf{\Delta}'$ vector. The direct solution of equation (9.52) yields the total answer by supplying all five vectors. For practical reasons, however, it is advisable to solve the system by partitioning. Thus we recall that

$$\mathbf{v} = \mathbf{Q}\mathbf{A}^t\mathbf{k} \tag{9.8}$$

$$\mathbf{k} = \mathbf{W}_e(-\mathbf{B}\mathbf{\Delta} + \mathbf{f}) \tag{9.9}$$

$$\mathbf{N} = \mathbf{B}^t\mathbf{W}_e\mathbf{B} \qquad \text{rank } (\mathbf{N}) = u \text{ because } u < c$$

$$\mathbf{t} = \mathbf{B}^t\mathbf{W}_e\mathbf{f}$$

$$\mathbf{\Delta}^0 = \mathbf{N}^{-1}\mathbf{t}$$

With these relationships, the third vector equation in (9.52) becomes

$$\mathbf{\Delta} = \mathbf{N}^{-1}(\mathbf{t} + \mathbf{D}_1^t\mathbf{k}_c) = \mathbf{\Delta}^0 + \mathbf{N}^{-1}\mathbf{D}_1^t\mathbf{k}_c \tag{9.53}$$

which reduces equation (9.52) to the form

$$\begin{bmatrix} (\mathbf{D}_1\mathbf{N}^{-1}\mathbf{D}_1^t) & \mathbf{D}_2 \\ \mathbf{D}_2^t & \mathbf{0} \end{bmatrix} \begin{bmatrix} \mathbf{k}_c \\ \mathbf{\Delta}' \end{bmatrix} = \begin{bmatrix} \mathbf{h} - \mathbf{D}_1\mathbf{\Delta}^0 \\ \mathbf{0} \end{bmatrix} \tag{9.54}$$

Equation (9.54) may be solved directly for both $\mathbf{k}_c$ and $\mathbf{\Delta}'$ or it may be reduced further

$$\underset{s,\,1}{\mathbf{k}_c} = \mathbf{P}^{-1}(\mathbf{h} - \mathbf{D}_1\mathbf{\Delta}^0 - \mathbf{D}_2\mathbf{\Delta}') \tag{9.55}$$

$$\underset{q,\,1}{\mathbf{\Delta}'} = \mathbf{R}^{-1}[\mathbf{D}_2^t\mathbf{P}^{-1}(\mathbf{h} - \mathbf{D}_1\mathbf{\Delta}^0)] \tag{9.56}$$

$$\underset{u,\,1}{\mathbf{\Delta}} = \mathbf{\Delta}^0 + \mathbf{N}^{-1}\mathbf{D}_1^t\mathbf{P}^{-1}(\mathbf{h} - \mathbf{D}_1\mathbf{\Delta}^0 - \mathbf{D}_2\mathbf{\Delta}') \tag{9.57}$$

where

$$\mathbf{P} = \mathbf{D}_1\mathbf{N}^{-1}\mathbf{D}_1^t \tag{9.58}$$

$$\mathbf{R} = \mathbf{D}_2^t\,\mathbf{P}^{-1}\mathbf{D}_2 \tag{9.59}$$

The computation would be carried out in this sequence, first $\mathbf{N}$ and $\mathbf{t}$ are computed and used to compute $\boldsymbol{\Delta}^0$. Next, with $\mathbf{D}_1$, $\mathbf{D}_2$, and $\mathbf{h}$ from the constraint equations, the vector of added parameters $\boldsymbol{\Delta}'$ is evaluated. Using $\boldsymbol{\Delta}'$ thus computed, the original parameters $\boldsymbol{\Delta}$ can then be computed.

The a posteriori cofactor matrices are

$$\mathbf{Q}_{\Delta'\Delta'} = \mathbf{R}^{-1} \tag{9.60}$$

$$\mathbf{Q}_{\Delta\Delta} = \mathbf{N}^{-1}[\mathbf{I} - \mathbf{D}_1^t\mathbf{P}^{-1}\mathbf{D}_1\mathbf{N}^{-1} + \mathbf{D}_1^t\mathbf{P}^{-1}\mathbf{D}_2\,\mathbf{R}^{-1}\mathbf{D}_2^t\,\mathbf{P}^{-1}\mathbf{D}_1\mathbf{N}^{-1}] \tag{9.61}$$

It is important to note that in case $\boldsymbol{\Delta}'$ is not present, equation (9.61) reduces to equation (9.15b) by enforcing $\mathbf{D}_2 = \mathbf{0}$.

An interesting case arises when the matrix $\mathbf{D}_1$ happens to be a square nonsingular matrix. In this case the first two terms of equation (9.61) cancel out and only the last term remains, or

$$\mathbf{Q}_{\Delta\Delta} = \mathbf{D}_1^{-1}\mathbf{D}_2\,\mathbf{R}^{-1}\mathbf{D}_2^t(\mathbf{D}_1^t)^{-1} \tag{9.62}$$

$$\mathbf{Q}_{\Delta\Delta'} = -\mathbf{N}^{-1}\mathbf{D}_1^t\mathbf{P}^{-1}\mathbf{D}_2\,\mathbf{R}^{-1} \tag{9.63}$$

Two useful computational checks are possible,

$$\mathbf{F} = \mathbf{Q}_{\Delta\Delta}\,\mathbf{D}_1^t + \mathbf{Q}_{\Delta\Delta'}\,\mathbf{D}_2^t = \mathbf{0} \tag{9.64}$$

$$\mathbf{G} = \mathbf{Q}_{\Delta\Delta'}\,\mathbf{D}_1^t + \mathbf{Q}_{\Delta'\Delta'}\,\mathbf{D}_2^t = \mathbf{0} \tag{9.65}$$

The cofactor matrices of $\mathbf{v}$ and $\hat{\boldsymbol{l}}$ are the same as those given in equations (9.17b) and (6.31).

Solution by Eliminating Constraints From q of the s' constraint equations ($q < s'$) eliminate the q added parameters $\boldsymbol{\Delta}'$. The remaining $(s' - q)$ constraint equations are then used to eliminate as many parameters from $\boldsymbol{\Delta}$. Partition equation (9.50b),

$$\underset{s,u}{\mathbf{D}_{11}}\ \underset{u,1}{\boldsymbol{\Delta}} + \underset{s,q}{\mathbf{D}_{12}}\ \underset{q,1}{\boldsymbol{\Delta}'} = \underset{s,1}{\mathbf{h}_1} \tag{9.66a}$$

$$\underset{q,u}{\mathbf{D}_{21}}\ \underset{u,1}{\boldsymbol{\Delta}} + \underset{q,q}{\mathbf{D}_{22}}\ \underset{q,1}{\boldsymbol{\Delta}'} = \underset{q,1}{\mathbf{h}_2} \tag{9.66b}$$

where

$$s = s' - q$$

Solve equation (9.66b) for Δ' in terms of Δ,

$$\underset{q,1}{\Delta'} = \mathbf{D}_{22}^{-1}(\mathbf{h}_2 - \mathbf{D}_{21}\Delta) \tag{9.66c}$$

making sure that $\mathbf{D}_{22}$ is nonsingular during the partitioning operation, and substitute into equation (9.66a)

$$(\mathbf{D}_{11} - \mathbf{D}_{12}\mathbf{D}_{22}^{-1}\mathbf{D}_{21})\Delta = \mathbf{h}_1 - \mathbf{D}_{12}\mathbf{D}_{22}^{-1}\mathbf{h}_2 \tag{9.66d}$$

Let

$$\underset{s,u}{\mathbf{C}} = \mathbf{D}_{11} - \mathbf{D}_{12}\mathbf{D}_{22}^{-1}\mathbf{D}_{21} \tag{9.67a}$$

$$\underset{s,1}{\mathbf{g}} = \mathbf{h}_1 - \mathbf{D}_{12}\mathbf{D}_{22}^{-1}\mathbf{h}_2 \tag{9.67b}$$

then equation (9.66d) combined with equation (9.50a) becomes the same as equations (9.3a) and (9.3b) and can be treated in a manner identical to that explained previously.

With respect to precision estimation, the cofactor matrix of the original parameters $\mathbf{Q}_{\Delta\Delta}$ may be computed from equation (9.29).

On the other hand $\mathbf{Q}_{\Delta'\Delta'}$ may be computed directly from equation (9.66c) as

$$\mathbf{Q}_{\Delta'\Delta'} = \mathbf{D}_{22}^{-1}\mathbf{D}_{21}\mathbf{Q}_{\Delta\Delta}\mathbf{D}_{21}^{t}(\mathbf{D}_{22}^{-1})^{t} \tag{9.68}$$

Example 9.5. The following case of adjustment is to estimate Δ, Δ', and their combined cofactor matrix:

$$\mathbf{A}\mathbf{v} + \mathbf{B}\Delta = \mathbf{f} \qquad \text{with } \mathbf{Q} = \mathbf{I}$$

$$\mathbf{D}_1\Delta + \mathbf{D}_2\Delta' = \mathbf{h}$$

with

$$\underset{3,3}{\mathbf{A}} = \mathbf{I} \qquad \underset{3,2}{\mathbf{B}} = \begin{bmatrix} 2 & -3 \\ -1 & 2 \\ 0 & 1 \end{bmatrix} \qquad \underset{3,1}{\mathbf{f}} = \begin{bmatrix} -1.1 \\ 1.2 \\ 1.0 \end{bmatrix} \qquad \underset{2,1}{\mathbf{h}} = \begin{bmatrix} -1 \\ 3 \end{bmatrix}$$

$$\underset{2,2}{\mathbf{D}_1} = \begin{bmatrix} 1 & -1 \\ 2 & -1 \end{bmatrix} \qquad \underset{2,1}{\mathbf{D}_2} = \begin{bmatrix} 1 \\ -2 \end{bmatrix} \qquad \underset{2,1}{\Delta} = \begin{bmatrix} x_1 \\ x_2 \end{bmatrix} \qquad \Delta' = x_3$$

Solution:

$$\mathbf{N} = \begin{bmatrix} 5 & -8 \\ -8 & 14 \end{bmatrix} \qquad \mathbf{t} = \begin{bmatrix} -3.4 \\ 6.7 \end{bmatrix} \qquad \Delta^0 = \begin{bmatrix} 1.00 \\ 1.05 \end{bmatrix}$$

Using equation (9.56)

$$\Delta' = -0.97922$$

Next from equation (9.57)

$$\Delta = \begin{bmatrix} x_1 \\ x_2 \end{bmatrix} = \begin{bmatrix} 1.00 \\ 1.05 \end{bmatrix} + \begin{bmatrix} 0.06233 \\ 0.03311 \end{bmatrix} = \begin{bmatrix} 1.06233 \\ 1.08311 \end{bmatrix}$$

hence the final vector is

$$\begin{bmatrix} x_1 \\ x_2 \\ x_3 \end{bmatrix} = \begin{bmatrix} 1.06233 \\ 1.08311 \\ -0.97922 \end{bmatrix}$$

To compute $\mathbf{Q}_{\Delta\Delta}$ for x_1 and x_2, use equation (9.62) because $\mathbf{D}_1$ is square and nonsingular. Thus

$$\mathbf{Q}_{\Delta\Delta} = \begin{bmatrix} 0.116883 & 0.155844 \\ \text{symmetric} & 0.207792 \end{bmatrix}$$

Next

$$\mathbf{Q}_{\Delta'\Delta'} = \mathbf{R}^{-1} = 1/77 = 0.012987$$

Finally, the crosscofactor matrix is computed from equation (9.63) as

$$\mathbf{Q}_{\Delta\Delta'} = [0.038961 \quad 0.051948]^t$$

leading to the final cofactor matrix for all three variables

$$\mathbf{Q}_{xx} = \begin{bmatrix} 0.116883 & 0.155844 & 0.038961 \\ & 0.207792 & 0.051948 \\ \text{symmetric} & & 0.012987 \end{bmatrix}$$

Solution by elimination is as follows:

$$\mathbf{D}_{11} = [1 \quad -1] \qquad \mathbf{D}_{12} = 1 \qquad \mathbf{h}_1 = -1$$

$$\mathbf{D}_{21} = [2 \quad -1] \qquad \mathbf{D}_{22} = -2 \qquad \mathbf{h}_2 = -3$$

$$\mathbf{C} = \mathbf{D}_{11} - \mathbf{D}_{12}\mathbf{D}_{22}^{-1}\mathbf{D}_{21} = [4 \quad -3]$$

$$\mathbf{g} = \mathbf{h}_1 - \mathbf{D}_{12}\mathbf{D}_{22}^{-1}\mathbf{h}_2 = 1$$

$$\mathbf{B}_1 = \begin{bmatrix} 2 \\ -1 \\ 0 \end{bmatrix} \qquad \mathbf{B}_2 = \begin{bmatrix} -3 \\ 2 \\ 1 \end{bmatrix} \qquad \mathbf{C}_{11} = 4 \qquad \mathbf{C}_{12} = -3$$

Thus

$$\bar{\mathbf{B}} = -\mathbf{B}_1\mathbf{C}_{11}^{-1}\mathbf{C}_{12} + \mathbf{B}_2 = [-1.50 \quad 1.25 \quad 1.00]^t$$

$$\bar{\mathbf{f}} = -\mathbf{B}_1\mathbf{C}_{11}^{-1}\mathbf{g} + \mathbf{f} = [-1.60 \quad 1.45 \quad 1.00]^t$$

from which, because both **A** and **Q** are identity matrices, we get,

$$\Delta_2 = x_2 = (\bar{\mathbf{B}}^t\bar{\mathbf{B}})^{-1}(\bar{\mathbf{B}}^t\bar{\mathbf{f}}) = 1.08311$$

Then

$$\Delta_1 = x_1 = \mathbf{C}_{11}^{-1}(\mathbf{g} - \mathbf{C}_{12}\,\Delta_2) = 1.06233$$

$$\Delta' = x_3 = \mathbf{D}_{22}^{-1}(\mathbf{h}_2 - \mathbf{D}_{21}\,\Delta) = -0.97933$$

thus the total vector is identical to that computed before. As for a posteriori cofactor matrices, suffice it to compute $\mathbf{Q}_{\Delta\Delta}$ for x_1 and x_2. Using equation (9.29),

$$\mathbf{Q}_{x1x1} = \mathbf{C}_{11}^{-1}\mathbf{C}_{12}\,\bar{\mathbf{N}}^{-1}\mathbf{C}_{12}^t\,\mathbf{C}_{11}^{-1} = 0.116883$$

$$\mathbf{Q}_{x1x2} = -\mathbf{C}_{11}^{-1}\mathbf{C}_{12}\,\bar{\mathbf{N}}^{-1} = -0.155844$$

$$\mathbf{Q}_{x2x2} = \bar{\mathbf{N}}^{-1} = 0.207792$$

with the final result the same as that evaluated previously.

Example 9.6

Given: Figure 9.3 shows essentially the same geometric layout of the problem in example 9.2, but expanded further to include four object points *A*, *B*, *C*, and *D*. The measurements for these four points on the three photographs are given in Table 9-1. The standard deviation for all the above measurements is 0.10 mm, the principal distance is $p = 100$ mm; $l_4 = 10$ m and $l_5 = 8$ m, and their standard deviation is 0.05 m (no correlation).

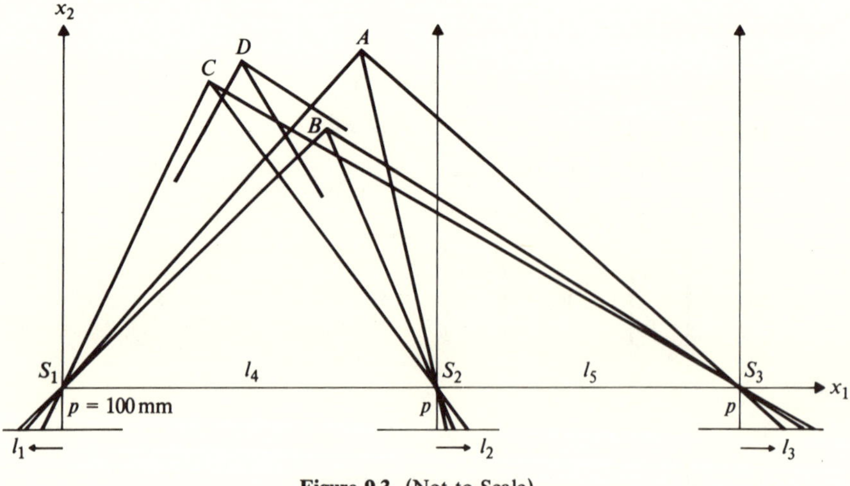

Figure 9.3. (Not to Scale)

TABLE 9-1

IMAGE POINT	$l_1 = \lvert x_1 \rvert$(mm) ON $S1$	$l_2 = \lvert x_2 \rvert$(mm) ON $S2$	$l_3 = \lvert x_3 \rvert$(mm) on $S3$
a	14.1	6.1	22.1
b	16.6	7.1	26.3
c	6.5	15.2	32.5
d	8.3	12.3	28.7

Required: (a) Compute the coordinates of the four points *A*, *B*, *C*, and *D*. (b) Compute a new set of coordinates for the four points when they are constrained to lie on a circle whose equation is

$$(x_{1i} - x_{10})^2 + (x_{2i} - x_{20})^2 - R^2 = 0 \qquad i = 1, \ldots, 4 \text{ (for } A, B, C, D)$$

where x_{10}, x_{20}, and R (representing the coordinates of the center of the circle and its radius) are three added parameters. Use the following approximations for the original and added parameters (the numbers between parenthesis represent the x_1 and x_2 coordinates, respectively).

$$A = (8, 51), \quad B = (7, 41), \quad C = (4, 47), \quad D = (4, 48)$$

$$x_{10} = 7, \quad x_{20} = 46, \quad R = 4$$

all values are in metres. Use both the direct procedure and the one by elimination of constraints.

Solution:

Part a. The number of observations is $n = 14$, the minimum number of variables necessary to define the geometry of the problem is $n_0 = 10$, redundancy $r = 4$; unknown parameters to be solved for are $u = 8$; and number of condition equations are $c = r + u = 12$. The condition equations are as follows

$$\begin{aligned}
&l_{1a}x_{2a} - px_{1a} = 0 && l_{1c}x_{2c} - px_{1c} = 0\\
&l_{2a}x_{2a} - pl_4 + px_{1a} = 0 && l_{2c}x_{2c} - p_4 + px_{1c} = 0\\
&l_{3a}x_{2a} - pl_4 - pl_5 + px_{1a} = 0 && l_{3c}x_{2c} - pl_4 - pl_5 + px_{1c} = 0\\
&l_{1b}x_{2b} - px_{1b} = 0 && l_{1d}x_{2d} - px_{1d} = 0\\
&l_{2b}x_{2b} - pl_4 + px_{1b} = 0 && l_{2d}x_{2d} - pl_4 + px_{1d} = 0\\
&l_{3b}x_{2b} - pl_4 - pl_5 + px_{1b} = 0 && l_{3d}x_{2d} - pl_4 - pl_5 + px_{1d} = 0
\end{aligned}$$

where l_i represent observational variables.

Linearization: For linearization purposes choose the sequence of the elements in the observation and parameter vectors as follows:

$$\underset{14,\,1}{\boldsymbol{l}} = [l_{1a} \quad l_{2a} \quad l_{3a} \quad l_{1b} \quad l_{2b} \quad l_{3b} \quad l_{1c} \quad l_{2c} \quad l_{3c} \quad l_{1d} \quad l_{2d} \quad l_{3d} \quad l_4 \quad l_5]^t$$

$$\underset{8,\,1}{\mathbf{x}} = [x_{1a} \quad x_{2a} \quad x_{1b} \quad x_{2b} \quad x_{1c} \quad x_{2c} \quad x_{1d} \quad x_{2d}]^t$$

$$
\underset{12,\,14}{\mathbf{A}} =
\begin{bmatrix}
x_{2a}^0 & 0 & \cdot & \cdot & \cdot & \cdot & \cdot & \cdot & \cdot & \cdot & \cdot & 0 & 0 & 0 \\
0 & x_{2a}^0 & \cdot & \cdot & \cdot & \cdot & \cdot & \cdot & \cdot & \cdot & \cdot & 0 & -p & 0 \\
0 & 0 & x_{2a}^0 & \cdot & \cdot & \cdot & \cdot & \cdot & \cdot & \cdot & \cdot & 0 & -p & -p \\
0 & & 0 & x_{2b}^0 & \cdot & \cdot & \cdot & \cdot & \cdot & \cdot & \cdot & 0 & 0 & 0 \\
0 & & & 0 & x_{2b}^0 & \cdot & \cdot & \cdot & \cdot & \cdot & \cdot & 0 & -p & 0 \\
0 & & & & 0 & x_{2b}^0 & \cdot & \cdot & \cdot & \cdot & \cdot & 0 & -p & -p \\
0 & & & & & 0 & x_{2c}^0 & \cdot & \cdot & \cdot & \cdot & 0 & 0 & 0 \\
0 & & & & & & 0 & x_{2c}^0 & \cdot & \cdot & \cdot & 0 & -p & 0 \\
0 & & & & & & & 0 & x_{2c}^0 & \cdot & \cdot & 0 & -p & -p \\
0 & & & & & & & & 0 & x_{2d}^0 & \cdot & 0 & 0 & 0 \\
0 & & & & & & & & & 0 & x_{2d}^0 & 0 & -p & 0 \\
0 & \cdot & \cdot & \cdot & & & & & & & 0 & x_{2d}^0 & -p & -p
\end{bmatrix}
$$

$$
\underset{12,\,8}{\mathbf{B}} =
\begin{bmatrix}
-p & l_{1a} & 0 & 0 & 0 & 0 & 0 & 0 \\
p & l_{2a} & 0 & 0 & \cdot & \cdot & & \\
p & l_{3a} & 0 & 0 & \cdot & \cdot & & \\
0 & 0 & -p & l_{1b} & \cdot & \cdot & \cdot & \cdot \\
 & & p & l_{2b} & & & \cdot & \cdot \\
\cdot & \cdot & p & l_{3b} & 0 & 0 & \cdot & \cdot \\
\cdot & \cdot & 0 & 0 & -p & l_{1c} & & \\
\cdot & \cdot & \cdot & \cdot & p & l_{2c} & & \\
 & & \cdot & \cdot & p & l_{3c} & 0 & 0 \\
 & & \cdot & \cdot & 0 & 0 & -p & l_{1d} \\
 & & & & 0 & 0 & p & l_{2d} \\
0 & 0 & 0 & 0 & 0 & 0 & p & l_{3d}
\end{bmatrix}
$$

$$
\underset{12,\,1}{\mathbf{f}} =
\begin{bmatrix}
px_{1a}^0 - l_{1a}x_{2a}^0 \\
pl_4 - l_{2a}x_{2a}^0 - px_{1a}^0 \\
pl_4 + pl_5 - l_{3a}x_{2a}^0 - px_{1a}^0 \\
px_{1b}^0 - l_{1b}x_{2b}^0 \\
pl_4 - l_{2b}x_{2b}^0 - px_{1b}^0 \\
pl_4 + pl_5 - l_{3b}x_{2b}^0 - px_{1b}^0 \\
px_{1c}^0 - l_{1c}x_{2c}^0 \\
pl_4 - l_{2c}x_{2c}^0 - px_{1c}^0 \\
pl_4 + pl_5 - l_{3c}x_{2c}^0 - px_{1c}^0 \\
px_{1d}^0 - l_{1d}x_{2d}^0 \\
pl_4 - l_{2d}x_{2d}^0 - px_{1d}^0 \\
pl_4 + pl_5 - l_{3d}x_{2d}^0 - px_{1d}^0
\end{bmatrix}
$$

The a priori cofactor matrix is diagonal (assuming $\sigma_0^2 = 1$)

$$\underset{14,\,14}{\mathbf{Q}} = \text{diag.}\{0.01, 0.01, 0.01, 0.01, 0.01, 0.01, 0.01, 0.01, 0.01, 0.01, 0.01, 0.01, 0.0025, 0.0025\}$$

Using the given approximations and observations, the above matrices may be evaluated and used to construct and solve the reduced normal equations. For purposes of checking, the values from the first and last iterations are given as follows:

$$\mathbf{\Delta}_{(1)} = [-1.00344 \quad -1.28713 \quad -0.01703 \quad -0.96588 \quad -1.00480 \quad -0.85048 \quad 0.03100 \quad 0.64280]^t(\text{m})$$

$$\mathbf{\Delta}_{(2)} = [0.00019 \quad -0.00029 \quad 0.00019 \quad -0.00026 \quad 0.00022 \quad -0.00027 \quad 0.00021 \quad -0.00028]^t(\text{m})$$

The elements of this vector are obviously sufficiently small that the process of iterations can be terminated giving the final required answer.

$$\hat{x}_{1a} = 6.997 \text{ m} \qquad \hat{x}_{1c} = 2.995 \text{ m}$$

$$\hat{x}_{2a} = 49.713 \text{ m} \qquad \hat{x}_{2c} = 49.149 \text{ m}$$

$$\hat{x}_{1b} = 6.983 \text{ m} \qquad \hat{x}_{1d} = 4.031 \text{ m}$$

$$\hat{x}_{2b} = 41.966 \text{ m} \qquad \hat{x}_{2d} = 48.643 \text{ m}$$

Part (b). First, reconsider the minimum number of variables that established the geometry of the problem (Table 9-2). Consequently, the minimum number of variables in this case is $n_0 = 9$, thus $r = 14 - 9 = 5$. Therefore

$$c + s = (r) + (u + q) = 5 + (8 + 3) = 16 \text{ equations}$$

TABLE 9-2

	NUMBER OF VARIABLES
Point S_1 is fixed by virtue of being the origin of the coordinate system	
Points S_2, S_3 being on the x_1 axis require one coordinate each to locate	2
Point A, two coordinates	2
Point B, two coordinates	2
Point C, two coordinates	2
(These three points would now determine a unique circle on which point D must lie. With this information about point D, only one more parameter is needed in order to fix its position)	1
Total	9

A logical choice would be to have $c = 12$ as in part (a) above and therefore write $s = 4$ constraints, one for each point. This simplifies the writing of both condition and constraint equations considerably, and satisfies all pertinent relations, particularly $s' > q$.

The constraint equations therefore take the following form:

$$(x_{1a} - x_{10})^2 + (x_{2a} - x_{20})^2 - R^2 = 0$$

$$(x_{1b} - x_{10})^2 + (x_{2b} - x_{20})^2 - R^2 = 0$$

$$(x_{1c} - x_{10})^2 + (x_{2c} - x_{20})^2 - R^2 = 0$$

$$(x_{1d} - x_{10})^2 + (x_{2d} - x_{20})^2 - R^2 = 0$$

Linearization:

$$\underset{4,8}{\mathbf{D}_1} = \begin{bmatrix} 2(x_{1a}^0 - x_{10}^0) & 2(x_{2a}^0 - x_{20}^0) & 0 & 0 & 0 & 0 & 0 & 0 \\ 0 & 0 & 2(x_{1b}^0 - x_{10}^0) & 2(x_{2b}^0 - x_{20}^0) & 0 & 0 & 0 & 0 \\ 0 & 0 & 0 & 0 & 2(x_{1c}^0 - x_{10}^0) & 2(x_{2c}^0 - x_{20}^0) & 0 & 0 \\ 0 & 0 & 0 & 0 & 0 & 0 & 2(x_{1d}^0 - x_{10}^0) & 2(x_{2d}^0 - x_{20}^0) \end{bmatrix}$$

$$\underset{4,3}{\mathbf{D}_2} = \begin{bmatrix} -2(x_{1a}^0 - x_{10}^0) & -2(x_{2a}^0 - x_{20}^0) & -2R^0 \\ -2(x_{1b}^0 - x_{10}^0) & -2(x_{2b}^0 - x_{20}^0) & -2R^0 \\ -2(x_{1c}^0 - x_{10}^0) & -2(x_{2c}^0 - x_{20}^0) & -2R^0 \\ -2(x_{1d}^0 - x_{10}^0) & -2(x_{2d}^0 - x_{20}^0) & -2R^0 \end{bmatrix}$$

$$\underset{4,1}{\mathbf{h}} = \begin{bmatrix} R^{02} - (x_{1a}^0 - x_{10}^0)^2 - (x_{2a}^0 - x_{20}^0)^2 \\ R^{02} - (x_{1b}^0 - x_{10}^0)^2 - (x_{2b}^0 - x_{20}^0)^2 \\ R^{02} - (x_{1c}^0 - x_{10}^0)^2 - (x_{2c}^0 - x_{20}^0)^2 \\ R^{02} - (x_{1d}^0 - x_{10}^0)^2 - (x_{2d}^0 - x_{20}^0)^2 \end{bmatrix}$$

Again, for checking purposes we have

First Iteration

$$\underset{8,1}{\boldsymbol{\Delta}} = [-1.00598 \quad -1.24835 \quad -0.01922 \quad -0.98883 \quad -1.02657 \quad -0.74275 \quad 0.05950 \quad 0.42954]^t$$

$$\underset{3,1}{\Delta'} = [-0.23980 \quad -0.15638 \quad -0.30651]^t$$

Third and Last Iteration

$$\Delta_{(3)} = [-0.00011 \quad 0.00019 \quad -0.00012 \quad 0.00124 \quad 0.00032$$
$$-0.00274 \quad 0.00008 \quad -0.00091]^t$$
$$\Delta'_{(3)} = [-0.00119 \quad 0.00063 \quad -0.00415]^t$$

The final answer is then

$\hat{x}_{1a} = 6.995$ m $\qquad \hat{x}_{1d} = 4.043$ m

$\hat{x}_{2a} = 49.774$ m $\qquad \hat{x}_{2d} = 48.552$ m

$\hat{x}_{1b} = 6.983$ m $\qquad \hat{x}_{10} = 6.879$ m

$\hat{x}_{2b} = 41.980$ m $\qquad \hat{x}_{20} = 45.876$ m

$\hat{x}_{1c} = 2.992$ m $\qquad \hat{R} = 3.898$ m

$\hat{x}_{2c} = 46.166$ m

The a posteriori cofactor matrices are (the differences between first iteration results, given here, and last iteration were insignificant) as follows.

a. *Without Circle Constraint*

$$\underset{8,8}{\mathbf{Q}_{\Delta\Delta}} = 10^{-3}\begin{bmatrix} 1.73 & 3.40 & 0.77 & 4.44 & 0.35 & 4.88 & 0.46 & 5.14 \\ & 77.27 & 5.25 & 31.87 & 2.22 & 35.05 & 3.00 & 36.94 \\ & & 1.39 & 3.41 & 0.35 & 4.87 & 0.46 & 5.13 \\ & & & 45.11 & 1.87 & 29.59 & 2.53 & 31.19 \\ & & & & 1.46 & -1.91 & 0.22 & 2.17 \\ & & & & & 61.46 & 2.79 & 34.29 \\ & & & & & & 1.45 & -0.71 \\ \text{symmetric} & & & & & & & 69.66 \end{bmatrix}$$

rank ($\mathbf{Q}_{\Delta\Delta} = 8 = u$)

$$
Q_{ll} = 10^{-3}\begin{bmatrix}
8.87 & -2.56 & 1.43 & 0.21 & 0.46 & -0.25 & 0.18 & 0.40 & -0.22 & 0.17 & 0.39 & -0.22 & 0.25 & -0.31 \\
 & 4.20 & 3.24 & 0.47 & 1.04 & -0.57 & 0.40 & 0.91 & -0.50 & 0.39 & 0.89 & -0.49 & 0.55 & -0.71 \\
 & & 8.19 & -0.26 & -0.58 & 0.32 & -0.22 & -0.51 & 0.28 & -0.22 & -0.50 & 0.28 & -0.31 & 0.40 \\
 & & & 8.93 & 2.39 & 1.32 & 0.22 & 0.50 & 0.28 & 0.22 & 0.49 & 0.27 & 0.32 & -0.39 \\
 & & & & 4.56 & 2.96 & 0.50 & 1.13 & -0.63 & 0.49 & 1.10 & -0.61 & 0.81 & -0.86 \\
 & & & & & 8.37 & -0.28 & -0.62 & 0.35 & 1.10 & -0.61 & 0.34 & -0.34 & 0.48 \\
 & & & & & & 8.89 & -2.51 & 1.40 & -0.19 & 0.43 & -0.24 & 0.27 & -0.34 \\
 & & & & & & & 4.35 & 3.15 & 0.43 & 0.96 & -0.54 & 0.61 & -0.77 \\
 & & & & & & & & 3.25 & -0.24 & -0.54 & 0.29 & -0.34 & 0.43 \\
 & & & & & & & & & 8.88 & -2.52 & 1.41 & 0.26 & -0.33 \\
 & & & & & & & & & & 4.30 & 3.17 & 0.59 & -0.75 \\
 & & & & & & & & & & & 8.28 & -0.33 & 0.42 \\
 & & & & & & & & & & & & 1.90 & 0.75 \\
\text{symmetric} & & & & & & & & & & & & & 1.57
\end{bmatrix}
$$

$$\text{rank}\ (\mathbf{Q}_{ll}) = 10 = n - r$$

$$
\mathbf{Q}_{pp} = 10^{-3}
\left[\begin{array}{cccccccc:ccc}
1.73 & 3.45 & 0.77 & 4.47 & 0.32 & 5.02 & 0.50 & 4.86 & -0.44 & 4.17 & -0.36 \\
 & 76.49 & 5.20 & 31.41 & 2.66 & 32.87 & 2.43 & 41.24 & 39.95 & 50.30 & 23.61 \\
 & & 1.39 & 3.43 & 0.33 & 4.99 & 0.49 & 4.89 & 1.51 & 4.29 & 1.07 \\
 & & & 44.83 & 2.13 & 28.30 & 2.19 & 33.73 & -4.08 & 38.98 & -7.32 \\
 & & & & 1.21 & -0.69 & 0.54 & -0.25 & 2.28 & 2.20 & 0.08 \\
 & & & & & 55.42 & 1.19 & 46.25 & -3.46 & 31.43 & 3.92 \\
 & & & & & & 1.02 & 2.45 & 1.01 & 2.26 & 0.08 \\
(\mathbf{Q}_{\Delta\Delta}) & & & & & & & 45.99 & 3.38 & 37.64 & 4.88 \\
\hdashline
 & & & & & & & & 38.35 & 14.06 & 22.67 \\
 & & & & & & & & & 43.65 & 5.84 \\
\text{symmetric} & & & & & & & & (\mathbf{Q}_{\Delta'\Delta'}) & & 16.46
\end{array}\right]
$$

The reference variance and reference standard deviation are

$$\hat{\sigma}_0^2 = 0.4852$$

$$\hat{\sigma}_0 = 0.70$$

b. *With Circle Constraint.* Because there are three additional parameters, $\mathbf{Q}_{\Delta\Delta}$, $\mathbf{Q}_{\Delta'\Delta'}$, and $\mathbf{Q}_{\Delta\Delta'}$ were computed separately and collected in a matrix of cofactors $\mathbf{Q}_{pp}$ for all 11 parameters. The value of $\mathbf{Q}_{pp}$ after the first iteration is on page 245. The computed rank of the 8×8 submatrix $\mathbf{Q}_{\Delta\Delta}$ is 7. This can be explained by recognizing that the number of independent primary parameters is $u - (s' - q) = 8 - (4 - 3) = 7$. The rank of $\mathbf{Q}_{\Delta'\Delta'}$ was also computed and found to be equal to 3 as expected.

The computational checks of equations (9.64) and (9.65) yield

$$\underset{4,8}{\mathbf{F}} = 10^{-11}\begin{bmatrix} 0.05 & 0.07 & 0.05 & -0.09 & -1.5 & 2.9 & -0.1 & -0.4 \\ -0.07 & 0.27 & 0 & -0.02 & 2.2 & 0 & 0.4 & 0.4 \\ -0.05 & 0.02 & -0.02 & 0 & -2.9 & 3.6 & -0.4 & -0.4 \\ 0.3 & -0.1 & 0.03 & 0.08 & 1.5 & 1.1 & 0.7 & 1.1 \end{bmatrix}$$

$$\underset{3,4}{\mathbf{G}} = 10^{-11}\begin{bmatrix} -1.8 & 0.68 & 0.82 & -0.43 \\ -1.1 & 1.8 & 0.41 & -0.73 \\ -1.5 & 0 & -0.05 & -0.50 \end{bmatrix}$$

both of which are effectively zero. Finally,

$$\hat{\sigma}_0^2 = 0.7724$$
$$\hat{\sigma}_0 = 0.8789$$

Solution by Elimination of Constraints: First, $s = s' - q = 1$. The partitioned matrices have the following submatrices:

$$\underset{1,8}{\mathbf{D}_{11}} = [2(x_{1a}^0 - x_{10}^0) \quad 2(x_{2a}^0 - x_{10}^0) \quad 0 \quad 0 \quad 0 \quad 0 \quad 0 \quad 0]$$

$$\underset{1,3}{\mathbf{D}_{12}} = [2(x_{1a}^0 - x_{10}^0) \quad -2(x_{2a}^0 - x_{20}^0) \quad -2R^0]$$

$$\underset{1,1}{\mathbf{h}_1} = (R^{02} - (x_{1a}^0 - x_{10}^0)^2 - (x_{2a}^0 - x_{20}^0)^2)$$

$$\underset{3,8}{\mathbf{D}_{21}} = \begin{bmatrix} 0 & 0 & 2(x_{1b}^0 - x_{10}^0) & 2(x_{2b}^0 - x_{20}^0) & 0 & 0 & 0 & 0 \\ 0 & 0 & 0 & 0 & 2(x_{1c}^0 - x_{10}^0) & 2(x_{2c}^0 - x_{20}^0) & 0 & 0 \\ 0 & 0 & 0 & 0 & 0 & 0 & 2(x_{1d}^0 - x_{10}^0) & 2(x_{2d}^0 - x_{20}^0) \end{bmatrix}$$

$$\underset{3,3}{\mathbf{D}_{22}} = \begin{bmatrix} -2(x_{1b}^0 - x_{10}^0) & -2(x_{2b}^0 - x_{20}^0) & -2R^0 \\ -2(x_{1c}^0 - x_{10}^0) & -2(x_{2c}^0 - x_{20}^0) & -2R^0 \\ -2(x_{1d}^0 - x_{10}^0) & -2(x_{2d}^0 - x_{20}^0) & -2R^0 \end{bmatrix}$$

$$\underset{3,1}{\mathbf{h}_2} = \begin{bmatrix} (R^{02} - (x_{1b}^0 - x_{10}^0)^2 - (x_{2b}^0 - x_{20}^0)^2) \\ (R^{02} - (x_{1c}^0 - x_{10}^0)^2 - (x_{2c}^0 - x_{20}^0)^2) \\ (R^{02} - (x_{1d}^0 - x_{10}^0)^2 - (x_{2d}^0 - x_{20}^0)^2) \end{bmatrix}$$

Using the given approximations, the numerical values of the matrices are

$$\mathbf{D}_{11} = [2 \quad 10 \quad 0 \quad 0 \quad 0 \quad 0 \quad 0 \quad 0]$$

$$\mathbf{D}_{12} = [-2 \quad -10 \quad -9]$$

$$\mathbf{h}_1 = -10$$

$$\mathbf{D}_{21} = \begin{bmatrix} 0 & 0 & 0 & -10 & 0 & 0 & 0 & 0 \\ 0 & 0 & 0 & 0 & -6 & 2 & 0 & 0 \\ 0 & 0 & 0 & 0 & 0 & 0 & -6 & 4 \end{bmatrix}$$

$$\mathbf{D}_{22} = \begin{bmatrix} 0 & 10 & -8 \\ 6 & -2 & -8 \\ 6 & -4 & -8 \end{bmatrix} \qquad \mathbf{h}_2 = \begin{bmatrix} -9 \\ 6 \\ 3 \end{bmatrix} \qquad \mathbf{c}_{11} = 2$$

$$\mathbf{C}_{12} = [10 \quad 0 \quad 13.33 \quad -74.0 \quad 24.67 \quad 72.0 \quad -48.0] \qquad \mathbf{g} = 40$$

The **B** matrix, accordingly partitioned, becomes

$$\mathbf{B}_1 = \begin{bmatrix} -100 \\ 100 \\ 100 \\ 0 \\ 0 \\ 0 \\ 0 \\ 0 \\ 0 \\ 0 \\ 0 \\ 0 \end{bmatrix} \qquad \mathbf{B}_2 = \begin{bmatrix} 14.1 & 0 & 0 & 0 & 0 & 0 & 0 \\ 6.1 & 0 & 0 & 0 & 0 & 0 & 0 \\ 22.1 & 0 & 0 & 0 & 0 & 0 & 0 \\ 0 & -100 & 16.6 & 0 & 0 & 0 & 0 \\ 0 & 100 & 7.1 & 0 & 0 & 0 & 0 \\ 0 & 100 & 26.3 & 0 & 0 & 0 & 0 \\ 0 & 0 & 0 & -100 & 6.5 & 0 & 0 \\ 0 & 0 & 0 & 100 & 15.2 & 0 & 0 \\ 0 & 0 & 0 & 100 & 32.5 & 0 & 0 \\ 0 & 0 & 0 & 0 & 0 & -100 & 8.3 \\ 0 & 0 & 0 & 0 & 0 & 100 & 12.3 \\ 0 & 0 & 0 & 0 & 0 & 100 & 28.7 \end{bmatrix}$$

Then

$$\underset{7,1}{\boldsymbol{\Delta}_2} = [-1.24835 \quad -0.01922 \quad 0.98883 \quad -1.02657 \quad -0.74275 \quad 0.05950 \quad 0.42954]^t$$

$$\boldsymbol{\Delta}_1 = -1.00598$$

$$\boldsymbol{\Delta}' = \begin{bmatrix} -0.23980 \\ -0.15638 \\ -0.30651 \end{bmatrix}$$

Adding these corrections to the given approximations, the solution is iterated until we get the final answer as

$$\begin{array}{lll} \hat{x}_{1a} = 6.994 \text{ m} & \hat{x}_{1b} = 6.983 \text{ m} & \hat{x}_{1c} = 2.992 \text{ m} \\ \hat{x}_{2a} = 49.774 \text{ m} & \hat{x}_{2b} = 41.980 \text{ m} & \hat{x}_{2c} = 46.166 \text{ m} \\ \hat{x}_{1d} = 4.043 \text{ m} & \hat{x}_{10} = 6.879 \text{ m} & \hat{R} = 3.898 \text{ m} \\ \hat{x}_{2d} = 48.552 \text{ m} & \hat{x}_{20} = 45.877 \text{ m} & \end{array}$$

As far as precision estimation is concerned, $\mathbf{Q}_{\Delta\Delta}$ was computed according to equation (9.29), and $\mathbf{Q}_{\Delta'\Delta'}$ according to equation (9.68), and both matrices were identical to those computed previously.

Example 9.7 Piecewise Polynomial Surface Fitting. In a practical application using side-looking airborne radar (SLAR) imagery to determine horizontal positions of terrain points, an interpolation method was used applying piecewise polynomials. The imagery covered a rectangular area with 21 control points. Two "pieces" of polynomial surfaces, the first covering 10 control points and the second 11 control points, were chosen. At the junction between pieces, the x coordinate is constant.

The polynomials in X have coefficients that are independent from those in Y and are also slightly different.

$$X = a_0 + a_1x + a_2x^2 + a_3y + a_4xy + a_5x^2y$$
$$Y = b_0 + b_1x + b_2x^2 + b_3y + b_4xy + b_5y^2$$

In these equations X and Y represent the control coordinates; x and y are the coordinates in the imagery either directly from measurements or after it has been subjected to preliminary transformation (for example, four parameter); and $a_0, \ldots, a_5$; $b_0, \ldots, b_5$ are the transformation parameters. The two surfaces represented by the two pieces must have the same value for the functions at the joining line. Since that line is of constant x, then for the X function (the subscript outside the parenthesis refers to the number of the "piece"),

$$(a_0 + a_1x + a_2x^2)_1 = (a_0 + a_1x + a_2x^2)_2 \qquad \text{for the constant term}$$
$$(a_3 + a_4x + a_5x^2)_1 = (a_3 + a_4x + a_5x^2)_2 \qquad \text{for the term in } y$$

and for the Y function

$$(b_0 + b_1x + b_2x^2)_1 = (b_0 + b_1x + b_2x^2)_2 \qquad \text{for the constant term}$$
$$(b_3 + b_4x)_1 = (b_3 + b_4x)_2 \qquad \text{for the term in } y$$
$$(b_5)_1 = (b_5)_2 \qquad \text{for the term in } y^2$$

with x in these equations being equal to the constant coordinate at the junction between the two pieces (see Tables 9-3 and 9-4).

The stochastic model for this problem will be rather simple and will be expressed by attaching a residual to each of the two polynomial equations for each point. This implies that the coordinates in the ground control system will be considered as observations. All observations will be assumed uncorrelated and of equal precision.

TABLE 9-3 Piece 1

		IMAGE COORDINATES		GROUND COORDINATES	
i	POINT NUMBER	x	y	X	Y
1	1	1042.9	1080.7	1556.2	4146.0
2	5	1011.9	1171.3	1016.6	7900.1
3	9	972.5	1261.7	21.9	11485.2
4	10	1080.7	1076.8	3269.6	3787.9
5	28	1069.1	1155.5	3443.0	6953.0
6	38	1052.4	1273.5	1273.5	11499.0
7	39	1137.8	1137.8	5826.8	3463.0
8	51	1192.2	1079.5	8293.9	3447.4
9	54	1132.0	1188.1	6504.1	7962.7
10	61	1139.0	1271.4	7430.7	11042.6

TABLE 9-4 Piece 2

		IMAGE COORDINATES		GROUND COORDINATES	
i	POINT NUMBER	x	y	X	Y
11	69	1176.9	1170.4	8302.0	7075.7
12	78	1195.7	1267.5	9840.7	10755.3
13	81	1298.0	1082.9	13130.4	3257.7
14	91	1279.5	1200.9	13165.4	7935.2
15	95	1282.9	1243.5	13677.5	9562.8
16	100	1371.3	1231.8	17771.5	8763.4
17	113	1405.7	1257.7	19662.7	9514.6
18	115	1380.8	1096.8	17113.6	3432.4
19	131	1458.5	1071.4	20383.2	2028.9
20	146	1444.0	1139.8	20366.9	4803.9
21	157	1404.4	1236.9	19393.5	8715.8

Solution: Because of the simplified stochastic assumptions, the pair of condition equations for each given point is directly linear and may be written in the following form:

For the x coordinate

$$v_{x_i} + \begin{bmatrix} 1 & x_i & x_i^2 & y_i & x_i y_i & x_i^2 y_i \end{bmatrix} \begin{bmatrix} a_0 \\ a_1 \\ a_2 \\ a_3 \\ a_4 \\ a_5 \end{bmatrix}_j = X_i$$

$j = 1$ (first piece)

$i = 1, 2, \ldots, 10$ (10 points in first piece)

For the y coordinate

$$v_{y_i} + \begin{bmatrix} 1 & x_i & x_i^2 & y_i & x_i y_i & y_i^2 \end{bmatrix} \begin{bmatrix} b_0 \\ b_1 \\ b_2 \\ b_3 \\ b_4 \\ b_5 \end{bmatrix}_j = Y_i$$

$j = 2$ (second piece)

$i = 11, 12, \ldots, 21$ (11 points in second piece)

both of which are of the form $\mathbf{v} + \mathbf{B}\boldsymbol{\Delta} = \mathbf{f}$.

To enforce the conditions at the joining line, two constraint equations are written for the X coordinate and three for the Y coordinate. Denoting by x_c $(= 1176)$ the constant coordinate at the junction, then

For the X coordinate

$$\begin{bmatrix} 1 & x_c & x_c^2 & 0 & 0 & 0 & -1 & -x_c & -x_c^2 & 0 & 0 & 0 \\ 0 & 0 & 0 & 1 & x_c & x_c^2 & 0 & 0 & 0 & -1 & -x_c & -x_c^2 \end{bmatrix} \begin{bmatrix} a_{01} \\ a_{11} \\ a_{21} \\ a_{31} \\ a_{41} \\ a_{51} \\ a_{02} \\ a_{12} \\ a_{22} \\ a_{32} \\ a_{42} \\ a_{52} \end{bmatrix} = \begin{bmatrix} 0 \\ 0 \end{bmatrix}$$

For the Y coordinate

$$\begin{bmatrix} 1 & x_c & x_c^2 & 0 & 0 & 0 & -1 & -x_c & -x_c^2 & 0 & 0 & 0 \\ 0 & 0 & 0 & 1 & x_c & 0 & 0 & 0 & 0 & -1 & -x_c & 0 \\ 0 & 0 & 0 & 0 & 0 & 1 & 0 & 0 & 0 & 0 & 0 & -1 \end{bmatrix} \begin{bmatrix} b_{01} \\ b_{11} \\ b_{21} \\ b_{31} \\ b_{41} \\ b_{51} \\ b_{02} \\ b_{12} \\ b_{22} \\ b_{32} \\ b_{42} \\ b_{52} \end{bmatrix} = \begin{bmatrix} 0 \\ 0 \\ 0 \end{bmatrix}$$

Both sets of constraint equations are of the general form $\mathbf{C\Delta} = \mathbf{g}$. Thus there are 20 condition equations for piece 1, 22 condition equations for piece 2, and 5 constraint equations. The total number of parameters is obviously 24, thus leading to the degrees of freedom,

$$r = (20 + 22) + 5 - 24 = 23$$

There are, of course, several ways in which this problem may be solved. Using any one of these [for example, the one carrying the Lagrange multipliers $\mathbf{k}_c$ as we used in programming this problem, see equation (9.33b)] leads to the same answer. Because of the relatively large numbers involved, it was found prudent to premultiply all coordinates x and y by 10^{-3}. The answers are

Piece 1 ($j = 1$)

$$X\ Coordinate \begin{bmatrix} a_0 \\ a_1 \\ a_2 \\ a_3 \\ a_4 \\ a_5 \end{bmatrix}_1 = \begin{bmatrix} -5.04 \\ -57.5 \\ 52.2 \\ -40.9 \\ 101.0 \\ -51.1 \end{bmatrix} \qquad Y\ Coordinate \begin{bmatrix} b_0 \\ b_1 \\ b_2 \\ b_3 \\ b_4 \\ b_5 \end{bmatrix}_1 = \begin{bmatrix} -30.5 \\ -27.4 \\ 8.09 \\ 56.0 \\ 4.37 \\ -9.46 \end{bmatrix}$$

Piece 2 ($j = 2$)

$$X\ Coordinate \begin{bmatrix} a_0 \\ a_1 \\ a_2 \\ a_3 \\ a_4 \\ a_5 \end{bmatrix}_2 = \begin{bmatrix} -150.0 \\ 204.0 \\ -65.2 \\ 99.3 \\ -151.0 \\ 62.0 \end{bmatrix} \qquad Y\ Coordinate \begin{bmatrix} b_0 \\ b_1 \\ b_2 \\ b_3 \\ b_4 \\ b_5 \end{bmatrix}_2 = \begin{bmatrix} -59.7 \\ 14.4 \\ -6.35 \\ 63.0 \\ -1.57 \\ -9.46 \end{bmatrix}$$

The values of these parameters may now be used in the original polynomial equations to compute new sets of coordinates X and Y for each point. These will in general be different from those originally given, and the differences can be used as an indication of how well the fitting process has been accomplished.

9.5. SUMMARY OF EQUATIONS

When parameters involved in the mathematical model of an adjustment are functionally dependent, constraint equations must be written to account for that dependency. Thus with n observations in $\mathbf{l}$ and u parameters in $\mathbf{\Delta}$, we write c conditions and s constraints.

$$\mathbf{Av} + \mathbf{B\Delta} = \mathbf{f} \tag{9.3a}$$

$$\mathbf{C\Delta} = \mathbf{g} \tag{9.3b}$$

$\mathbf{f}$ is $c \times 1$ and $\mathbf{g}$ is $s \times 1$, both are vectors of constants.
$\mathbf{A}$ is $c \times n$ with rank $= c$.
$\mathbf{v}$ is $n \times 1$ vector of residuals.
$\mathbf{B}$ is $c \times u$ with rank $=$ smaller of c and u.
$\mathbf{\Delta}$ is $u \times 1$ vector of parameters having $(u - s)$ independent elements.
$\mathbf{C}$ is $s \times u$ with rank $= s$.

9.5.1. When $u < c$

Direct Solution:

$$\mathbf{Q}_e = \mathbf{AQA}^t \qquad \mathbf{W}_e = \mathbf{Q}_e^{-1} \tag{9.9}$$

$$\mathbf{N} = \mathbf{B}^t\mathbf{W}_e\mathbf{B} \qquad \mathbf{t} = \mathbf{B}^t\mathbf{W}_e\mathbf{f}$$

$$\boldsymbol{\Delta}^0 = \mathbf{N}^{-1}\mathbf{t} \qquad \mathbf{M} = (\mathbf{C}\mathbf{N}^{-1}\mathbf{C}^t)$$

$$\boldsymbol{\Delta} = \boldsymbol{\Delta}^0 + \delta\boldsymbol{\Delta} \qquad \delta\boldsymbol{\Delta} = \mathbf{N}^{-1}\mathbf{C}^t\mathbf{M}^{-1}(\mathbf{g} - \mathbf{C}\boldsymbol{\Delta}^0)$$

$$\hat{\sigma}_0^2 = (\mathbf{v}^t\mathbf{W}\mathbf{v})/r$$

$$\mathbf{Q}_{\Delta\Delta} = \mathbf{N}^{-1}[\mathbf{I} - \mathbf{C}^t(\mathbf{C}\mathbf{N}^{-1}\mathbf{C}^t)^{-1}\mathbf{C}\mathbf{N}^{-1}] \tag{9.15b}$$

$$\mathbf{Q}_{\Delta\Delta}\mathbf{C}^t = \mathbf{0} \tag{9.16}$$

$$\mathbf{v} = \mathbf{Q}\mathbf{A}^t\mathbf{W}_e(-\mathbf{B}\boldsymbol{\Delta} + \mathbf{f}) \tag{9.8), (9.9}$$

$$\hat{\boldsymbol{l}} = \boldsymbol{l} + \mathbf{v}$$

$$\mathbf{Q}_{vv} = \mathbf{Q}\mathbf{A}^t\mathbf{W}_e\mathbf{A}\mathbf{Q} - \mathbf{Q}\mathbf{A}^t\mathbf{W}_e\mathbf{B}\mathbf{Q}_{\Delta\Delta}\mathbf{B}^t\mathbf{W}_e\mathbf{A}\mathbf{Q} \tag{9.17b}$$

$$\mathbf{Q}_{\hat{l}\hat{l}} = \mathbf{Q} - \mathbf{Q}_{vv}$$

9.5.2. Solution by Constraint Elimination

$$a = u - s$$

Partition

$$\underset{c,u}{\mathbf{B}} = \begin{bmatrix} \underset{c,s}{\mathbf{B}_1} & \underset{c,a}{\mathbf{B}_2} \end{bmatrix} \quad \underset{s,u}{\mathbf{C}} = \begin{bmatrix} \underset{s,s}{\mathbf{C}_{11}} & \underset{s,a}{\mathbf{C}_{12}} \end{bmatrix} \qquad \text{with } \mathbf{C}_{11} \text{ nonsingular} \tag{9.19b}$$

$$\underset{c,a}{\bar{\mathbf{B}}} = -\mathbf{B}_1\mathbf{C}_{11}^{-1}\mathbf{C}_{12} + \mathbf{B}_2 \qquad \underset{c,1}{\bar{\mathbf{f}}} = \mathbf{f} - \mathbf{B}_1\mathbf{C}_{11}^{-1}\mathbf{g} \tag{9.21}$$

$$\bar{\mathbf{N}} = \bar{\mathbf{B}}^t\mathbf{W}_e\bar{\mathbf{B}} \qquad \bar{\mathbf{t}} = \bar{\mathbf{B}}^t\mathbf{W}_e\bar{\mathbf{f}} \tag{9.24}$$

$$\boldsymbol{\Delta}_2 = \bar{\mathbf{N}}^{-1}\bar{\mathbf{t}} \tag{9.23}$$

$$\boldsymbol{\Delta}_1 = \mathbf{C}_{11}^{-1}(\mathbf{g} - \mathbf{C}_{12}\boldsymbol{\Delta}_2) \tag{9.20}$$

$$\boldsymbol{\Delta} = \begin{bmatrix} \boldsymbol{\Delta}_1 \\ \boldsymbol{\Delta}_2 \end{bmatrix}$$

$$\mathbf{Q}_{\Delta\Delta} = \begin{bmatrix} \mathbf{C}_{11}^{-1}\mathbf{C}_{12}\bar{\mathbf{N}}^{-1}\mathbf{C}_{12}^t(\mathbf{C}_{11}^{-1})^t & -\mathbf{C}_{11}^{-1}\mathbf{C}_{12}\bar{\mathbf{N}}^{-1} \\ -\bar{\mathbf{N}}^{-1}\mathbf{C}_{12}^t(\mathbf{C}_{11}^{-1})^t & \bar{\mathbf{N}}^{-1} \end{bmatrix} \tag{9.29}$$

9.5.3. When $u > c$

$\mathbf{N} = \mathbf{B}^t\mathbf{W}_e\mathbf{B}$ is here a singular matrix,
$\mathbf{t} = \mathbf{B}^t\mathbf{W}_e\mathbf{f}$.

$$\begin{bmatrix} \boldsymbol{\Delta} \\ \mathbf{k}_c \end{bmatrix} = \begin{bmatrix} -\mathbf{N} & \mathbf{C}^t \\ \mathbf{C} & \mathbf{0} \end{bmatrix}^{-1} \begin{bmatrix} -\mathbf{t} \\ \mathbf{g} \end{bmatrix} = \begin{bmatrix} \boldsymbol{\alpha} & \boldsymbol{\beta}^t \\ \boldsymbol{\beta} & \boldsymbol{\gamma} \end{bmatrix} \begin{bmatrix} -\mathbf{t} \\ \mathbf{g} \end{bmatrix} \tag{9.33b), (9.34}$$

($\mathbf{k}_c$ is $s \times 1$ vector of Lagrange multipliers for the constraints.)

$$\mathbf{Q}_{\Delta\Delta} = -\boldsymbol{\alpha} \tag{9.42}$$

All other relations for the remainder of the direct solution as well as the solution by constraint elimination for the case of $u < c$ apply equally here.

9.5.4. Constraints with Added Parameters If in addition to the u parameters in $\boldsymbol{\Delta}$, a set of q added parameters in $\boldsymbol{\Delta}'$ appear in the constraints, the number of constraints becomes s' such that

$$c + s' = r + (u + q) \qquad (r = \text{redundancy}) \tag{9.48}$$

$$q < s' < (u + q) \qquad (9.47), (9.49)$$

The conditions and constraints are of the form

$$\mathbf{A}\mathbf{v} + \mathbf{B}\boldsymbol{\Delta} = \mathbf{f}$$

$$\mathbf{D}_1\boldsymbol{\Delta} + \mathbf{D}_2\boldsymbol{\Delta}' = \mathbf{h}$$

$\mathbf{h}$ is $s' \times 1$ vector of constants
$\mathbf{D}_1$ is $s' \times u$ with rank $= s$
$\mathbf{D}_2$ is $s' \times q$ with rank $= q$

When $u < c$ all relations up to computing $\boldsymbol{\Delta}^0$ apply here. Then

$$\mathbf{P} = \mathbf{D}_1\mathbf{N}^{-1}\mathbf{D}_1^t \tag{9.58}$$

$$\mathbf{R} = \mathbf{D}_2^t\mathbf{P}^{-1}\mathbf{D}_2 \tag{9.59}$$

$$\boldsymbol{\Delta}' = \mathbf{R}^{-1}[\mathbf{D}_2^t\mathbf{P}^{-1}(\mathbf{h} - \mathbf{D}_1\boldsymbol{\Delta}^0)] \tag{9.56}$$

$$\boldsymbol{\Delta} = \boldsymbol{\Delta}^0 + \mathbf{N}^{-1}\mathbf{D}_1^t\mathbf{P}^{-1}(\mathbf{h} - \mathbf{D}_1\boldsymbol{\Delta}^0 - \mathbf{D}_2\boldsymbol{\Delta}') \tag{9.57}$$

$$\mathbf{Q}_{\Delta'\Delta'} = \mathbf{R}^{-1} \tag{9.60}$$

$$\mathbf{Q}_{\Delta\Delta} = \mathbf{N}^{-1}[\mathbf{I} - \mathbf{D}_1^t\mathbf{P}^{-1}\mathbf{D}_1\mathbf{N}^{-1} + \mathbf{D}_1^t\mathbf{P}^{-1}\mathbf{D}_2\mathbf{R}^{-1}\mathbf{D}_2^t\mathbf{P}^{-1}\mathbf{D}_1\mathbf{N}^{-1}] \tag{9.61}$$

$$\mathbf{Q}_{\Delta\Delta'} = -\mathbf{N}^{-1}\mathbf{D}_1^t\mathbf{P}^{-1}\mathbf{D}_2\mathbf{R}^{-1} \tag{9.64}$$

$\mathbf{Q}_{vv}$ and $\mathbf{Q}_{ll}$ are the same as given in Section 9.5.1.

9.5.5. Elimination of Constraints with Added Parameters With $s = s' - q$ partition the constraints into

$$\underset{s',u}{\mathbf{D}_1} = \begin{bmatrix} \underset{s,u}{\mathbf{D}_{11}} \\ \underset{q,u}{\mathbf{D}_{21}} \end{bmatrix} \qquad \underset{s',q}{\mathbf{D}_2} = \begin{bmatrix} \underset{s,q}{\mathbf{D}_{12}} \\ \underset{q,q}{\mathbf{D}_{22}} \end{bmatrix} \qquad \underset{s',u}{\mathbf{h}} = \begin{bmatrix} \underset{s,1}{\mathbf{h}_1} \\ \underset{q,1}{\mathbf{h}_2} \end{bmatrix} \tag{9.66}$$

such that $\mathbf{D}_{22}$ is nonsingular.

$$\mathbf{C} = (\mathbf{D}_{11} - \mathbf{D}_{12}\mathbf{D}_{22}^{-1}\mathbf{D}_{21}) \tag{9.67a}$$

$$\mathbf{g} = (\mathbf{h}_1 - \mathbf{D}_{12}\mathbf{D}_{22}^{-1}\mathbf{h}_2) \tag{9.67b}$$

With **C** and **g**, partition and solve as in Section 9.5.2 to get $\mathbf{\Delta}_1$ and $\mathbf{\Delta}_2$, thus the total parameter vector $\mathbf{\Delta}$. Then

$$\mathbf{\Delta}' = \mathbf{D}_{22}^{-1}(\mathbf{h}_2 - \mathbf{D}_{21}\mathbf{\Delta}) \tag{9.66c}$$

$\mathbf{Q}_{\Delta\Delta}$ as in equation (9.29)

$$\mathbf{Q}_{\Delta'\Delta'} = \mathbf{D}_{22}^{-1}\mathbf{D}_{21}\mathbf{Q}_{\Delta\Delta}\mathbf{D}_{21}^{t}\mathbf{D}_{22}^{-1} \tag{9.68}$$

10

Adjustment with Derived Observations and Adjustment in Steps (of Observations Only)

10.1. INTRODUCTION

In earlier discussions it was shown that observational variables need not be the direct numerical outcomes of the physical operations of measurement. In fact, as was shown by Example 7.5, we could use in the adjustment other stochastic vaiables that are derived from the originally available variables. The outcome is exactly the same *provided the stochastic model is modified in such a way that the stochastic properties of the derived observations are also properly obtained.* This rather important concept will be discussed further and demonstrated by examples in the first part of this chapter.

The second part of the chapter will deal with another related concept. It involves what is termed "adjustment in steps" when the technique applied is that of adjusting observations only. In many practical situations a set of observations is obtained that belongs to a certain (geometrical) model. This first set is then adjusted and estimates for the observations are obtained from the adjustment. At a later stage, a second set of observations may be acquired that belongs to either the same model or to an expanded form of it. Thus the adjustment of the second step would involve in it not only the second set of observations but also some or all of the *new estimates* of the observations in the first set.

By necessity, then, the adjustment is performed in two steps and the final result must be the same as would have been obtained had we performed the total adjustment of all sets of observations together. This will be demonstrated both algebraically and by numerical example.

10.2. ADJUSTMENT WITH DERIVED OBSERVATIONS

Consideration is limited to linear functions (which may be originally linear or may have been linearized). Also only the case of adjustment of observations is considered here, whereas other possibilities, which include parameters, will be treated in Part III of this book.

With n observations and r redundancy the conditions may be written as

$$\underset{r,n}{\mathbf{A}_l}\ \underset{n,1}{\mathbf{v}_l} = (-\mathbf{A}_l\boldsymbol{l} + \mathbf{d}_l) = \underset{r,1}{\mathbf{f}_l} \tag{10.1}$$

with $\mathbf{d}_l$ and $\mathbf{f}_l$ being constant term vectors. Associated with $\boldsymbol{l}$ is the cofactor matrix $\mathbf{Q}_{ll}$.† If an adjustment is carried out using these conditions, the following relations may be written (refer to Chapter 7):

$$\mathbf{k}_l = (\mathbf{A}_l\mathbf{Q}_{ll}\mathbf{A}_l^t)^{-1}\mathbf{f}_l = \mathbf{W}_{e_l}\mathbf{f}_l \tag{7.4}$$

$$\mathbf{v}_l = \mathbf{Q}_{ll}\mathbf{A}_l^t\mathbf{k}_l = \mathbf{Q}_{ll}\mathbf{A}_l^t\mathbf{W}_{e_l}\mathbf{f}_l \tag{7.5}$$

$$\mathbf{Q}_{vvl} = \mathbf{Q}_{ll}\mathbf{A}_l^t\mathbf{W}_{e_l}\mathbf{A}_l\mathbf{Q}_{ll} \tag{7.7}$$

$$\mathbf{Q}_{\hat{l}\hat{l}} = \mathbf{Q}_{ll} - \mathbf{Q}_{vvl} \tag{6.31}$$

Suppose now that we are to *derive* a new set of observations $\mathbf{m}_{p,1}$ from the original set $\boldsymbol{l}$, such that $p \le n$. This new set can be used in the adjustment in place of the original set with results consistent with those that would have been obtained from $\boldsymbol{l}$, provided the following assertions are considered:

1. The number of derived observations is less than or equal to the number of original observations, that is, $p \le n$.
2. The cofactor matrix $\underset{p,p}{\mathbf{Q}_{mm}}$ for the new set $\mathbf{m}$ must be properly evaluated.

   Thus if

$$\underset{p,1}{\mathbf{m}} = \underset{p,n}{\mathbf{D}}\ \underset{n,1}{\boldsymbol{l}} \qquad \text{and} \qquad \mathbf{v}_m = \mathbf{D}\mathbf{v}_l \tag{10.2}$$

then

$$\mathbf{Q}_{mm} = \mathbf{D}\mathbf{Q}_{ll}\mathbf{D}^t \tag{10.3}$$

must be computed.

† The subscript l is reimposed here to facilitate the comparison with the derived set applying the subscript m.

3. Redundancy, or the number of degrees of freedom r, *must remain the same* after obtaining the derived observations. In other words the model must not change while switching over from one set of observations to another if we are to get consistent adjustment results.

Because of point (1) above, we may write the conditions in terms of the derived observations $\mathbf{m}$ as follows:

$$\underset{r,p}{\mathbf{A}_m}\ \underset{p,1}{\mathbf{v}_m} = (-\mathbf{A}_m\,\mathbf{m} + \mathbf{d}_m) = \underset{r,1}{\mathbf{f}_m} \tag{10.4}$$

Using a vector $\mathbf{f}_m$ that is different from $\mathbf{f}_l$ allows for the possibility of using a different set of conditions, even though the number of conditions is the same. For the obvious reason that any r conditions chosen are independent, then

$$\underset{r,1}{\mathbf{f}_m} = \underset{r,r}{\mathbf{C}}\ \underset{r,1}{\mathbf{f}_l} \tag{10.5}$$

with $\mathbf{C}$ being nonsingular.

Applying least squares to equation (10.4) leads directly to the following relations:

$$\underset{r,1}{\mathbf{k}_m} = \mathbf{W}_{e_m}\,\mathbf{f}_m = (\mathbf{A}_m\,\mathbf{Q}_{mm}\,\mathbf{A}_m^t)^{-1}\mathbf{f}_m \tag{10.6a}$$

$$\mathbf{v}_m = \mathbf{Q}_{mm}\,\mathbf{A}_m^t\,\mathbf{k}_m = \mathbf{Q}_{mm}\,\mathbf{A}_m^t\,\mathbf{W}_{e_m}\,\mathbf{f}_m \tag{10.6b}$$

$$\mathbf{Q}_{v_m v_m} = \mathbf{Q}_{mm}\,\mathbf{A}_m^t\,\mathbf{W}_{e_m}\,\mathbf{A}_m\,\mathbf{Q}_{mm} \tag{10.6c}$$

$$\mathbf{Q}_{\hat{m}\hat{m}} = \mathbf{Q}_{mm} - \mathbf{Q}_{v_m v_m} \tag{10.6d}$$

Equations (10.6) give the least squares solution in terms of the new set of observations $\mathbf{m}$ and their a priori cofactor matrix $\mathbf{Q}_{mm}$. We must now compute the residuals $\mathbf{v}_l$ and the a posteriori matrix $\mathbf{Q}_{v_l v_l}$ from these, and also prove that they are the same values that would have been obtained if $\boldsymbol{l}$ and $\mathbf{Q}_{ll}$ were used directly.

From equations (10.1), (10.2), (10.4), and (10.5) it follows that

$$\mathbf{A}_m\,\mathbf{D}\mathbf{v}_l = \mathbf{C}\mathbf{f}_l$$

$$(\mathbf{C}^{-1}\mathbf{A}_m\,\mathbf{D})\mathbf{v}_l = \mathbf{f}_l$$

which leads to

$$\mathbf{A}_l = \mathbf{C}^{-1}\mathbf{A}_m\,\mathbf{D} \tag{10.7}$$

From equations (7.4, 10.3, 10.5, and 10.6a) we write

$$\begin{aligned}
\mathbf{k}_l &= (\mathbf{A}_l\,\mathbf{Q}_{ll}\,\mathbf{A}_l^t)^{-1}\mathbf{f}_l \\
&= [(\mathbf{C}^{-1}\mathbf{A}_m\,\mathbf{D})\mathbf{Q}_{ll}(\mathbf{D}^t\mathbf{A}_m(\mathbf{C}^{-1})^t)]^{-1}\mathbf{f}_l \\
&= [\mathbf{C}^{-1}\mathbf{A}_m\,\mathbf{Q}_{mm}\,\mathbf{A}_m^t(\mathbf{C}^t)^{-1}]^{-1}\mathbf{f}_l \\
&= \mathbf{C}^t(\mathbf{A}_m\,\mathbf{Q}_{mm}\,\mathbf{A}_m^t)^{-1}\mathbf{C}\mathbf{f}_l \\
&= \mathbf{C}^t\mathbf{W}_{e_m}\,\mathbf{f}_m
\end{aligned}$$

or

$$\mathbf{k}_l = \mathbf{C}^t\mathbf{k}_m \tag{10.8}$$

Now, to derive a relation for $\mathbf{v}_l$ in terms of $\mathbf{v}_m$,

$$\begin{aligned}\mathbf{v}_l &= \mathbf{Q}_{ll}\,\mathbf{A}_l^t\,\mathbf{k}_l\\ &= \mathbf{Q}_{ll}\,\mathbf{D}^t\mathbf{A}_m^t(\mathbf{C}^{-1})^t\mathbf{C}^t\mathbf{k}_m\end{aligned}$$

or

$$\mathbf{v}_l = \mathbf{Q}_{ll}\,\mathbf{D}^t\mathbf{A}_m^t\,\mathbf{k}_m \tag{10.9a}$$

or, with reference to equation (10.6b)

$$\mathbf{v}_l = \mathbf{Q}_{ll}\,\mathbf{D}^t\mathbf{Q}_{mm}^{-1}\,\mathbf{v}_m \tag{10.9b}$$

It should be straightforward to show that premultiplying both sides of equation (10.9b) by $\mathbf{D}$ yields equation (10.2), taking equation (10.3) into account.

Using equation (10.9b) we can derive a relationship for the a posteriori cofactor matrix $\mathbf{Q}_{v_lv_l}$.

$$\mathbf{Q}_{v_lv_l} = (\mathbf{Q}_{ll}\,\mathbf{D}^t\mathbf{Q}_{mm}^{-1})\mathbf{Q}_{v_mv_m}(\mathbf{Q}_{ll}\,\mathbf{D}^t\mathbf{Q}_{mm}^{-1})^t$$

and from equation (10.6c),

$$\mathbf{Q}_{v_lv_l} = \mathbf{Q}_{ll}\,\mathbf{D}^t\mathbf{Q}_{mm}^{-1}\mathbf{Q}_{mm}\,\mathbf{A}_m^t\,\mathbf{W}_{e_m}\,\mathbf{A}_m\,\mathbf{Q}_{mm}\,\mathbf{Q}_{mm}^{-1}\,\mathbf{D}\mathbf{Q}_{ll}$$

or

$$\mathbf{Q}_{v_lv_l} = \mathbf{Q}_{ll}\,\mathbf{D}^t\mathbf{A}_m^t\,\mathbf{W}_{e_m}\,\mathbf{A}_m\,\mathbf{D}\mathbf{Q}_{ll} \tag{10.10}$$

Finally, it must be shown that equation (10.10) is equivalent to equation (7.7).

$$\mathbf{Q}_{v_lv_l} = \mathbf{Q}_{ll}\,\mathbf{D}^t\mathbf{A}_m^t(\mathbf{A}_m\,\mathbf{D}\mathbf{Q}_{ll}\,\mathbf{D}^t\mathbf{A}_m^t)^{-1}\mathbf{A}_m\,\mathbf{D}\mathbf{Q}_{ll}$$

and from equation (10.7),

$$\begin{aligned}\mathbf{Q}_{v_lv_l} &= \mathbf{Q}_{ll}\,\mathbf{D}^t\mathbf{A}_m^t(\mathbf{C}\mathbf{A}_l\,\mathbf{Q}_{ll}\,\mathbf{A}_l^t\,\mathbf{C}^t)^{-1}\mathbf{A}_m\,\mathbf{D}\mathbf{Q}_{ll}\\ &= \mathbf{Q}_{ll}[\mathbf{D}^t\mathbf{A}_m^t(\mathbf{C}^{-1})^t](\mathbf{A}_l\,\mathbf{Q}_{ll}\,\mathbf{A}_l^t)^{-1}(\mathbf{C}^{-1}\mathbf{A}_m\,\mathbf{D})\mathbf{Q}_{ll}\end{aligned}$$

or

$$\mathbf{Q}_{v_lv_l} = \mathbf{Q}_{ll}\,\mathbf{A}_l^t\,\mathbf{W}_{e_l}\,\mathbf{A}_l^t\,\mathbf{Q}_{ll}$$

which is identical to equation (7.7). To show that

$$\mathbf{v}_l^t\,\mathbf{W}_{ll}\,\mathbf{v}_l = \mathbf{v}_m^t\,\mathbf{W}_{mm}\,\mathbf{v}_m$$

we refer to equations (10.9b) and (10.3).

$$\begin{aligned}\mathbf{v}_l^t\,\mathbf{W}_{ll}\,\mathbf{v}_l &= (\mathbf{v}_m^t\,\mathbf{Q}_{mm}^{-1}\,\mathbf{D}\mathbf{Q}_{ll})\mathbf{W}_{ll}(\mathbf{Q}_{ll}\,\mathbf{D}^t\mathbf{Q}_{mm}^{-1}\,\mathbf{v}_m)\\ &= \mathbf{v}_m^t\,\mathbf{Q}_{mm}^{-1}(\mathbf{D}\mathbf{Q}_{ll}\,\mathbf{D}^t)\mathbf{Q}_{mm}^{-1}\,\mathbf{v}_m\\ &= \mathbf{v}_m^t\,\mathbf{Q}_{mm}^{-1}\,\mathbf{v}_m = \mathbf{v}_m^t\,\mathbf{W}_{mm}\,\mathbf{v}\end{aligned}$$

To demonstrate the relations derived above some numerical examples follow.

Example 10.1. As a first example consider the simple case of a plane triangle *ABC* in Figure 10.1. Six directions are measured, two at each corner. From these six directions, which may be considered as the original observations, three angles may be computed as derived observations. It is required to show that from the adjustment

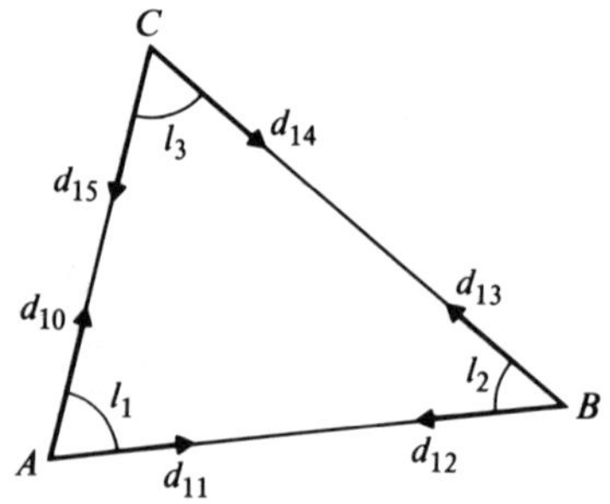

Figure 10.1

with the derived observations (angles), residuals for the original directions may be computed as well as their cofactor matrix which are the same as those obtained from adjustment with the directions themselves. Assume that the cofactor matrix for the directions is the identity matrix and that the triangle's misclosure (contradiction or discrepancy) is w seconds of arc.

Solution: For the sake of later checks, the triangle is first adjusted using directions. Since there are three zero directions, one at each corner, and since two variables are necessary for determining the shape of the triangle, then $r = 1$. This is the same as adjusting by angles, which satisfies requirement (3) in Section 10.2. From this it is obvious that there is only one **f**, or $\mathbf{f}_l = \mathbf{f}_m = w$ which leads to $\mathbf{C} = \mathbf{I}$, in equation (10.5).

The condition for adjustment by directions is

$$[-1 \quad 1 \quad -1 \quad 1 \quad -1 \quad 1]\mathbf{v}_d = w \qquad \text{or} \qquad \underset{1,6}{\mathbf{A}_d}\ \underset{6,1}{\mathbf{v}_d} = \underset{1,1}{\mathbf{f}}$$

with

$$\mathbf{Q}_{dd} = \mathbf{I} \qquad \mathbf{Q}_{e_d} = 6 \qquad k_d = k = \frac{w}{6} \qquad \mathbf{v}_d^t = [-1 \quad 1 \quad -1 \quad 1 \quad -1 \quad 1]\frac{w}{6}$$

$$\mathbf{Q}_{v_dv_d} = \tfrac{1}{6}\begin{bmatrix} 1 & -1 & 1 & -1 & 1 & -1 \\ & 1 & -1 & 1 & -1 & 1 \\ & & 1 & -1 & 1 & -1 \\ & & & 1 & -1 & 1 \\ & & & & 1 & -1 \\ \text{symmetric} & & & & & 1 \end{bmatrix}$$

To get the angles

$$\begin{bmatrix} l_1 \\ l_2 \\ l_3 \end{bmatrix} = \begin{bmatrix} -1 & 1 & 0 & 0 & 0 & 0 \\ 0 & 0 & -1 & 1 & 0 & 0 \\ 0 & 0 & 0 & 0 & -1 & 1 \end{bmatrix} \begin{bmatrix} d_{10} \\ d_{11} \\ d_{12} \\ d_{13} \\ d_{14} \\ d_{15} \end{bmatrix} \qquad \text{or} \qquad \mathbf{l} = \mathbf{Dd}$$

and

$$\mathbf{Q}_{ll} = \mathbf{D}\mathbf{Q}_{dd}\,\mathbf{D}^t = 2\mathbf{I}_3$$

The condition equation is

$$[1 \quad 1 \quad 1]\mathbf{v}_l = w$$

$$\mathbf{Q}_{e_m} = 6 \qquad k_l = k = \frac{w}{6}$$

$$\mathbf{V}_l^t = \frac{w}{3}[1 \quad 1 \quad 1]$$

$$\mathbf{Q}_{v_l v_l} = \tfrac{2}{3}\begin{bmatrix} 1 & 1 & 1 \\ & 1 & 1 \\ \text{symmetric} & & 1 \end{bmatrix}$$

Now compute $\mathbf{v}_d$ and $\mathbf{Q}_{v_d v_d}$ from $\mathbf{v}_l$, $\mathbf{Q}_{v_l v_l}$, and other matrices belonging to the adjustment. From equation (10.9a),

$$\mathbf{v}_d = \mathbf{Q}_{dd} D^t \mathbf{A}_l^t k_l = \begin{bmatrix} -1 & 0 & 0 \\ 1 & 0 & 0 \\ 0 & -1 & 0 \\ 0 & 1 & 0 \\ 0 & 0 & -1 \\ 0 & 0 & 1 \end{bmatrix} \begin{bmatrix} 1 \\ 1 \\ 1 \end{bmatrix} \frac{w}{6} = w/6 \begin{bmatrix} -1 \\ 1 \\ -1 \\ 1 \\ -1 \\ 1 \end{bmatrix}$$

Then from equation (10.10),

$$\mathbf{Q}_{v_d v_d} = \mathbf{Q}_{dd}\,\mathbf{D}^t\mathbf{A}_l^t\,\mathbf{W}_{e_l}\,\mathbf{A}_l\,\mathbf{D}\mathbf{Q}_{dd} = \tfrac{1}{6}\begin{bmatrix} 1 & -1 & 1 & -1 & 1 & -1 \\ & 1 & -1 & 1 & -1 & 1 \\ & & 1 & -1 & 1 & -1 \\ & & & 1 & -1 & 1 \\ & & & & 1 & -1 \\ \text{symmetric} & & & & & 1 \end{bmatrix}$$

Both $\mathbf{v}_d$ and $\mathbf{Q}_{v_d v_d}$ are the same as was obtained previously. This verifies the relations derived in the present section.

Example 10.2. In this example the results are used from the adjustment in Example 7.5 of a quadrilateral by angles and directions in order to verify further the applicability of equations (10.9) and (10.10).

Solution: Using equation (10.9b)

$$\underset{12,12}{\mathbf{v}_d} = \underset{12,12}{\mathbf{Q}_{dd}}\ \underset{12,8}{\mathbf{D}^t}\ \underset{8,8}{\mathbf{Q}_{ll}^{-1}}\ \underset{8,1}{\mathbf{v}_l}$$

where

$$\mathbf{Q}_{dd} = \mathbf{I}_{12} \qquad \text{and} \qquad \mathbf{D} = \text{matrix } \underset{8,12}{\mathbf{F}} \text{ in Example 7.5}$$

$$\underset{8,8}{\mathbf{Q}_{ll}^{-1}} = \frac{1}{3}\begin{bmatrix} 2 & 1 & 0 & & & \text{(zeros)} & & \\ 1 & 2 & 0 & 0 & & & & \\ 0 & 0 & 2 & 1 & 0 & & & \\ & 0 & 1 & 2 & 0 & & & \\ & & 0 & 0 & 2 & 1 & 0 & \\ & & & 0 & 1 & 2 & 0 & 0 \\ & & & & 0 & 0 & 2 & 1 \\ & \text{(zeros)} & & & & 0 & 1 & 2 \end{bmatrix}$$

$$\mathbf{v}_l = [-1.0122 \quad 0.9822 \quad -0.3378 \quad 1.7137 \quad 1.7620 \quad -0.3722 \quad 1.1065 \quad -1.1022]^t$$

$$\mathbf{v}_d = \frac{1}{3}\begin{bmatrix} -2 & -1 & 0 & 0 & 0 & 0 & 0 & 0 \\ 1 & -1 & 0 & 0 & 0 & 0 & 0 & 0 \\ 1 & 2 & 0 & 0 & 0 & 0 & 0 & 0 \\ 0 & 0 & -2 & -1 & 0 & 0 & 0 & 0 \\ 0 & 0 & 1 & 1 & 0 & 0 & 0 & 0 \\ 0 & 0 & 1 & 2 & 0 & 0 & 0 & 0 \\ 0 & 0 & 0 & 0 & 2 & -1 & 0 & 0 \\ 0 & 0 & 0 & 0 & 1 & -1 & 0 & 0 \\ 0 & 0 & 0 & 0 & 1 & 2 & 0 & 0 \\ 0 & 0 & 0 & 0 & 0 & 0 & -2 & -1 \\ 0 & 0 & 0 & 0 & 0 & 0 & 1 & -1 \\ 0 & 0 & 0 & 0 & 0 & 0 & 1 & 2 \end{bmatrix}\begin{bmatrix} -1.0122 \\ 0.9822 \\ -0.3378 \\ 1.7137 \\ 1.7620 \\ -0.3722 \\ 1.1065 \\ -1.1022 \end{bmatrix} = \begin{bmatrix} 0.3474 \\ -0.6648 \\ 0.3174 \\ -0.3460 \\ -0.6838 \\ 1.0299 \\ -1.0506 \\ 0.7114 \\ 0.3392 \\ -0.3703 \\ 0.7362 \\ -0.3660 \end{bmatrix}$$

which is exactly the same vector obtained from direct adjustment of directions (see Example 7.5). With regard to equation (10.10), compute $\mathbf{Q}_{v_d v_d}$ as follows:

$$\mathbf{Q}_{v_d v_d} = \mathbf{Q}_{ll}\,\mathbf{D}^t\mathbf{A}_l^t\,\mathbf{W}_{e_l}\,\mathbf{A}_l\,\mathbf{D}\mathbf{Q}_{ll} = \underset{12,8}{\mathbf{D}}\ \underset{8,4}{\mathbf{A}_l^t}\ \underset{4,4}{\mathbf{W}_{el}}\ \underset{4,8}{\mathbf{A}_l}\ \underset{8,12}{\mathbf{D}}$$

where

$$\underset{4,8}{\mathbf{A}_l} = \begin{bmatrix} 1 & 1 & 1 & 0 & 0 & 0 & 0 & 1 \\ 0 & 0 & 0 & 1 & 1 & 1 & 1 & 0 \\ 1 & 0 & 0 & 0 & 0 & 1 & 1 & 1 \\ 2.4716 & -3.3244 & 0.6457 & -2.8156 & 0.0602 & -0.6062 & 3.7025 & -0.1037 \end{bmatrix}$$

$$\underset{4,4}{\mathbf{W}_{e_l}} = \begin{bmatrix} 0.2773 & 0.1490 & -0.1765 & 0.0195 \\ & 0.2712 & -0.1703 & 0.0171 \\ & & 0.3471 & -0.0367 \\ \text{symmetric} & & & 0.0139 \end{bmatrix}$$

The value of $\mathbf{D}^t\mathbf{A}_l^t$ is first computed

$$\mathbf{D}^t\mathbf{A}_l^t = \begin{bmatrix} -1 & 0 & -1 & -2.4716 \\ 0 & 0 & 1 & 5.7960 \\ 1 & 0 & 0 & -3.3244 \\ 1 & 0 & 0 & -0.6457 \\ 1 & -1 & 0 & 3.4613 \\ 0 & 1 & 0 & -2.8156 \\ 0 & 1 & 0 & -0.0602 \\ 0 & 0 & -1 & 0.6644 \\ 0 & 1 & 1 & -0.6062 \\ 0 & -1 & -1 & -3.7026 \\ -1 & 1 & 0 & 3.8062 \\ 1 & 0 & 1 & -0.1037 \end{bmatrix}$$

This is equal to $\mathbf{A}_d^t$, and since $\mathbf{W}_{e_d} = \mathbf{W}_{e_l}$, the computed $\mathbf{Q}_{v_d v_d}$ will be exactly the same as that computed from the adjustment of directions and need not be repeated.

Example 10.3. Refer to Figure 10.1 of the plane triangle and consider that the angle A is measured twice (l_1 and l_2), whereas angles B and C are each measured once l_3 and l_4, respectively. The four measurements, which are uncorrelated and of equal precision, are

$$l_1 = 60°00'$$

$$l_2 = 60°02'$$

$$l_3 = 60°00'$$

$$l_4 = 60°01'$$

It is required to adjust the triangle, first using all four observations and then using derived observations, and to check the consistency of both results.

Solution: The model for this problem requires a minimum of two variables; thus $r = 2$, and two conditions are needed as follows:

$$l_1 - l_2 = 0$$

$$l_1 + l_3 + l_4 - \pi = 0$$

or

$$\begin{bmatrix} 1 & -1 & 0 & 0 \\ 1 & 0 & 1 & 1 \end{bmatrix} \mathbf{v}_l = \begin{bmatrix} 2' \\ -1' \end{bmatrix} \qquad \text{with } \mathbf{Q}_{ll} = \mathbf{I}_4$$

Thus

$$\mathbf{k}_l = \begin{bmatrix} 2 & 1 \\ 1 & 3 \end{bmatrix}^{-1} \begin{bmatrix} 2 \\ -1 \end{bmatrix} = \tfrac{1}{5} \begin{bmatrix} 7 \\ -4 \end{bmatrix}$$

$$\mathbf{v}_l = \tfrac{1}{5} \begin{bmatrix} 1 & 1 \\ -1 & 0 \\ 0 & 1 \\ 0 & 1 \end{bmatrix} \begin{bmatrix} 7 \\ -4 \end{bmatrix} = \tfrac{1}{5} \begin{bmatrix} 3' \\ -7' \\ -4' \\ -4' \end{bmatrix} \qquad \mathbf{Q}_{v_l v_l} = \tfrac{1}{5} \begin{bmatrix} 3 & -2 & 1 & 1 \\ -2 & 3 & 1 & 1 \\ 1 & 1 & 2 & 2 \\ 1 & 1 & 2 & 2 \end{bmatrix}$$

It is possible to first derive one "observation" from l_1 and l_2. Since both are of equal precision, then the derived observations are

$$\begin{bmatrix} m_1 \\ m_2 \\ m_3 \end{bmatrix} = \begin{bmatrix} \frac{1}{2} & \frac{1}{2} & 0 & 0 \\ 0 & 0 & 1 & 0 \\ 0 & 0 & 0 & 1 \end{bmatrix} \begin{bmatrix} l_1 \\ l_2 \\ l_3 \\ l_4 \end{bmatrix} \qquad \text{or} \qquad \mathbf{m} = \mathbf{D}\boldsymbol{l}$$

where m_1 is the arithmetic mean of l_1 and l_2. With these three observations, we may now adjust the triangle. However, with only three observations, we would have only one redundancy. This means that in the process of obtaining the derived observations we also have reduced the redundancy from two to one. This makes it no longer a technique using derived observations, and thus a treatment using adjustment in steps is warranted instead. First adjust l_1 and l_2, and in the second part use the adjusted values as derived observations. This is an adjustment in steps, as will be explained in the next section.

Adjusting l_1, l_2 first we get

$$v_1 = +1' \qquad v_2 = -1'$$

$$\hat{l}_1 = 60°01' \qquad \hat{l}_2 = 60°01'$$

and

$$\hat{\mathbf{Q}} = \begin{bmatrix} \frac{1}{2} & \frac{1}{2} \\ \frac{1}{2} & \frac{1}{2} \end{bmatrix}$$

Now we work with the "derived" observations $\mathbf{m}$ noting that we should use $\hat{l}_1$ and $\hat{l}_2$ instead of l_1 and l_2.

Hence

$$\underset{3,\,3}{\mathbf{Q}_{mm}} = \begin{bmatrix} \frac{1}{2} & \frac{1}{2} & 0 & 0 \\ 0 & 0 & 1 & 0 \\ 0 & 0 & 0 & 1 \end{bmatrix} \begin{bmatrix} \frac{1}{2} & \frac{1}{2} & 0 & 0 \\ \frac{1}{2} & \frac{1}{2} & 0 & 0 \\ 0 & 0 & 1 & 0 \\ 0 & 0 & 0 & 1 \end{bmatrix} \begin{bmatrix} \frac{1}{2} & 0 & 0 \\ \frac{1}{2} & 0 & 0 \\ 0 & 1 & 0 \\ 0 & 0 & 1 \end{bmatrix} = \begin{bmatrix} \frac{1}{2} & 0 & 0 \\ 0 & 1 & 0 \\ 0 & 0 & 1 \end{bmatrix}$$

and the condition $\mathbf{A}_m \mathbf{V}_m = \mathbf{f}_m$ is

$$[1 \quad 1 \quad 1]\mathbf{V}_m = 180 - 60°01' - 60°00' - 60°01' = -2'$$

$$k_m = \frac{2}{5}(-2) = -\frac{4'}{5}$$

$$\mathbf{v}_m = \begin{bmatrix} \frac{1}{2} & 0 & 0 \\ 0 & 1 & 0 \\ 0 & 0 & 1 \end{bmatrix} \begin{bmatrix} 1 \\ 1 \\ 1 \end{bmatrix} (-\tfrac{4}{5}) = \tfrac{1}{5} \begin{bmatrix} -2 \\ -4 \\ -4 \end{bmatrix}$$

Now, in order to compute $\mathbf{v}_{l'}$ apply equation (10.9a)

$$\mathbf{v}_{l'} = \mathbf{Q}_{l'l'}\,\mathbf{D}^t\mathbf{A}_m^t k_m = \begin{bmatrix} \frac{1}{2} & \frac{1}{2} & 0 & 0 \\ \frac{1}{2} & \frac{1}{2} & 0 & 0 \\ 0 & 0 & 1 & 0 \\ 0 & 0 & 0 & 1 \end{bmatrix} \begin{bmatrix} \frac{1}{2} & 0 & 0 \\ \frac{1}{2} & 0 & 0 \\ 0 & 1 & 0 \\ 0 & 0 & 1 \end{bmatrix} \begin{bmatrix} 1 \\ 1 \\ 1 \end{bmatrix} (-\tfrac{4}{5}) = \tfrac{1}{5} \begin{bmatrix} -2 \\ -2 \\ -4 \\ -4 \end{bmatrix}$$

Note the first two corrections $-\frac{2}{5}$ and $-\frac{2}{5}$ are to be applied to $\hat{l}_1$ and $\hat{l}_2$, respectively, after the first step of adjustment, and *not* to the original observations l_1 and l_2. In order to get the total residuals, add the contribution from the first step to that from the adjustment with reduced observations, or

$$\mathbf{v}_l = \begin{bmatrix} 1' \\ -1' \\ 0 \\ 0 \end{bmatrix} + \tfrac{1}{5} \begin{bmatrix} -2' \\ -2' \\ -4' \\ -4' \end{bmatrix} = \tfrac{1}{5} \begin{bmatrix} 3' \\ -7' \\ -4' \\ -4' \end{bmatrix}$$

which is identical to that obtained directly from the adjustment of all four observations. Next, we should evaluate and check the cofactor matrices of the estimated observations. (Check the derivations in the next section to ascertain this point.) Thus first compute

$$\underset{4,\,4}{\mathbf{Q}_{ll}} \text{ from } \mathbf{Q}_{v_l v_l}$$

evaluated in the first part

$$\mathbf{Q}_{ll} = \mathbf{I} - \tfrac{1}{5} \begin{bmatrix} 3 & -2 & 1 & 1 \\ & 3 & 1 & 1 \\ & & 2 & 2 \\ \text{symmetric} & & & 2 \end{bmatrix} = \tfrac{1}{5} \begin{bmatrix} 2 & 2 & -1 & -1 \\ & 2 & -1 & -1 \\ & & 3 & -2 \\ \text{symmetric} & & & 3 \end{bmatrix}$$

Then compute $\mathbf{Q}_{v_{l'}v_{l'}}$ from the second part (with derived observations) using equation (10.10), or

$$\mathbf{Q}_{v_{l'}v_{l'}} = \mathbf{Q}_{l'l'}\,\mathbf{D}^t\mathbf{A}_m^t\,\mathbf{W}_{e_m}\,\mathbf{A}_m\,\mathbf{D}\mathbf{Q}_{l'l'}$$

$$\mathbf{Q}_{v_{l'}v_{l'}} = \tfrac{1}{10} \begin{bmatrix} 1 & 1 & 2 & 2 \\ & 1 & 2 & 2 \\ & & 4 & 4 \\ \text{symmetric} & & & 4 \end{bmatrix}$$

Finally,

$$\mathbf{Q}_{ll} = \frac{1}{3}\begin{bmatrix} 2 & 2 & -1 & -1 \\ & 2 & -1 & -1 \\ & & 3 & -2 \\ \text{symmetric} & & & 3 \end{bmatrix}$$

This value is identical to that computed by the other procedure. A last and *important* remark: It is worth noting that although it may not have been readily evident, it was possible to propagate backward from a smaller set of variables (for example, angles) to a larger set of variables (directions).

10.3. ADJUSTMENT IN STEPS

We begin adjustment in steps by having a total set of $(r + s)$ condition equations

$$\underset{(r+s),\,n}{\mathbf{A}}\left(\boldsymbol{l} + \underset{n,\,1}{\mathbf{v}}\right) = \underset{(r+s),\,1}{\mathbf{d}}$$

amongst the observations $\boldsymbol{l}_{n,\,1}$, the cofactor matrix of which is $\mathbf{Q}$. The solution for this total system is given by equations (7.4), (7.5), and (7.7) in Section 10.2. Due to the manner in which the conditions arise, the total system of conditions is partitioned into two systems, the first being

$$\underset{r,\,n}{\mathbf{A}_1}\left(\boldsymbol{l} + \underset{n,\,1}{\mathbf{v}_1}\right) = \underset{r,\,1}{\mathbf{d}_1} \tag{10.11}$$

Using the equations just referred to, we get $\mathbf{v}_1$, $\mathbf{Q}_{v_1v_1}$, and $\mathbf{Q}_2 = \mathbf{Q} - \mathbf{Q}_{v_1v_1}$ as the solution for the first system in the first step. Next, the second system of conditions is

$$\underset{s,\,n}{\mathbf{A}_2}\left(\boldsymbol{l} + \underset{n,\,1}{\mathbf{v}_1} + \mathbf{v}_2\right) = \underset{s,\,1}{\mathbf{d}_2} \tag{10.12}$$

for which the cofactor matrix is now $\mathbf{Q}_2$. Again using the same equations, solve for $\mathbf{v}_2$ and $\mathbf{Q}_{v_2v_2}$ and compute

$$\bar{\mathbf{Q}}_{ll} = \mathbf{Q}_2 - \mathbf{Q}_{v_2v_2} \tag{10.13}$$

The answer obtained from the total system of conditions may be obtained from the adjustment in steps as follows:

$$\mathbf{v} = \mathbf{v}_1 + \mathbf{v}_2 \tag{10.14a}$$

$$\mathbf{Q}_{ll} = \bar{\mathbf{Q}}_{ll} \tag{10.14b}$$

$$\phi = \phi_1 + \phi_2 \qquad \text{(the quadratic forms)} \tag{10.14c}$$

10.3.1. Proof that $\mathbf{v} = \mathbf{v}_1 + \mathbf{v}_2$

The total system may be partitioned to

$$\begin{bmatrix} \mathbf{A}_1 \\ \mathbf{A}_2 \end{bmatrix} (\boldsymbol{l} + \mathbf{v}) = \begin{bmatrix} \mathbf{d}_1 \\ \mathbf{d}_2 \end{bmatrix}$$

or

$$\mathbf{A}_1 \mathbf{v} = \mathbf{d}_1 - \mathbf{A}_1 \boldsymbol{l} = \mathbf{f}_1 \tag{10.15a}$$

$$\mathbf{A}_2 \mathbf{v} = \mathbf{d}_2 - \mathbf{A}_2 \boldsymbol{l} = \mathbf{f}_2 \tag{10.15b}$$

Now with regard to the adjustment in steps, from the first step:

$$\mathbf{A}_1 \mathbf{v}_1 = \mathbf{d}_1 - \mathbf{A}_1 \boldsymbol{l} = \mathbf{f}_1 \tag{10.16a}$$

and

$$\mathbf{Q}_{v_1 v_1} = \mathbf{Q}\mathbf{A}_1^t(\mathbf{A}_1\mathbf{Q}\mathbf{A}_1^t)^{-1}\mathbf{A}_1\mathbf{Q} \tag{10.16b}$$

$$\mathbf{Q}_2 = \mathbf{Q} - \mathbf{Q}\mathbf{A}_1^t(\mathbf{A}_1\mathbf{Q}\mathbf{A}_1^t)^{-1}\mathbf{A}_1\mathbf{Q} \tag{10.16c}$$

For the second step

$$\mathbf{A}_2(\boldsymbol{l} + \mathbf{v}_1 + \mathbf{v}_2) = \mathbf{d}_2$$

$$\mathbf{A}_2(\mathbf{v}_1 + \mathbf{v}_2) = \mathbf{d}_2 - \mathbf{A}_2 \boldsymbol{l} \tag{10.17}$$

From equations (10.17) and (10.15) it follows that

$$\mathbf{v} = \mathbf{v}_1 + \mathbf{v}_2$$

provided that the consistency between equations (10.15a) and (10.16a) is ascertained. For such a consistency to stand we must prove that

$$\mathbf{A}_1 \mathbf{v}_2 = \mathbf{0}$$

Solving equation (10.17) for $\mathbf{v}_2$

$$\begin{aligned} \mathbf{v}_2 &= \mathbf{Q}_2 \mathbf{A}_2^t(\mathbf{A}_2 \mathbf{Q}_2 \mathbf{A}_2^t)^{-1}(\mathbf{d}_2 - \mathbf{A}_2 \mathbf{l} - \mathbf{A}_2 \mathbf{v}_1) \\ &= \mathbf{Q}_2 \mathbf{A}_2^t(\mathbf{A}_2 \mathbf{Q}_2 \mathbf{A}_2^t)^{-1}\mathbf{f}_2' \end{aligned}$$

and using equation (10.16c)

$$\mathbf{v}_2 = [\mathbf{Q} - \mathbf{Q}\mathbf{A}_1^t(\mathbf{A}_1\mathbf{Q}\mathbf{A}_1^t)^{-1}\mathbf{A}_1\mathbf{Q}]\mathbf{A}_2^t(\mathbf{A}_2\mathbf{Q}_2\mathbf{A}_2^t)^{-1}\mathbf{f}_2'$$

Now premultiplying by $\mathbf{A}_1$

$$\begin{aligned} \mathbf{A}_1\mathbf{v}_2 &= [\mathbf{A}_1\mathbf{Q} - (\mathbf{A}_1\mathbf{Q}\mathbf{A}_1^t)(\mathbf{A}_1\mathbf{Q}\mathbf{A}_1^t)^{-1}\mathbf{A}_1\mathbf{Q}]\mathbf{A}_2^t(\mathbf{A}_2\mathbf{Q}_2\mathbf{A}_2^t)^{-1}\mathbf{f}_2' \\ &= \mathbf{0} \end{aligned}$$

thus proving the validity of equation (10.14a).

10.3.2. Proof that $\mathbf{Q}_{ll} = \bar{\mathbf{Q}}_{ll}$ We first evaluate $\mathbf{Q}_{ll}$ from the total system after it is partitioned.

$$\mathbf{Q}_{ll} = \mathbf{Q} - \mathbf{QA}^t(\mathbf{AQA}^t)^{-1}\mathbf{AQ}$$

for which we expand

$$(\mathbf{AQA}^t)^{-1} = \left\{\begin{bmatrix}\mathbf{A}_1\\ \mathbf{A}_2\end{bmatrix}\mathbf{Q}[\mathbf{A}_1^t\mathbf{A}_2^t]\right\}^{-1} = \begin{bmatrix}\mathbf{A}_1\mathbf{QA}_1^t & \mathbf{A}_1\mathbf{QA}_2^t\\ \mathbf{A}_2\mathbf{QA}_1^t & \mathbf{A}_2\mathbf{QA}_2^t\end{bmatrix}^{-1}$$

$$= \begin{bmatrix}\dot{\mathbf{M}} & \bar{\mathbf{M}}\\ \bar{\mathbf{M}}^t & \ddot{\mathbf{M}}\end{bmatrix}^{-1} = \begin{bmatrix}\mathbf{E} & \mathbf{G}\\ \mathbf{G}^t & \mathbf{H}\end{bmatrix}$$

where

$$\mathbf{H} = (\ddot{\mathbf{M}} - \bar{\mathbf{M}}^t\dot{\mathbf{M}}^{-1}\bar{\mathbf{M}})^{-1} \tag{10.18a}$$

$$\mathbf{G} = -\dot{\mathbf{M}}^{-1}\bar{\mathbf{M}}\mathbf{H} \tag{10.18b}$$

$$\mathbf{E} = \dot{\mathbf{M}}^{-1} - \dot{\mathbf{M}}^{-1}\bar{\mathbf{M}}\mathbf{G}^t \tag{10.18c}$$

Thus

$$\mathbf{Q}_{ll} = \mathbf{Q} - \mathbf{Q}[\mathbf{A}_1^t\mathbf{A}_2^t]\begin{bmatrix}\mathbf{E} & \mathbf{G}\\ \mathbf{G}^t & \mathbf{H}\end{bmatrix}\begin{bmatrix}\mathbf{A}_1\\ \mathbf{A}_2\end{bmatrix}\mathbf{Q}$$

or

$$\begin{aligned}\mathbf{Q}_{ll} &= \mathbf{Q} - \mathbf{QA}_1^t\mathbf{EA}_1\mathbf{Q} - \mathbf{QA}_2^t\mathbf{G}^t\mathbf{A}_1\mathbf{Q} - \mathbf{QA}_1^t\mathbf{GA}_2\mathbf{Q} - \mathbf{QA}_2^t\mathbf{HA}_2\mathbf{Q}\\ &= \mathbf{Q} - \mathbf{QA}_1^t(\dot{\mathbf{M}}^{-1} - \dot{\mathbf{M}}^{-1}\bar{\mathbf{M}}\mathbf{G}^t)\mathbf{A}_1\mathbf{Q} - \mathbf{QA}_2^t\mathbf{G}^t\mathbf{A}_1\mathbf{Q}\\ &\quad - \mathbf{QA}_1^t\mathbf{GA}_2\mathbf{Q} - \mathbf{QA}_2^t\mathbf{HA}_2\mathbf{Q}\end{aligned}$$

or

$$\begin{aligned}\mathbf{Q}_{ll} &= \mathbf{Q} - \underset{(2)}{(\mathbf{QA}_1^t\dot{\mathbf{M}}^{-1}\mathbf{A}_1\mathbf{Q})} + \underset{(3)}{(\mathbf{QA}_1^t\dot{\mathbf{M}}^{-1}\bar{\mathbf{M}}\mathbf{G}^t\mathbf{A}_1\mathbf{Q})}\\ &\quad - \underset{(4)}{(\mathbf{QA}_2^t\mathbf{G}^t\mathbf{A}_1\mathbf{Q})} - \underset{(5)}{(\mathbf{QA}_1^t\mathbf{GA}_2\mathbf{Q})}\\ &\quad - \underset{(6)}{(\mathbf{QA}_2^t\mathbf{HA}_2\mathbf{Q})}\end{aligned} \tag{10.19}$$

Then we also evaluate $\bar{\mathbf{Q}}_{ll}$ as follows:

$$\begin{aligned}\bar{\mathbf{Q}}_{ll} &= \mathbf{Q}_2 - \mathbf{Q}_2\mathbf{A}_2^t(\mathbf{A}_2\mathbf{Q}_2\mathbf{A}_2^t)^{-1}\mathbf{A}_2\mathbf{Q}_2\\ &= \underset{(1)}{\mathbf{Q}} - \underset{(2)}{\mathbf{QA}_1^t\dot{\mathbf{M}}^{-1}\mathbf{A}_1\mathbf{Q}}\\ &\quad - (\mathbf{Q} - \mathbf{QA}_1^t\dot{\mathbf{M}}^{-1}\mathbf{A}_1\mathbf{Q})\mathbf{A}_2^t(\mathbf{A}_2\mathbf{Q}_2\mathbf{A}_2^t)^{-1}\mathbf{A}_2(\mathbf{Q} - \mathbf{QA}_1^t\dot{\mathbf{M}}^{-1}\mathbf{A}_1\mathbf{Q})\end{aligned} \tag{10.20}$$

We evaluate the following term first:

$$\begin{aligned}(\mathbf{A}_2\mathbf{Q}_2\mathbf{A}_2^t)^{-1} &= [\mathbf{A}_2(\mathbf{Q}-\mathbf{Q}\mathbf{A}_1^t\dot{\mathbf{M}}^{-1}\mathbf{A}_1\mathbf{Q})\mathbf{A}_2^t]^{-1}\\ &= [(\mathbf{A}_2\mathbf{Q}\mathbf{A}_2^t)-(\mathbf{A}_2\mathbf{Q}\mathbf{A}_1^t)\dot{\mathbf{M}}^{-1}(\mathbf{A}_1\mathbf{Q}\mathbf{A}_2^t)]^{-1}\\ &= (\ddot{\mathbf{M}}-\bar{\mathbf{M}}^t\dot{\mathbf{M}}^{-1}\bar{\mathbf{M}})^{-1}\\ &= \mathbf{H} \qquad (10.21)\end{aligned}$$

Since the first two terms are the same in both equation (10.19) and equation (10.20), we consider the last term in equation (10.20), which is denoted by **t**, using the just derived result.

$$\begin{aligned}\mathbf{t} = &-\underset{(6)}{\mathbf{Q}\mathbf{A}_2^t\mathbf{H}\mathbf{A}_2\mathbf{Q}} + \mathbf{Q}\mathbf{A}_2^t\mathbf{H}\mathbf{A}_2\mathbf{Q}\mathbf{A}_1^t\dot{\mathbf{M}}^{-1}\mathbf{A}_1\mathbf{Q} + \mathbf{Q}\mathbf{A}_1^t\dot{\mathbf{M}}^{-1}\mathbf{A}_1\mathbf{Q}\mathbf{A}_2^t\\ &\times \mathbf{H}\mathbf{A}_2\mathbf{Q} - \mathbf{Q}\mathbf{A}_1^t\dot{\mathbf{M}}^{-1}\mathbf{A}_1\mathbf{Q}\mathbf{A}_2^t\mathbf{H}\mathbf{A}_2\mathbf{Q}\mathbf{A}_1^t\dot{\mathbf{M}}^{-1}\mathbf{A}_1\mathbf{Q}\end{aligned}$$

Dropping the first term since it is the same as the sixth term of equation (10.19), we continue with

$$\begin{aligned}\mathbf{t}' &= \mathbf{Q}\mathbf{A}_2^t\mathbf{H}\bar{\mathbf{M}}^t\dot{\mathbf{M}}^{-1}\mathbf{A}_1\mathbf{Q} + \mathbf{Q}\mathbf{A}_1^t\dot{\mathbf{M}}^{-1}\bar{\mathbf{M}}\mathbf{H}\mathbf{A}_2\mathbf{Q}\\ &\quad - \mathbf{Q}\mathbf{A}_1^t\dot{\mathbf{M}}^{-1}\bar{\mathbf{M}}\mathbf{H}\bar{\mathbf{M}}^t\dot{\mathbf{M}}^{-1}\mathbf{A}_1\mathbf{Q}\\ &= -\underset{(4)}{\mathbf{Q}\mathbf{A}_2^t\mathbf{G}^t\mathbf{A}_1\mathbf{Q}} - \underset{(5)}{\mathbf{Q}\mathbf{A}_1^t\mathbf{G}\mathbf{A}_2\mathbf{Q}} + \underset{(3)}{\mathbf{Q}\mathbf{A}_1\dot{\mathbf{M}}^{-1}\bar{\mathbf{M}}\mathbf{G}^t\mathbf{A}_1\mathbf{Q}} \qquad (10.22)\end{aligned}$$

The first term of equation (10.21) is the same as the fourth term in equation (10.19), the second is the same as the fifth, and the last term is the same as the third, respectively. This proves that both $\mathbf{Q}_{ll}$ and $\bar{\mathbf{Q}}_{ll}$ can be expanded to six terms that are identical. Hence

$$\mathbf{Q}_{ll} = \bar{\mathbf{Q}}_{ll}$$

10.3.3. Proof that $\phi = \phi_1 + \phi_2$ In order to compute an a posteriori estimate of the reference variance, the quadratic form $\phi = \mathbf{v}^t\mathbf{W}\mathbf{v}$ must be computed. For the computation of ϕ we use equation (7.6) or

$$\phi = \mathbf{k}^t\mathbf{f} = \mathbf{f}^t(\mathbf{A}\mathbf{Q}\mathbf{A}^t)^{-1}\mathbf{f}$$

From equations (10.15b) and (10.17) we can readily write

$$\mathbf{f}_2' = (\mathbf{d}_2 - \mathbf{A}_2\boldsymbol{l}) - \mathbf{A}_2\mathbf{v}_1 = \mathbf{f}_2 - \mathbf{A}_2\mathbf{v}_1 \qquad (10.23)$$

and we also recall that

$$\mathbf{v}_1 = \mathbf{Q}\mathbf{A}_1^t(\mathbf{A}_1\mathbf{Q}\mathbf{A}_1^t)^{-1}\mathbf{f}_1 = \mathbf{Q}\mathbf{A}_1^t\dot{\mathbf{M}}^{-1}\mathbf{f}_1 \qquad (10.24)$$

$$\begin{aligned}\phi &= \mathbf{f}^t\begin{bmatrix}\dot{\mathbf{M}} & \bar{\mathbf{M}}\\ \bar{\mathbf{M}}^t & \ddot{\mathbf{M}}\end{bmatrix}^{-1}\mathbf{f} = [\mathbf{f}_1^t \quad \mathbf{f}_2^t]\begin{bmatrix}\mathbf{E} & \mathbf{G}\\ \mathbf{G}^t & \mathbf{H}\end{bmatrix}\begin{bmatrix}\mathbf{f}_1\\ \mathbf{f}_2\end{bmatrix}\\ &= (\underset{(1)}{\mathbf{f}_1^t\dot{\mathbf{M}}^{-1}\mathbf{f}_1} - \underset{(2)}{\mathbf{f}_1^t\dot{\mathbf{M}}^{-1}\bar{\mathbf{M}}\mathbf{G}^t\mathbf{f}_1}) + \underset{(3)}{2\mathbf{f}_1^t\mathbf{G}\mathbf{f}_2} + \underset{(4)}{\mathbf{f}_2^t\mathbf{H}\mathbf{f}_2} \qquad (10.25)\end{aligned}$$

and

$$\phi_1 = \mathbf{f}_1^t(\mathbf{A}_1\mathbf{Q}\mathbf{A}_1^t)^{-1}\mathbf{f}_1 = \mathbf{f}_1^t\dot{\mathbf{M}}^{-1}\mathbf{f}_1 \tag{10.26}$$

finally,

$$\phi_2 = \mathbf{f}_2'^t(\mathbf{A}_2\,\mathbf{Q}_2\,\mathbf{A}_2^t)^{-1}\mathbf{f}_2' \tag{10.27}$$

which according to equations (10.21), (10.23), and (10.24) becomes

$$\begin{aligned}\phi_2 &= \mathbf{f}_2'^t\mathbf{H}\mathbf{f}_2' \\ &= (\mathbf{f}_2^t - \mathbf{v}_1^t\mathbf{A}_2^t)\mathbf{H}(\mathbf{f}_2 - \mathbf{A}_2\,\mathbf{v}_1) \\ &= \mathbf{f}_2^t\,\mathbf{H}\mathbf{f}_2 - 2\mathbf{v}_1^t\mathbf{A}_2^t\,\mathbf{H}\mathbf{f}_2 + \mathbf{v}_1^t\mathbf{A}_2^t\,\mathbf{H}\mathbf{A}_2\,\mathbf{v}_1 \\ &= \mathbf{f}_2^t\,\mathbf{H}\mathbf{f}_2 - 2\mathbf{f}_1^t\dot{\mathbf{M}}^{-1}(\mathbf{A}_1\mathbf{Q}\mathbf{A}_2^t)\mathbf{H}\mathbf{f}_2 \\ &\quad + \mathbf{f}_1^t\dot{\mathbf{M}}^{-1}(\mathbf{A}_1\mathbf{Q}\mathbf{A}_2^t)\mathbf{H}(\mathbf{A}_2\,\mathbf{Q}\mathbf{A}_1^t)\dot{\mathbf{M}}^{-1}\mathbf{f}_1 \\ &= \mathbf{f}_2^t\,\mathbf{H}\mathbf{f}_2 - 2\mathbf{f}_1^t\dot{\mathbf{M}}^{-1}\bar{\mathbf{M}}\mathbf{H}\mathbf{f}_2 + \mathbf{f}_1^t\dot{\mathbf{M}}^{-1}\bar{\mathbf{M}}(\mathbf{H}\bar{\mathbf{M}}^t\dot{\mathbf{M}}^{-1})\mathbf{f}_1\end{aligned}$$

$$\phi_2 = -\mathbf{f}_1^t\dot{\mathbf{M}}^{-1}\bar{\mathbf{M}}\mathbf{G}^t\mathbf{f}_1 + 2\mathbf{f}_1^t\mathbf{G}\mathbf{f}_2 + \mathbf{f}_2^t\,\mathbf{H}\mathbf{f}_2 \tag{10.28}$$

It is obvious from equations (10.25), (10.26), and (10.28) that

$$\phi = \phi_1 + \phi_2$$

Example 10.4. Figure 10.2 shows two adjacent plane triangles where six angles are measured. These are uncorrelated and of equal precision and have the following values:

$l_1 = 60°00'03''$
$l_2 = 60°00'02''$
$l_3 = 60°00'01''$
$l_4 = 120°00'05''$
$l_5 = 120°00'05''$
$l_6 = 60°00'01''$

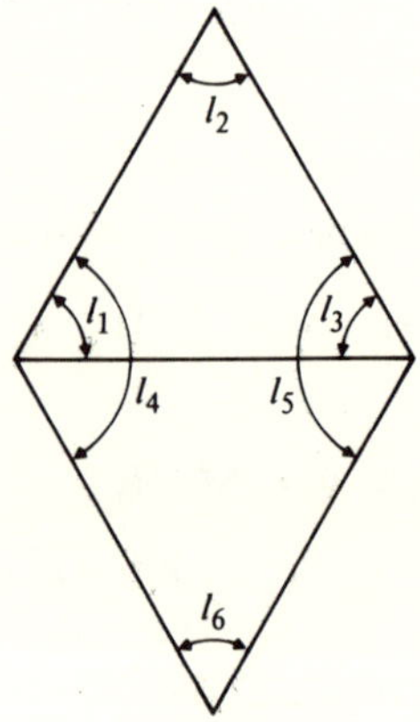

Figure 10.2

It is required to adjust the whole figure with all six observations together. We adjust the figure in two steps: First, using angles l_1, l_2, and l_3, and in the second step, using the remaining data. Then we verify the consistency of the answers from both procedures.

Solution

The Direct Adjustment. There are 2 degrees of freedom (the reader should verify this) for which we write the following two independent conditions:

$$l_1 + l_2 + l_3 - 180 = 0$$

$$l_2 + l_4 + l_5 + l_6 - 360 = 0$$

or

$$\begin{bmatrix} 1 & 1 & 1 & 0 & 0 & 0 \\ 0 & 1 & 0 & 1 & 1 & 1 \end{bmatrix} \mathbf{v} = \begin{bmatrix} 180 - l_1 - l_2 - l_3 \\ 360 - l_2 - l_4 - l_5 - l_6 \end{bmatrix} = \begin{bmatrix} -6'' \\ -12 \end{bmatrix}$$

$$\mathbf{k} = \begin{bmatrix} 3 & 1 \\ 1 & 4 \end{bmatrix}^{-1} \begin{bmatrix} -6 \\ -12 \end{bmatrix} = \begin{bmatrix} -1.09'' \\ -2.73'' \end{bmatrix}$$

$$\mathbf{v} = \mathbf{QA}^t\mathbf{k} = \begin{bmatrix} 1 & 0 \\ 1 & 1 \\ 1 & 0 \\ 0 & 1 \\ 0 & 1 \\ 0 & 1 \end{bmatrix} \begin{bmatrix} -1.09'' \\ -2.73'' \end{bmatrix} = \begin{bmatrix} -1.09'' \\ -3.82'' \\ -1.09'' \\ -2.73'' \\ -2.73'' \\ -2.73'' \end{bmatrix}$$

$$\mathbf{Q}_{vv} = \tfrac{1}{11} \begin{bmatrix} 4 & 3 & 4 & -1 & -1 & -1 \\ & 5 & 3 & 2 & 2 & 2 \\ & & 4 & -1 & -1 & -1 \\ & & & 3 & 3 & 3 \\ & & & & 3 & 3 \\ \text{symmetric} & & & & & 3 \end{bmatrix}$$

$$\mathbf{Q}_{ll} = \mathbf{I} - \mathbf{Q}_{vv} = \tfrac{1}{11} \begin{bmatrix} 7 & -3 & -4 & 1 & 1 & 1 \\ & 6 & -3 & -2 & -2 & -2 \\ & & 7 & 1 & 1 & 1 \\ & & & 8 & -3 & -3 \\ & & & & 8 & -3 \\ \text{symmetric} & & & & & 8 \end{bmatrix}$$

Adjustment in Steps

First Step: Here there is only one condition,

$$l_1 + l_2 + l_3 - 180 = 0$$

$$[1 \quad 1 \quad 1 \quad 0 \quad 0 \quad 0]\mathbf{v}_1 = -6''$$

$$k_1 = 2'' \qquad \mathbf{v}_1 = [-2'' \quad -2'' \quad -2'' \quad 0 \quad 0 \quad 0]^t$$

$$\mathbf{Q}_{v_1v_1} = \tfrac{1}{3}\begin{bmatrix} 1 & 1 & 1 & 0 & 0 & 0 \\ & 1 & 1 & 0 & 0 & 0 \\ & & 1 & 0 & 0 & 0 \\ & & & 0 & 0 & 0 \\ & & & & 0 & 0 \\ \text{symmetric} & & & & & 0 \end{bmatrix}$$

$$\mathbf{Q}_2 = \mathbf{Q} - \mathbf{Q}_{v_1v_1} = \mathbf{I} - \mathbf{Q}_{v_1v_1} = \tfrac{1}{3}\begin{bmatrix} 2 & -1 & -1 & 0 & 0 & 0 \\ & 2 & -1 & 0 & 0 & 0 \\ & & 2 & 0 & 0 & 0 \\ & & & 3 & 0 & 0 \\ & & & & 3 & 0 \\ \text{symmetric} & & & & & 3 \end{bmatrix}$$

Second Step: The vector of observations in this step is not the one originally given at the beginning of the problem. It is updated by the residuals from the first step. Thus

$$l'_1 = 60°00''01' \qquad l'_2 = 60°00''00' \qquad l'_3 = 59°59''39'$$

$$l'_4 = 120°00''05' \qquad l'_5 = 120°00''04' \qquad l'_6 = 60°00''01'$$

The one condition for this step is

$$l'_2 + l'_4 + l'_5 + l'_6 - 360 = 0$$

$$[0 \quad 1 \quad 0 \quad 1 \quad 1 \quad 1]\mathbf{v}_2 = 360 - l'_2 - l'_4 - l'_5 - l'_6 = -10''$$

Note that $f'_2 = -10''$ is not the same as $f_2 = -12''$. The difference as given by equation (10.23) is $(-\mathbf{A}_2\mathbf{v}_1) = +2''$, which is correct. Thus

$$\mathbf{k}_2 = (\mathbf{A}_2\mathbf{Q}_2\mathbf{A}_2)^{-1}(-10) = -\left(\frac{30}{11}\right)''$$

$$\mathbf{v}_2 = \mathbf{Q}_2\mathbf{A}_2\mathbf{k}_2$$

$$= [0.91 \quad -1.82 \quad 0.91 \quad -2.73 \quad -2.73 \quad -2.73]^t \qquad \text{(sec of arc)}$$

and

$$\mathbf{v}_1 + \mathbf{v}_2 = [-1.09 \quad -3.82 \quad -1.09 \quad -2.73 \quad -2.73 \quad -2.73]^t \qquad \text{(sec of arc)}$$

which is identical to **v** computed in the direct adjustment.

$$\mathbf{Q}_{v2v2} = \tfrac{1}{33}\begin{bmatrix} 1 & -2 & 1 & -3 & -3 & -3 \\ & 4 & -2 & 6 & 6 & 6 \\ & & 1 & -3 & -3 & -3 \\ & & & 9 & 9 & 9 \\ & & & & 9 & 9 \\ \text{symmetric} & & & & & 9 \end{bmatrix}$$

$$\bar{\mathbf{Q}}_{ll} = \mathbf{Q}_2 - \mathbf{Q}_{v2v2} = \tfrac{1}{11}\begin{bmatrix} 7 & -3 & -4 & 1 & 1 & 1 \\ & 6 & -3 & -2 & -2 & -2 \\ & & 7 & 1 & 1 & 1 \\ & & & 8 & -3 & -3 \\ \text{symmetric} & & & & & 8 \end{bmatrix}$$

which is again identical to $\mathbf{Q}_{ll}$ computed in the direct adjustment.

It may be concluded that both adjustment with derived observations and adjustment in steps are feasible provided we make sure of satisfying the requirements of each concept. Most important of these are not changing the model when adjusting with derived observations and making the necessary modification of the stochastic model for both concepts of adjustment.

11

Numerical and Statistical Considerations in Adjustment

This chapter is devoted to several topics that were mentioned but not discussed in detail in the foregoing chapters. Such topics pertain mostly to practical matters of adjustment and include selection of approximations for model variables, criteria for termination of iterations for nonlinear cases, data editing for gross errors, and computational aspects of adjustment, and so on. The problem of linearization of nonlinear equations is discussed first.

11.1. NONLINEARITY OF THE EQUATIONS

Nonlinear equations arise more often in practice than those that are directly linear. Therefore the adjustment problem quite often involves nonlinear estimation. The least squares criterion can, in principle, be applied directly to nonlinear equations but with the consequence that the resulting normal equations would also be nonlinear. Such nonlinear equations may pose considerable difficulties in seeking a unique solution. Furthermore, in cases of large adjustment problems, computational efforts may become so extensive as to render the solution impractical. To demonstrate this point and the possible complexity of this approach, a simple example is given.

Example 11.1. In this example reconsider the problem of Example 7.2 and refer to Figure 6.1 for the layout of the three terrestrial photographs. In that example there was 1 degree of freedom, leading to one condition equation in terms of the observations only, of the form

$$\mathbf{F}(l) = -l_1 l_5 - l_2 l_4 - l_2 l_5 + l_3 l_4 = 0$$

where l_i represents the five measurements as shown in Figure 6.1. In Example 7.2 (as in all other treatments) this nonlinear equation was first linearized by Taylor series expansion before the least squares criterion was applied. Here we go directly into the criterion with the nonlinear condition, thus in effect performing nonlinear least squares adjustment. In this case the quadratic form to be minimized is

$$\phi = \mathbf{v}^t\mathbf{W}\mathbf{v} - 2\mathbf{k}_n^t(-l_1 l_5 - l_2 l_4 - l_2 l_5 + l_3 l_4)$$

where $\mathbf{k}_n$ is one Lagrange multiplier for the nonlinear condition. Replacing l_i by $(l_i + v_i)$ and remembering (from Example 7.2) that $\mathbf{W}$ is a 5×5 diagonal matrix with elements w_i, the quadratic form ϕ may be expanded to

$$\begin{aligned}\phi = {} & (w_1 v_1^2 + w_2 v_2^2 + w_3 v_3^2 + w_4 v_4^2 + w_5 v_5^2) \\ & + 2k_n[(l_1 + v_1)(l_5 + v_5) \\ & + (l_2 + v_2)(l_4 + v_4) + (l_2 + v_2)(l_5 + v_5) \\ & - (l_3 + v_3)(l_4 + v_4)]\end{aligned}$$

To fulfill the minimum criterion, differentiate ϕ with respect to each v_i and equate each partial derivative to zero. Thus

$$\begin{aligned} & w_1 v_1 + k_n(l_5 + v_5) && = 0 \\ & w_2 v_2 + k_n[(l_4 + v_4) + (l_5 + v_5)] && = 0 \\ & w_3 v_3 - k_n(l_4 + v_4) && = 0 \\ & w_4 v_4 + k_n[(l_2 + v_2) - (l_3 + v_3)] && = 0 \\ & w_5 v_5 + k_n[(l_1 + v_1) + (l_2 + v_2)] && = 0 \end{aligned}$$

These are five equations that, when combined with the one condition, result in *six nonlinear* normal equations in six unknowns; the unknowns being the five residuals and k_n. It must be recognized that the solution of this set of nonlinear equations is not as easily obtainable as for linear equations, as their solution requires more involved numerical schemes. Another point that is worth mentioning is the fact that here there are *six* equations to solve simultaneously, whereas in the linearized case the v's were easily eliminated from the six equations and *one* normal equation remained in terms of k. Finally, if the weight matrix were a full matrix, the expression for ϕ would have included many more terms thus resulting in more complicated nonlinear equations. This change in weight matrix, on the other hand, would have less effect on the normal equations in the linearized case.

These remarks point out the relative complexity of nonlinear least squares when compared to the linear case. This is the reason that in practice we invariably linearize first, then apply the least squares method.

11.1.1. The Linearization Concept Section 5.4 included a brief description of linearization by Taylor series expansion using zero and first-order terms and dropping second- and higher-order terms. In Appendix B the complete series for one-dimensional and multidimensional functions are given. These are introduced in terms of functional variables without regard to their classification in different groups when used in adjustment. Here, on the other hand, linearization is discussed as it is applied in adjustment techniques.

It has been customary in practice to linearize the conditions at the given observations and at approximations for the parameters. This means that for cases of adjustment of observations only the solution is not iterated, and for all other cases the solution is iterated on the parameters only. Although this procedure may be totally practical in most cases, it is not theoretically sound. Of course, we must admit that in many situations the given a priori values of the observations are sufficiently close to the final estimates that no iterations are needed. Notwithstanding such cases we should, in principle, treat all variables in the model equally and perform the linearization and iteration on all. There are situations in which failure to linearize and iterate on the observable variables leads to incorrect results. As an example, in a unified approach to least squares (see Chapter 12 in Part III) when parameters are treated strictly as observations, lack of proper iteration causes the method to fail. Although the given observations will, as usual, be denoted by $\boldsymbol{l}$, an approximate vector that varies from one iteration to another will be denoted by $\boldsymbol{l}^0$. Furthermore, the vector of residuals will be retained as $\mathbf{v}$, but a vector of corrections $\boldsymbol{\Delta l}$ will also be necessary. For the parameters $\mathbf{x}$, the customary $\mathbf{x}^0$ will be used for approximations and $\boldsymbol{\Delta}$ for corrections. On the basis of these symbols the linearization schemes for each of the different techniques of adjustment are given.

11.1.2. Linearization Form for General Case of Adjustment with Conditions Only $F(\boldsymbol{l}, \mathbf{x}) = \mathbf{0}$ The original nonlinear conditions are expressed functionally by

$$f(\boldsymbol{l}, \mathbf{x}) = \mathbf{0} \tag{11.1}$$

or

$$F[(\boldsymbol{l}^0 + \boldsymbol{\Delta l}), (\mathbf{x}^0 + \boldsymbol{\Delta})] = \mathbf{0} \tag{11.2}$$

Applying Taylor's series to equation (11.2), dropping second- and higher-order terms, and using the auxiliaries

$$\mathbf{A} = \left.\frac{\partial F}{\partial l}\right|_{\boldsymbol{l}^0,\, \mathbf{x}^0} \qquad \text{and} \qquad \mathbf{B} = \left.\frac{\partial F}{\partial \mathbf{x}}\right|_{\boldsymbol{l}^0,\, \mathbf{x}^0} \tag{11.3}$$

the linearized form of equation (11.2) is

$$F(\boldsymbol{l}^0, \mathbf{x}^0) + \mathbf{A}\,\boldsymbol{\Delta l} + \mathbf{B}\boldsymbol{\Delta} = \mathbf{0} \tag{11.4}$$

The final estimate of the observations $\hat{l}$ must be equal to the observed vaues plus the least squares residuals, as well as the approximate values plus the corrections, or

$$\hat{l} = l + \mathbf{v} = l^0 + \Delta l \tag{11.5}$$

from which

$$\Delta l = \mathbf{v} + (l - l^0) \tag{11.6}$$

Using this relation in equation (11.4) and letting

$$\mathbf{f}^0 = -[F(l^0, \mathbf{x}^0) + \mathbf{A}(l - l^0)] \tag{11.7}$$

we get

$$\mathbf{Av} + \mathbf{B\Delta} = \mathbf{f}^0 \tag{11.8}$$

Equation (11.8) is of the same form as the basic equation (6.5). The only apparent difference is in the term $\mathbf{f}^0$ on the right-hand side. When the conditions are originally linear, $\mathbf{f}$ as given by equation (6.6) is used. If the conditions are nonlinear and we do not wish to iterate on the observations, $l = l^0$ and $\mathbf{f}^0$ would reduce to

$$\mathbf{f} = -F(l, \mathbf{x}^0) \tag{11.9}$$

In the general case of iterating on all variables equation (11.8) is used. The details of iterating by updating approximations will be treated in a subsequent section after giving the linearization forms for the remaining techniques.

11.1.3. Linearization Form for Adjustment of Observations Only $F(l) = \mathbf{0}$ In a manner similar to the general case above, we begin with the nonlinear conditions

$$F(l) = \mathbf{0} \tag{11.10}$$

which in this case are in terms of observations only (no parameters). Therefore equations (11.7) and (11.8) become

$$\mathbf{f}^0 = -[F(l^0) + \mathbf{A}(l - l^0)] \tag{11.11}$$

$$\mathbf{Av} = \mathbf{f}^0 \tag{11.12}$$

If we choose to iterate on the observations, the constant term would be as in equation (11.11). Otherwise the term $\mathbf{A}(l - l^0)$ drops out and $\mathbf{f}^0 = -F(l^0)$ for nonlinear conditions, but the solution is performed only once.

11.1.4. Linearization Form for Adjustment of Indirect Observations $l + F(\mathbf{x}) = \mathbf{0}$ The nonlinear conditions for this technique of adjustment take the form

$$l + F(\mathbf{x}) = \mathbf{0} \tag{11.13}$$

which is obviously linear in the observations. Therefore we may write the linearized form directly as

$$\mathbf{v} + \mathbf{B}\boldsymbol{\Delta} = \mathbf{f}^0 \tag{11.14}$$

$$\mathbf{f}^0 = -[\boldsymbol{l} + F(\mathbf{x}^0)] \tag{11.15}$$

Note that no iteration on the observations is necessary here since the conditions in them are originally linear.

11.1.5. Linearization Form for Adjustment with Conditions and Constraints Only the general case is considered here, as other specialized cases can be readily treated in a straightforward manner. The original nonlinear conditions and constraints are functionally expressed as

$$F(\boldsymbol{l}, \mathbf{x}) = \mathbf{0} \tag{11.16a}$$

$$G(\mathbf{x}) = \mathbf{0} \tag{11.16b}$$

If

$$\mathbf{C} = \left.\frac{\partial G}{\partial \mathbf{x}}\right|_{\mathbf{x}^0} \tag{11.17}$$

$$\mathbf{g}^0 = -G(\mathbf{x}^0) \tag{11.18}$$

then with equations (11.3) and (11.7) the linearized form of the pair of equations in equation (11.16) becomes

$$\mathbf{A}\mathbf{v} + \mathbf{B}\boldsymbol{\Delta} = \mathbf{f}^0 \tag{11.19a}$$

$$\mathbf{C}\boldsymbol{\Delta} = \mathbf{g}^0 \tag{11.19b}$$

The different cases involving $\mathbf{f}^0$ and $\mathbf{f}$ have already been discussed in Section 11.1.2. As regards $\mathbf{g}^0$, it would simply become a vector of numerical constants $\mathbf{g}$, given a priori, when the constraints are originally linear.

Before leaving this section it should be emphasized that linearization, if it is necessary, must be applied to *both* the conditions and constraints, and at the same approximation vector $\mathbf{x}^0$, even if either set is originally linear.

11.1.6. Scheme of Iteration With the approximate vectors $\boldsymbol{l}_0^0$ and $\mathbf{x}_0^0$ we would compute the coefficient matrices $\mathbf{A}_0$ and $\mathbf{B}_0$ (either or both depending on the case of adjustment involved). Also the constant term vector $\mathbf{f}_0^0$ (or $\mathbf{f}_0$ depending again on the adjustment case) is evaluated. With these numerical values, normal equations are formed and solved after which we may directly obtain first values for $\boldsymbol{\Delta}$ and $\mathbf{v}$, which may be denoted by $\boldsymbol{\Delta}_1$ and $\mathbf{v}_1$. This essentially concludes the computations for one iteration. In order to begin the second iteration, updated approximations are required. These are computed from

$$\mathbf{x}_1^0 = \mathbf{x}_1^0 + \mathbf{\Delta}_1 \tag{11.20a}$$

$$\boldsymbol{l}_1^0 = \boldsymbol{l} + \mathbf{v}_1 \tag{11.20b}$$

Given $\boldsymbol{l}_1^0$ and $\mathbf{x}_1^0$, the matrices $\mathbf{A}_1$, $\mathbf{B}_1$, and $\mathbf{f}_1^0$ (or $\mathbf{f}_1$) are evaluated anew. Another least squares solution is effected yielding $\mathbf{\Delta}_2$ and $\mathbf{v}_2$ as answer vectors. Then

$$\mathbf{x}_2^0 = \mathbf{x}_1^0 + \mathbf{\Delta}_2 \tag{11.21a}$$

$$\boldsymbol{l}_2^0 = \boldsymbol{l} + \mathbf{v}_2 \tag{11.21b}$$

Note that although $\mathbf{\Delta}_2$ is added to the previous approximation vector $\mathbf{x}_1^0$, the vector $\mathbf{v}_2$ is added to the original observational vector. This is a direct consequence of equation (11.5). In general, then, the approximations $\boldsymbol{l}_i^0$ and $\mathbf{x}_i^0$ at the end of the ith iteration may be computed from

$$\mathbf{x}_i^0 = \mathbf{x}_{i-1}^0 + \mathbf{\Delta}_i \tag{11.22a}$$

$$\boldsymbol{l}_i^0 = \boldsymbol{l} + \mathbf{v}_i \tag{11.22b}$$

At the end of each iteration a check is performed to see if it is necessary to continue with another iteration or to terminate the process. This check is usually performed according to a present criterion of tolerance. This aspect is treated in Section 11.4.

Example 11.2. Recall the problem of the three terrestrial photographs in Example 7.2, where it was solved by the technique of adjustment of observations only. In that example the one nonlinear condition was linearized at the given observations and the solution performed only once. If we now extend the solution such that iterations are performed on observations, equations (11.11) and (11.12) should be applied. In doing so, four iterations were carried out. The vectors of residuals after the first and final iterations are

$$\mathbf{v}_1 = \begin{bmatrix} 0.043\ 642\ 1495\ \text{(mm)} \\ 0.098\ 194\ 8365\ \text{(mm)} \\ -0.054\ 552\ 6869\ \text{(mm)} \\ -0.022\ 639\ 3650\ \text{(m)} \\ 0.027\ 685\ 4886\ \text{(m)} \end{bmatrix} \qquad \mathbf{v}_f = \begin{bmatrix} 0.043\ 775\ 4674\ \text{(mm)} \\ 0.098\ 182\ 5901\ \text{(mm)} \\ -0.054\ 407\ 1226\ \text{(mm)} \\ -0.022\ 421\ 6799\ \text{(m)} \\ 0.027\ 867\ 1859\ \text{(m)} \end{bmatrix}$$

The vector $\mathbf{v}_1$ would be identical to that obtained in Example 7.2, if that many digits were carried in the latter. In fact, 10 digits to the right of the decimal were carried there for the purpose of illustrating the exactness of the procedures derived above. To ascertain the consistency of these procedures, the same problem was solved by the method of adjustment of indirect observations. Remember that in such a case we compute only one value for the vector $\mathbf{v}$ after terminating the iterations on the parameters. A criterion for iteration termination was selected such that the last corrections $\mathbf{\Delta}$ must all be less than 10^{-10}. With this tolerance criterion met, *all* five residuals agreed with the elements of $\mathbf{v}_f$ to *all* 10 decimal places.

As a further illustration the scheme of direct nonlinear least squares discussed in Example 11.1 was also used. Instead of directly solving the six nonlinear equations, a much simpler procedure was applied in order to verify the correctness of $\mathbf{v}_f$. The values of the residuals (in $\mathbf{v}_f$) are substituted in the five equations (obtained by partial differentiation in Example 11.1) and five values for k_n are computed as follows:

$$\begin{aligned} k_n &= -0.545\ 293\ 867\ 41 \\ &= -0.545\ 293\ 867\ 44 \\ &= -0.545\ 293\ 867\ 41 \\ &= -0.545\ 293\ 867\ 43 \\ &= -0.545\ 293\ 867\ 44 \end{aligned}$$

These values agree with each other to the tenth decimal place. When the elements of $\mathbf{v}_1$ instead of $\mathbf{v}_f$ were used (thus representing the noniterative procedure of Example 7.2) that agreement was not attained.

To conclude this example it should be mentioned that the intention here was to carry enough digits in order to verify the theoretical considerations given earlier. In this particular example, we would not really need to iterate, since the differences between $\mathbf{v}_1$ and $\mathbf{v}_f$ are of no significance in view of the given σ's. However, there may be situations in which failure to iterate on the observations results in serious problems.

Example 11.3. As a further illustration consider the problem of the two-parameter transformation of Example 7.3. Without undue repetition, the solution of this problem was done using iterations on the observations with the following two vectors being the residuals after the first and the final iterations:

$$\mathbf{v}_1 = \begin{bmatrix} 0.009\ 296\ 1919 \\ 0.080\ 495\ 5370 \\ 0.014\ 389\ 2837 \\ 0.011\ 362\ 0128 \\ -0.061\ 013\ 1615 \\ -0.063\ 078\ 9819 \end{bmatrix} \qquad \mathbf{v}_f = \begin{bmatrix} 0.009\ 340\ 9445 \\ 0.078\ 360\ 1453 \\ 0.017\ 125\ 0649 \\ 0.010\ 378\ 8272 \\ -0.053\ 450\ 9601 \\ -0.054\ 488\ 8428 \end{bmatrix}$$

The problem was also programmed in the form $\mathbf{v} + \mathbf{B}\boldsymbol{\Delta} = \mathbf{f}^0$, the solution iterated on the parameters, and the final residual vector computed and found to agree with $\mathbf{v}_f$ above to even more digits than those given.

A possibility may be considered in which the conventional one-run solution is used to obtain a vector of estimates $\hat{\mathbf{l}}$ and a corresponding cofactor matrix $\mathbf{Q}_{\hat{l}\hat{l}}$. We may feel that with these new values a relinearization can be performed and another adjustment solution carried out. Such a procedure is incorrect and would fail because $\mathbf{Q}_{\hat{l}\hat{l}}$ is always a singular matrix (see Appendix C). This point can be demonstrated using the data of the present example. For the first iteration

$\mathbf{A}_1$ is a 4×6 matrix with a rank 4.

$\mathbf{Q}$ is a 6×6 matrix with a rank 6. (Original observations are given as functionally independent, thus a priori cofactor matrix is nonsingular.)

$\mathbf{Q}_{e_1} = \mathbf{A}_1\mathbf{Q}\mathbf{A}_1^t$ is a 4×4 nonsingular coefficient matrix of partially reduced normal equations (of rank 4).

After the first iteration, the computed $\hat{l}$ will be functionally dependent, which leads to a singular $\mathbf{Q}_{ll}$ matrix of a rank 2. Consequently, $\mathbf{Q}_{e_2}$ for the second iteration will have rank 2, and we cannot iterate the solution with this scheme.

11.2. APPROXIMATE VALUES FOR MODEL VARIABLES

With a nonlinear least squares problem, the procedure is to linearize the functional relationships and to apply the linear least squares algorithm repeatedly on the different levels of linearization. In order to linearize or expand a given function into power series, approximate values for all the variables in the function are required. The given a priori estimates of the observations are the best approximations possible when linearizing with respect to the observation variables. In the past, providing parameter approximations was not considered a problem. This is because in conventional applications in geodesy and astronomy it was always possible to arrive at such approximations by performing preliminary computations with a selected small subset of the measurements. At the present time, however, the problem of approximations is not at all trivial, particularly when we consider present least squares applications. They are usually fully automated, carried out on a computer, and involve extensive amounts of data. Obtaining approximations is therefore a problem of a general nature for which there is, unfortunately, no general algorithm available.

11.2.1. General Criteria for Parameter Approximations There are two main reasons for seeking close approximations: First, nonlinear problems may have several real solutions. With coarse initial approximate values, iterative relinearizations may converge, but to a solution that does not correspond to the minimum criterion of least squares. Second, it is, in general, true that the closer the approximations, the lesser is the number of iterations needed to reach the final solution. This obviously means a more economical solution, particularly for large problems.

Both of the above criteria figure more prominantly in cases in which gross data errors confound the solutions. But, in general, we usually strive to satisfy them both. However, with modern computer solutions the tendency is toward automated methods for obtaining approximate values. This added criterion of automation may in certain cases lead to the sacrifice of the second item above, at least to some extent. Of course, the ultimate result of automation is overall economy which obviously offsets any increase in the number of iterations.

11.2.2. Methods for Deriving Approximations Since there is no general algorithm for obtaining approximate values, it is usually the analysis of each particular adjustment problem that leads to an appropriate procedure for deriving approximations. As illustration, the following are several possibilities.

1. The procedure very frequently used in geodesy in the past was to perform computation with a minimum number of the given observations. This often allowed the calculation in a sequential manner of one (or more) approximate value after another. A good example is the case of a trigonometric net where we proceed to solve one triangle at a time until approximations for all the coordinates in the net are computed. In such procedures, partial adjustments may even be possible, as, for instance, in adjusting triangle closures.

This classically successful procedure can be formulated in a general principle. Redundancy is reduced to such a degree that it becomes possible to evaluate sequentially groups of unknowns as explicit functions of observations or other unknowns (or both) for which approximations have already been calculated. Such a scheme can often work quite well. In fact, for some problems it may yield approximations so close to the final estimates that only one linearized adjustment will be needed. The principal weakness of this method is that it relies on subjective judgment rather than an automated approach. In some cases, however, the subjective part can be automated.

For example, the hand method for geodetic nets can be simulated by computer techniques. In the simplest case it means operating on a few subprograms, such as intersection, resection, or traverse computation, where the sequence of operation is fed into the computer by a human operator, perhaps by making use of a graph technique. Another possibility is working out a special search program by which the computer searches by itself as to how to proceed from one point to the next, trying several possibilities in case difficulty is encountered. Here we can distinguish between two cases: whether the computer is at least told where to start, or whether it is to start at random or according to certain features that it would systematically search for.

Although such procedures work satisfactorily in most practical cases, they are by no means general. Depending on the geometrical structure of a system and on redundancy, such systems may break down in unusual cases.

A second weakness is the sensitivity of such systems to blunders in the data. Because the philosophy of such strategies is to find a straightforward way through the system, without actual adjustment, the success depends on the assumption that the observations can be relied on. If a blunder exists in the system, it can make many of the computed approximations grossly erroneous. Even if the program can detect blunders, it will usually not be able to circumvent them. All this leads to the conclusion that in well-determined (well-conditioned) systems the described strategy of automatic procedure for determination of approximate values works very well. As a general algorithm it can break down, however, or lead to unacceptable results.

With the recent progress in computer applications, the requirements for generalization and automation of a program are continuously increasing.

The ultimate goal is to look for a completely automatic system of obtaining approximate values.

2. In many cases, notably in photogrammetry, it is often sufficient simply to assume certain initial values as approximations for some of the unknowns. Such values are easily recognized because of the specific physical and geometric circumstances of the problem. As an example consider block triangulation: (a) for regular aerial photography, zero tilts are quite frequently assumed and the azimuth taken from the flight plan; (b) particularly with high-altitude photography, the flying height may be approximated by its nominally known value; (c) approximate scale may readily be evaluated from the nominal flying height and principal distance; (d) terrain may be taken as being flat, particularly for small-scale photography.

3. It is often sufficient, in certain cases, to have approximations for only one group of unknowns and compute the other group automatically in the adjustment program. A good example of this is block triangulation of bundles using the collinearity equations in analytical photogrammetry. In some systems we need only introduce input approximations pertaining to the exposure station parameters. Built into the triangulation program would be routines for computing approximations for object point coordinates. It is also possible to accept approximations for even a part of the exposure station elements and compute the rest in the program. For example, knowing the nominal direction of flight and overlap, approximations for the position of one or a few camera stations would be sufficient to compute the others in a strip, given that the angular elements are approximately zero (or known).

4. Sometimes enough information may be available to solve part of the problem first. The results from that step may then serve as approximate values for the rest of the problem. As an example, consider the situation in which auxiliary data are available for the orientation parameters of aerial photographs. With such data, preliminary block triangulation may be carried out in order to obtain approximations for the joint adjustment (of photo data and auxiliary data).

As another example, some block triangulation programs contain special distance conditions. It is a good practice to perform a preliminary block adjustment first to obtain approximate coordinates for linearizing the conditions arising from distance measurement.

Yet another example concerns orbital photography. Given orbit tracking data and time, we can perform preliminary computations in order to derive approximations for the camera parameters. Using these, the rigorous adjustment of the photographic block, together with orbital constraints, may be performed.

5. It is sometimes possible to change to a simpler mathematical model in order to obtain approximations. A good example of this is performing planimetric block triangulation using four-parameter transformation at the plate

scale. This reduces a relatively involved nonlinear case to one that is linear and does not require approximations. In fact, in some systems a sequential set of such approximate procedures is applied so that the final rigorous adjustment is only performed once or twice.

Another example concerns the eight-parameter transformation equations that represent the projectivity between two planes. This pair of fractional linear equations may be reduced to two linear equations by disregarding or otherwise reducing the denominator to unity. A unique one-iteration solution can then be effected to compute approximations for six of the eight parameters. Using these and some or all of the observations, approximations for the remaining two can be obtained. This procedure has often been termed "quasi least squares."

This method is rather important when the adjustment problem is extensive. Here, determining approximations is not necessarily a simple computational effort. It can, in fact, become a sizable effort in its own right. But it needs to be said that such an effort for acquiring approximations is justified only if it is more than offset by thee savings that would accrue from having good approximations. This is one reason why, in many instances, the operation of detecting gross data errors (data editing) is combined with deriving approximations. It has a twofold advantage: to use efficiently the apparently extensive computation effort and to detect outliers and gross errors at an early stage. A very good example of this is performing sequential triangulation prior to a large block reduction by bundles. Forming models, then strips, and adjusting a block by strips using polynomials often will form a sequential set of steps in order both to detect and eliminate gross errors and blunders and to obtain the best approximations for use in a subsequent block adjustment by bundles. Experiences with such a system indicate that the extensive block triangulations by bundles, which require long computer time, are usually complete after only one iteration.

11.2.3. Recapitulation All the examples given above illustrate several points that may be summarized as follows.

First, each problem or family of problems usually has its own suitable possibilities for either obtaining approximations or reducing the model such that approximations are not needed.

Second, the degree of generality of the method selected for computing approximations determines the extent of its suitability for rendering adequate approximations; that is, a relatively general procedure may succeed in the majority of cases to yield adequate approximations but may fail badly in certain specific cases. Consequently, whenever generality is sought, we must be aware that problems may arise in some special cases.

Third, whenever consideration is for large and extensive adjustment problems, automated techniques should be sought. Whenever possible such methods should also include the possibility for blunder detection.

It has been said on occasion that poor initial approximations are acceptable since adjustment is performed on a computer. The reasoning goes that in such a case all that is necessary is to let the solution continue for a large number of iterations. Although this approach may succeed for some problems, we must view this reasoning with much caution. If, for instance, the topological structure of the problem is changed due to poor initial approximations, the procedure might not converge to the proper solution.

Finally, we cannot push the system to extremes. For instance, it is not possible to select zero approximations as valid for all coordinates of a geodetic system of points or for, say, the sines and cosines of the same angle in a transformation.

11.3. A POSTERIORI ESTIMATE OF THE REFERENCE VARIANCE ($\hat{\sigma}_0^2$)

Whether the problem is originally linear, or is linearized, an estimate of the reference variance $\hat{\sigma}_0^2$ can be computed from the observational residuals and the a priori weight matrix of the observations, or by other formulas, as was given in all techniques of least squares presented. The reference variance is an essential element of each adjustment problem. It is useful for a number of purposes, including the assessment of the adjustment results (see Section 11.5) and termination of the iterations (see Section 11.4). Although equations for its computations have been given, it is useful to give the corresponding derivations, which is the purpose of this section. Before proceeding, let us enumerate a number of useful relationships.

1. A useful matrix relationship from Appendix A may be recalled first.

$$\mathbf{x}^t\mathbf{A}\mathbf{x} = tr(\mathbf{x}\mathbf{x}^t\mathbf{A}) \qquad (\mathbf{A} \text{ symmetric}) \tag{A.67}$$

2. Next, it is recalled that the covariance matrix of a random vector $\tilde{\mathbf{x}}$ is $\boldsymbol{\Sigma}_{xx}$ and its mean or expectation is $\boldsymbol{\mu}_x = E(\tilde{\mathbf{x}})$. Then by definition,

$$\begin{aligned}\boldsymbol{\Sigma}_{xx} &= E[(\tilde{\mathbf{x}} - \boldsymbol{\mu}_x)(\tilde{\mathbf{x}} - \boldsymbol{\mu}_x)^t] \\ &= E[\tilde{\mathbf{x}}\tilde{\mathbf{x}}^t] - E[\tilde{\mathbf{x}}\boldsymbol{\mu}_x^t] - E[\boldsymbol{\mu}_x \tilde{\mathbf{x}}^t] + E[\boldsymbol{\mu}_x \boldsymbol{\mu}_x^t] \\ &= E[\tilde{\mathbf{x}}\tilde{\mathbf{x}}^t] - 2\boldsymbol{\mu}_x \boldsymbol{\mu}_x^t + \boldsymbol{\mu}_x \boldsymbol{\mu}_x^t\end{aligned}$$

or

$$\boldsymbol{\Sigma}_{xx} = E[\tilde{\mathbf{x}}\tilde{\mathbf{x}}^t] - \boldsymbol{\mu}_x \boldsymbol{\mu}_x^t \tag{11.23}$$

and alternatively

$$E[\tilde{\mathbf{x}}\tilde{\mathbf{x}}^t] = \boldsymbol{\Sigma}_{xx} + \boldsymbol{\mu}_x\boldsymbol{\mu}_x^t \tag{11.24}$$

3. The elements of the observational vector $\boldsymbol{l}$ are such that their residuals have an expectation or mean vector of zero, or

$$\boldsymbol{\mu}_v = \mathbf{0} \tag{11.25}$$

This means that there are no biases in the observations.

4. By definition, the relations between covariance, cofactor, and weight matrices are

$$\mathbf{W} = \mathbf{Q}^{-1} = \sigma_0^2 \boldsymbol{\Sigma}^{-1} \tag{11.26}$$

where $\mathbf{W}$ is the weight matrix, $\mathbf{Q}$ is the cofactor matrix, and σ_0^2 is the reference variance.

With these general remarks, we may now proceed to derive the relations for an unbiased a posteriori estimate for the reference variance. Each technique of adjustment is treated separately.

11.3.1. General Case of Adjustment with Conditions (Observations and Parameters) For completeness, the pertinent relations are recalled

$$\underset{c,n}{\mathbf{A}}\ \underset{n,1}{\mathbf{v}} + \underset{c,u}{\mathbf{B}}\ \underset{u,1}{\boldsymbol{\Delta}} = \underset{c,1}{\mathbf{f}}$$

$$\underset{c,c}{\mathbf{Q}_e} = \mathbf{AQA}^t = \mathbf{W}_e^{-1} \qquad \underset{u,u}{\mathbf{N}} = \mathbf{B}^t\mathbf{W}_e\mathbf{B} \qquad \mathbf{t} = \mathbf{B}^t\mathbf{W}_e\mathbf{f}$$

and from equation (11.26), the following relations may be introduced

$$\boldsymbol{\Sigma}_e^{-1} = \frac{1}{\sigma_0^2}\mathbf{W}_e = (\mathbf{A\Sigma A}^t)^{-1} \qquad \mathbf{M} = \mathbf{B}^t\boldsymbol{\Sigma}_e^{-1}\mathbf{B}$$

Further, recalling the relations

$$\mathbf{Q}_{ff} = \mathbf{Q}_e \qquad \mathbf{Q}_{tt} = \mathbf{N} \qquad \mathbf{Q}_{\Delta\Delta} = \mathbf{N}^{-1}$$

and factoring the reference variance leads to

$$\boldsymbol{\Sigma}_{ff} = \boldsymbol{\Sigma}_e \qquad \boldsymbol{\Sigma}_{tt} = \sigma_0^4\mathbf{M} \qquad \boldsymbol{\Sigma}_{\Delta\Delta} = \mathbf{M}^{-1}$$

Finally, we compute the expectation or mean of $\mathbf{f}$ as

$$\boldsymbol{\mu}_f = E[\mathbf{f}] = \mathbf{A}E[\mathbf{v}] + \mathbf{B}E[\boldsymbol{\Delta}]$$

or

$$\boldsymbol{\mu}_f = \mathbf{B}\boldsymbol{\mu}_\Delta$$

because

$$E[\mathbf{v}] = \boldsymbol{\mu}_v = \mathbf{0}$$

The quadratic form of the residuals is

$$\mathbf{v}^t\mathbf{W}\mathbf{v} = \sigma_0^2(\mathbf{v}^t\boldsymbol{\Sigma}^{-1}\mathbf{v}) \tag{11.27}$$

and from equation (6.27)

$$\mathbf{v}^t\mathbf{W}\mathbf{v} = \mathbf{f}^t\mathbf{W}_e\mathbf{f} - \boldsymbol{\Delta}^t\mathbf{t} = \mathbf{f}^t\mathbf{W}_e\mathbf{f} - \boldsymbol{\Delta}^t\mathbf{N}\boldsymbol{\Delta}$$

and

$$\mathbf{v}^t\mathbf{\Sigma}^{-1}\mathbf{v} = \mathbf{f}^t\mathbf{\Sigma}_e^{-1}\mathbf{f} - \mathbf{\Delta}^t\mathbf{M}\mathbf{\Delta}$$

$$E[\mathbf{v}^t\mathbf{\Sigma}^{-1}\mathbf{v}] = E[\mathbf{f}^t\mathbf{\Sigma}_e^{-1}\mathbf{f}] - E[\mathbf{\Delta}^t\mathbf{M}\mathbf{\Delta}]$$

or from equation (A67)

$$\begin{aligned}
E[\mathbf{v}^t\mathbf{\Sigma}^{-1}\mathbf{v}] &= E[tr(\mathbf{f}\mathbf{f}^t\mathbf{\Sigma}_e^{-1})] - E[tr(\mathbf{\Delta}\mathbf{\Delta}^t\mathbf{M})] \\
&= tr(E[\mathbf{f}\mathbf{f}^t]\mathbf{\Sigma}_e^{-1}) - tr(E[\mathbf{\Delta}\mathbf{\Delta}^t]\mathbf{M}) \\
&= tr([\mathbf{\Sigma}_{ff} + \boldsymbol{\mu}_f\boldsymbol{\mu}_f^t]\mathbf{\Sigma}_e^{-1}) - tr([\mathbf{\Sigma}_{\Delta\Delta} + \boldsymbol{\mu}_\Delta\boldsymbol{\mu}_\Delta^t]\mathbf{M}) \\
&= tr(\mathbf{I}_c + \boldsymbol{\mu}_f\boldsymbol{\mu}_f^t\mathbf{\Sigma}_e^{-1}) - tr(\mathbf{I}_u + \boldsymbol{\mu}_\Delta\boldsymbol{\mu}_\Delta^t\mathbf{M}) \\
&= tr(\mathbf{I}_c - \mathbf{I}_u) + tr(\boldsymbol{\mu}_f\boldsymbol{\mu}_f^t\mathbf{\Sigma}_e^{-1} - \boldsymbol{\mu}_\Delta\boldsymbol{\mu}_\Delta^t\mathbf{M}) \\
&= (c - u) + \boldsymbol{\mu}_f^t\mathbf{\Sigma}_e^{-1}\boldsymbol{\mu}_f - \boldsymbol{\mu}_\Delta^t\mathbf{M}\boldsymbol{\mu}_\Delta
\end{aligned}$$

which, in view of the fact that $\boldsymbol{\mu}_f = \mathbf{B}\boldsymbol{\mu}_\Delta$, becomes

$$\begin{aligned}
E[\mathbf{v}^t\mathbf{\Sigma}^{-1}\mathbf{v}] &= (c - u) + \boldsymbol{\mu}_\Delta^t\mathbf{B}^t\mathbf{\Sigma}_e^{-1}\mathbf{B}\boldsymbol{\mu}_\Delta - \boldsymbol{\mu}_\Delta^t\mathbf{M}\boldsymbol{\mu}_\Delta \\
&= (c - u) + \boldsymbol{\mu}_\Delta^t\mathbf{M}\boldsymbol{\mu}_\Delta - \boldsymbol{\mu}_\Delta^t\mathbf{M}\boldsymbol{\mu}_\Delta \\
&= (c - u) \\
&= r + u - u = r = \text{the degree of freedom (or redundancy)}
\end{aligned}$$

Thus according to equation (11.27),

$$E[\mathbf{v}^t\mathbf{W}\mathbf{v}] = \sigma_0^2 E[\mathbf{v}^t\mathbf{\Sigma}^{-1}\mathbf{v}] = \sigma_0^2 r$$

$$\sigma_0^2 = \frac{E[\mathbf{v}^t\mathbf{W}\mathbf{v}]}{r} \tag{11.28}$$

Consequently, an unbiased estimate of the reference variance may be computed a posteriori of the least squares adjustment from

$$\hat{\sigma}_0^2 = \frac{\mathbf{v}^t\mathbf{W}\mathbf{v}}{c - u} = \frac{\mathbf{v}^t\mathbf{W}\mathbf{v}}{r} \tag{11.29}$$

11.3.2. Case of Adjustment of Observations Only In this case the relations are

$$\mathbf{A}\mathbf{v} = \mathbf{f}, \qquad \mathbf{Q}_e = \mathbf{A}\mathbf{Q}\mathbf{A}^t, \qquad \mathbf{\Sigma}_e = \mathbf{A}\mathbf{\Sigma}\mathbf{A}^t, \qquad \mathbf{Q}_{ff} = \mathbf{Q}_e,$$

and

$$\mathbf{\Sigma}_{ff} = \mathbf{\Sigma}_e$$

Since $\boldsymbol{\mu}_v = \mathbf{0}$, then $\boldsymbol{\mu}_f = \mathbf{0}$ and from equation (11.23)

$$E[\mathbf{f}\mathbf{f}^t] = \mathbf{\Sigma}_{ff} = \mathbf{\Sigma}_e$$

In a manner similar to the derivation above, it can be shown that

$$E(\mathbf{v}^t\boldsymbol{\Sigma}^{-1}\mathbf{v}) = r$$

and an unbiased a posteriori estimate would be

$$\hat{\sigma}_0^2 = \frac{\mathbf{v}^t\mathbf{W}\mathbf{v}}{r} \tag{11.30}$$

11.3.3. Case of Adjustment of Indirect Observations Earlier relations are recalled,

$$\mathbf{v} + \mathbf{B}\boldsymbol{\Delta} = \mathbf{f}, \qquad \mathbf{N} = \mathbf{B}^t\mathbf{W}\mathbf{B}, \qquad \mathbf{t} = \mathbf{B}^t\mathbf{W}\mathbf{f} \qquad \mathbf{M} = \mathbf{B}^t\boldsymbol{\Sigma}^{-1}\mathbf{B}$$

and

$$\mathbf{Q}_{ff} = \mathbf{Q}$$

thus

$$\boldsymbol{\Sigma}_{ff} = \boldsymbol{\Sigma}$$

$$\mathbf{Q}_{\Delta\Delta} = \mathbf{N}$$

hence

$$\boldsymbol{\Sigma}_{\Delta\Delta} = \mathbf{M}$$

Again because $\boldsymbol{\mu}_v = \mathbf{0}$, then $\boldsymbol{\mu}_f = \mathbf{B}\boldsymbol{\mu}_\Delta$. Next

$$\mathbf{v}^t\mathbf{W}\mathbf{v} = \mathbf{f}^t\mathbf{W}\mathbf{f} - \boldsymbol{\Delta}^t\mathbf{t} = \mathbf{f}^t\mathbf{W}\mathbf{f} - \boldsymbol{\Delta}^t\mathbf{N}\boldsymbol{\Delta}$$

from which it can be shown that

$$E(\mathbf{v}^t\boldsymbol{\Sigma}^{-1}\mathbf{v}) = (n - u) = r = \text{degrees of freedom}$$

and an unbiased estimate is therefore

$$\hat{\sigma}_0^2 = \frac{\mathbf{v}^t\mathbf{W}\mathbf{v}}{r} \tag{11.31}$$

The above derivations show the generality of the relation for the a posteriori estimate of the reference variance. It is always equal to the value of the quadratic form of the residuals divided by the number of degrees of freedom. Because of this rather general relation, no attempt is made to prove the cases of adjustment with functionally dependent parameters, as the same formula will always hold.

The reader must be aware of the assumptions made during the derivations. Of these, two important ones need to be mentioned: the assumption that $\boldsymbol{\mu}_v = \mathbf{0}$ or no bias in the observations, and the assumption that one has a priori the population (or distribution) covariance (or cofactor) matrix. The inclusion of these two assumptions provides for obtaining an unbiased estimate of the reference variance.

11.4. ITERATION TERMINATION WITH LINEARIZED CONDITIONS

When setting up an iterative adjustment program, criteria for terminating the iterations must be incorporated in the program. In order to select the appropriate criterion, or criteria, we must recognize that iteration termination is the mechanism by which we ascertain that the solution has converged sufficiently "close" to the theoretical nonlinear solution. The "closeness" of the solution to be accepted is ultimately limited by the word "length" in the computer—that is, the number of digits carried in the computation unless special techniques are applied. Consequently, practical considerations are important when we design an iterative adjustment system.

Before addressing the actual iteration termination subject, it is useful to mention a point of general nature regarding convergence. In principle, not all iterative processes necessarily converge. Some may diverge, others may oscillate or repeat themselves in certain cycles, and yet others may converge but not to the desired solution. It is not the intention here to dwell on all types of iterative procedures, which is a topic for numerical analysis to be found in the corresponding literature. Instead, consideration is given only to one type: repeated relinearization of the nonlinear condition equations and the performance of a complete linear least squares solution within each linearization cycle. This procedure belongs to a class of iterative methods, called "Newton processes," the convergence properties of which are known by general theory. Such processes have a positive definite coefficient matrix and will converge to the correct (minimum) solution. This is always the case for least squares normal equations' coefficient matrix; hence the desired answer will be attained, given certain conditions concerning the approximate values. Convergence in such methods is known to be of second order and is relatively fast. All this implies that unless the system is degenerate or the adjustment model is incorrectly established, convergence would be quickly obtained provided that problems of numerical acuity are not encountered.

The problem of limited numerical acuity is therefore another practical matter that merits consideration. It would make very little sense to continue with meaningless iterations beyond the level dictated by the number of significant digits and accumulating rounding-off errors.

In addition to the two elements—striving to reach as close an answer to the theoretical solution as possible, and the practical limitations due to digital considerations—there is a third element, that of economy. Although we may wish to get the answer with the highest precision, it is unwise to carry on numerous iterations for a relatively small gain in accuracy. Therefore whenever a criterion for terminating iterations is selected, this rather important economic factor must be considered. It actually refers to proper experimental design of balancing accuracy levels versus cost, which in turn affects iteration termination.

11.4.1. Criteria for Iteration Termination The selected criteria should be representative indicators of convergence. They should have high discriminative power to signalize the attainment of convergence. Several criteria will therefore be given, together with examples, and a discussion as to their power.

FIRST GROUP OF CRITERIA The first group of criteria involves the quadratic form, $\mathbf{v}^t\mathbf{W}\mathbf{v}$, the reference variance, $\hat{\sigma}_0^2 = (1/r)\mathbf{v}^t\mathbf{W}\mathbf{v}$, or the reference standard deviation (the so-called standard error of unit weight) $\hat{\sigma}_0$, which are obviously all related. We may test the absolute value against a preselected tolerance value, or we may test the numerical change from one iteration to the next against a threshold. It is also possible to test the rate of change, rather than just the change. For instance, if we consider $\hat{\sigma}_0$, then we may test

$$\hat{\sigma}_0 < \delta_1 \tag{11.32a}$$

$$|\hat{\sigma}_{0_j} - \hat{\sigma}_{0_{j+1}}| < \delta_2 \tag{11.32b}$$

$$\left|\frac{\hat{\sigma}_{0_j} - \hat{\sigma}_{0_{j+1}}}{\hat{\sigma}_{0_j}}\right| = \left|1 - \frac{\hat{\sigma}_{0_{j+1}}}{\hat{\sigma}_{0_j}}\right| < \delta_3 \tag{11.32c}$$

where δ_i is the appropriate tolerance value and j the iterative index. Equations (11.32) apply equally to $\hat{\sigma}_0^2$ and to $(\mathbf{v}^t\mathbf{W}\mathbf{v})$. Through the iterative procedure, the quadratic form or the reference variance (or both) can be, and often are, computed and listed; but these criteria are not very discriminative. In fact, one of the general difficulties is to select the numerical value of the threshold, such that "sufficient" convergence is reached. This bears directly on the discriminative power of the criterion and no general scheme is available for the choice of δ. Correspondingly, each case or group of cases, as in a particular field of application, needs to be studied separately.

The use of equation (11.32a) has been found to be unadvisable. The value of $\hat{\sigma}_0$ either does not reach δ_1 at all, or when the iterations are terminated at it, the convergence is not actually reached. It is usually better to use equation (11.32c) as it is more indicative of convergence, although still not with much power. At any rate it is quite difficult to determine a priori the proper ratio that would signify convergence. The value of δ_3 in equation (27.32c) can be as small as 10^{-5} and in some cases even smaller. In other cases there may be constant convergence ratios established experimentally through a number of simulated studies. In these latter cases these constants may be directly used for problems of the same type.

The use of relations depending on the reference variance (or reference standard deviation) can be nearly useless in cases of adjustment problems with low numerical stability. Such cases are encountered when nearly indeterminate (or poorly determinate) situations are involved in which the determinant of the normal equations coefficient matrix approaches zero. An example of this case involves a photogrammetric strip adjustment by independent models with ground control only at the beginning and end of fairly long strips (30–50 models). This represents a case of low determinacy.

Using σ_0 as a criterion, the Newton process converged after three iterations to the fifth decimal place. By contrast, the method of conjugate gradiant used to reach the least squares minimum required 230 iterations (which is here not directly relevant) to reach σ_0 within 10^{-4} μm. *Nevertheless* the coordinates in the center of the strip were still different from the check values by some 30 μm in the negative scale. This means that in cases of such poor control, the use of σ_0 with the conjugate gradiant method is nearly useless for iteration termination. Even with the very large number of iterations, correct convergence was not reached. It is because of situations like this that the criterion of terminating iterations after a certain prefixed number of iterations is used, as is discussed in the section to follow on the third criterion.

SECOND GROUP OF CRITERIA The second group of criteria is related to parameter increments or functions of such increments. There are a number of possibilities.

1. The value of *each* increment (or parameter correction) Δ, should approach zero. In this case a threshold ε which is insignificantly small may be used for the test,

$$|\Delta| < \varepsilon \tag{11.33a}$$

2. Instead of having each parameter increment approach zero, it is often sufficient to check on the maximum increment, or,

$$|\Delta_{\max}| < \varepsilon \tag{11.33b}$$

3. In many problems the parameters involved may be in the form of several distinct classes or subsets, each of which contains parameters of the same type (angular, linear, and so on). In such a case the criterion would check the maximum value within each subset i of parameters of a kind, or

$$|\Delta_{i\,\max}| < \varepsilon \tag{11.33c}$$

As a simple example, consider a geodetic net: We may stop iterations when the maximum alteration in coordinates is less than, say, 1 mm, and the maximum change in angular parameters is less than 10^{-6} radians. The same would apply to aerial photography with a threshold for the camera position and another for the camera orientation parameters.

4. Another possibility for a criterion is to use the rate of change of the parameter increments, rather than the increments themselves, or

$$\left| \frac{\Delta_{i-1} - \Delta_i}{\Delta_i} \right| < \varepsilon \tag{11.33d}$$

In some cases, a different form is used

$$\left|\frac{\Delta_{i-1} - \Delta_i}{\sigma_\Delta}\right| < \varepsilon \tag{11.33e}$$

The use of equation (11.33d) or (11.33e) as a test criterion for terminating iterations is somewhat more general than the use of equation (11.33a), (11.33b), or (11.33c) because of being not too dependent on the type of parameters. For example, the rate of change of angular parameters of an aerial photograph may be

$$\left|\frac{\alpha_{i-1} - \alpha_i}{\sigma}\right| < \frac{0.1^{cc}}{10^{cc}} < 10^{-2}$$

5. Yet another group of criteria utilizes a function of the parameters or functions of subsets of them instead of the parameters themselves. A very simple example would be to test on

$$\omega^2 + \phi^2 + \kappa^2 < \varepsilon^2 \tag{11.33f}$$

instead of each of the rotation elements (ω, ϕ, and κ) of an aerial photograph. A more sophisticated example for criteria of this kind concerns photogrammetric block adjustment. Instead of testing on the parameters (exterior orientation elements of the photographs, both positional and rotational), we may test on the maximum changes in point coordinates at the negative scale, or

$$\Delta x_{max}, \Delta y_{max}, \Delta z_{max} < \varepsilon \tag{11.33g}$$

in which ε may take a small value of say 0.1 μm.

THIRD CRITERION A criterion that is useful to use in addition to other criteria is to specify that iterations must not exceed a certain number which is input to the adjustment program. Such a criterion is used usually as a safeguard against excessive iterations in case other criteria fail. The failure of the built-in criteria could be due to either the presence of large mistakes or the poor numerical condition of the system. Consequently, the choice of the number of iterations after which cutoff is effected should not be taken lightly. Although it obviously should be small when blunders contaminate the data, it must be large enough to allow for the suitable number of iterations in case of quasistable systems. In other words the threshold number of iterations in the latter case should not lead to stopping the iterations too soon.

It is evident from the preceding paragraph that gross data errors affect the iterations and consequently the criterion for their termination. In the following section, this point will be briefly discussed.

11.4.2. Iteration Termination and Gross Errors The question of terminating iterations is, in practice, always intertwined with the problem of gross data errors—their detection and elimination. In large systems these error types usually occur and therefore the editing of the data is a necessary operation in order for the solution to converge to the appropriate answer with a reasonable number of iterations. Gross errors and blunders can disturb the iterative process considerably, to the extent that the system may diverge. This is one reason for using a criterion based on σ_0. Here, a check that $\sigma_{0(i+1)} < \sigma_{0i}$ will assure that at least the solution is not diverging. Of course, it has little power in pointing out the rate of convergence. Thus with some criterion on σ_0, very large blunders in the data may be intercepted and eliminated, hence avoiding the possibility of divergence. Other criteria may also be incorporated to detect blunders of moderate size which may slow down, but not impede, convergence, and which may also lead to an unacceptable minimum solution.

In concluding this section, it is worth stressing that iteration termination is an important operation both technically and economically. Well-conditioned and designed adjustment problems, with no gross errors, are, in principle, quite reasonable to handle using the criteria given above. As an example, a quotation† may be given regarding two block adjustments of large-scale photography, summarized in Table 11-1. The method of triangulation is called PAT-M-43 and uses the method of independent models. The first block, Block *A*, contained 50 models of photo scale 1 : 3400, and the second, Block *B*, had 129 models of photo scale 1 : 14000. Counting a plan and height adjustment sequence as one iteration, the maximum alteration against the previous iteration were reduced to 6 mm for *A*, and 0 mm for *B*, after three steps, and after the first iteration the coordinates changed 1758 m and 10229 m, respectively. It can be seen that only three iterations were more than sufficient for convergence under the apparently good conditions of the problem.

TABLE 11-1 Rates of Convergence of Plane and Height Iterations of Three-Dimensional Block Adjustments with PAT-M-43

BLOCK	COORDINATE DIFFERENCES	ITERATION STEP 1		ITERATION STEP 2		ITERATION STEP 3	
		PLAN (m)	HEIGHT (m)	PLAN (m)	HEIGHT (m)	PLAN (m)	HEIGHT (m)
A	x_{max}	1501	8	0.78	0.18	0.005	0.001
	y_{max}	1758	23	0.93	0.19	0.006	0.001
	z_{max}	318	147	0.47	0.21	0.005	0.001
B	x_{max}	10,229	30	0.92	0.52	0.000	0.000
	y_{max}	7,303	85	2.68	0.35	0.000	0.000
	z_{max}	389	527	0.78	0.65	0.000	0.000

† See Ackermann (1973) in the Bibliography.

By contrast to well-conditioned systems, the termination tolerances may have to be quite fine for poorly conditioned problems. In such cases there may be large numbers of iterations and we may have to compare the nth iteration not only with $(n - 1)$ but perhaps also with the $(n - 10)$ iteration. Another point is always to ascertain that iterations do not proceed beyond the point at which the numerical acuity of the system is reached. Finally, then, when designing a large adjustment system for nonlinear problems, it is advisable to incorporate several different criteria for iteration termination—as well as, in some cases, automatic data editing. The extensive programs for photogrammetric block triangulation by DBA systems are good examples for both automatic data editing and iteration termination schemes.†

11.5. A POSTERIORI STATISTICAL ANALYSIS

The results of least squares adjustment are composed of (point) estimates for one or more sets of random variables (parameters or observations, or both) as well as estimates of the corresponding cofactor or covariance matrices. Once the adjustment is performed, statistical techniques may be used to analyze and assess the results obtained. Although least squares estimation does not require a priori knowledge of the distributions of the random variables involved, the a posteriori statistical analysis does require such distributions. Therefore it is usually assumed that the observations are normally distributed and that the least squares estimation is performed on a linear functional model. Consequently, for nonlinear cases it is assumed that after iteration termination, the linearized form is sufficiently close to make the statistical testing valid.

11.5.1. Tests on the Reference Variance One of the statistical tests commonly applied to least squares data concerns the a posteriori estimate of the reference variance $\hat{\sigma}_0^2$. This value may be compared to the a priori value σ_0^2 using a χ^2 test (see Section 2.8.3), since

$$\chi_r^2 = \frac{r\hat{\sigma}_0^2}{\sigma_0^2}$$

has a χ^2 distribution with r degrees of freedom (r = redundancy in the adjustment). There are two possible hypothesis tests: First,

$$H_0: \sigma^2 = \sigma_0^2 \qquad \text{vs.} \qquad H_1: \sigma^2 \neq \sigma_0^2$$

and reject H_0 when

$$\chi_r^2 < \chi_{1-\alpha/2,\,r}^2 \qquad \text{or} \qquad \chi_r^2 > \chi_{\alpha/2,\,r}^2$$

† See Davis (1967) in the Bibliography.

The value χ_r^2 is computed from the adjustment results as

$$\chi_r^2 = \frac{\phi}{\sigma_0^2} = \frac{\mathbf{v}^t\mathbf{W}\mathbf{v}}{\sigma_0^2}$$

Second,

$$H_0\colon \sigma^2 = \sigma_0^2 \qquad \text{vs.} \qquad H_1\colon \sigma^2 > \sigma_0^2$$

and reject H_0 when

$$\chi_r^2 > \chi_{\alpha, r}^2$$

The second hypothesis test, which is a one-tail upper-bound test, is more frequently employed in practice. However, the first two-tail test may be useful, particularly in the cases of the design of new adjustment programs.

The use of the χ^2 test above required the a priori knowledge of the reference variance σ_0^2. In some situations the best available value for the unit variance may itself be an estimate with its own statistical fluctuations. In such instances we may compare the estimate from the adjustment to that a priori estimate and apply the F test for the equality of two variances (see Section 2.8.3).

Let $\hat{\sigma}_0^2$ with r degrees of freedom be the result from the least squares adjustment to be tested, and $\hat{\sigma}_{00}^2$ while r_0 degrees of freedom be the a priori known values. Denote the larger value of the two variances ($\hat{\sigma}_0^2$ or $\hat{\sigma}_{00}^2$) by $\hat{\sigma}_1^2$ and the smaller by $\hat{\sigma}_2^2$, with corresponding degrees of freedom r_1, r_2, and compute $F_{r_1, r_2} = \hat{\sigma}_1^2/\hat{\sigma}_2^2$. The hypothesis to be tested is $H_0\colon \sigma_1^2 = \sigma_2^2$ versus the alternative $H_1\colon \sigma_1^2 > \sigma_2^2$. Thus we would reject H_0 if $F_{r_1, r_2} > F_{\alpha, r_1, r_2}$. In practice, the F ratio is often evaluated as $\hat{\sigma}_0^2$ (from the adjustment) divided by $\hat{\sigma}_{00}^2$ (a priori). If this ratio is less than one, then we usually do not need to test a hypothesis and $\hat{\sigma}_0^2$ is taken to be insignificantly different from $\hat{\sigma}_{00}^2$. Obviously, if the ratio is larger than one, the test proceeds as specified.

If the test on $\hat{\sigma}_0^2$ fails, a poor adjustment is indicated and we must investigate the possible cause. There are several possibilities.

First, we must ascertain that the computations were performed correctly before proceeding to determine other possible sources of difficulty.

Second, we need to investigate how adequately the systematic effects were accounted for. This also relates to the adequacy of the selected model since systematic effects are best interpreted as model deficiencies.

Third, it is very important that the *proper number* of condition (and constraint) equations be determined. Furthermore, the actual functional relationships forming such equations must be correct in the sense of reflecting the physical and geometric situation of the adjustment problem. If both of these points are met, then the linearized form (if nonlinear problem) must also be checked, particularly with the selected approximations, to make sure that numerical instabilities did not occur. All these factors pertain to the functional model.

Finally, the proper stochastic model is equally important. In some instances the a priori cofactor matrix may be accurately known but σ_0^2 is estimated. Here we should apply the F test in order to account for the fact that the a priori value of the reference variance is an estimate and not a numerical constant (representing the population variance). If the a priori variances and covariances are estimated through repeated observations, the number of repetitions should be sufficiently large, that is, 20 to 30.

Example 11.4. Refer to the three-photo problem of Example 6.1 and test if the a posteriori reference variance $\hat{\sigma}_0^2$ is significantly different from its a priori value. The a posteriori value is $\hat{\sigma}_0^2 = 1.9634$ with 1 degree of freedom and the a priori value is $\sigma_0^2 = 1.0000$ (with infinite degrees of freedom). Thus according to Section 2.8.3,

$$\chi_m^2 = \frac{ms^2}{\sigma^2} \qquad \text{or} \qquad \chi_1^2 = \frac{(1)\hat{\sigma}_0^2}{\sigma_0^2} = \frac{(1)(1.9634)}{(1)} = 1.9634$$

and

$$H_0\colon \sigma^2 = \sigma_0^2 \qquad \text{vs.} \qquad H_1\colon \sigma^2 > \sigma_0^2$$

The null hypothesis (H_0) would be rejected if $\chi_1^2 > \chi_{\alpha,1}^2$. At the level of significance

$$\alpha = 0.05, \qquad \chi_{0.05,1}^2 = 3.84 > \chi_1^2$$

and H_0 cannot be rejected. In fact, H_0 would still be accepted at $\alpha = 0.15$ level of significance.

11.5.2. Confidence Regions It has been shown in Section 2.6 that the covariance matrix may be used to establish regions around the estimated values, each of which having a specified probability. In order to review this concept we begin by a numerical example before discussing its application to least squares results.

Example 11.5. The position of point A, in Figure 11.1 is determined by measuring r, α whose values are

$$r = 206\,264 \text{ cm} \qquad \sigma_r = 2 \text{ cm}$$

$$\alpha = 60°00'00'' \qquad \sigma_\alpha = 1''$$

with $\sigma_{r\alpha} = 0$ (that is, r and α are uncorrelated). If $\mathbf{v} = [r \quad \alpha]^t$, then

$$\mathbf{\Sigma}_{vv} = \begin{bmatrix} 4 \text{ cm}^2 & 0 \\ 0 & 1/\rho^2 \end{bmatrix}$$

where ρ is for conversion from seconds of arc to radian and is equal to 206,264 or r. The position of point A may be expressed by the cartesian coordinates x_1 and x_2 and the covariance matrix $\mathbf{\Sigma}_{xx}$. This matrix is obtained from $\mathbf{\Sigma}_{vv}$ by propagation rule, or

$$\mathbf{\Sigma}_{xx} = \mathbf{J}_{xv}\mathbf{\Sigma}_{vv}\mathbf{J}_{xv}^t$$

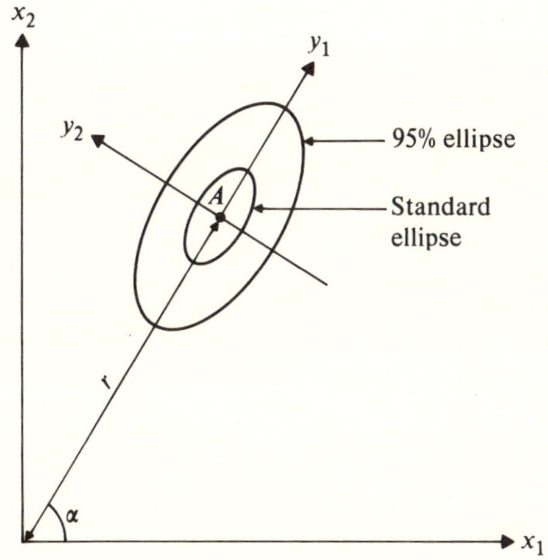

Figure 11.1

To evaluate the Jacobian $\mathbf{J}_{xv}$, the cartesian coordinates

$$\mathbf{x} = [r \cos \alpha \quad r \sin \alpha]^t$$

are differentiated with respect to r and α. Thus

$$\mathbf{J} = \begin{bmatrix} \cos \alpha & -r \sin \alpha \\ \sin \alpha & r \cos \alpha \end{bmatrix} \quad \text{and} \quad \mathbf{\Sigma}_{xx} = \frac{1}{4} \begin{bmatrix} 7 & 3\sqrt{3} \\ 3\sqrt{3} & 13 \end{bmatrix}$$

Suppose we are interested in obtaining the standard ellipse associated with $\mathbf{\Sigma}_{xx}$. First the eigenvalues are computed from the characteristic polynomial

$$\lambda^2 - 5\lambda + 4 = 0$$

or

$$\lambda_1 = 4 \quad \text{and} \quad \lambda_2 = 1$$

Second, the eigenvectors are evaluated as

$$\mathbf{y}_1 = [1 \quad \sqrt{3}]^t \quad \text{for } \lambda_1$$

and

$$\mathbf{y}_2 = [1 \quad -\sqrt{3}/2]^t \quad \text{for } \lambda_2$$

Consequently, the semimajor axis of the standard ellipse is $a = \sqrt{\lambda_1} = 2$ cm and is in the direction of $\mathbf{y}_1$, and the semiminor axis is $b = \sqrt{\lambda_2} = 1$ cm in the direction of $\mathbf{y}_2$ (see Figure 11.1). These results are consistent with the original data of the problem since

$$a = \sigma_{y_1} = \sigma_r \quad \text{and} \quad b = \sigma_{y_2} = \rho\sigma_\alpha$$

Furthermore, the diagonalized form of $\mathbf{\Sigma}_{xx}$ reverts back to $\mathbf{\Sigma}_{vv}$.

Although the probability is 0.394 for the point A to lie inside the standard ellipse, a 95% confidence ellipse can be established. Referring to Table 2.1, for $P = 0.95$ the value of $k = 2.447$. Therefore the 95% confidence ellipse has a semimajor axis

$$a = 2.447 \times 2 = 4.894 \text{ cm}$$

and a semiminor axis

$$b = 2.447 \times 1 = 2.447 \text{ cm}$$

Other confidence ellipses can be constructed through the use of the appropriate values of k from Table 2.1. An equivalent circular confidence region may also be established using the techniques presented in Section 2.6. For example, the circular standard error is

$$\sigma_c = \simeq \tfrac{1}{2}(\sigma_{y_1} + \sigma_{y_2}) \simeq 1.5 \text{ cm}$$

and the circular probable error is

$$\text{CPE} \simeq 0.589(\sigma_{y_1} + \sigma_{y_2}) \simeq 1.767 \text{ cm}$$

The probability for the point to lie within the circle with radius σ_c is approximately equal to 0.394 (same as the standard ellipse), and that inside the circle with radius of circular probable error is approximately equal to 0.500.

The preceding example dealt with a direct point determination and associated error propagation. The same techniques used for getting the confidence regions apply equally to the results of least squares adjustment. Suppose $\hat{\mathbf{x}}_{u,1}$ is the vector of least squares estimates of a set of parameters $\mathbf{x}$, where $\mathbf{x}$ is conventionally known as the "true" values, or in more modern terminology it is the mean vector of the multidimensional distribution of these random variables. Again, from the information in Section 2.6 the quadratic form

$$\phi = (\hat{\mathbf{x}} - \mathbf{x})^t \boldsymbol{\Sigma}_{xx}^{-1} (\hat{\mathbf{x}} - \mathbf{x})$$

represents an equation of a hyperellipsoid in u-dimensional space, centered at $\hat{\mathbf{x}}$. It also has a χ^2 distribution with u degrees of freedom. Thus for $\phi = 1$ the region of constant probability will be that of the standard hyperellipsoid. For other numerical values a corresponding region may be established and its associated probability determined. For example, consider the spatial determination in three-dimensional space. If we are interested in the 95% confidence ellipsoid, then with $\chi^2_{0.05,3} = 7.81$ (see Table D3), its equation would be

$$(\hat{\mathbf{x}} - \mathbf{x})^t \underset{3,3}{\boldsymbol{\Sigma}_{xx}^{-1}} (\hat{\mathbf{x}} - \mathbf{x}) = 7.81$$

Let the eigenvalues of $\boldsymbol{\Sigma}$ be denoted, in decreasing numerical value, by $\lambda_1, \lambda_2, \lambda_3$, and the corresponding eigenvectors by $\mathbf{y}_1, \mathbf{y}_2, \mathbf{y}_3$. The 95% confidence ellipsoid would have the semimajor axes: $a = \sqrt{7.81\lambda_1}$ oriented along $\mathbf{y}_1$,

$b = \sqrt{7.81\lambda_2}$ oriented along $\mathbf{y}_2$, and $c = \sqrt{7.81\lambda_3}$ oriented along $\mathbf{y}_3$. The interpretation of this confidence ellipsoid is that if the determination of $\hat{\mathbf{x}}$ is repeated a large number of times, then we would expect on the average that 95% of the confidence ellipsoids thus determined would include the "true" point.

In the preceding discussion the covariance matrix $\boldsymbol{\Sigma}_{xx}$ is assumed to be known. This implies that σ_0^2 is known a priori and used to obtain $\boldsymbol{\Sigma}_{xx}$ from the inverse of the reduced normal equations' coefficient matrix $\mathbf{N}^{-1}$. If instead, the reference variance is estimated by $\hat{\sigma}_0^2$ from the least squares adjustment, then

$$\frac{1}{u\hat{\sigma}_0^2}(\hat{\mathbf{x}} - \mathbf{x})^t\mathbf{N}(\hat{\mathbf{x}} - \mathbf{x}) = F_{u,\,r} \tag{11.34}$$

and hypothesis tests can be performed using the tabulated F values. For example, for

$$H_0\colon \mathbf{x} = \mathbf{x}_0 \qquad \text{vs.} \qquad H_1\colon \mathbf{x} \neq \mathbf{x}_0$$

we would reject H_0 if $F_{u,\,r}$ computed from equation (11.34) exceeds $F_{\alpha,\,u,\,r}$ at the α level of significance (r = redundancy in the adjustment). In a similar manner confidence regions can be established using the appropriate F values instead of χ^2 when the estimate $\hat{\sigma}_0^2$ is used.

Example 11.6. Let us continue the analysis of the three-photo problem (Examples 6.1 and 11.4) by considering the a posteriori covariance matrix of the parameters x_1 and x_2, that is, the position of point P (see Figure 6.1). Since $\hat{\sigma}_0^2$ was found to be consistent with the a priori value (Example 11.5), then the covariance matrix $\boldsymbol{\Sigma}_{xx}$ is ($\mathbf{Q}_{\Delta\Delta(f)}$ in Example 6.1)

$$\boldsymbol{\Sigma}_{xx} = 10^{-3}\begin{bmatrix} 1.86 & 5.03 \\ 5.03 & 71.33 \end{bmatrix} \; (\text{m}^2)$$

The eigenvalues of this matrix may be computed to be $\lambda_1 = 0.0729$ and $\lambda_2 = 0.0016$, thus $a = 0.27$ m and $b = 0.04$ m for the standard ellipse. The major axis deviates only slightly from the x_2 axis (by about 4°07′). The rather flattened shape of the ellipse is due to the geometry of the intersection of the three rays. Due to the small angles of intersection, the point P is relatively poorly determined in the direction of the ellipse major axis which is here very nearly the x_2 axis.

The shape of the ellipse is independent of the actual values of the observations. Rather, it depends on the geometry and the a priori covariance matrix. It can therefore be studied without the actual measurements. Such studies are performed in the planning phase so that we may decide beforehand whether the planned observations will yield the required results, or whether additional observations or more precise observations are needed.

11.5.3. Correlation Analysis The concept of correlation has been introduced and explained in Chapter 2 of Part I, and a relationship for the correlation coefficient ρ between two random variables has been presented.

The numerical value of ρ_{xy}, which by definition varies between $+1$ and -1, reflects the degree of correlation between the two random variables $\tilde{x}$ and $\tilde{y}$. When $|\rho| = 1$, there exists perfect linear correlation, and plotted observations (x as abscissa and y as ordinate) will lie on a straight line. If the plotted points fall in a narrow band about a straight line, linear correlation of moderate value is said to exist. The value of the correlation coefficient will tend toward zero when the plotted points either fall on a curved line (nonlinear correlation) or are widely scattered, thus implying lack of correlation.

An estimate of the correlation coefficient $\hat{\rho}$ may be computed from a sample in the same way as would the sample variance s_x^2 and sample covariance s_{xy}, or

$$\hat{\rho} = \frac{s_{xy}}{s_x s_y} \tag{11.35}$$

where

$$s_{xy} = \sum \frac{(x - \bar{x})(y - \bar{y})}{n - 1} \tag{11.36a}$$

$$s_x^2 = \sum \frac{(x - \bar{x})^2}{n - 1} \tag{11.36b}$$

$$s_y^2 = \sum \frac{(y - \bar{y})^2}{n - 1} \tag{11.36c}$$

with n being the sample size. There are two possible tests on $\hat{\rho}$ under the null hypothesis, $H_0: \rho = 0$ against the alternative $H_1: \rho \neq 0$. The first applies the t distribution by computing

$$t_{n-2} = \left[\frac{(n-2)\hat{\rho}^2}{1 - \hat{\rho}^2}\right]^{1/2} \tag{11.37}$$

and accepting H_0 when

$$-t_{\alpha/2,\, n-2} < t_{n-2} < t_{\alpha/2,\, n-2}$$

The second (for large n) applies the normal distribution by computing

$$z = \frac{\sqrt{n-3}}{2} \ln \left[\frac{1 + \hat{\rho}}{1 - \hat{\rho}}\right] \tag{11.38}$$

and accepting H_0 when

$$-z_{\alpha/2} < z < z_{\alpha/2}$$

When H_0 is rejected, the two random variables in question are said to be significantly correlated at the selected α level. These tests assume that the random variables involved are normally distributed.

If the problem to be analyzed concerns *only* two random variables (jointly normally distributed), then $\hat{\rho}$ as computed by equation (11.35) and either of the two tests above will be appropriate. On the other hand if a multinormal set of random variables are involved, it is better to use the *partial correlation coefficient*.

For example, if $\mathbf{\Sigma}_{xx}$ is the covariance matrix of the random vector $\tilde{\mathbf{x}}$ (which is $u \times 1$) and $\mathbf{W}_{xx} = \mathbf{\Sigma}_{xx}^{-1}$, then the correlation coefficient between x_1 and x_2 when the remaining $u - 2$ are fixed is

$$\rho_{12|3\cdots u} = \frac{-w_{12}}{(w_{11}w_{22})^{1/2}} \tag{11.39}$$

where w_{ij} are elements of $\mathbf{\Sigma}_{xx}^{-1}$. The estimate $\hat{\rho}_{12|3\cdots u}$ is defined the same way as in equation (11.39) using the covariance matrix evaluated from the sample values. Note that when $u = 2$, the two correlation coefficients from both equation (11.35) and (11.39) become identical since the inverse of the covariance matrix is

$$\begin{bmatrix} w_{11} & w_{12} \\ w_{12} & w_{22} \end{bmatrix} = \frac{1}{s_1^2 s_2^2 - s_{12}^2} \begin{bmatrix} s_2^2 & -s_{12} \\ -s_{12} & s_1^2 \end{bmatrix}$$

Similar to equation (11.37) the test of hypothesis H_0: $\rho_{12|3\cdots u} = 0$ versus the alternative H_1: $\rho_{12|3\cdots u} \neq 0$ is performed by computing

$$t_{n-u} = \left[\frac{(n-u)\hat{\rho}_{12|3\cdots u}^2}{1 - \hat{\rho}_{12|3\cdots u}^2}\right]^{1/2} \tag{11.40}$$

and accepting H_0 when

$$-t_{\alpha/2,\,n-u} < t_{n-u} < t_{\alpha/2,\,n-u}$$

In the test, n is the number of repeated measurements on the $u \times 1$ vector of observations, and u is the number of parameters to be determined (the size of the random vector). In case of linear least squares, n becomes the number of observations and u the number of parameters estimated in the vector $\mathbf{\Delta}$. In this case, $n - u$ is the redundancy r, and the partial correlation coefficient may be evaluated directly from the coefficient matrix $\mathbf{N}$ of the partially reduced normal equations (being the inverse of the scaled covariance matrix). Hence $\hat{\rho}_{12|3\cdots u}$ for the correlation between δ_1 and δ_2 of $\mathbf{\Delta}$ is evaluated from

$$\hat{\rho}_{12|3\cdots u} = \frac{-n_{12}}{(n_{11}n_{22})^{1/2}} \tag{11.41}$$

where n_{ij} are elements of $\mathbf{N}$. The computed t value is

$$t_r = \left[\frac{r\hat{\rho}_{12|3\cdots u}^2}{1 - \hat{\rho}_{12|3\cdots u}^2}\right]^{1/2} \tag{11.42}$$

and the test is carried out as before.

Example 11.7. Consider the x_1 and x_2 coordinates of point P determined from three photographs (see Example 6.1). The covariance matrix is given in Example 11.6, and we may test if the correlation between x_1 and x_2 is significant. First,

$$\hat{\rho}_{12} = \frac{5.03}{(1.86 \times 71.33)^{1/2}} = 0.437$$

and since $r = 1$, then

$$t_1 = \frac{0.437}{[1 - (.437)^2]^{1/2}} = 0.49$$

At $\alpha = 0.05$, $t_{0.025,\,1} = 12.706$, and thus x_1 and x_2 can safely be assumed to be uncorrelated.

Example 11.8. For this example recall the problem of the five-point level net solved by the method of indirect observations in Example 7.11. Recalling $\mathbf{Q}_{xx}$ as $\mathbf{N}^{-1}$, first compute the estimates of the (marginal) correlation coefficients $\hat{\rho}_{ij}$ according to equation (11.35),

$$\mathbf{Q}_{xx} = \mathbf{N}^{-1} = \begin{bmatrix} 0.806 & 0.445 & 0.505 & 0.333 \\ & 0.645 & 0.504 & 0.414 \\ & & 0.999 & 0.505 \\ \text{symmetric} & & & 0.742 \end{bmatrix} \quad \text{and}$$

$$\{\hat{\rho}_m\} = \begin{bmatrix} . & 0.617 & 0.563 & 0.436 \\ & . & 0.628 & 0.606 \\ & & . & 0.594 \\ & & & . \end{bmatrix}$$

The inverse of $\mathbf{Q}_{xx}$ is $\mathbf{N}$ from which the partial correlation coefficients $\hat{\rho}_p$ may be computed according to equation (11.41),

$$N = \begin{bmatrix} 2.185 & -1.064 & -0.568 & 0 \\ & 3.492 & -0.714 & -1.010 \\ & & 2.023 & -0.741 \\ \text{symmetric} & & & 2.476 \end{bmatrix} \quad \text{and}$$

$$\{\hat{\rho}_p\} = \begin{bmatrix} . & 0.385 & 0.270 & 0 \\ & . & 0.269 & 0.344 \\ & & . & 0.331 \\ & & & . \end{bmatrix}$$

The redundancy for the problem is $r = 4$. For the largest element (0.628) in $\{\hat{\rho}_m\}$, the computed t is 1.614; and for the largest element (0.385) in $\{\hat{\rho}_p\}$, the computed t is 0.834. Since $t_{4,\,0.025} = 2.776$, no significant correlation exists between the elevations of the four points determined in the adjustment of the level net. It is worth noting that the correlation coefficients in $\{\hat{\rho}_m\}$ are generally higher than those in $\{\hat{\rho}_p\}$. As an example consider $\hat{\rho}_{m12} = 0.617$; it reflects not only the direct correlation between x_1 and x_2, but also the indirect effect of the correlation between x_1 and x_2, and x_3 and x_4. On the other hand $\hat{\rho}_{p12} = 0.385$ is correlation between x_1 and x_2 when x_3 and x_4

remain fixed. Therefore in this particular case the inclusion of the indirect effect (with x_3, x_4) increases the correlation between x_1 and x_2, though still not to a significant level at $\alpha = 0.05$.

In addition to the statistical analyses presented in this section, several others may be employed. These may include: (a) plotting the residuals versus the estimated observations to determine trends, (b) performing regression analysis on the residuals, (c) performing analysis of variance techniques on discrepancies at withheld control to compare different methods of analysis; and so forth. These and other advanced statistical techniques are beyond the scope of this book.

11.6. COMPUTATIONAL AND NUMERICAL CONSIDERATIONS

Once the mathematical model of an adjustment problem is designated and the appropriate equations formulated and linearized, the technique of least squares becomes a sequence of matrix operations. The algorithms written to perform these operations are usually automated in the sense of being carried out on a computer. Consequently, the application of least squares reduces at the end to computational matters. The larger the adjustment problem, the more important is the efficient performance of the computational algorithms. In fact, the full potential of the least squares method lies in applications involving very large adjustment systems. Although the treatment of computational methods is in the field of numerical analysis, a concise discussion of pertinent aspects is given in this section. More detailed information can be obtained from textbooks on numerical methods.

11.6.1. Selection of Adjustment Technique and Parameters The technique of adjustment of observations only ($\mathbf{Av} = \mathbf{f}$) was frequently chosen in the past for the adjustment of survey problems. The main reason lay in the small number of condition and normal equations (r) and the limited capacity of the computational aids available. With present computers the size of the normal equations is not such a totally overriding factor as it used to be. Instead, the adjustment with conditions containing parameters, particularly the technique of indirect observations wherever possible, is preferred. These techniques make the systematic handling of the adjustment data much easier, even though the resulting normal equations may frequently be larger in number than with the procedure of adjustment of observations only.

Another point that bears on the computational effort is the number of parameters carried in the adjustment. This aspect has been discussed in several places in previous chapters, mainly in regard to whether the parameters are required or not. Here the general remark is made that the ability to set up the computational algorithm systematically and optimally is more important than attempting to carry the fewest number of parameters in the adjustment. The prime reason is the fact that in most cases the normal equations exhibit a special structure which when exploited leads to using considerably more economical computational techniques. Another consideration is to develop and apply computer adjustment programs that cover a wide range of problems. The effort involved in setting up good programs is such that one program should be applicable to a whole family of cases, even at the expense of not being fully optimized for smaller or simpler special cases. Separate programming of limited algorithms for special cases is not economical.

11.6.2. Computational Organization and Grouping of Variables Because the mechanics of matrix operations for least squares is known beforehand, it is usually not necessary to set up the matrices of the condition and constraint equations. Instead, we proceed to compute directly the elements of the normal equations. In fact, the partially reduced normal equations after eliminating the vectors $\mathbf{v}$, $\mathbf{k}$, and $\mathbf{k}_c$ (if it is present in the adjustment) are usually set up and solved for the unknown parameters. Although this approach requires, in general, special programming effort for certain types of adjustment problems, it leads to increased efficiency and savings in computer storage. This is particularly true when the observations, or certain subsets of the observations, are uncorrelated. The careful arrangement of the variables results in having zero submatrices and other submatrices that may be of diagonal or block diagonal form. The computational algorithm takes advantage of these particular characteristics (see the following section) particularly by not operating on the zero submatrices.

11.6.3. Structure of Normal Equations' Matrices When speaking of matrix structure of normal equations, we refer to the arrangement of zero and nonzero elements. It has been shown repeatedly before that the total system of normal equations contains a large number of zero submatrices. On the basis of such structure it has been possible to solve separately for the vectors $\mathbf{v}$, $\mathbf{k}$, $\boldsymbol{\Delta}$, and so on. In a similar manner the partially reduced normal equations, that is, those in terms of $\boldsymbol{\Delta}$, either have or can be made to have a useful pattern of zero and nonzero submatrices. An excellent example of this is the photogrammetric block triangulation by units (either bundles or models). For instance, Figure 11.2 depicts the structure of a partially reduced normal equations' coefficient matrix for bundle triangulation of a 4×5 block of aerial photographs. It can be readily seen that the matrix is sparse indeed, since the nonzero elements are much smaller in numbers than the zero elements. Furthermore, both submatrices $\dot{\mathbf{N}}$ and $\ddot{\mathbf{N}}$ are block diagonal, whereas $\bar{\mathbf{N}}$ has a banded structure. This particular pattern, which is obtained when cross-strip ordering of both photographs and points is done, is brought about by grouping the photoparameters together and separate from

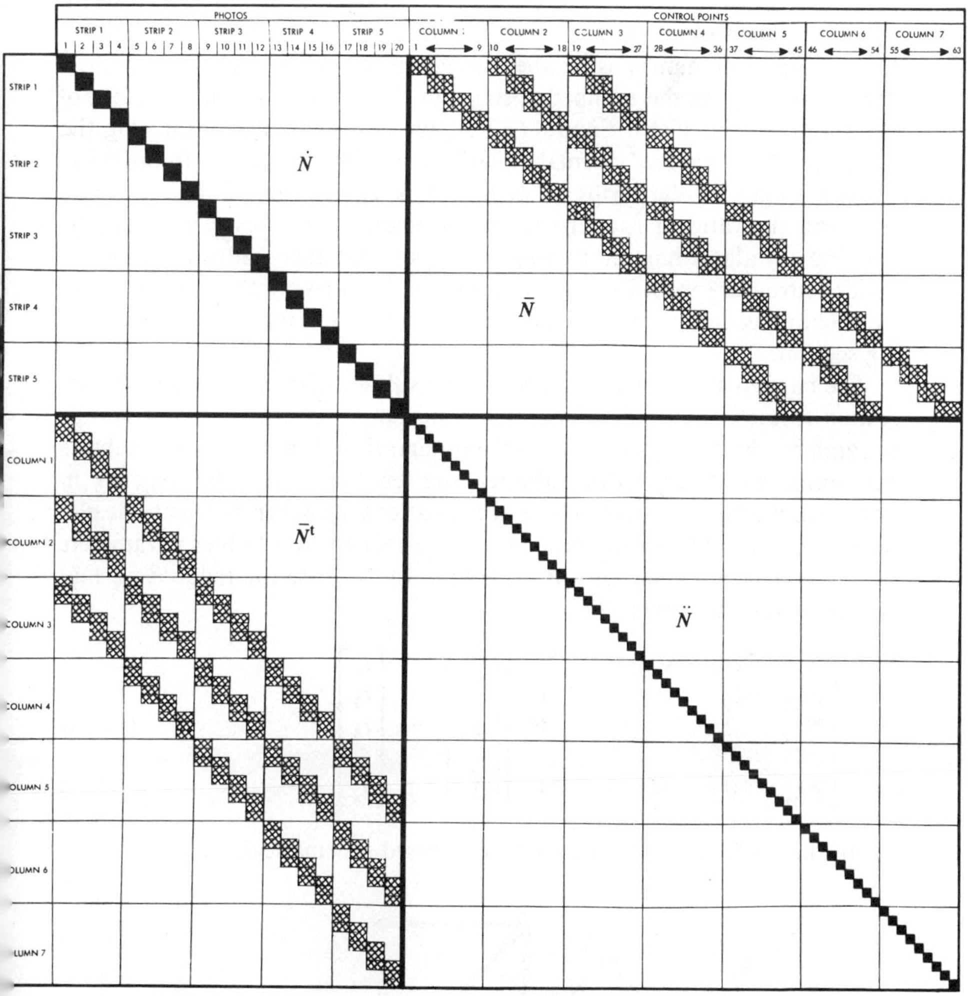

Figure 11.2. Structure of Normal Equations from Bundle Adjustment of 4 × 5 Photo Block of Photographs (From Brown, 1974. See Bibliography.)

the object point parameters. It therefore allows for the elimination of either of the two groups of unknowns by further reducing the normal equations. For example, the reduced normals in terms of photo parameters are ($\dot{\mathbf{N}} - \bar{\mathbf{N}}\ddot{\mathbf{N}}^{-1}\bar{\mathbf{N}}^t$), and in terms of object point parameters are ($\ddot{\mathbf{N}} - \bar{\mathbf{N}}^t\dot{\mathbf{N}}^{-1}\bar{\mathbf{N}}$). Not only does either reduction reduce the size of system to be solved, but it also exploits the fact that both $\dot{\mathbf{N}}$ and $\ddot{\mathbf{N}}$ are block diagonal matrices with submatrices of small order. (For $\dot{\mathbf{N}}$, the submatrix is usually of order 6, whereas $\ddot{\mathbf{N}}$ has 3 × 3 submatrices along the main diagonal).

Although the reduction of the normal equations to only one group of parameters yields significant savings in the computational effort, further study of the structure of reduced normals indicates that there are still a significant number of zero submatrices. In fact the reduced normal equations' coefficient matrix is of the banded type where the band width is dependent only on the number of strips in the block and is independent of the length of the strips. Figure 11.3 shows four examples illustrating the structure of the reduced normal equations' system for cross-strip ordering. As the length of the strips is increased, keeping the number of strips constant, the ratio of nonzero to zero elements decreases. Consequently, devising an algorithm for solving a system that takes advantage of the banded structure would make the triangulation of very large blocks both feasible and economical. Such an algorithm will be discussed in the following section.

The banded structure arises when small subsets of the parameters are not related to each other either by actual condition or constraint equations, or through stochastical correlation of the observations. The size of such subsets determines the band width although the length of the band may be, in principle, indefinitely large. There are situations, however, in which the elements of one subset of the parameters are related to all other parameters through condition equations or otherwise. In this case the reduced normal equations take the form

$$\begin{bmatrix} \mathbf{N}_{11} & \mathbf{N}_{12} & \mathbf{N}_{13} & \cdots & \mathbf{N}_{1n} \\ \mathbf{N}_{12}^t & \mathbf{N}_{22} & \mathbf{0} & \cdots & \mathbf{0} \\ \mathbf{N}_{13}^t & \mathbf{0} & \mathbf{N}_{33} & \cdots & \mathbf{0} \\ \vdots & \vdots & \vdots & & \vdots \\ \mathbf{N}_{1n}^t & \mathbf{0} & \mathbf{0} & \cdots & \mathbf{N}_{nn} \end{bmatrix} \begin{bmatrix} \boldsymbol{\Delta}_1 \\ \boldsymbol{\Delta}_2 \\ \boldsymbol{\Delta}_3 \\ \vdots \\ \boldsymbol{\Delta}_n \end{bmatrix} = \begin{bmatrix} \mathbf{t}_1 \\ \mathbf{t}_2 \\ \mathbf{t}_3 \\ \vdots \\ \mathbf{t}_n \end{bmatrix} \tag{11.43}$$

where the coefficient matrix may be expressed schematically as†

N ~ 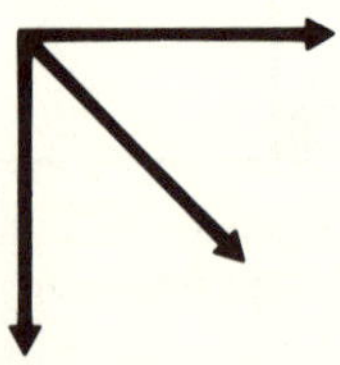

The sense of this scheme is that the horizontal and vertical bars generally represent nonzero elements, the diagonal bar represents block diagonal nonzero submatrices, and remaining region (blank) is formed of totally zero elements. The arrows indicate that the total matrix **N** can extend indefinitely, keeping the character the same as depicted. This system is referred to as *banded bordered*. Brown† also names it *first-order partitioned* system because

† See Brown and Trotter in the Bibliography.

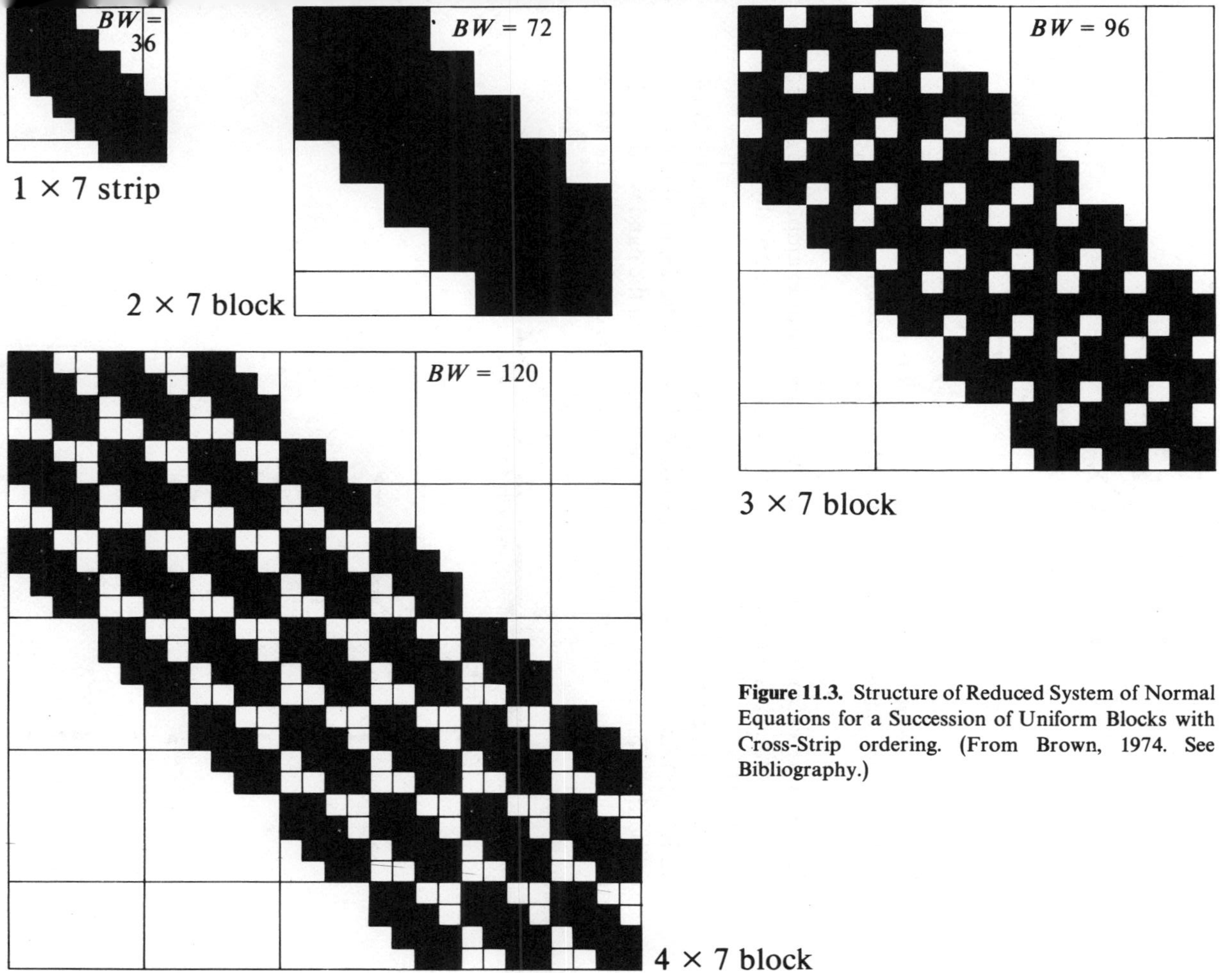

Figure 11.3. Structure of Reduced System of Normal Equations for a Succession of Uniform Blocks with Cross-Strip ordering. (From Brown, 1974. See Bibliography.)

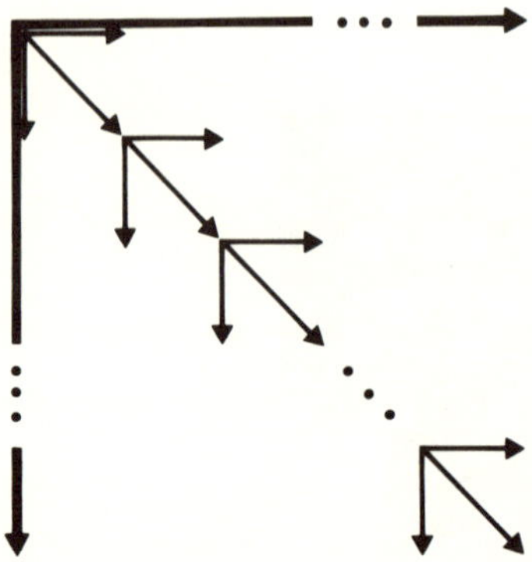

Figure 11.4. Schematic of Coefficient Matrix Characteristic of a Second-Order Partitioned System of Normal Equations (From Brown and Trotter, See Bibliography.)

of the way he extends it to higher-order systems. For example, the second- and third-order partitioned systems are depicted schematically in Figures 11.4 and 11.5.

When setting up the equations and arranging the parameters, the objective is to obtain as narrow a band as possible. As will be seen in the following section the computational time will depend mainly on the band width. As for the border, it may exist at the outset because of the conditions of the adjust-

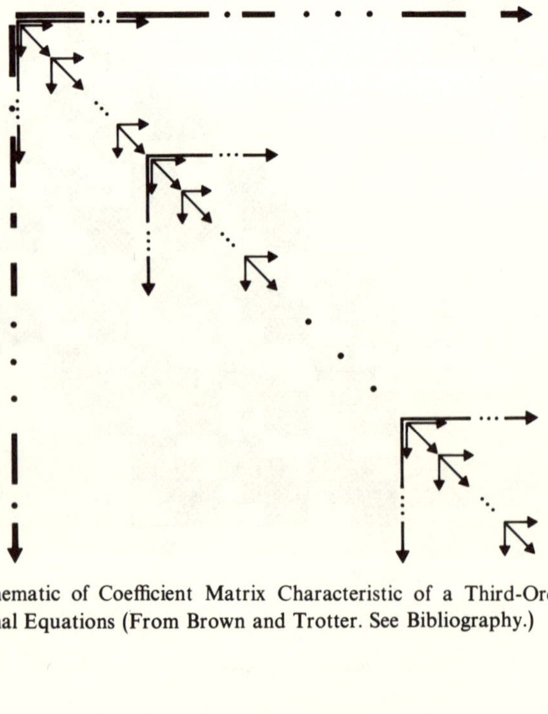

Figure 11.5. Schematic of Coefficient Matrix Characteristic of a Third-Order Partitioned System of Normal Equations (From Brown and Trotter. See Bibliography.)

ment problem or it may actually be introduced in order to minimize the band width. The latter case is common when constraints exist between parameters that are widely separated in the numbering scheme.

11.6.4. Algorithms for Solution of Normal Equations All adjustment problems eventually come down to the solution of a system of linear equations—the reduced normal equations. In general, the larger the adjustment problem, the larger the system of equations to be solved. In addition to arranging the problem such that special matrix structure results, the selected algorithm for the solution must both exploit the structure and be as efficient as possible. In fact, the capacity and efficiency of the algorithm selected are often the limiting factors on the size of the adjustment system to be solved by a given computer configuration. Therefore it is quite important that the most suitable algorithm be selected for a given adjustment problem with a specific structure of normal equations.

Algorithms for the solution of systems of linear equations constitute a large segment of the activities of the numerical analysis field. We can only briefly comment therefore on the topic here and leave the details to the appropriate sources. There are, in general, two basic classes of algorithms: one group applying iterative processes and the other direct solution procedures. Of the iterative methods the following may be mentioned: block Gauss-Seidel, block Gauss-Seidel with acceleration, block successive over relaxation (BSOR), and conjugate gradient method. Many of these methods have been applied successfully in large adjustment programs, particularly BSOR and the conjugate gradient methods. Their stated advantages refer to required limited core storage, numerical acuity, and relative simplicity in programming. However, as with most iterative procedures, the rate of convergence is an important factor. It often depends on the particular circumstances of a given problem. For example, using BSOR for adjustment of photogrammetric blocks, it was found that the rate of convergence is dependent on the amount and disposition of available ground control. Although its convergence is satisfactory with suitable distribution of control, it is quite slow in cases of sparse control. For this reason and others, the tendency has been lately to apply direct closed solution of the system of reduced normal equations.

There are also many direct solution algorithms. In Appendix A an explanation is given of the classical procedure of Gauss elimination and the Gauss-Jordon method. Other direct techniques include the Cholesky method and the partitioning scheme which is often called the "bordering" technique. Some of these methods take advantage of the symmetry of the reduced normal equations' coefficient matrix. A relatively recent procedure, termed *recursive partitioning*, exploits the special structure of the matrices and is efficiently applied to both banded and banded-bordered matrices. It is therefore selected for further discussion.

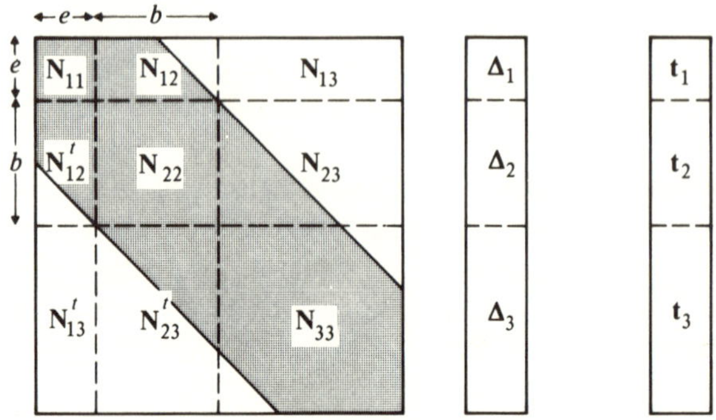

Figure 11.6

11.6.5. Recursive Partitioning† Consider the banded system of Figure 11.6 with the triple partitioning shown, for which $e \leq b$. The number of elements in the second partition is b or the bandwidth. As shown in the figure this partitioning yields $\mathbf{N}_{13}$ (and $\mathbf{N}_{13}^t$) as a null matrix. The vector $\mathbf{\Delta}_1$ may be eliminated using the method of partitioning (see Appendix A) thus

$$\left[\begin{bmatrix}\mathbf{N}_{22} & \mathbf{N}_{23}\\ \mathbf{N}_{23}^t & \mathbf{N}_{33}\end{bmatrix} - \begin{bmatrix}\mathbf{N}_{12}^t\\ \mathbf{N}_{13}^t\end{bmatrix}\mathbf{N}_{11}^{-1}[\mathbf{N}_{12}\ \mathbf{N}_{13}]\right]\begin{bmatrix}\mathbf{\Delta}_2\\ \mathbf{\Delta}_3\end{bmatrix} = \begin{bmatrix}\mathbf{t}_2\\ \mathbf{t}_3\end{bmatrix} - \begin{bmatrix}\mathbf{N}_{12}^t\\ \mathbf{N}_{13}^t\end{bmatrix}\mathbf{N}_{11}^{-1}\mathbf{t}_1 \tag{11.44}$$

But since $\mathbf{N}_{13} = \mathbf{0}$, then

$$\begin{bmatrix}\mathbf{N}_{22} - \mathbf{N}_{12}^t\,\mathbf{N}_{11}^{-1}\mathbf{N}_{12} & \mathbf{N}_{23}\\ \mathbf{N}_{23}^t & \mathbf{N}_{33}\end{bmatrix}\begin{bmatrix}\mathbf{\Delta}_2\\ \mathbf{\Delta}_3\end{bmatrix} = \begin{bmatrix}\mathbf{t}_2 - \mathbf{N}_{12}^t\,\mathbf{N}_{11}^{-1}t_1\\ \mathbf{t}_3\end{bmatrix} \tag{11.45}$$

Comparing equation (11.45) to the original system (Figure 11.6) reveals that both the banded form and the bandwidth are preserved. The elimination of $\mathbf{\Delta}_1$ affects only the $\mathbf{N}_{22}$ portion of the coefficient matrix which lies entirely within the band. The system may be triply partitioned further and the procedure repeated recurrently, hence the name "recursive partitioning." At the end there will remain a system that is sufficiently small to be solved directly for the remaining unknowns. Once the elements of the last unknown vector are computed, a backward procedure can be carried out to obtain each of the unknown subvectors (the $\mathbf{\Delta}_i$) eliminated at the different steps of the forward process. As in the forward reduction, the bandwidth is preserved throughout the backward procedure. The computational efficiency of recursive partitioning stems from confining all the numerical operations to a submatrix ($\mathbf{N}_{22}$) which is totally inside the band. The number of arithmetic operations required for the reduction of a $u \times u$ banded system of bandwidth b is proportional to b^2u. By comparison, conventional Guass elimination requires a number of operations which is proportional to u^3. Thus

† See Brown, 1974, in the Bibliography.

recursive partitioning leads to considerable savings in computational effort particularly when the bandwidth b is much smaller than the matrix order u.

The method of recursive partitioning applies equally to banded-bordered systems such that the numerical operations are restricted to the nonzero elements within the band and border. In fact, the number of arithmetic operations required in the reduction of a system of bandwidth b and border-width b' is proportional to $(b + b')^2u$, where u is the order of the system. Therefore, whenever possible, attempts should be made to keep both the bandwidth and border width as narrow as possible through prudent arrangement of parameters. This leads to the most economical computational algorithm.

It is frequently the practice that the selected algorithm is used to solve the system of reduced normal equations and not invert the coefficient matrix. This is particularly true when the original equations are nonlinear and relinearization is applied. During the last relinearization, however, we often seek the inverse of the coefficient matrix for purposes of a posteriori precision estimation. The reason for not seeking the inverse repeatedly is that it normally requires more computational time. For the particular algorithm of recursive partitioning it is possible to *generate only that portion of the inverse that corresponds to the original band* (which is primarily what is needed for a posteriori precision estimation). Nevertheless, the computation of only a portion of the inverse was found to require approximately twice the time needed for the solution of the system alone.

Problems for Part II

1. The angle α was measured with three different instruments as indicated below. Determine the "best value" of α and its standard deviation.

INSTRUMENT	α	σ
WILD T1 AE	45°27′20″	6″
WILD T16 ED	45°27′24″	9″
WILD T2	45°27′23″	3″

2. The three angles of a triangle were measured with three different instruments as indicated below. Determine the estimates of the angles that satisfy the condition that $\alpha + \beta + \gamma = 180°$.

INSTRUMENT	ANGLE	σ ANGLE
WILD T1 AE	α = 30°00′05″	6″
WILD T16 ED	β = 60°00′07″	9″
WILD T2	γ = 90°00′10″	3″

3. A baseline was measured with four different measurement systems, yielding the following results:

MEASUREMENT VALUE (m)	STANDARD DEVIATION (cm)
1000.19	12
1000.40	13
1000.50	14
1000.28	15

a. Compute the weighted mean.
b. Compute the standard deviation of the weighted mean.

4. Given the simple level net where A is a reference bench mark whose elevation is, for simplicity, zero, the observed differences in elevation are given in *metres* as

$$l_1 = 4 \qquad l_2 = 2 \qquad l_3 = -6.3 \qquad l_4 = 5$$

Assuming that all the observations are uncorrelated and of equal weight, compute the elevation of point D and its cofactor.

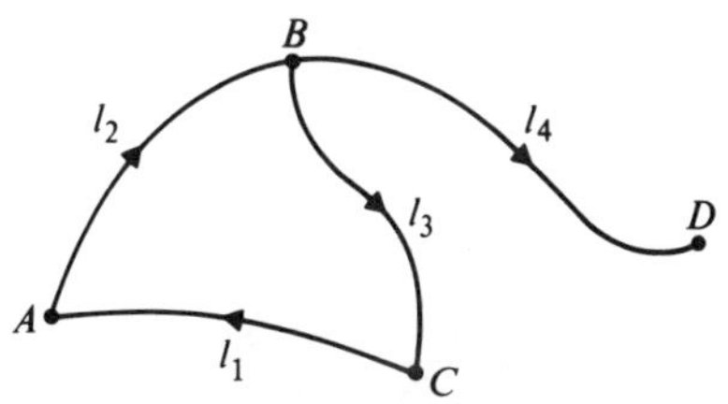

5. Assume that a gravimeter has been used to determine gravity differences along the lines indicated in the diagram below. Assume further, for simplicity, that all systematic corrections have been accounted for and that the terrain is level. The absolute value of gravity at point A has been established from pendulum observations as $g_A = 980000.0$ mgal. The observations (Δg) are uncorrelated and it has been determined empirically that the weight of Δg is given by

$$W_{\Delta g_i}\left(\frac{1}{\text{mgal}^2}\right) = \frac{1}{d} \qquad (d \text{ in 10 km units})$$

Thus the following observations are available:

LINE	DISTANCE (km)	Δg (mgal)
1	18.1	+25.42
2	9.4	+10.34
3	14.2	−35.20
4	17.6	−15.54
5	13.5	+21.32
6	9.9	+ 4.82
7	13.8	−31.02
8	14.0	−26.11

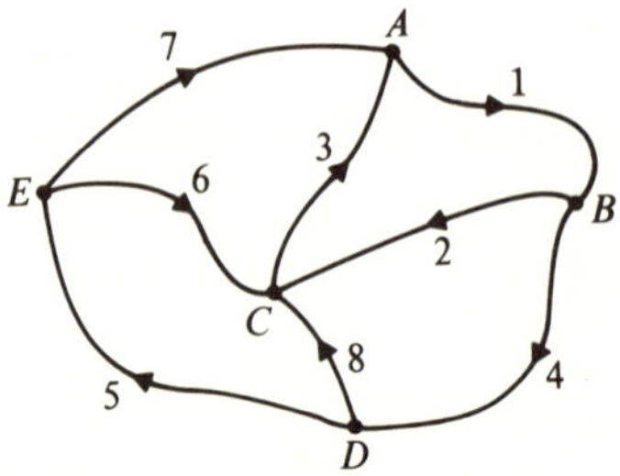

a. Using least squares adjustment of indirect observations compute the gravity values at points B, C, D, and E and their standard deviations.
b. Check the values obtained in (a) above using the method of least squares adjustment of observations only.

6. The figure shows the diagram for a system of measured angles about position 0. The azimuth of line $0A$ is considered known and fixed. It is desired to determine the adjusted azimuth of lines $0B$, $0C$, and $0D$.

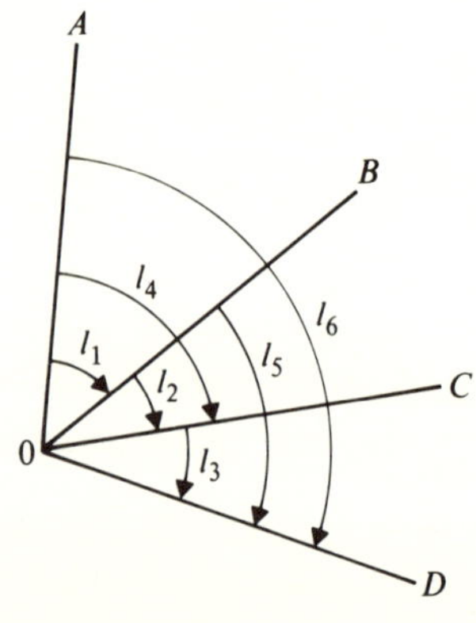

a. Considering application of the method of adjustment of indirect observations (i) How many condition equations should be written? (ii) Write all the condition equations, specifying the parameters.
b. Considering application of the method of adjustment of observations only: (i) How many condition equations should be written? (ii) Write all the condition equations and explain how the required azimuths and their covariance matrix will be determined.

7. Given the following observations with respect to the figure

$$
\begin{aligned}
l_1 &= A\hat{0}B = 30^\circ 15' 15'' & \sigma_1 &= 1'' \\
l_2 &= A\hat{0}C = 50^\circ 37' 51'' & \sigma_2 &= 2'' \\
l_3 &= C\hat{0}D = 15^\circ 31' 17'' & \sigma_3 &= 1'' \\
l_4 &= B\hat{0}D = 35^\circ 54' 00'' & \sigma_4 &= 1'' \\
l_5 &= C\hat{0}E = 45^\circ 31' 18'' & \sigma_5 &= 2''
\end{aligned}
$$

Compute the angle θ ($D\hat{0}E$) and its standard deviation σ_θ.

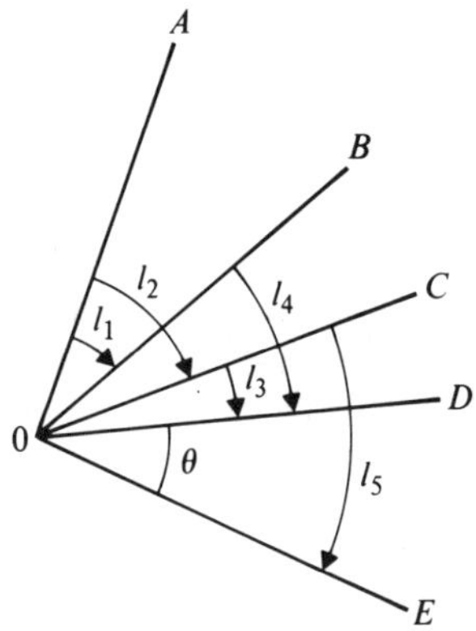

8. Given the following data with respect to the figure:

$l_1 = A\hat{0}B = 30°15'15''$

$l_2 = B\hat{0}C = 20°00'00''$

$l_3 = A\hat{0}C = 50°15'18''$

$l_4 = C\hat{0}D = 30°00'00''$

$l_5 = B\hat{0}E = 70°00'01''$

Assuming that all observations are *uncorrelated* and of equal precision, compute the least squares estimate of the angle $D0E$ *and its cofactor.*

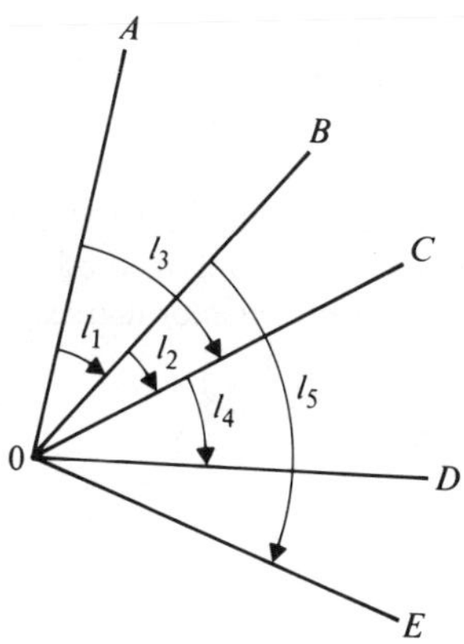

9. Given the following information with respect to the figure:

Direction $l_1 = 10°$ $\quad \sigma_1 = 1''$

Direction $l_2 = 20°$ $\quad \sigma_2 = 1''$

Direction $l_3 = 30°$ $\quad \sigma_3 = 1''$

Direction $l_4 = 60°$ $\quad \sigma_4 = 1''$

Angle $l_5 = 20°$ $\quad \sigma_5 = 2''$

No correlation exists between the observations. Compute the angle θ and its standard deviation.

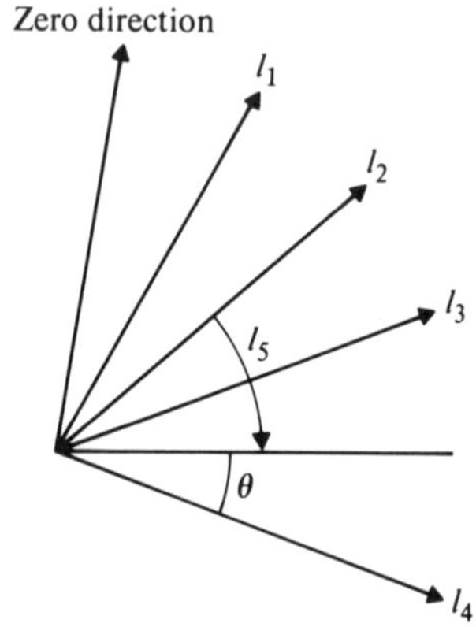

10. In the level circuit shown, the following elevation differences were observed:

Station 1 42.107 m above Station 5

Station 1 12.424 m above Station 2

Station 2 42.251 m above Station 3

Station 4 8.464 m above Station 3

Station 5 4.138 m above Station 4

Station 1 46.269 m above Station 4

Station 2 33.802 m above Station 4

Station 5 is a benchmark with known (assumed fixed) elevation of 500.000 m AMSL. Using the adjustment of indirect observations method compute the elevations of stations 1 through 4 and their a posteriori cofactor matrix. Assume that the variances

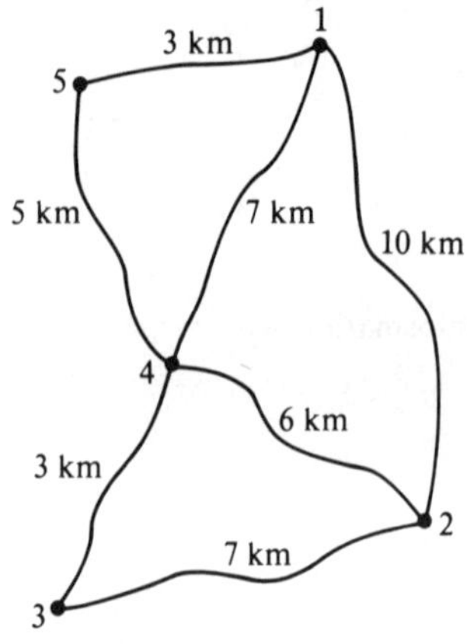

in the observations are directly proportional to the distance leveled (as indicated on the sketch).

11. Solve Problem 10 using the method of adjustment of observations only.

12. Set up each of the following problems such that a least squares solution can be determined. Show the steps used to get to the final equations, specifying the dimensions on all matrices used.

a. The following 11 angles have been observed in a pentagon as shown.

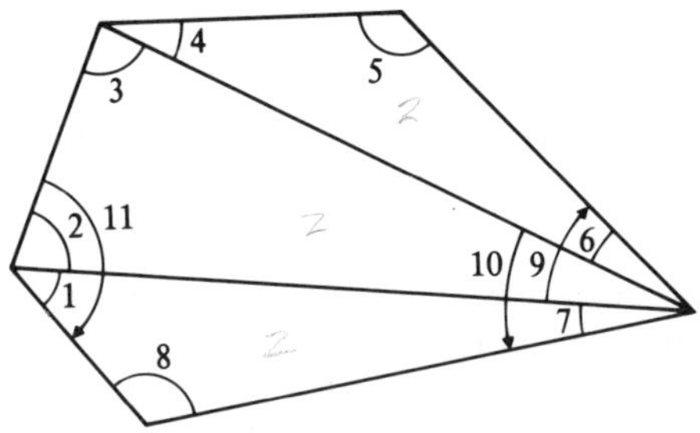

b. The following 16 elevation differences have been observed in a level net as shown. Stations A, B, and C are considered to be of known elevation.

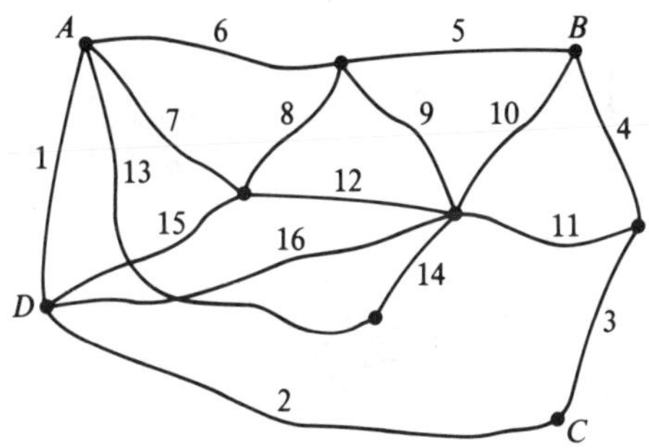

13. The figure depicts a simplified problem of intersection in a plane. The three points a, b, and c are perfectly known control with x y coordinates as follows:

POINT	x	y
a	2.00	19.32
b	2.00	2.00
c	19.32	2.00

An *unknown* point P is to be determined such that its x coordinate is the same as that

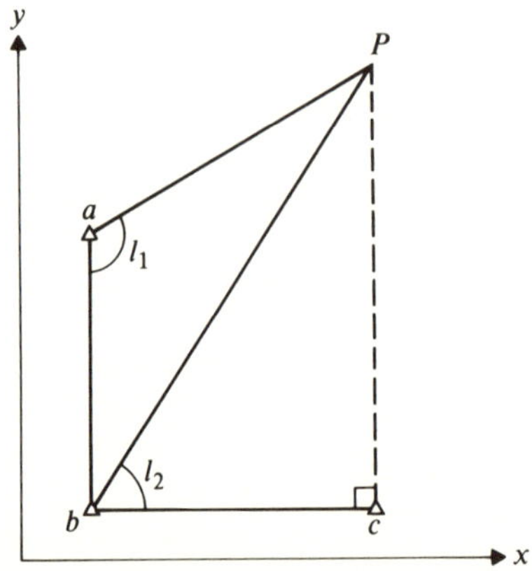

for point c (that is, angle bcP is perfectly known to be 90 degrees). The observations are the two angles shown in the figure,

$$l_1 = 120° \quad \text{and} \quad l_2 = 57°40'$$

where l_1 and l_2 are uncorrelated and of equal weight. Compute the position of point P. (If an approximate value is needed use $y_p^\circ = 29.00$.)

14. In the figure, A and B are *known* horizontal control points. It is desired to determine the position of point P using the three observations shown. Give the *linearized* mathematical model for the adjustment of the problem to obtain the coordinates of point P.

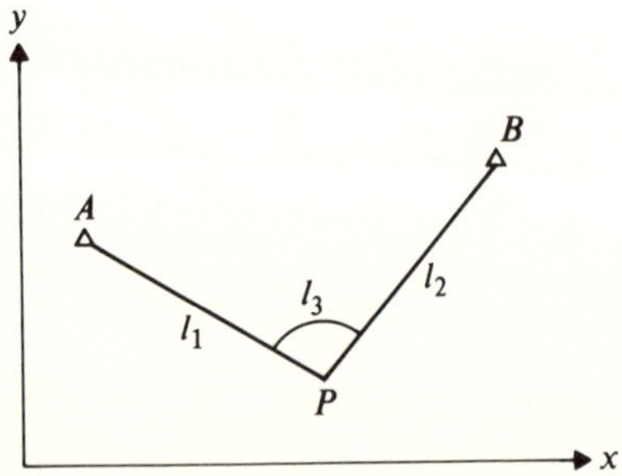

15. The figure shows two *known* horizontal control points A and B. To determine the coordinates of point P, the two lengths l_1 and l_3 and the two angles l_2 and l_4 are

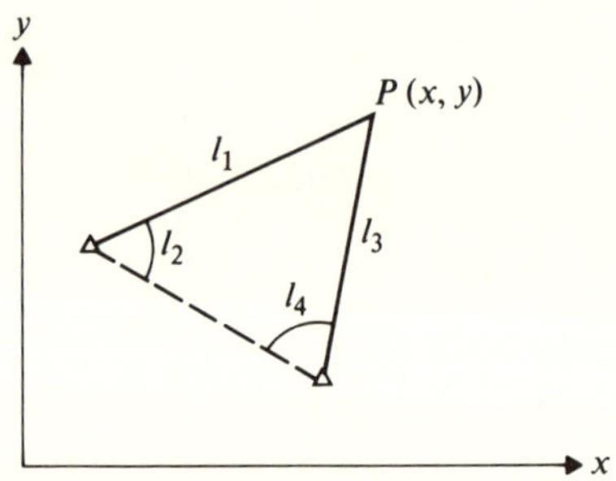

measured. Set this problem up for an adjustment, showing in detail the makeup of the appropriate matrices.

16. For the plane triangle shown, the following quantities are measured:

Angle $A = l_1 = 59°59'58''$ $\sigma_1 = 5''$

Angle $B = l_2 = 59°59'51''$ $\sigma_2 = 5''$

Angle $C = l_3 = 59°59'54''$ $\sigma_3 = 5''$

side $AC = l_4 = 1000$ m $\sigma_4 = 0.05$ m.

Compute the coordinates of point B with respect to the coordinate system shown, and determine their covariance matrix.

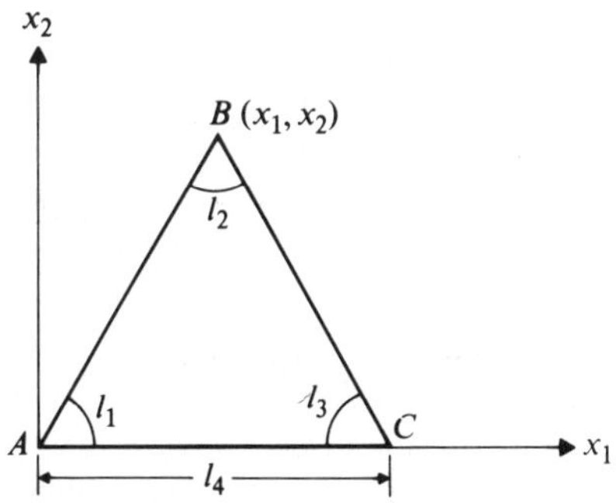

17. The three sides and three angles of a plane triangle were observed as follows:

Sides $l_1 = 1400$ ft $\sigma_1 = 0.01$ ft

Side $l_2 = 1100$ ft $\sigma_2 = 0.02$ ft

Side $l_3 = 1700$ ft $\sigma_3 = 0.01$ ft

Angle $l_4 = 39°23'40''$ $\sigma_4 = 1''$

Angle $l_5 = 88°33'05''$ $\sigma_5 = 1''$

Angle $l_6 = 52°03'17''$ $\sigma_6 = 1''$

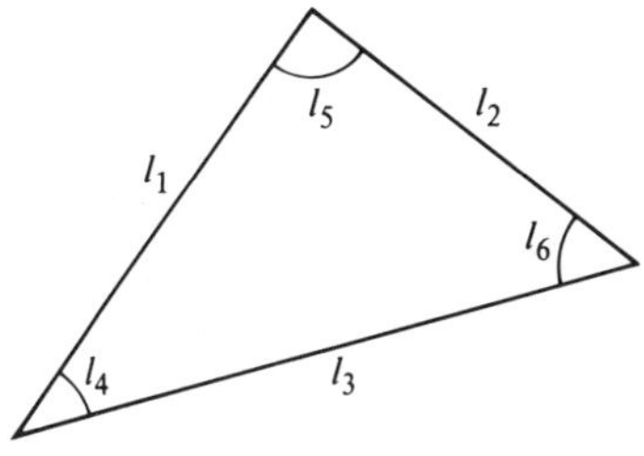

Compute the least squares estimates of the six observations, their a posteriori cofactor matrix, and the a posteriori estimate of the reference variance.

18. The figure shows a simple intersection problem to determine the horizontal position $(X, Y)_p$ of the unknown point P. The coordinates $(X, Y)_a$, $(X, Y)_b$, and $(X, Y)_c$ of the three horizontal control points A, B, and C are *known*. If the four angles l_1, l_2, l_3, and l_4 are observed, give the steps for determining the coordinates of point P by least squares.

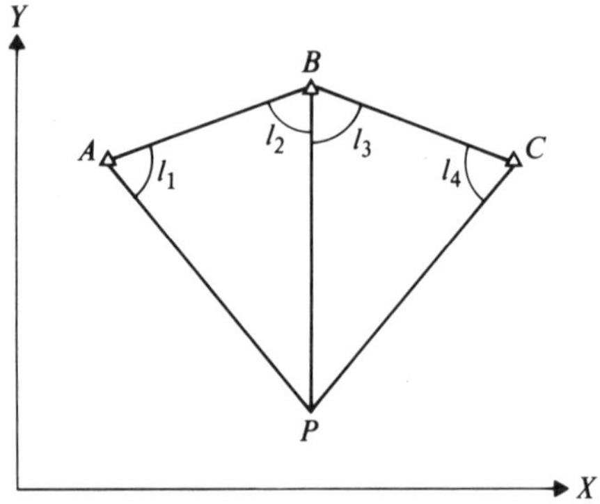

19. The figure shows four (4) positions separated by six (6) measured distances. Assuming station 1 fixed and the Y coordinate of 2 equal to the Y coordinate of 1, *develop* the appropriate equations for a least squares solution to determine the coordinates of points 2, 3, 4, and their covariance matrix. State clearly any assumptions you make.

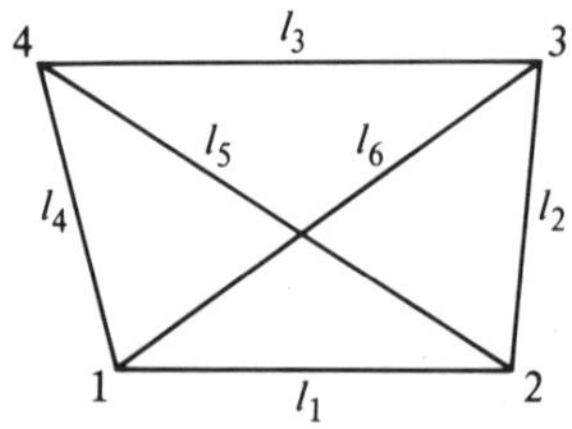

20. Given the triangle shown in the figure with four observed values as follows:

Side AB $= l_1 = 1.1$ m

Side BC $= l_2 = 0.9$ m

Side CA $= l_3 = 1.0$ m

Angle $BAC = l_4 = 60°$

Assuming that the approximate value of the angle ABC is 60 degrees, compute the least squares estimate of the correction δx. Assume that all observations are uncorrelated and of equal weight.

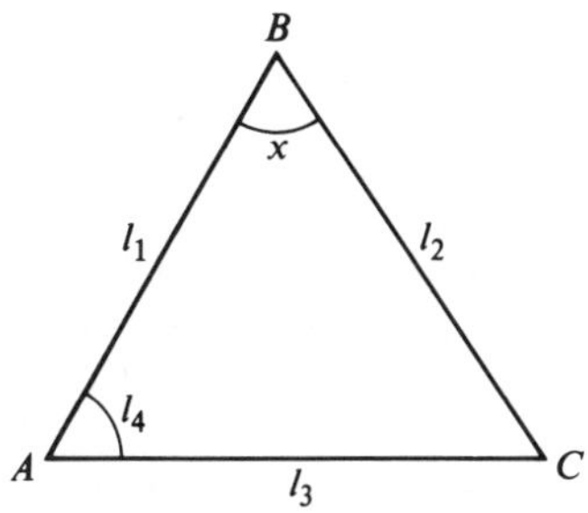

21. Instead of the information given in Problem 20, you are given the information shown in the figure where

Angle $ABC = l_1 = 60°02'$

Angle $BCA = l_2 = 60°01'$

Side $CA \quad = l_3 = 1.0$ m

Angle $BAC = l_4 = 60°00'$

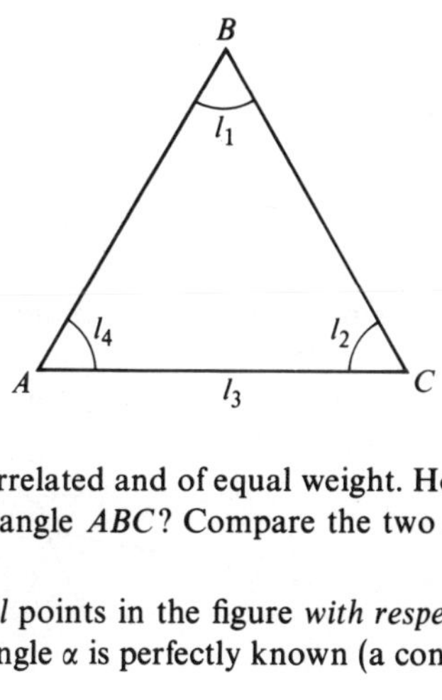

All observations are uncorrelated and of equal weight. How would you compute the least squares estimate of angle ABC? Compare the two adjustment procedures for the two problems.

22. The *positions of all* points in the figure *with respect to the coordinate system shown* are required. The angle α is perfectly known (a constant). All three sides l_1, l_2,

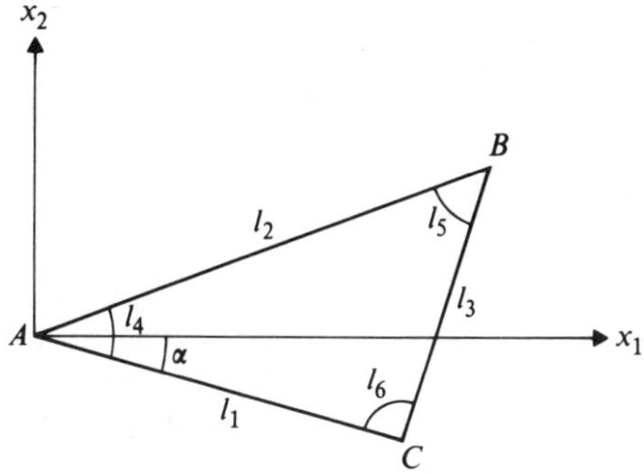

and l_3 and all three interior angles l_4, l_5, and l_6 are measured with an a priori cofactor matrix $\underset{6,6}{\mathbf{Q}}$.

a. Set up the mathematical model giving all pertinent information (for example, n, n_0, r, u, c, …, etc.). Write the equations in detail. If they are nonlinear, explain how you would linearize them. Show how the desired coordinates will be obtained.
b. What would change in the problem if α is also an observation?
c. What would change in the problem if α is not given, that is, if you only have $l_1, \ldots, l_6$?

23. The figure presents a simple triangle in which the vertices Q and S are horizontal control stations with known X, Y coordinates. As shown, the measurements are l_1, l_2, l_3, l_4, l_5, and l_6. Set up the least squares adjustment solution for the determination of the horizontal position of station R.

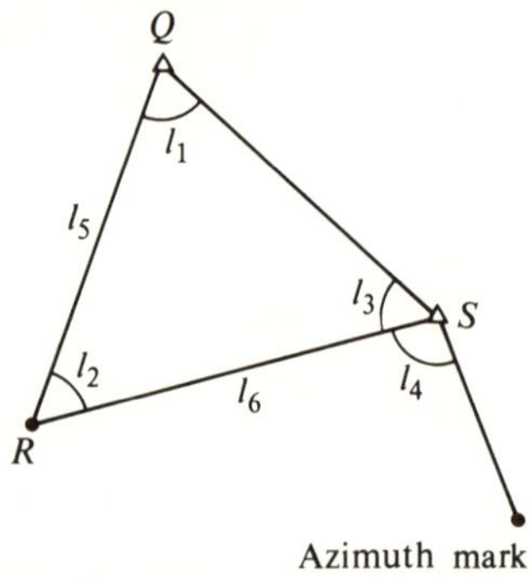

24. a. In the figure the four observations shown are given. Explain the adjustment procedure you would use to determine the *shape* of the figure.
b. If the interior angle at C, which is l_5, is also a given observation, show in detail any change to your answer to (a) above.

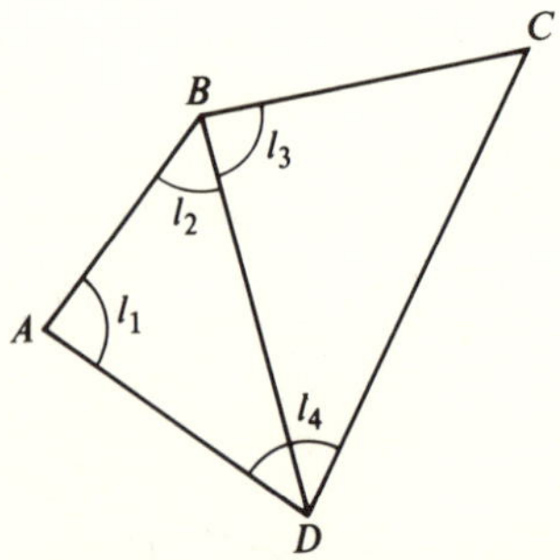

25. The figure shows a *cartesian* coordinate system xy with origin at point B. Points A and C are *known control* points on the y and x axes, respectively. Point D is of unknown coordinates. The observations are the three angles l_1, l_2, and l_3 shown in the figure. Determine n, n_0, r, and then

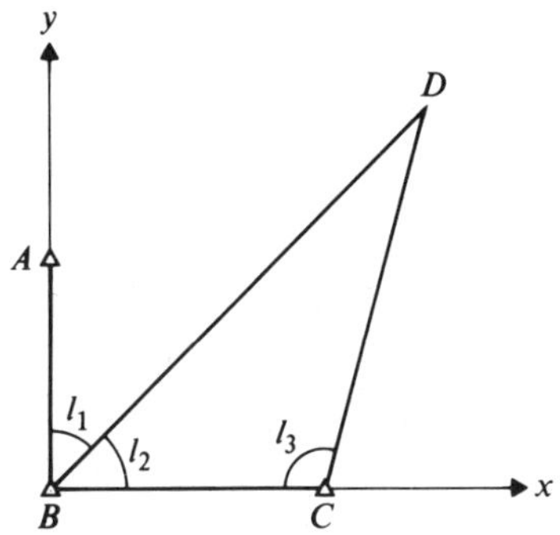

a. For $\mathbf{Av} = \mathbf{f}$ write the condition equations (no linearization).
b. For $\mathbf{Av} + \mathbf{B\Delta} = \mathbf{f}$ write the condition equations necessary for obtaining the estimates of the coordinates of point D. (No linearization.) State explicitly any assumptions you make.

26. The figure shows a circle for which the line AD is a *diameter*, the estimate of which is desired. Because point D could not be occupied, two other points (B and C) are selected on the circumference. The angles l_1 and l_2 and the distances l_3 and l_4 are measured. *Show in complete detail*, giving the appropriate condition equations, how the method of least squares may be used to estimate $\overline{AD}$ and its variance. Specify any assumptions you may make.

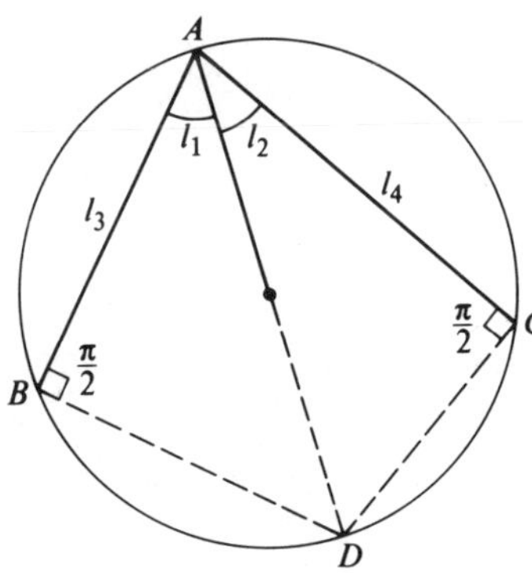

27. The figure shows a photogrammetric model that has been leveled and approximately scaled. All that remains is final scaling. Two distances are *measured* in the model which are $l_1 = 300$ mm and $l_2 = 400$ mm. The corresponding *known* control distances are $D_1 = 305$ mm and $D_2 = 410$ mm, at the nominal model scale. The current value of the base, that is, that corresponding to l_1 and l_2, is $b = 200$ mm.
a. Compute the final value of the base B that, when introduced, would complete the scaling of the model.
b. Compute the observational residuals in l_1 and l_2.
c. Check to verify that the mathematical model is satisfied after the adjustment. Assume all observations to be uncorrelated and of equal weight.

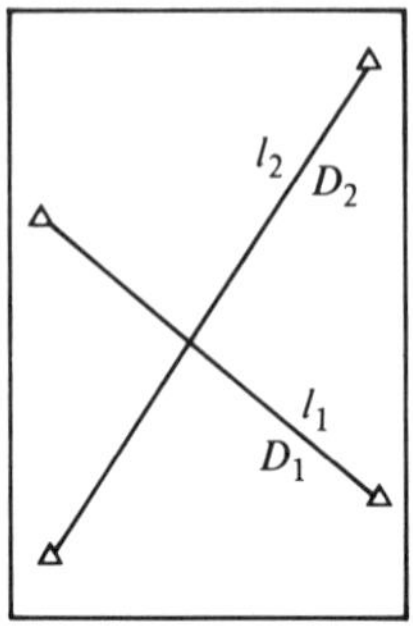

28. The figure shows a photogrammetric model that is scaled and approximately leveled. The four points 3, 4, 5, and 6 are given vertical control points. If Z denotes ground elevation in feet, and z denotes model elevation in feet, let the following values represent the *observations*:

$$l_3 = Z_3 - z_3 = +4 \text{ ft}$$

$$l_4 = Z_4 - z_4 = +2 \text{ ft}$$

$$l_5 = Z_5 - z_5 = +1 \text{ ft}$$

$$l_6 = Z_6 - z_6 = -5 \text{ ft}$$

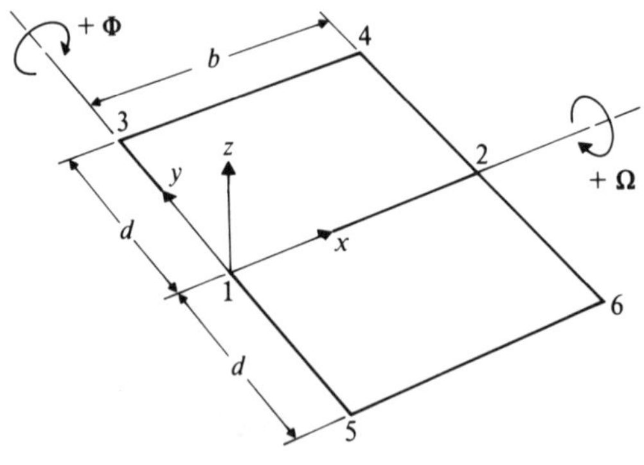

If these observations are uncorrelated and of equal weight, compute the estimated observations using the method of least squares adjustment of observations only.

29. In a chemistry lab a student desired to find the radius of the inside of a glass sphere. He filled the glass sphere with mercury and carefully poured the mercury into a graduated vial from which he read the volume in cubic centimetres. He observed readings of 31.9, 32.2, and 32.0. He considered that the second measurement was about one half as good as the first and the third measurement about one third as good as the first. Considering these as weights, determine what the best estimate of the radius is *and* how good that estimate is.
(Hint: Adjustment of observations only would probably be easiest.)

30. a. Find the best-fitting straight line to the following data:

X (CONSTANT)	Y (OBSERVATIONS)
-4	-1.9
-2	-0.2
0	2.0
2	4.2
4	6.1

The observations are assumed to be independent and have variances of 0.05.

b. Does the a posteriori estimate of reference variance seem reasonable?

31. The equation of a plane is usually given as

$$Ax + By + Cz + D = 0$$

We know that three points determine a plane. Therefore only three of these parameters are necessary. Thus we can write the equation as

$$ax + by + cz + 1 = 0$$

Given the x, y, z coordinates of the four corners of a building from a stereo pair of photographs, and assuming that these four points must lie in a plane,

a. Show that the normal equations for the determination of the best-fitting plane can be formed by a summation process. Be sure to dimension all matrices. Assume

$$\underset{12,\,12}{\mathbf{Q}} = \mathbf{I}$$

b. Using as approximations $a^0 = 1$, $b^0 = 1$, and $c^0 = -1$, find the best equation for that plane. Do *one* iteration only.

	MODEL COORDINATES		
POINT	X	Y	Z
1	1.1	-1.0	0.9
2	-2.0	2.0	1.0
3	2.0	-2.0	1.0
4	-1.1	1.0	0.9

32. Two rectangular coordinate systems, A and B, are related through translation and rotation. For the two-dimensional case in transforming from system A to system B, the following expressions apply:

$$x_b = a_0 + a_1 x_a - a_2 y_a$$

$$y_b = a_3 + a_2 x_a + a_1 y_a$$

Measurements on five points in each coordinate system were recorded as follows:

POINT	x_a	y_a	x_b	y_b
1	2.020	4.107	8.457	16.740
2	5.132	1.098	12.472	15.292
3	0.080	6.204	5.863	17.865
4	7.483	0.109	15.155	15.367
5	4.206	8.128	8.818	21.333

Solve for the transformation coefficients a_0, a_1, a_2, and a_3 assuming that all measurements are uncorrelated and of equal precision.

33. The outcome of an experiment is to be plotted with respect to an x-y system of coordinates. The values of the ordinate y are *observations* taken at equal intervals of x. The following table lists the data:

x	y	σ_y
2	2.1	0.05
4	3.9	0.05
6	6.2	0.05
8	8.1	0.05

If a straight line with the equation:

$$y = ax + b$$

is to be fitted to the data, compute the least squares estimates for a and b and their covariance matrix.

34. A parabola with the equation $y^2 = ax$ is to be fitted to the two points (1, 2) and (2, 3). Considering that the y values are the observations, which are uncorrelated and of equal precision, compute the *first* correction to the approximate value of a, $a^0 = 4.0$.

35. The figure presents a traverse situation in which the endpoints A and D are horizontal control stations with known X, Y coordinates. B and E are the given

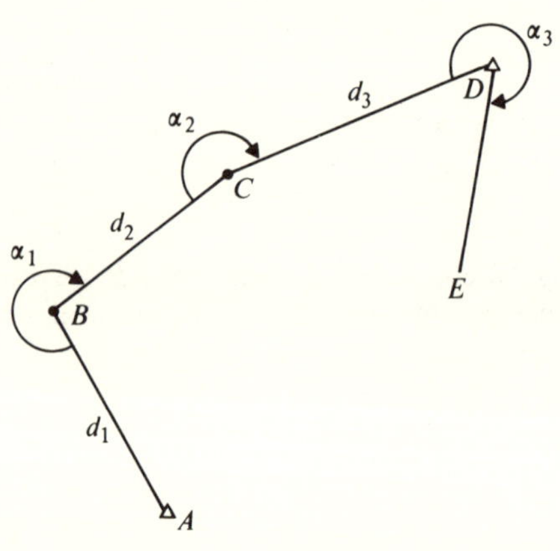

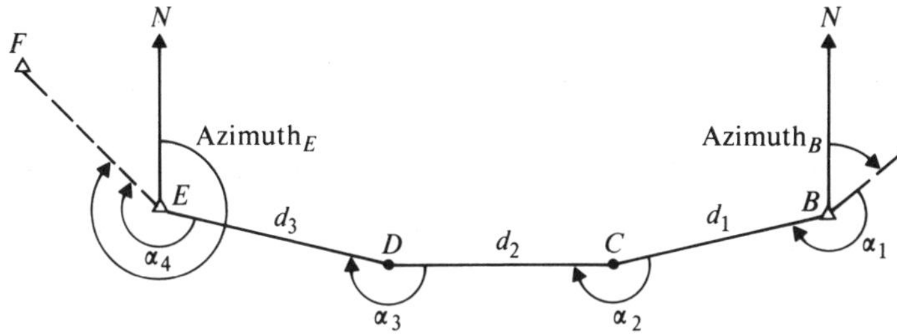

azimuth marks for stations A and D, respectively. Set up the least squares adjustment solution for the determination of the horizontal position of stations B and C, given that stations B, C, and D are occupied to measure α_1, α_2, α_3, d_1, d_2, and d_3.

36. For the traverse shown, the following angle and distance measurements were recorded with their associated standard deviations.

ANGLE	σ	DISTANCE	σ
$\alpha_1 = 172°53'34''$	2″	$d_1 = 281.832$ m	0.016 m
$\alpha_2 = 185°22'14''$	2″	$d_2 = 271.300$ m	0.016 m
$\alpha_3 = 208°26'19''$	2″	$d_3 = 274.100$ m	0.016 m
$\alpha_4 = 205°13'51''$	2″		

POINT	COORDINATES		
	X (m)	Y (m)	AZIMUTH
B	8478.139	2483.826	68°15′20.″7
E	7709.336	2263.411	300°11′30.″5

a. Compute the coordinates of points C and D and their respective covariance matrices.
b. Compute the adjusted estimates of the observed angles and distances.
c. Compute the estimate $\hat{\sigma}_0^2$.
d. Compute the covariance matrix for the estimated observations.

37. In the figure $b\hat{0}c = \theta_2$ and $c\hat{0}d = \theta_1$ are unknown parameters. The measurements l_1, l_2, l_3, and l_4 are given by

$$\boldsymbol{l} = \begin{bmatrix} d\hat{0}e \\ a\hat{0}b \\ c\hat{0}e \\ a\hat{0}c \end{bmatrix} = \begin{bmatrix} 51° \\ 38° \\ 80° \\ 98° \end{bmatrix}$$

Let $\theta_1^0 = 29$ and $\theta_2^0 = 60$. Take the covariance matrix of the observations to be the identity matrix
a. Find θ_1 and θ_2.
b. It is known that angle $b0d$ is a right angle. Find θ_1 and θ_2 under this *constraint*.

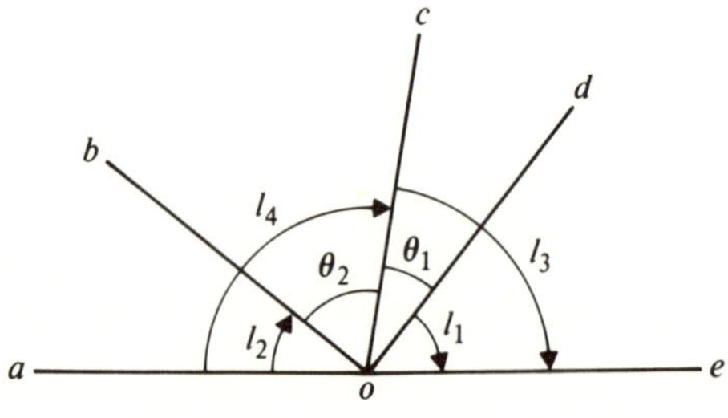

38. A parabola with the equation $y^2 = ax$ is to be fitted to the following data:

POINT	(1)	(2)	(3)
x	1	2	3
y	2	3	3.5

Consider the *y values as observations* that are uncorrelated and of equal weight.
a. With an approximate value $a^0 = 4$ compute the correction δa from *first iteration only*.
b. What would the solution be if the parabola is *constrained* to pass through point (1) precisely?

39. The figure shows two *perfectly* known points: $A(0, 2)$ and $B(0, 5)$. Two other points 1 and 2 are to be located in the x, y plane through measuring the three distances: $l_1 = \overline{B1} = 10.1$; $l_2 = \overline{B2} = 5.1$; $l_3 = \overline{A2} = 4.0$ and $\mathbf{Q} = \mathbf{I}$. Also, point 1 has a fixed y coordinate $= 5.0$.
a. Compute the three coordinates x_1, x_2, y_2.
b. If points 1 and 2 are constrained to lie on the line $y = mx$, write the *conditions* and *constraints* giving the numerical values of all matrices. Then eliminate the constraints and give the numerical values of the matrices $\bar{\mathbf{B}}$ and $\bar{\mathbf{f}}$ of the modified conditions. Use reasonable approximations such as $x_1^0 = 10$, $x_2^0 = 4$, and $y_2^0 = 2$.

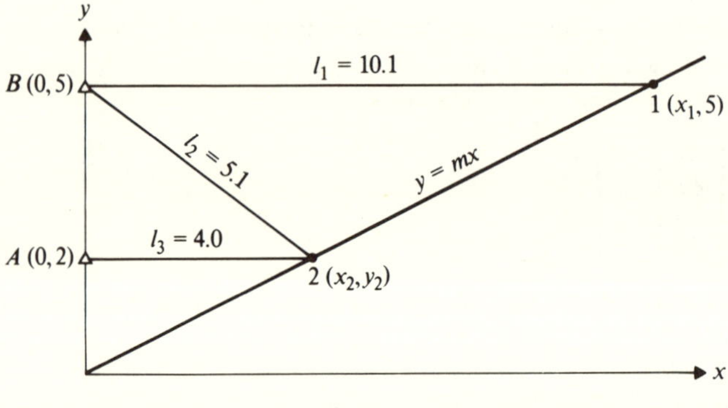

40. Given two coordinate systems $\mathbf{x} = [x_1\, x_2]^t$, $\mathbf{y} = [y_1\, y_2]^t$, and the following three points with measured coordinates in both systems (take a priori $\mathbf{Q} = \mathbf{I}$)

POINT	x_1	x_2	y_1	y_2
1	576.889	286.144	738.284	372.541
2	598.213	449.326	745.184	561.693
3	608.554	597.625	741.096	732.570

The transformation between the two systems is

$$\begin{bmatrix} y_1 \\ y_2 \end{bmatrix} = s \begin{bmatrix} a & b \\ c & d \end{bmatrix} \begin{bmatrix} x \\ y \end{bmatrix} + \begin{bmatrix} t_1 \\ t_2 \end{bmatrix}$$

where s is a scale factor and a, b, c, d, t_1, t_2 are six parameters. Compute the least squares estimates for these seven parameters applying appropriate constraints such that the transformation is in effect the four-parameter linear conformal in two dimensions [equation (8.6)]. Perform the adjustment with the four-parameter transformation and ascertain that the results from both procedures are the same.

41. In the figure, the four observations shown are given. If the four points A, B, C, and D fall on the circumference of a circle, write the elements of the functional model for the least squares adjustment for determining the shape of the figure.

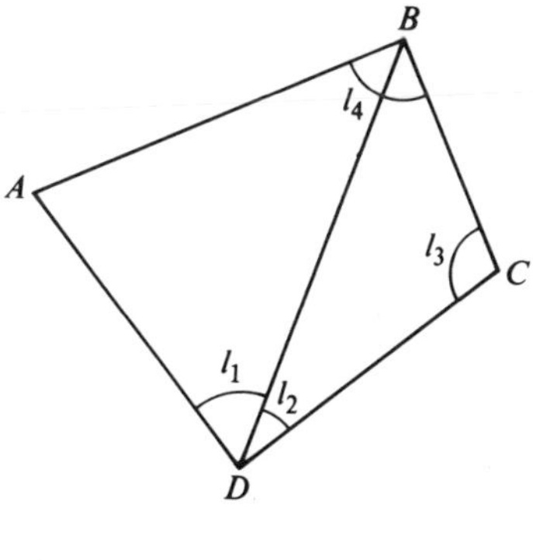

42. The figure shows a large *rectangular* area divided into two *rectangular* areas (that is, all angles are perfectly known). The observations, which are uncorrelated and of equal precision, are

$$l_1 = 2.00 \qquad l_2 = 3.00 \qquad l_3 = 3.10 \qquad l_4 = 1.00 \qquad l_5 = 1.94$$

a. Compute the least squares estimates of the areas X_1 and X_2 and their cofactors.
b. Under the added information of constraining X_1 to be $3X_2$: (i) Write all the elements of the functional model giving the condition and constraint equations; if necessary, linearize these equations and give the numerical values of all matrices involved. (ii) Eliminate the constraints and give the numerical values of the matrices

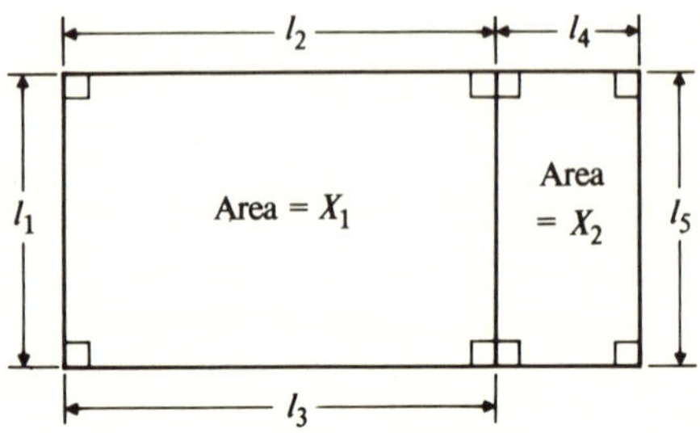

involved in remaining equations. (iii) Eliminate the parameters left after part (ii) above and give the numerical values of the matrices involved in the remaining equations.

PART III

12

A Unified Approach to Least Squares Adjustment

12.1. INTRODUCTION

In the different chapters of Part II of this book several techniques of least squares were introduced. In all these techniques a distinction was made between different groups of variables appearing in the mathematical formulation. Observations were designated as those variables for which covariance matrices are available a priori, and the parameters were the variables usually unknown in the adjustment. It was specified further that condition equations are those written as functions of both observations and parameters, and constraint equations are written only in terms of the parameters. Such classification of variables and equations follows what happens in reality and is therefore quite adequate for formulating many of the adjustment problems encountered in practice. However, recent rapid advances in the fields of geodesy, surveying, and photogrammetry have made it desirable to have an adjustment approach that can handle a mixture of input data in a more general and unified manner. As an example, computational orbital photogrammetry problems may involve regular photogrammetric data, orbital constraints, time measurements, auxiliary sensor information such as stellar photomeasurements, laser altitude and doppler information, and gravity information. Although it would be possible to handle such complex adjustment problems by techniques developed in Part II, an approach in which varying types of data can be handled with minimum effort and specialization would certainly be more appropriate. It is the purpose of this chapter to introduce and discuss such an approach, which is being progressively incorporated in many of modern photogrammetric and geodetic computational systems.

The most important underlying factor to this approach is the assumption at the outset that *all variables involved in the mathematical formulation are observations.*

With this assumption no individual group classification becomes necessary as long as we introduce into the concept some kind of mechanism by which a differentiation can be made whenever necessary. Such a mechanism is conveniently given by the covariance, or weight, matrix of the *observations*, which in this case are all the variables in the model. To demonstrate the practicality of this approach, consider the following two extreme cases:

1. If an observation (in this case any variable in the model) is given an infinitely large variance, that is, its weight is $w = 0$, then it is allowed to vary freely in the adjustment and will therefore assume the role of an unknown parameter in the classical sense.
2. If on the other hand the observation is given a zero variance, or a weight that approaches infinity, $w \to \infty$, it is simply not allowed to change in the adjustment, with the consequence that its residual will be zero and it would assume the classical meaning of a constant.

Between the above two extremes lies a large set of possibilities within which actual observations (in the classical sense) fit. Of course, one of the most important points to watch is that when the covariances of a heterogeneous set of variables are all collected together, care must be exercised in regard to units and dimensions. This is particularly important if we choose to use cofactor matrices instead of covariance matrices. In such a case only one common factor must be chosen to relate the different covariance matrices together.

Having lifted up the distinction between observations and parameters, the differentiation between condition equations and constraint equations used in Part II no longer applies. This is because the parameters are now considered as observations, thus rendering the constraints as conditions. However, because the constraints usually refer to strict geometric or functional relations (such as points lying on a straight line, circle, and so on), their treatment is deferred until later. But first we shall consider the case of adjustment of observations and independent parameters treated in Chapter 6 (Part II).

12.2. FORMULATION

We begin by limiting consideration to linear functions. Let the classically designated observational vector be $\boldsymbol{l}$ with a priori cofactor matrix $\mathbf{Q}$ and a vector of residuals $\mathbf{v}$. The variables, usually referred to by parameters, will in the present case be treated as observations. For this purpose we use the vector $\mathbf{x}$ with a priori cofactor matrix $\mathbf{Q}_{xx}$ and a vector of "corrections" $\boldsymbol{\Delta}$. There are n elements (observations) in $\boldsymbol{l}$ (and of course in $\mathbf{v}$) and u elements (parameters) in $\mathbf{x}$ (and in $\boldsymbol{\Delta}$). In the classical approach of Part II the redundancy is $r = n - n_0$ and the total number of conditions is $c = r + u$, remembering that n_0 is the minimum number of variables necessary to specify the problem. In the unified concept all variables are observations, thus a total of $(n + u)$ observed values. This is equivalent to adjustment of observations only, given in Chapter 7, where the number of equations is equal to the redundancy. In the present case the redundancy is equal to the total number of observations $(n + u)$ minus n_0, or

$$(n + u - n_0) = r + u = c$$

Thus the new considerations under the unified approach lead to the same situation as would be the case when the classical procedure is followed.

Let us write the c originally linear equations in the form

$$\mathbf{A}(\boldsymbol{l} + \mathbf{v}) + \mathbf{B}(\mathbf{x} + \boldsymbol{\Delta}) = \mathbf{d} \tag{12.1}$$

where

$\mathbf{A}$ is $c \times n$, rank $(\mathbf{A}) = c$

$\mathbf{B}$ is $c \times u$, rank $(\mathbf{B}) = u$

$\mathbf{d}$ is $c \times 1$

all of which are coefficient matrices. Rearranging equation (12.1), gives

$$\mathbf{A}\mathbf{v} + \mathbf{B}\boldsymbol{\Delta} = \mathbf{d} - \mathbf{A}\boldsymbol{l} - \mathbf{B}\mathbf{x} = \bar{\mathbf{f}} \tag{12.2}$$

Equation (12.2) may be rewritten more concisely as

$$\bar{\mathbf{A}}\bar{\mathbf{v}} = \bar{\mathbf{f}} \tag{12.3}$$

where

$$\bar{\mathbf{A}} = [\mathbf{A} \quad \mathbf{B}] \tag{12.4a}$$

$$\bar{\mathbf{v}} = [\mathbf{v}^t \quad \boldsymbol{\Delta}^t]^t \tag{12.4b}$$

and the corresponding a priori cofactor matrix is

$$\bar{\mathbf{Q}} = \begin{bmatrix} \mathbf{Q} & \mathbf{0} \\ \mathbf{0} & \mathbf{Q}_{xx} \end{bmatrix} \tag{12.4c}$$

where no correlation is assumed between the two vectors $\boldsymbol{l}$ and $\mathbf{x}$. The least squares solution for equation (12.3) is obtained by applying equations (7.4) and (7.5) as follows:

$$\bar{\mathbf{v}} = \bar{\mathbf{Q}}\bar{\mathbf{A}}^t(\bar{\mathbf{A}}\bar{\mathbf{Q}}\bar{\mathbf{A}}^t)^{-1}\bar{\mathbf{f}}$$
$$= \bar{\mathbf{Q}}\bar{\mathbf{A}}^t(\mathbf{AQA}^t + \mathbf{BQ}_{xx}\mathbf{B}^t)^{-1}\bar{\mathbf{f}}$$

or

$$\begin{bmatrix} \mathbf{v} \\ \mathbf{\Delta} \end{bmatrix} = \begin{bmatrix} \mathbf{Q} & \mathbf{0} \\ \mathbf{0} & \mathbf{Q}_{xx} \end{bmatrix} \begin{bmatrix} \mathbf{A}^t \\ \mathbf{B}^t \end{bmatrix} (\mathbf{Q}_e + \mathbf{BQ}_{xx}\mathbf{B}^t)^{-1}\bar{\mathbf{f}}$$

from which

$$\mathbf{v} = \mathbf{QA}^t(\mathbf{Q}_e + \mathbf{BQ}_{xx}\mathbf{B}^t)^{-1}\bar{\mathbf{f}} \tag{12.5a}$$

$$\mathbf{\Delta} = \mathbf{Q}_{xx}\mathbf{B}^t(\mathbf{Q}_e + \mathbf{BQ}_{xx}\mathbf{B}^t)^{-1}\bar{\mathbf{f}} \tag{12.5b}$$

We first reduce equation (12.5b), by substituting for the inverse in it by its equivalent as given in equations (A68) and (A69) (Appendix A):

$$\mathbf{\Delta} = \mathbf{Q}_{xx}\mathbf{B}^t\mathbf{Q}_e^{-1}[\mathbf{I} - \mathbf{B}(\mathbf{B}^t\mathbf{Q}_e^{-1}\mathbf{B} + \mathbf{Q}_{xx}^{-1})^{-1}\mathbf{B}^t\mathbf{Q}_e^{-1}]\bar{\mathbf{f}} \tag{12.6a}$$

Realizing that $\mathbf{W}_e = \mathbf{Q}_e^{-1}$ and referring to the auxiliaries used regularly,

$$\mathbf{\Delta} = \mathbf{Q}_{xx}\mathbf{t} - \mathbf{Q}_{xx}\mathbf{N}(\mathbf{N} + \mathbf{W}_{xx})^{-1}\mathbf{t} \tag{12.6b}$$

Now applying equation (A70) (see Appendix A) to the inverse in equation (12.6b), then

$$\mathbf{\Delta} = \mathbf{Q}_{xx}\mathbf{t} - \mathbf{Q}_{xx}\mathbf{N}\mathbf{N}^{-1}(\mathbf{N}^{-1} + \mathbf{Q}_{xx})^{-1}\mathbf{Q}_{xx}\mathbf{t}$$
$$= [\mathbf{Q}_{xx} - \mathbf{Q}_{xx}(\mathbf{N}^{-1} + \mathbf{Q}_{xx})^{-1}\mathbf{Q}_{xx}]\mathbf{t}$$

which from equations (A69) and (A68) again leads to

$$\mathbf{\Delta} = (\mathbf{N} + \mathbf{Q}_{xx}^{-1})^{-1}\mathbf{t} = (\mathbf{N} + \mathbf{W}_{xx})^{-1}\mathbf{t} \tag{12.7}$$

Equation (12.7) shows that the only difference between the classical approach and the unified treatment is that the a priori weight matrix of the parameters $\mathbf{W}_{xx}$ is added to $\mathbf{N}$ and in computing $\mathbf{t}$ the vector $\bar{\mathbf{f}}$ of equation (12.2) is used instead of the usual $\mathbf{f}$ of equation (6.6). When $\mathbf{W}_{xx} = \mathbf{0}$ and a priori $\mathbf{x} = \mathbf{0}$, that is, when $\mathbf{\Delta}$ represents strictly unknown variables (parameters), the unified approach immediately reduces to the "classical" one. Once $\mathbf{\Delta}$ is computed from equation (12.7), the final estimate of the parameter vector would be $\hat{\mathbf{x}} = \mathbf{x} + \mathbf{\Delta}$, where $\mathbf{x}$ is the a priori value.

Returning to equation (12.5a) we may reduce it in a manner similar to that used above. Thus

$$\mathbf{v} = \mathbf{QA}^t(\mathbf{Q}_e + \mathbf{BQ}_{xx}\mathbf{B}^t)^{-1}\bar{\mathbf{f}}$$
$$= \mathbf{QA}^t\mathbf{Q}_e^{-1}[\mathbf{I} - \mathbf{B}(\mathbf{B}^t\mathbf{Q}_e^{-1}\mathbf{B} + \mathbf{Q}_{xx}^{-1})^{-1}\mathbf{B}^t\mathbf{Q}_e^{-1}]\bar{\mathbf{f}}$$
$$= \mathbf{QA}^t\mathbf{W}_e[\bar{\mathbf{f}} - \mathbf{B}(\mathbf{N} + \mathbf{Q}_{xx}^{-1})^{-1}\mathbf{t}]$$

or

$$\mathbf{v} = \mathbf{QA}^t\mathbf{W}_e(\bar{\mathbf{f}} - \mathbf{B}\boldsymbol{\Delta}) \tag{12.8}$$

Equation (12.8) is similar to the result obtained in Chapter 6 (Part II) given by equations (6.13) and (6.17). Therefore the relation for computing $\mathbf{v}$ remains the same except for the fact that here $\bar{\mathbf{f}}$ and $\boldsymbol{\Delta}$ are computed from equations (12.2) and (12.7) instead of equation (6.24).

12.2.1. Alternative Derivation In the foregoing development we treated the case of having the classically designated parameters considered directly as observations, and applied the results of adjustment of observations only to derive equations (12.7) and (12.8). The same results can be obtained by applying the minimum criterion of least squares to equation (12.2). Of basic importance here is the fact that $\boldsymbol{\Delta}$ in the present case plays a role equivalent to that of "residuals." In other words, although $\mathbf{v}$ is the vector of residuals for $\boldsymbol{l}$, $\boldsymbol{\Delta}$ is the vector of "residuals" for the "observations" $\mathbf{x}$. Thus the scalar to be minimized in this case is

$$\phi = \mathbf{v}^t\mathbf{W}\mathbf{v} + \boldsymbol{\Delta}^t\mathbf{W}_{xx}\,\boldsymbol{\Delta} - 2\mathbf{k}^t(\mathbf{A}\mathbf{v} + \mathbf{B}\boldsymbol{\Delta} - \bar{\mathbf{f}}) \tag{12.9}$$

and then,

$$\frac{\partial\phi}{\partial\mathbf{v}} = 2\mathbf{v}^t\mathbf{W} - 2\mathbf{k}^t\mathbf{A} = \mathbf{0}^t$$

and

$$\frac{\partial\phi}{\partial\boldsymbol{\Delta}} = 2\boldsymbol{\Delta}^t\mathbf{W}_{xx} - 2\mathbf{k}^t\mathbf{B} = \mathbf{0}^t$$

or

$$\mathbf{W}\mathbf{v} - \mathbf{A}^t\mathbf{k} = 0 \tag{12.10a}$$

$$\mathbf{W}_{xx}\boldsymbol{\Delta} - \mathbf{B}^t\mathbf{k} = 0 \tag{12.10b}$$

From equation (12.10a)

$$\mathbf{v} = \mathbf{QA}^t\mathbf{k} \tag{12.10c}$$

and using it in equation (6.5) and solving for $\mathbf{k}$

$$\mathbf{k} = (\mathbf{AQA}^t)^{-1}(\bar{\mathbf{f}} - \mathbf{B}\boldsymbol{\Delta}) = \mathbf{W}_e(\bar{\mathbf{f}} - \mathbf{B}\boldsymbol{\Delta}) \tag{12.10d}$$

Substituting equation (12.10d) into (12.10b) leads to

$$\mathbf{W}_{xx}\,\boldsymbol{\Delta} - \mathbf{B}^t\mathbf{W}_e(\bar{\mathbf{f}} - \mathbf{B}\boldsymbol{\Delta}) = \mathbf{0}$$

or

$$(\mathbf{N} + \mathbf{W}_{xx})\boldsymbol{\Delta} = \mathbf{t} \tag{12.11}$$

The solution of equation (12.11) for $\mathbf{\Delta}$ gives equation (12.7), thus proving that either of the two procedures given above leads to the same results. Equation (12.8) can also be obtained from combining equations (12.10c) and (12.10d).

12.2.2. Parameter/Observation Consideration The two methods given above dealt with the case when the parameters, as classically designated, are treated strictly as observations. At the end of each derivation we ended up with a relation for computing the vector $\mathbf{\Delta}$. In practice, even though the unified approach does not distinguish between observations and parameters, the groups of variables do in fact present themselves in specific categories, and the parameters are most often those variables that are required. On the basis of this premise it is possible to have a third approach to the unified concept.

We can begin by considering that $\boldsymbol{l}$ is an $n \times 1$ vector of observations with a cofactor matrix $\mathbf{Q}$ and a residual vector $\mathbf{v}$, and $\mathbf{x}$ is a $u \times 1$ vector of parameters with a correction vector $\mathbf{\Delta}$. Thus if $r = n - n_0$ is the redundancy, we may as usual write $c = r + u$ conditions in terms of $\mathbf{v}$, $\boldsymbol{l}$, $\mathbf{x}$, and $\mathbf{\Delta}$. We may further consider that there are u "observations" $\mathbf{x}$ with a cofactor matrix $\mathbf{Q}_{xx}$ and a residual vector $\mathbf{v}_x$. This makes the total number of observations $(n + u)$, which with the given minimum number of required variables n_0 changes the redundancy to $(n + u - n_0)$ or $(r + u)$. Given u parameters in addition leads to a required total number of condition equations of $(r + 2u)$ or $(c + u)$.

The first set of c conditions is that usually available from the geometric or physical conditions of the problem and formulated in equation (12.2). Added to that we must have u more "conditions" to account for having a priori "observations" on the parameters. These may be formulated by recognizing that at the end of the adjustment, the value of the estimated parameters $\hat{\mathbf{x}}$ must be identical to the "estimated observations" for the same variables, or

$$\hat{\mathbf{x}} = \mathbf{x} + \mathbf{\Delta} = \mathbf{x} + \mathbf{v}_x$$

or

$$\mathbf{v}_x - \mathbf{\Delta} = \mathbf{0} \tag{12.12}$$

Combining equations (12.2) and (12.12) gives

$$\begin{bmatrix} \mathbf{A} & \mathbf{0} \\ \mathbf{0} & \mathbf{I} \end{bmatrix} \begin{bmatrix} \mathbf{v} \\ \mathbf{v}_x \end{bmatrix} + \begin{bmatrix} \mathbf{B} \\ -\mathbf{I} \end{bmatrix} \mathbf{\Delta} = \begin{bmatrix} \mathbf{f} \\ \mathbf{0} \end{bmatrix} \tag{12.13a}$$

or

$$\dot{\mathbf{A}}\dot{\mathbf{v}} + \dot{\mathbf{B}}\mathbf{\Delta} = \dot{\mathbf{f}} \tag{12.13b}$$

for which the total cofactor matrix is $\bar{\mathbf{Q}}$ given by equation (12.4c). Least squares may now be applied directly to equation (12.13b). Thus

$$\dot{\mathbf{Q}}_e = \dot{\mathbf{A}}\bar{\mathbf{Q}}\dot{\mathbf{A}}^t = \begin{bmatrix} \mathbf{AQA}^t & \mathbf{0} \\ \mathbf{0} & \mathbf{Q}_{xx} \end{bmatrix} = \begin{bmatrix} \mathbf{Q}_e & \mathbf{0} \\ \mathbf{0} & \mathbf{Q}_{xx} \end{bmatrix} \tag{12.14a}$$

$$\dot{\mathbf{N}} = \dot{\mathbf{B}}^t\dot{\mathbf{W}}_e\,\dot{\mathbf{B}} = [\mathbf{B}^t\mathbf{W}_e\,\mathbf{B} + \mathbf{W}_{xx}] = (\mathbf{N} + \mathbf{W}_{xx}) \tag{12.14b}$$

$$\dot{\mathbf{t}} = \dot{\mathbf{B}}^t\dot{\mathbf{W}}_e\,\dot{\mathbf{f}} = \mathbf{B}^t\mathbf{W}_e\,\bar{\mathbf{f}} = \mathbf{t} \tag{12.14c}$$

and thus

$$\boldsymbol{\Delta} = \dot{\mathbf{N}}^{-1}\dot{\mathbf{t}} = (\mathbf{N} + \mathbf{W}_{xx})^{-1}\mathbf{t} \tag{12.14d}$$

Equation (12.14d) is obviously identical to (12.7) and hence all three methods of derivation are identical. Again, $\hat{\mathbf{x}} = \mathbf{x} + \boldsymbol{\Delta}$.

In this linear case it is possible to assume that $\boldsymbol{\Delta}$ is the total parameter vector as has been the practice in Part II. Thus equation (12.12) becomes

$$\mathbf{v}_x - \boldsymbol{\Delta} = -\mathbf{x} \tag{12.15}$$

with $\mathbf{x}$ being the a priori "observational values" on the parameters, and $\dot{\mathbf{f}}$ changes to

$$\dot{\mathbf{f}} = \begin{bmatrix} \mathbf{f} \\ -\mathbf{x} \end{bmatrix} \tag{12.16a}$$

Hence,

$$\dot{\mathbf{t}} = \dot{\mathbf{B}}^t\dot{\mathbf{W}}_e\,\dot{\mathbf{f}} = [\mathbf{B}^t\mathbf{W}_e\,\mathbf{f} + \mathbf{W}_{xx}\,\mathbf{x}] = (\mathbf{t} + \mathbf{W}_{xx}\,\mathbf{x}) \tag{12.16b}$$

$$\boldsymbol{\Delta} = (\mathbf{N} + \mathbf{W}_{xx})^{-1}(\mathbf{t} + \mathbf{W}_{xx}\,\mathbf{x}) \tag{12.16c}$$

Note that the value of $\mathbf{t}$ in equation (12.16c) is computed on the basis of $\mathbf{f} = \mathbf{d} - \mathbf{A}l$ and not $\bar{\mathbf{f}}$ given in equation (12.2).

Example 12.1. Consider a very simple problem (Figure 12.1) of determining the elevation of a point P by measuring the difference in elevation between it and two reference points A, B, the elevations of which are zero. The observations are $l_1 = 1.10$ m and $l_2 = 1.12$ m with a covariance matrix $\boldsymbol{\Sigma} = 10^{-4}\mathbf{I}_2$ m^2. Let the a priori value for the elevation of point P be 1.00 m with a variance of 0.01 m^2.

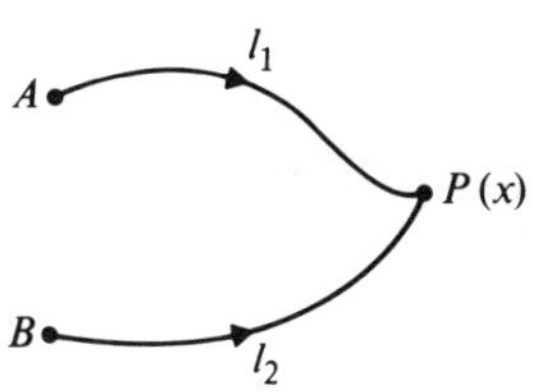

Figure 12.1

The first procedure of solving the problem is to consider having three observations 1.10, 1.12, and 1.00 m with the covariance matrix

$$\boldsymbol{\Sigma} = \text{diag. } \{10^{-4}, 10^{-4}, 10^{-2}\} = 10^{-4} \text{ diag. } \{1, 1, 100\} = \sigma_0^2 \mathbf{Q}$$

Since there is only one elevation to be determined, we write two conditions

$$v_1 - v_x = (x - l_1) = -0.10 \text{ m}$$

$$v_2 - v_x = (x - l_2) = -0.12 \text{ m}$$

$$\begin{bmatrix} 1 & 0 & -1 \\ 0 & 1 & -1 \end{bmatrix} \mathbf{v} = \begin{bmatrix} -0.10 \\ -0.12 \end{bmatrix}$$

$$\mathbf{Q}_e = \begin{bmatrix} 101 & 100 \\ 100 & 101 \end{bmatrix} \qquad \mathbf{W}_e = \tfrac{1}{201} \begin{bmatrix} 101 & -100 \\ -100 & 101 \end{bmatrix} \qquad \mathbf{k} = \begin{bmatrix} 0.0094527 \\ -0.0105472 \end{bmatrix}$$

(the k's are computed to many decimal places for purposes of checking later). Next

$$[v_1 \quad v_2 \quad v_x] = [0.0094527 \quad -0.0105472 \quad 0.10945] \text{ m}$$

Thus

$$\hat{l}_1 = 1.109453 \text{ m} \qquad \hat{l}_2 = 1.109453 \text{ m} \qquad \text{and} \qquad \hat{x} = 1.10945 \text{ m}$$

which show the consistency of the results even to five decimal places. Practically these values should be rounded to two decimal places only.

A second procedure would be to apply equation (12.7). However, we first write the conditions

$$l_1 + v_1 - x - \Delta = 0 \qquad \text{and} \qquad l_2 + v_2 - x - \Delta = 0$$

$$\mathbf{v} + \begin{bmatrix} -1 \\ -1 \end{bmatrix} \Delta = \begin{bmatrix} x - l_1 \\ x - l_2 \end{bmatrix} = \begin{bmatrix} -0.10 \text{ m} \\ -0.12 \text{ m} \end{bmatrix} \qquad \text{or} \qquad \mathbf{v} + \mathbf{B}\Delta = \bar{\mathbf{f}}$$

Then with

$$\mathbf{Q} = \mathbf{W} = \mathbf{I} \qquad \text{and} \qquad \mathbf{W}_{xx} = (100)^{-1} = 0.01$$

$$\mathbf{N} = 2 \qquad \mathbf{t} = 0.22 \text{ m} \qquad \text{and} \qquad \boldsymbol{\Delta} = (2 + 0.01)^{-1}(0.22) = 0.109453 \text{ m}$$

Hence

$$\hat{\mathbf{x}} = \mathbf{x} + \boldsymbol{\Delta} = 1.109453 \text{ m}$$

The third technique allows for solving for the parameters directly rather than for a correction vector $\boldsymbol{\Delta}$ since the equations are linear. In this case

$$\mathbf{f} = [-1.10 \qquad -1.12]^t$$

$$\mathbf{t} = 2.22 \text{ m}$$

and

$$\mathbf{t} = \mathbf{t} + \mathbf{W}_{xx}\,\mathbf{x} = 2.22 + 0.01 = 2.23 \text{ m}$$

Applying equation (12.16c) yields

$$\boldsymbol{\Delta} = (2 + 0.01)^{-1}(2.23) = 1.109453 \text{ m}$$

which is identical to those obtained before.

12.3. THE UNIFIED APPROACH AND NONLINEAR FUNCTIONS

In this section the following nonlinear conditions are considered

$$F(\boldsymbol{l}, \mathbf{x}) = \mathbf{0} \tag{12.17}$$

Denoting by $\boldsymbol{l}^0$ and $\mathbf{x}^0$ two approximation vectors for the observations and parameters, the linearization of equation (12.17) follows the procedure given in Section 11.1.2 or

$$\mathbf{Av} + \mathbf{B\Delta} = \mathbf{f}^0 \tag{11.8}$$

in which $\mathbf{f}^0$ is given by

$$\mathbf{f}^0 = -[F(\boldsymbol{l}^0, \mathbf{x}^0) + \mathbf{A}(\boldsymbol{l} - \boldsymbol{l}^0)] \tag{11.7}$$

In addition to the two approximation vectors, there are also two vectors of observations, $\boldsymbol{l}$ and $\mathbf{x}$, with a priori cofactor matrices $\mathbf{Q}$ and $\mathbf{Q}_{xx}$, respectively. The estimate $\hat{\mathbf{x}}$ of $\mathbf{x}$ is

$$\hat{\mathbf{x}} = \mathbf{x}^0 + \mathbf{\Delta} = \mathbf{x} + \mathbf{v}_x$$

from which

$$\mathbf{v}_x - \mathbf{\Delta} = \mathbf{x}^0 - \mathbf{x} = \mathbf{f}_x \tag{12.18}$$

Equation (12.18) differs from equation (12.15) in that the latter does not contain $\mathbf{x}^0$, because no approximations are needed if the conditions are originally linear. Therefore if we replace $\mathbf{x}$ by $(-\mathbf{f}_x)$ in equation (12.16c), then the solution for equations (11.8) and (12.18) is

$$\mathbf{\Delta} = (\mathbf{N} + \mathbf{W}_{xx})^{-1}(\mathbf{t} - \mathbf{W}_{xx}\,\mathbf{f}_x) \tag{12.19}$$

In evaluating $\mathbf{t}$ in equation (12.19) the appropriate value of $\mathbf{f}^0$ must be used.

It is important to say a word about the value of $\mathbf{f}_x$ during different iterations of the nonlinear solution. At the beginning of the first iteration, it is quite practical and convenient to take $\mathbf{x}^0 = \mathbf{x}$, or to assume that the approximations for the parameters are equal to their a priori "observational" estimates. This will obviously lead to the first $\mathbf{f}_x$ being equal to zero. After the first iteration is completed a value $\mathbf{\Delta}_1$ is computed, which when added to $\mathbf{x}^0$ gives an updated value for the vector of approximations. In the meantime the "observational" values $\mathbf{x}$ do not change all through the iterative process. Consequently, for the second iteration, $\mathbf{f}_x$ is no longer zero but is equal to $\mathbf{x}^0 + \mathbf{\Delta}_1 - \mathbf{x} = \mathbf{\Delta}_1$, since $\mathbf{x}^0 = \mathbf{x}$. At the beginning of the third iteration, $\mathbf{f}_x = \mathbf{\Delta}_1 + \mathbf{\Delta}_2$, and so on.

The same discussion applies also for the term $\mathbf{A}(\boldsymbol{l} - \boldsymbol{l}^0)$ in the computation $\mathbf{f}^0$ from equation (11.7), when we iterate on the observations. Although it may be zero at the beginning, it acquires finite values in succeeding iterations.

Example 12.2. It is required to fit a straight line with the equation $y - mx = 0$ through the two points (1.1, 2.1) and (2.1, 4.0). The cofactor matrix for these coordinates is the identity matrix. An a priori estimate for the one parameter (the slope of the line) is $m = 1.0$, and its a priori cofactor is $q_m = 100$. Compute the final estimate m for the line slope from this data.

Solution: It would take one point with its two coordinates, in addition to the origin, to determine the line. The second point is intended to lie on the line and therefore *only one coordinate* is needed to fix its position on the line. Hence there is *one additional observation* more than the minimum necessary to determine the geometry of the problem, that is, 1 degree of freedom. In total, $n = 4$, $n_0 = 3$, and $r = 1$. With $u = 1$ for the parameter m, then $c = 2$ conditions, one line equation for each of the two given points,

$$y_i - mx_i = 0 \qquad \text{with } i = 1, 2$$

Denoting by m^0 the approximate value for the parameter, the linearized form of this equation becomes (we choose not to iterate on the observed coordinates in this problem)

$$-m^0 v_{x_i} + v_{y_i} - x_i\ \delta = m^0 x_i - y_i$$

or

$$[-m^0 \quad 1]\mathbf{v}_i + [-x_i]\boldsymbol{\Delta} = \mathbf{f}$$

Using the given data and choosing $m^0 = m = 1$, the matrices for the two condition equations become

$$\mathbf{A} = \begin{bmatrix} -1 & 1 & 0 & 0 \\ 0 & 0 & -1 & 1 \end{bmatrix} \qquad \mathbf{B} = \begin{bmatrix} -1.1 \\ -2.1 \end{bmatrix} \qquad \mathbf{f} = \begin{bmatrix} -1.0 \\ -1.9 \end{bmatrix}$$

with

$$\mathbf{W} = \mathbf{I}, \qquad \mathbf{W}_e = \mathbf{AA}^t = \begin{bmatrix} 2 & 0 \\ 0 & 2 \end{bmatrix}, \qquad \mathbf{N} = 2.81, \qquad \text{and} \qquad \mathbf{t} = 2.545$$

Because $m^0 = m$, the value of $\mathbf{f}_x$ for the first iteration is zero according to equation (12.18). Thus with $\mathbf{W}_{xx} = 0.01$ then

$$\boldsymbol{\Delta}_1 = (2.81 + 0.01)^{-1}(2.545) = 0.902482$$

$$m_1^0 = m^0 + \boldsymbol{\Delta}_1 = 1.902482$$

$$f_{x_1} = 0.902482$$

Using these values, the second iteration becomes

$$\boldsymbol{\Delta}_2 = (1.216598 + 0.01)^{-1}(-0.005118) = -0.004172$$

For the third iteration,

$$m_2^0 = 1.898310, \qquad f_{x_2} = 0.898310, \qquad \text{and} \qquad \boldsymbol{\Delta}_3 = 0.000025$$

which is considered to be sufficiently small to terminate the iterations. The final estimate of the line slope is

$$m = 1.898310 + 0.000025 = 1.898335$$

12.4. THE UNIFIED APPROACH AND PARAMETER CONSTRAINTS

Any constraint is in a real sense a function reflecting some geometric or physical relation between a number of variables. Therefore they can practically be viewed as "conditions" containing some type of "observations." For example, constraining a distance to a given value D can be done as closely as the known value of D. In the treatments of Chapter 9 (Part II), in which equations of the type $\mathbf{C\Delta} = \mathbf{g}$ were used, the distance D would be assumed known perfectly, or with infinite weight (zero variance). In the present situation under the unified concept, we would take D as an observed quantity and choose a value for its weight commensurate with the reliability of its known value. The higher the value of its weight, the closer we get to perfect satisfaction of the constraint.

Although the example of a distance can be easily visualized, there are many other instances in which constraints do not include variables that are amenable to being considered as observations. As an example we refer to the case of constraining a set of points to lie on a line or a circle. All variables in such functions are what have been conventionally considered parameters. The only other variable may simply be the numerical zero, which would obviously be a constant. It is that constant which we now convert to an "observation," regardless of its numerical value. Consequently, in each equation conventionally known as a "constraint," the constant is replaced by an observational variable that is denoted by $\boldsymbol{l}_c$. (This would replace the distance D in the first example, and the "zero" in the second example.) In addition to the total vector of "constraint observations" $\boldsymbol{l}_c$, the parameters $\mathbf{x}$ would be the only other variables in the set of equations replacing the constraints. These equations may in general be nonlinear and therefore take the form

$$F_c(\boldsymbol{l}_c, \mathbf{x}) = \mathbf{0} \tag{12.20}$$

which is the same form as equation (12.17). Thus we can appreciate how the unified concept allows for only one general type of condition equations. In fact, with $\mathbf{x}$ also treated as a vector of observable quantities, we may combine equations (12.17) and (12.20) into one equation,

$$F_t(\boldsymbol{l}_t) = \mathbf{0} \tag{12.21}$$

in which $\boldsymbol{l}_t$ represents the total vector variables. Equation (12.21) is essentially the same as that given by equation (11.10). The only difference is that in equation (11.10) the vector $\boldsymbol{l}$ denoted only the one set of variables conventionally known as observations, whereas in (12.21) the vector $\boldsymbol{l}_t$ includes several subvectors each referring to one class of conventional variables (that is, observations, parameters, and so on). It would be possible to develop a solution directly from equation (12.21). However, as has been shown on several occasions such a direct treatment invariably leads to undue complexities in the derivation. Therefore for practical reasons as well as for convenience, it is prudent to partition the system of equation (12.21) and proceed with the development such that groups of variables usually encountered together in practice are treated as separate subsets.

Considering equations (12.17) and (12.20) together, the variables are

- $\boldsymbol{l}$ $n \times 1$ vector of conventionally designated observations ($\mathbf{Q}$ is its a priori cofactor matrix.)
- $\boldsymbol{l}_c$ $t \times 1$ vector of observations arising with the constraints ($\mathbf{Q}_{cc}$ is its a priori cofactor matrix.)
- $\mathbf{x}$ $u \times 1$ vector of variables conventionally known as parameters ($\mathbf{Q}_{xx}$ is its a priori cofactor matrix.)

There are c equations in (12.17) and s equations in (12.20). If $\boldsymbol{l}^0$, $\boldsymbol{l}_c^0$, and $\mathbf{x}^0$ denote vectors of approximations, $\mathbf{v}$, $\mathbf{v}_c$, and $\mathbf{v}_x$ denote vectors of residuals, and $\boldsymbol{\Delta}_l$, $\boldsymbol{\Delta}_c$, and $\boldsymbol{\Delta}$ represent vector of corrections to the approximate vectors, then

$$\boldsymbol{l} + \mathbf{v} = \boldsymbol{l}^0 + \boldsymbol{\Delta}_l \tag{12.22a}$$

$$\boldsymbol{l}_c + \mathbf{v}_c = \boldsymbol{l}_c^0 + \boldsymbol{\Delta}_c \tag{12.22b}$$

For the linearization of equations (12.17) and (12.20) we use the auxiliaries

$$\mathbf{A} = \frac{\partial F}{\partial \boldsymbol{l}} \quad \text{is } c \times n \text{ matrix of rank equal to } c$$

$$\mathbf{B} = \frac{\partial F}{\partial \mathbf{x}} \quad \text{is } c \times u \text{ matrix of rank equal to } u$$

$$\mathbf{A}_c = \frac{\partial F_c}{\partial \boldsymbol{l}_c} \quad \text{is } s \times t \text{ matrix of rank equal to } s$$

$$\mathbf{C} = \frac{\partial F_c}{\partial \mathbf{x}} \quad \text{is } s \times u \text{ matrix of rank equal to } s$$

Thus

$$\mathbf{A}\mathbf{v} + \mathbf{B}\boldsymbol{\Delta} = \mathbf{f}^0 \tag{12.23a}$$

$$\mathbf{A}_c\mathbf{v}_c + \mathbf{C}\boldsymbol{\Delta} = \mathbf{f}_c^0 \tag{12.23b}$$

in which

$$\mathbf{f}^0 = -[F(\boldsymbol{l}^0, \mathbf{x}^0) - \mathbf{A}(\boldsymbol{l}^0 - \boldsymbol{l})] \tag{12.24a}$$

$$f_c^0 = -[F_c(\boldsymbol{l}_c^0, \mathbf{x}^0) - \mathbf{A}_c(\boldsymbol{l}_c^0 - \boldsymbol{l}_c)] \tag{12.24b}$$

with

$$\mathbf{f}_x = \mathbf{x}^0 - \mathbf{x} \tag{12.24c}$$

then

$$\begin{bmatrix} \mathbf{A} & \mathbf{0} & \mathbf{0} \\ \mathbf{0} & \mathbf{A}_c & \mathbf{0} \\ \mathbf{0} & \mathbf{0} & \mathbf{I} \end{bmatrix} \begin{bmatrix} \mathbf{v} \\ \mathbf{v}_c \\ \mathbf{v}_x \end{bmatrix} + \begin{bmatrix} \mathbf{B} \\ \mathbf{C} \\ -\mathbf{I} \end{bmatrix} \boldsymbol{\Delta} = \begin{bmatrix} \mathbf{f}^0 \\ \mathbf{f}_c^0 \\ \mathbf{f}_x \end{bmatrix} \tag{12.25}$$

Letting, as has been customary,

$$\mathbf{W}_e = (\mathbf{AQA}^t)^{-1} \tag{12.26a}$$

$$\mathbf{W}_{ec} = (\mathbf{A}_c \mathbf{Q}_{cc} \mathbf{A}_c^t)^{-1} \tag{12.26b}$$

the normal equations for $\boldsymbol{\Delta}$ from the conditions of equation (12.25) would be

$$[\mathbf{B}^t \quad \mathbf{C}^t \quad -\mathbf{I}] \begin{bmatrix} \mathbf{W}_e & \mathbf{0} & \mathbf{0} \\ \mathbf{0} & \mathbf{W}_{ec} & \mathbf{0} \\ \mathbf{0} & \mathbf{0} & \mathbf{W}_{xx} \end{bmatrix} \begin{bmatrix} \mathbf{B} \\ \mathbf{C} \\ -\mathbf{I} \end{bmatrix} \boldsymbol{\Delta}$$

$$= [\mathbf{B}^t \quad \mathbf{C}^t \quad -\mathbf{I}] \begin{bmatrix} \mathbf{W}_e & \mathbf{0} & \mathbf{0} \\ \mathbf{0} & \mathbf{W}_{ec} & \mathbf{0} \\ \mathbf{0} & \mathbf{0} & \mathbf{W}_{xx} \end{bmatrix} \begin{bmatrix} \mathbf{f}^0 \\ \mathbf{f}_c^0 \\ \mathbf{f}_x \end{bmatrix}$$

or

$$[\mathbf{B}^t\mathbf{W}_e \mathbf{B} + \mathbf{C}^t\mathbf{W}_{ec} \mathbf{C} + \mathbf{W}_{xx}]\boldsymbol{\Delta}$$

$$= [\mathbf{B}^t\mathbf{W}_e \mathbf{f}^0 + \mathbf{C}^t\mathbf{W}_{ec} \mathbf{f}_c^0 - \mathbf{W}_{xx} \mathbf{f}_x] \tag{12.27}$$

Introducing new auxiliaries and repeating earlier ones for the sake of completeness,

$$\mathbf{N} = \mathbf{B}^t\mathbf{W}_e \mathbf{B} \tag{12.28a}$$

$$\mathbf{t} = \mathbf{B}^t\mathbf{W}_e \mathbf{f}^0 \tag{12.28b}$$

$$\mathbf{N}_c = \mathbf{C}^t\mathbf{W}_{ec} \mathbf{C} \tag{12.28c}$$

$$\mathbf{t}_c = \mathbf{C}^t\mathbf{W}_{ec} \mathbf{f}_c^0 \tag{12.28d}$$

equation (12.27) takes the shorter form,

$$[\mathbf{N} + \mathbf{N}_c + \mathbf{W}_{xx}]\boldsymbol{\Delta} = [\mathbf{t} + \mathbf{t}_c - \mathbf{W}_{xx} \mathbf{f}_x] \tag{12.29a}$$

from which

$$\boldsymbol{\Delta} = (\mathbf{N} + \mathbf{N}_c + \mathbf{W}_{xx})^{-1}(\mathbf{t} + \mathbf{t}_c - \mathbf{W}_{xx} \mathbf{f}_x) \tag{12.29b}$$

Equation (12.29b) gives the vector of estimates for corrections to parameter approximations when all variables have a priori values as well as a priori cofactor matrices. The terms $\mathbf{N}$ and $\mathbf{t}$ reflect the contribution of the conventionally known observations; the terms $\mathbf{N}_c$ and $\mathbf{t}_c$, the contribution of constraints; and the remaining terms, the contribution due to a priori knowledge of the parameters themselves.

In the forgoing development allowance was made for linearization and iteration on *all variables in the model* to make the analysis as general as possible. It should now be shown how the iterative process proceeds on this general case, and then the possible simplifications for special cases should be pointed out.

Let $\mathbf{x}^0_{(0)}$, $\boldsymbol{l}^0_{(0)}$, and $\boldsymbol{l}^0_{c(0)}$ denote the initial approximation vectors, with, of course, $\mathbf{x}$, $\boldsymbol{l}$, and $\boldsymbol{l}_c$ being the a priori (observational) values that do not change during the adjustment. With these values the matrices and vectors $\mathbf{A}$, $\mathbf{B}$, $\mathbf{A}_c$, $\mathbf{C}$, $\mathbf{f}^0$, $\mathbf{f}^0_c$, and $\mathbf{f}_x$ are numerically evaluated and used to form $\mathbf{N}$, $\mathbf{t}$, $\mathbf{N}_c$, and $\mathbf{t}_c$, which in turn are substituted in equation (12.29b) to yield the first vector $\boldsymbol{\Delta}_{(1)}$. This vector $\boldsymbol{\Delta}_{(1)}$ is then added to $\mathbf{x}^0_{(0)}$ to obtain an updated vector of approximation $\mathbf{x}^0_{(1)}$. In order to get updated vectors of approximations $\boldsymbol{l}^0_{(1)}$ and $\boldsymbol{l}^0_{c(1)}$, we need to compute the two vectors $\mathbf{v}$ and $\mathbf{v}_c$. From equation (12.8) we may easily write

$$\mathbf{v} = \mathbf{Q}\mathbf{A}^t\mathbf{W}_e(\mathbf{f}^0 - \mathbf{B}\boldsymbol{\Delta}) \tag{12.29c}$$

and in a similar manner

$$\mathbf{v}_c = \mathbf{Q}_{cc}\,\mathbf{A}^t_c\,\mathbf{W}_{ec}(\mathbf{f}^0_c - \mathbf{C}\boldsymbol{\Delta}) \tag{12.30a}$$

finally,

$$\mathbf{v}_x = \mathbf{x}^0 - \mathbf{x} + \boldsymbol{\Delta} = \mathbf{f}_x + \boldsymbol{\Delta} \tag{12.30b}$$

Using the value $\boldsymbol{\Delta}_{(1)}$ in equations (12.30a) and (12.30b), the first two values $\mathbf{v}_{(1)}$ and $\mathbf{v}_{c(1)}$ can be computed. These may be used in equations (12.22a) and (12.22b) to compute the correction vectors $\boldsymbol{\Delta}_{l(1)}$ and $\boldsymbol{\Delta}_{c(1)}$ for the corresponding approximation vectors $\boldsymbol{l}^0_{(0)}$ and $\boldsymbol{l}^0_{c(0)}$. These relations in recursive form are

$$\boldsymbol{\Delta}_{l(i+1)} = \boldsymbol{l} + \mathbf{v}_{(i+1)} - \boldsymbol{l}^0_{(i)} \tag{12.30c}$$

$$\boldsymbol{\Delta}_{c(i+1)} = \boldsymbol{l}_c + \mathbf{v}_{c(i+1)} - \boldsymbol{l}^0_{c(i)} \tag{12.30d}$$

in which we must note that $\boldsymbol{l}$ and $\boldsymbol{l}_c$ are the a priori "observations" that remain constant all through the iterations. Equations (12.30c) and (12.30d) may be rearranged to give a more convenient form from which the updated approximations $\boldsymbol{l}^0$ and $\boldsymbol{l}^0_c$ may be obtained directly.

$$\boldsymbol{l}^0_{(i+1)} = \boldsymbol{l}^0_{(i)} + \boldsymbol{\Delta}_{l(i+1)} = \boldsymbol{l} + \mathbf{v}_{(i+1)} \tag{12.30e}$$

$$\boldsymbol{l}^0_{c(i+1)} = \boldsymbol{l}^0_{c(i)} + \boldsymbol{\Delta}_{c(i+1)} = \boldsymbol{l}_c + \mathbf{v}_{c(i+1)} \tag{12.30f}$$

Thus with $\mathbf{v}_{(1)}$ and $\mathbf{v}_{c(1)}$ we can readily get, from equations (12.30e) and (12.30f) the vectors $\boldsymbol{l}^0_{(1)}$ and $\boldsymbol{l}^0_{c(1)}$, respectively. Given these, together with $\mathbf{x}^0_{(1)}$ computed before, we may proceed to re-evaluate all matrices and vectors leading to computing a new vector $\boldsymbol{\Delta}_{(2)}$. The procedure of obtaining a new $\mathbf{x}^0_{(2)}$, $\boldsymbol{l}^0_{(2)}$, and $\boldsymbol{l}^0_{c(2)}$ is repeated in its entirety. At the end of each iteration a criterion, established a priori, is checked to see if the process is to be continued or stopped (see section 11.4). When such criterion is satisfied and the iterations terminated, we get the final estimates,

$$\hat{\mathbf{x}} = \mathbf{x}^0_{(0)} + \sum_j \boldsymbol{\Delta}_{(j)} \tag{12.31a}$$

$$\hat{l} = l + \mathbf{v}_{(f)} \tag{12.31b}$$

$$\hat{l}_c = l_c + \mathbf{v}_{c(f)} \tag{12.31c}$$

(f) denotes final value.

12.4.1. Special Cases 1. If all functions are originally *linear*, then

$$l^0 = l^0_c = \mathbf{x}^0 = \mathbf{0} \tag{12.32a}$$

$$\mathbf{f}^0 = \mathbf{f} = \mathbf{d} - \mathbf{A}l \tag{12.32b}$$

$$\mathbf{f}^0_c = \mathbf{f}_c = \mathbf{d}_c - \mathbf{A}_c l_c \tag{12.32c}$$

$$\mathbf{f}_x = -\mathbf{x} \tag{12.32d}$$

$$\boldsymbol{\Delta} = (\mathbf{N} + \mathbf{N}_c + \mathbf{W}_{xx})^{-1}(\mathbf{t} + \mathbf{t}_c + \mathbf{W}_{xx}\,\mathbf{x}) \tag{12.32e}$$

noting that in computing $\mathbf{t}$ and $\mathbf{t}_c$, we use $\mathbf{f}^0$ and $\mathbf{f}^0_c$ as evaluated from equations (12.32b) and (12.32c).

2. If the functions are *linear* and there are no *constraints*, then equation (12.32e) reduces to equation (12.16c), or

$$\boldsymbol{\Delta} = (\mathbf{N} + \mathbf{W}_{xx})^{-1}(\mathbf{t} + \mathbf{W}_{xx}\,\mathbf{x}) \tag{12.16c}$$

by dropping the two terms $\mathbf{N}_c$ and $\mathbf{t}_c$.

3. For *nonlinear* functions *without constraints*, equation (12.33b) becomes the same as equation (12.19), or

$$\boldsymbol{\Delta} = (\mathbf{N} + \mathbf{W}_{xx})^{-1}(\mathbf{t} - \mathbf{W}_{xx}\,\mathbf{f}_x) \tag{12.19}$$

If we iterate on the observations, we compute $\mathbf{t}$ using $\mathbf{f}^0$ from equation (12.24a). If we do not iterate on the observations, a situation that occurs often in practice, then

$$\mathbf{f} = -\mathbf{F}(l, \mathbf{x}^0) \tag{11.9}$$

must be used in evaluating $\mathbf{t}$. The iterative process would involve computing successive values for $\boldsymbol{\Delta}$ only, whereas $\mathbf{v}$ may be computed only at the last iteration, unless it is needed for the criterion of termination of the iterations.

Example 12.3. Consider the problem of Example 9.5 and rework it to show how the unified approach accommodates parameter constraints by changing them to condition equations with appropriate weight matrices. For convenience first recall the data,

$$\mathbf{A}\mathbf{v} + \mathbf{B}\boldsymbol{\Delta} = \mathbf{f} \qquad \text{with } \mathbf{Q} = \mathbf{I}$$

$$\mathbf{D}_1\boldsymbol{\Delta} + \mathbf{D}_2\boldsymbol{\Delta}' = \mathbf{h}$$

where

$$\mathbf{A} = \mathbf{I}_3 \qquad \underset{3,\,2}{\mathbf{B}} = \begin{bmatrix} 2 & -3 \\ -1 & 2 \\ 0 & 1 \end{bmatrix} \qquad \underset{3,\,1}{\mathbf{f}} = \begin{bmatrix} -1.1 \\ 1.2 \\ 1.0 \end{bmatrix} \qquad \mathbf{h} = \begin{bmatrix} -1 \\ 3 \end{bmatrix}$$

$$\mathbf{D}_1 = \begin{bmatrix} 1 & -1 \\ 2 & -1 \end{bmatrix} \qquad \mathbf{D}_2 = \begin{bmatrix} 1 \\ -2 \end{bmatrix} \qquad \mathbf{\Delta} = \begin{bmatrix} x_1 \\ x_2 \end{bmatrix} \qquad \mathbf{\Delta}' = x_3$$

Rewrite the equations in the general form

$$\mathbf{A}\mathbf{v} + [\mathbf{B} \quad \mathbf{0}]\begin{bmatrix} \mathbf{\Delta} \\ \mathbf{\Delta}' \end{bmatrix} = \mathbf{f} \qquad \text{or} \qquad \mathbf{A}\mathbf{v} + \tilde{\mathbf{B}}\tilde{\mathbf{\Delta}} = \mathbf{f}$$

$$\mathbf{A}_c\,\mathbf{v}_c + [\mathbf{D}_1 \quad \mathbf{D}_2]\begin{bmatrix} \mathbf{\Delta} \\ \mathbf{\Delta}' \end{bmatrix} = \mathbf{h} = \mathbf{f}_c \qquad \text{or} \qquad \mathbf{A}_c\,\mathbf{v}_c + \tilde{\mathbf{C}}\tilde{\mathbf{\Delta}} = \mathbf{f}_c$$

with obvious correspondence in terms. The numerical values of these matrices are

$$\mathbf{A}_c = \mathbf{I}$$

$$\underset{3,\,3}{\tilde{\mathbf{B}}} = \begin{bmatrix} 2 & -3 & 0 \\ -1 & 2 & 0 \\ 0 & 1 & 0 \end{bmatrix} \qquad \underset{2,\,3}{\tilde{\mathbf{C}}} = \begin{bmatrix} 1 & -1 & 1 \\ 2 & -1 & -2 \end{bmatrix}$$

and $\mathbf{f}$ and $\mathbf{f}_c$ ($= \mathbf{h}$) are as given above. Realizing that this is a linear case, the solution for $\tilde{\mathbf{\Delta}} = [x_1 \quad x_2 \quad x_3]'$ may be obtained by applying equation (12.32e), modified slightly to accommodate the fact that there were no a priori values for those three variables. Thus with $\mathbf{W}_{xx} = \mathbf{0}$ and $\mathbf{x} = \mathbf{0}$, we have

$$\mathbf{\Delta} = (\mathbf{N} + \mathbf{N}_c)^{-1}(\mathbf{t} + \mathbf{t}_c)$$

which, because $\mathbf{A} = \mathbf{A}_c = \mathbf{I}$, may be expanded to

$$\tilde{\mathbf{\Delta}} = (\tilde{\mathbf{B}}'\mathbf{W}\tilde{\mathbf{B}} + \tilde{\mathbf{C}}'\mathbf{W}_{cc}\,\tilde{\mathbf{C}})^{-1}(\tilde{\mathbf{B}}'\mathbf{W}\mathbf{f} + \tilde{\mathbf{C}}'\mathbf{W}_{cc}\,\mathbf{f}_c)$$

with

$$\mathbf{W} = \mathbf{Q}^{-1} \qquad \text{and} \qquad \mathbf{W}_{cc} = \mathbf{Q}_{cc}^{-1}$$

The relative magnitudes of $\mathbf{W}$ and $\mathbf{W}_{cc}$ comprise the mechanism by which we can insure that the constraints are treated properly. Since $\mathbf{W} = \mathbf{I}$ as given in the data of the problem, we are free to choose a suitable value for $\mathbf{W}_{cc}$. The ideal case is when $\mathbf{W}_{cc}$ is infinitely large relative to $\mathbf{W}$. In order to determine which value of $\mathbf{W}_{cc}$ is "practically" infinite, consider several possibilities. Just for the sake of establishing a trend, begin by the rather unrealistic value of $\mathbf{W}_{cc} = \mathbf{I}$ as well. With these two values, the above equation gives

$$[x_1 \quad x_2 \quad x_3]_1 = [1.04486 \quad 1.07383 \quad -0.98785]$$

The answer is clearly different from the correct one obtained in Example 9.5, a fact that should have been expected. Next, use $\mathbf{W}_{cc} = 10\mathbf{I}$ and get

$$[x_1 \quad x_2 \quad x_3]_2 = [1.06000 \quad 1.08188 \quad -0.98038]$$

which is already closer to the correct answer. Further, with $\mathbf{W}_{cc} = 100\mathbf{I}$

$$[x_1 \quad x_2 \quad x_3]_3 = [1.06210 \quad 1.08299 \quad -0.97934]$$

which may be considered as essentially the same as the correct answer. If, however, a closer set of values is desired, we may increase $\mathbf{W}_{cc}$ to $1000\mathbf{I}$ or even more, depending on the accuracy of the required final answer.

12.5. PRECISION ESTIMATION

Applying the propagation principle to equation (12.29b), it can be readily shown that

$$\mathbf{Q}_{\Delta\Delta} = (\mathbf{N} + \mathbf{N}_c + \mathbf{W}_{xx})^{-1} \tag{12.33a}$$

The fact that $\mathbf{Q}_{\Delta\Delta}$ is the inverse of the sum of three weight matrices should seem reasonable, since each matrix represents the contribution of one of the three sources of information. The matrix $\mathbf{W}_{xx}$ expresses what is already known about the parameters before the adjustment, and $\mathbf{N}$ and $\mathbf{N}_c$ express the contributions of the conditions and constraints, respectively.

Equation (12.33a) reduces to already known relationships when different restrictions are enforced. For example, when $\mathbf{W}_{xx} = \mathbf{N}_c = \mathbf{0}$, which means that there are no a priori parameter information and no constraints, then

$$\mathbf{Q}_{\Delta\Delta} = \mathbf{N}^{-1} \tag{12.33b}$$

which is the same as equation (6.29).

If there are no constraints, that is, if $\mathbf{N}_c = \mathbf{0}$, but the a priori weight matrix for the parameters $\mathbf{W}_{xx}$ is given, then

$$\mathbf{Q}_{\Delta\Delta} = (\mathbf{N} + \mathbf{W}_{xx})^{-1} \tag{12.33c}$$

A third interesting case is when $\mathbf{W}_{xx} = \mathbf{0}$ but $\mathbf{N}_c \neq 0$, for which

$$\mathbf{Q}_{\Delta\Delta} = (\mathbf{N} + \mathbf{N}_c)^{-1} \tag{12.33d}$$

This equation can also be shown to reduce to an earlier form given the proper assumption. Rewriting equation (12.33d) in an expanded form in view of equation (12.28c), then

$$\mathbf{Q}_{\Delta\Delta} = (\mathbf{N} + \mathbf{C}^t\mathbf{W}_{ec}\,\mathbf{C})^{-1} \tag{12.34a}$$

which may be expanded further, by applying equations (A68) and (A69) (Appendix A), to

$$\mathbf{Q}_{\Delta\Delta} = \mathbf{N}^{-1} - \mathbf{N}^{-1}\mathbf{C}^t(\mathbf{Q}_{ec} + \mathbf{C}\mathbf{N}^{-1}\mathbf{C}^t)^{-1}\mathbf{C}\mathbf{N}^{-1} \tag{12.34b}$$

In equation (12.34b), the matrix $\mathbf{Q}_{ec}$ is the inverse of $\mathbf{W}_{ec}$, which is from equation (12.26b)

$$\mathbf{Q}_{ec} = \mathbf{A}_c \mathbf{Q}_{cc} \mathbf{A}_c^t \tag{12.34c}$$

Conventional treatment of constraints requires those equations to be strictly satisfied. This is equivalent to saying that $\mathbf{Q}_{cc} = \mathbf{0}$ (or equivalently $\mathbf{W}_{cc} \to \infty$), which leads to having $\mathbf{Q}_{ec} = \mathbf{0}$ from equation (12.34c). With the vanishing of $\mathbf{Q}_{ec}$, equation (12.34b) may be rewritten as

$$\mathbf{Q}_{\Delta\Delta} = \mathbf{N}^{-1}[\mathbf{I} - \mathbf{C}^t(\mathbf{C}\mathbf{N}^{-1}\mathbf{C}^t)^{-1}\mathbf{C}\mathbf{N}^{-1}] \tag{12.34d}$$

which is identical to equation (9.15b) derived for the conventional case.

The next cofactor matrix to evaluate is that for *all* the observational residuals $\mathbf{v}_t$, which is denoted by $\mathbf{Q}_{v_t v_t}$. To derive this matrix, we first rewrite equation (12.25) in a more compact form

$$\mathbf{A}_t \mathbf{v}_t + \mathbf{B}_t \Delta = \mathbf{f}_t \tag{12.35a}$$

with associated total cofactor matrix,

$$\mathbf{Q}_{tt} = \begin{bmatrix} \mathbf{Q} & \mathbf{0} & \mathbf{0} \\ \mathbf{0} & \mathbf{Q}_{cc} & \mathbf{0} \\ \mathbf{0} & \mathbf{0} & \mathbf{Q}_{xx} \end{bmatrix} \tag{12.35b}$$

Applying equation (6.30)

$$\mathbf{Q}_{v_t v_t} = \mathbf{Q}_{tt} \mathbf{A}_t^t \mathbf{W}_{et} \mathbf{A}_t \mathbf{Q}_{tt} - \mathbf{Q}_{tt} \mathbf{A}_t^t \mathbf{W}_{et} \mathbf{B}\mathbf{Q}_{\Delta\Delta} \mathbf{B}^t \mathbf{W}_{et} \mathbf{A}_t \mathbf{Q}_{tt} \tag{12.36}$$

in which

$$\mathbf{W}_{et} = (\mathbf{A}_t \mathbf{Q}_{tt} \mathbf{A}_t^t)^{-1} = \begin{bmatrix} (\mathbf{A}\mathbf{Q}\mathbf{A}^t)^{-1} & \mathbf{0} & \mathbf{0} \\ \mathbf{0} & (\mathbf{A}_c \mathbf{Q}_{cc} \mathbf{A}_c^t)^{-1} & \mathbf{0} \\ \mathbf{0} & \mathbf{0} & \mathbf{Q}_{xx}^{-1} \end{bmatrix}$$

$$= \begin{bmatrix} \mathbf{W} & \mathbf{0} & \mathbf{0} \\ \mathbf{0} & \mathbf{W}_{cc} & \mathbf{0} \\ \mathbf{0} & \mathbf{0} & \mathbf{W}_{xx} \end{bmatrix} \tag{12.37}$$

Thus

$$\mathbf{Q}_{v_t v_t} = \begin{bmatrix} \mathbf{Q}_{vv} & \mathbf{Q}_{vv_c} & \mathbf{Q}_{vv_x} \\ & \mathbf{Q}_{v_c v_c} & \mathbf{Q}_{v_c v_x} \\ \text{symmetric} & & \mathbf{Q}_{v_x v_x} \end{bmatrix}$$

$$= \begin{bmatrix} \begin{matrix} \mathbf{Q}\mathbf{A}^t\mathbf{W}_e\mathbf{A}\mathbf{Q} \\ -\mathbf{Q}\mathbf{A}^t\mathbf{W}_e\mathbf{B}\mathbf{Q}_{\Delta\Delta}\mathbf{B}^t\mathbf{W}_e\mathbf{A}\mathbf{Q}] \end{matrix} & [-\mathbf{Q}\mathbf{A}^t\mathbf{W}_e\mathbf{B}\mathbf{Q}_{\Delta\Delta}\mathbf{C}^t\mathbf{W}_{ec}\mathbf{A}_c\mathbf{Q}_{cc}] & [\mathbf{Q}\mathbf{A}^t\mathbf{W}_e\mathbf{B}\mathbf{Q}_{\Delta\Delta}] \\ & \begin{matrix} [\mathbf{Q}_{cc}\mathbf{A}_c^t\mathbf{W}_{ec}\mathbf{A}_c\mathbf{Q}_{cc} \\ -\mathbf{Q}_{cc}\mathbf{A}_c^t\mathbf{W}_{ec}\mathbf{C}\mathbf{Q}_{\Delta\Delta}\mathbf{C}^t\mathbf{W}_{ec}\mathbf{A}_c\mathbf{Q}_{cc}] \end{matrix} & [\mathbf{Q}_{cc}\mathbf{A}_c^t\mathbf{W}_{ec}\mathbf{C}\mathbf{Q}_{\Delta\Delta}] \\ \text{symmetric} & & [\mathbf{Q}_{xx} - \mathbf{Q}_{\Delta\Delta}] \end{bmatrix} \tag{12.38}$$

Equation (12.38) is rather general in as much as it gives not only autocofactor matrices for all three residual vectors, $\mathbf{v}$, $\mathbf{v}_c$, and $\mathbf{v}_x$, but also their crosscofactor matrices (for example, $\mathbf{Q}_{vv_c}$, $\mathbf{Q}_{vv_x}$, and so on). The total cofactor matrix of the estimated "observations" may readily be obtained from

$$\hat{\mathbf{Q}}_{tt} = \mathbf{Q}_{tt} - \mathbf{Q}_{v_tv_t} = \begin{bmatrix} \mathbf{Q}_{\hat{l}\hat{l}} & \mathbf{Q}_{\hat{l}\hat{l}_c} & \mathbf{Q}_{\hat{l}\hat{x}} \\ & \mathbf{Q}_{\hat{l}_c\hat{l}_c} & \mathbf{Q}_{\hat{l}_c\hat{x}} \\ \text{symmetric} & & \mathbf{Q}_{\hat{x}\hat{x}} \end{bmatrix}$$

$$= \begin{bmatrix} (\mathbf{Q} - \mathbf{Q}_{vv}) & -\mathbf{Q}_{vv_c} & -\mathbf{Q}_{vv_x} \\ & (\mathbf{Q}_{cc} - \mathbf{Q}_{v_cv_c}) & -\mathbf{Q}_{v_cv_x} \\ \text{symmetric} & & (\mathbf{Q}_{xx} - \mathbf{Q}_{v_xv_x}) \end{bmatrix} \tag{12.39}$$

where the terms such as $\mathbf{Q}_{vv}$, $\mathbf{Q}_{vv_c}$, and so on, are obtainable from equation (12.38). Similar to equation (12.38), equation (12.39) is also of general nature such that various relationships derived earlier can be extracted from it. From example, the (1, 1) term (on the first row and first column) in equation (12.39) gives directly the familiar relation $\mathbf{Q}_{\hat{l}\hat{l}} = \mathbf{Q} - \mathbf{Q}_{vv}$. Another example is the crosscofactor $\mathbf{Q}_{\hat{l}\hat{x}}$ in the (1, 3) term. Recognizing that $\hat{\mathbf{x}} = \mathbf{\Delta}$ for the conventional treatment of linear functions given in Chapter 6 (Part II), then

$$\mathbf{Q}_{\hat{l}\Delta} = \mathbf{Q}_{\hat{l}\hat{x}} = -\mathbf{Q}_{vv_x} = -\mathbf{Q}\mathbf{A}^t\mathbf{W}_e\,\mathbf{B}\mathbf{Q}_{\Delta\Delta}$$

which, in view of equation (6.29) becomes

$$\mathbf{Q}_{\hat{l}\Delta} = -\mathbf{Q}\mathbf{A}^t\mathbf{W}_e\,\mathbf{B}\mathbf{N}^{-1} \tag{12.40}$$

Equation (12.40) gives a result that is identical to that given by equation (6.32).

Example 12.4. Compute the a posteriori cofactor matrices for the numerical Examples 12.1, 12.2, and 12.3.

Solution:

1. *The Problem of Example 12.1:* The cofactor of the elevation of point p is

$$\mathbf{Q}_{\Delta\Delta} = (\mathbf{N} + \mathbf{W}_{xx})^{-1} = (2.01)^{-1} = 0.498$$

2. *The Problem of Example 12.2:* The problem here was of fitting a straight line through the origin and two given points. A priori value for the one parameter, the slope of the line, was given. Thus this is a case of a nonlinear problem with a priori parameter data but no constraints. Given a priori $\mathbf{W}_{xx} = 0.01$, the solution was iterated three times in Example 12.2, for which we give the corresponding $\mathbf{W}_{\Delta\Delta}$,

$$\mathbf{W}_{\Delta\Delta(1)} = 2.81 + 0.01 = 2.82 \qquad \text{after the first iteration}$$

$$\mathbf{W}_{\Delta\Delta(2)} = 1.216598 + 0.01 = 1.226598 \qquad \text{after the second iteration}$$

$$\mathbf{W}_{\Delta\Delta(3)} = 1.220788 + 0.01 = 1.230788 \qquad \text{after the third iteration}$$

We have given the computed values for all three iterations in order to illustrate a point. It is quite important when iterating a nonlinear problem to make sure that $\mathbf{Q}_{\Delta\Delta}$ or $\mathbf{W}_{\Delta\Delta}$ is computed from the last set of normal equations. Otherwise although the estimate of the vector itself, $\boldsymbol{\Delta}$, may be correct, the estimate of the associated cofactor or weight matrix can be considerably in error as shown by the above set of values.

3. *The Problem of Example 12.3* Here we had linear conditions, linear constraints, but no a priori $\mathbf{W}_{xx}$ for the parameters. According to the symbols in Example 12.3, then,

$$\mathbf{Q}_{xx} = \mathbf{Q}_{\bar{\Delta}\bar{\Delta}} = [\tilde{\mathbf{B}}^t(\mathbf{W})\tilde{\mathbf{B}} + \tilde{\mathbf{C}}^t(\mathbf{W}_{cc})\tilde{\mathbf{C}}]^{-1}$$

with

$$\tilde{\mathbf{B}} = \begin{bmatrix} 2 & -3 & 0 \\ -1 & 2 & 0 \\ 0 & 1 & 0 \end{bmatrix} \qquad \tilde{\mathbf{C}} = \begin{bmatrix} 1 & -1 & 1 \\ 2 & -1 & -2 \end{bmatrix} \qquad \text{and} \qquad \mathbf{W} = \mathbf{I}$$

The computed result will depend on the choice for a value of $\mathbf{W}_{cc}$. If we write a general expression as $\mathbf{W}_{cc} = \alpha\mathbf{I}$, then

$$\mathbf{Q}_{xx} = \left[\begin{bmatrix} 5 & -8 & 0 \\ -8 & 14 & 0 \\ 0 & 0 & 0 \end{bmatrix} + \alpha \begin{bmatrix} 5 & -3 & -3 \\ -3 & 2 & 1 \\ -3 & 1 & 5 \end{bmatrix}\right]^{-1}$$

With an $\alpha = 10$,

$$\mathbf{Q}_{xx(1)} = \begin{bmatrix} 505 & -308 & -300 \\ -308 & 214 & 100 \\ -300 & 100 & 500 \end{bmatrix}^{-1} = \begin{bmatrix} 0.125488 & 0.160414 & 0.043208 \\ & 0.210219 & 0.054204 \\ \text{symmetric} & & 0.017084 \end{bmatrix}$$

With an $\alpha = 100$,

$$\mathbf{Q}_{xx(2)} = \begin{bmatrix} 5005 & -3008 & -3000 \\ & 2014 & 1000 \\ \text{symmetric} & & 5000 \end{bmatrix}^{-1}$$

$$= \begin{bmatrix} 0.117745 & 0.156300 & 0.039886 \\ & 0.208033 & 0.052173 \\ \text{symmetric} & & 0.013397 \end{bmatrix}$$

Finally, with an $\alpha = 1000$,

$$\mathbf{Q}_{xx(3)} = \begin{bmatrix} 50005 & -30008 & -30000 \\ -30008 & 20014 & 10000 \\ -30000 & 10000 & 50000 \end{bmatrix}^{-1}$$

$$= \begin{bmatrix} 0.117133 & 0.156108 & 0.039058 \\ & 0.208107 & 0.052043 \\ \text{symmetric} & & 0.013046 \end{bmatrix}$$

The values in $\mathbf{Q}_{xx(1)}$ agree with the proper values computed in Example 9.5 only to the first decimal place. As the value of α increases to bring $\mathbf{W}_{cc}$ closer to what it should be (infinity in the limit), the values computed from the present unified approach get closer and closer to the answer by conventional means.

12.6. THE REFERENCE VARIANCE

The computation of the a posteriori estimate of the reference variance is similar here to the other cases covered in Part II. We need to compute the total quadratic form ($\mathbf{v}_t^t \mathbf{W}_{tt} \mathbf{v}_t$) and divide it by the number of degrees of freedom in the adjustment. Since all the variables are considered as observations, the situation may be treated in the same way as the limited case of adjustment of observations only, covered in Chapter 7 (Part II). The quadratic form in the present case, however, is formed of several parts corresponding to the natural division of variables as they are presented in practice. For example, referring to equation (12.25) the quadratic form may be written as

$$\phi = \mathbf{v}^t\mathbf{W}\mathbf{v} + \mathbf{v}_c^t \mathbf{W}_{cc} \mathbf{v}_c + \mathbf{v}_x^t \mathbf{W}_{xx} \mathbf{v}_x \tag{12.41}$$

As regards to the number of degrees of freedom, this is often determined prior to the writing of the condition (and constraint) equations. A direct way to compute it is often to take the difference between the total number of equations and the number of unknowns.

Example 12.5. Consider the simple problem of the one-loop level net given in Example 12.1. If we assume for the moment that there was no a priori value for point P, the problem may be solved quickly by the method of adjustment of observations only:

$$[1 \quad -1]\mathbf{v} = 0.02$$

from which

$$\mathbf{v} = [0.01 \quad -0.01]^t$$

Then

$$\phi_1 = \mathbf{v}^t\mathbf{W}\mathbf{v} = \mathbf{v}^t\mathbf{v} = 2 \times 10^{-4}$$

and with 1 degree of freedom

$$\hat{\sigma}_{01}^2 = \phi_1 = 2 \times 10^{-4} \text{ m}^2$$

From the unified approach solution

$$[v_1 \quad v_2 \quad v_x] = [0.0095 \quad -0.0105 \quad 0.1095] \text{ m}$$

and the quadratic form becomes

$$\phi_2 = \mathbf{v}^t\mathbf{W}\mathbf{v} + \mathbf{v}_x^t \mathbf{W}_{xx} \mathbf{v}_x = 2.005 \times 10^{-4} + 1.199 + 10^{-4} = 3.204 \times 10^{-4} \text{ m}^2$$

The degrees of freedom in the case of the unified treatment is equal to 2. Consequently, the estimate of the reference variance would be

$$\hat{\sigma}_{02}^2 = \frac{\phi_2}{2} = 1.602 \times 10^{-4} \text{ m}^2$$

The value of $\hat{\sigma}_{02}^2$ is not substantially different from $\hat{\sigma}_{01}^2$ because the a priori estimates of the elevation of point P and its variance were consistent. Therefore their treatment as observations was quite realistic. If on the other hand the a priori estimates were selected in such a way as to allow for the convenience of applying the unified approach, then we must be careful in computing and interpreting the a posteriori reference variance. For example, if $\mathbf{W}_{xx} = 10^{-6}$ were selected, then

$$\phi_2 = 2.005 \times 10^{-4} + 0.012 \times 10^{-6} = 2.00512 \times 10^{-4}$$

This value is almost identical to ϕ_1, but if we use 2 degrees of freedom, then

$$\hat{\sigma}_{02}^2 = 1.003 \times 10^{-4} \text{ m}^2$$

which is considerably different from $\hat{\sigma}_{01}^2$. Therefore it is advisable that we carefully analyze the given data, particularly when evaluating the degrees of freedom.

12.7. REVIEW EXAMPLES

In this section some of the problems used in the examples in Part II will be solved by applying the techniques of the unified approach developed in this chapter. In each of the examples to follow, several possibilities will be selected for the weight matrices of the parameters and the constraints, whenever applicable. The objective of these variations is to show how the solution vector behaves and when we can assume it to be the same as that obtained from the classical procedures.

Example 12.6. In this example, consider the problem of the three photos with two points given in Example 9.2, but without the distance constraint. Refer to Figure 9.1 and the data given, which consist of eight observations and their corresponding covariance matrix. The four parameters, representing the coordinates of the two given points A, B should have a priori observational values. Let these values be equal to the approximations used in Example 9.2 in order to be able to compare the results directly,

$$x_{1a} = 8.00 \text{ m} \qquad x_{2a} = 51.00 \qquad x_{1b} = 7.00 \text{ m} \qquad x_{2b} = 41.00 \text{ m}$$

Associated with these values we should have an a priori cofactor matrix $\mathbf{Q}_{xx}$. We shall choose different values for this matrix and compare the results. These results are summarized in Table 12-1

Included in the table are four cases in which a priori $\mathbf{Q}_{xx}$ takes values from $10\mathbf{I}$ to $10^4\mathbf{I}$, as well as the answers from the classical solution of Example 9.2. Note that with

TABLE 12-1

(m)	$\mathbf{Q}_{xx} = 10\mathbf{I}$	$\mathbf{Q}_{xx} = 10^2\mathbf{I}$	$\mathbf{Q}_{xx} = 10^3\mathbf{I}$	$\mathbf{Q}_{xx} = 10^4\mathbf{I}$	CLASSICAL $\mathbf{Q}_{xx} = \infty\mathbf{I}$
$\hat{x}_{1a}$	6.9954	6.9952	6.9952	6.9952	6.9952
$\hat{x}_{2a}$	49.7243	49.7181	49.7175	49.7174	49.7174
$\hat{x}_{1b}$	6.9820	6.9817	6.9816	6.9816	6.9816
$\hat{x}_{2b}$	41.9698	41.9698	41.9698	41.9698	41.9698

$\mathbf{Q}_{xx} = 10\mathbf{I}$ the estimates of the coordinates agree within less than 1 cm with those from the ultimate classical solution. This should be expected since a $\mathbf{Q}_{xx} = 10\mathbf{I}$ is already several order of magnitude larger than the a priori $\mathbf{Q}$, and since the elements of the latter corresponding to measurements in the object space are 0.0025 m^2. This is a ratio of 10/0.0025 or 4000 : 1, which is certainly large enough to yield such close answers. As the magnitude of $\mathbf{Q}_{xx}$ is increased relative to $\mathbf{Q}$, the answers get closer and closer to those from the classical procedure. It should be noted though that the gain in proximity of the results is not proportional to the increase in the value of $\mathbf{Q}_{xx}$. Finally, the choice of $\mathbf{Q}_{xx} = 10^4\mathbf{I}$ gives identical results to within 0.1 mm in the object. The implication of this is that $(10^4\mathbf{I})$ is practically a set of infinite values as far as this particular example is concerned.

As regards to the a posteriori estimate of the cofactor matrix of the parameters, that is, $\mathbf{Q}_{\Delta\Delta}$, the values obtained behaved in much the same way as the vector of estimates $\hat{\mathbf{x}}$ in Table 12-1, with $\mathbf{Q}_{\Delta\Delta}$ being identical to $\mathbf{Q}_{\Delta\Delta(f)}$ in Example 9.2 when $\mathbf{Q}_{xx} = 10^4\mathbf{I}$.

Example 12.7. In this example consider the same problem of the preceding example after introducing the distance constraint. This constraint seeks to enforce the distance between the two points A and B to be equal to 7.80 m. Here we need to select two a priori cofactor matrices, $\mathbf{Q}_{xx}$ for the four coordinates and $\mathbf{Q}_{cc}$ for the constraints. Running through five values of $\mathbf{Q}_{xx}$, from $10\mathbf{I}$ to $10^5\mathbf{I}$, a value $\mathbf{Q}_{cc} = 0.0001$ associated with the linearized constraint equation was chosen. This means an a priori variance of 0.0001 m^2 (since a priori σ_0^2 is 1) to the given distance. The results are summarized in Table 12-2. These results show that the choice of $\mathbf{Q}_{cc} = 0.0001$ m^2 is quite realistic since the answers by classical means are obtained to within less than 0.1 mm in the object space when $\mathbf{Q}_{xx}$ reached the value of $10^3\mathbf{I}$.

TABLE 12-2

	$\mathbf{Q}_{cc} = 0.0001$					$\mathbf{Q}_{cc} = 0$
$\mathbf{Q}_{xx} =$	$10\mathbf{I}$	$10^2\mathbf{I}$	$10^3\mathbf{I}$	$10^4\mathbf{I}$	$10^5\mathbf{I}$	CLASSICAL $\mathbf{Q}_{xx} = \infty\mathbf{I}$
$\hat{x}_{1a}$	6.99463	6.99437	6.99434	6.99434	6.99434	6.99434
$\hat{x}_{2a}$	49.75862	49.75716	49.75701	49.75700	49.75700	49.75700
$\hat{x}_{1b}$	6.98352	6.98334	6.98333	6.98332	6.98332	6.98332
$\hat{x}_{2b}$	41.95863	41.95717	41.95702	41.95700	41.95700	41.95700

A posteriori cofactor matrix of the parameters $\mathbf{Q}_{\Delta\Delta}$ was also monitored in this example, and the conclusions obtained are essentially the same as those given in the preceding example.

Example 12.8. In this example we rework, using the unified approach, the problem of three photos containing four points given in Example 9.6, without applying the circle constraints. Figure 9.3 shows the schematic of the problem followed by the pertinent data. For the a priori "observational" values for the eight parameters (the coordinates of the four points), select for simplicity the approximations used in Example 9.6, which are

$$x_{1a} = 8.00 \text{ m} \qquad x_{2a} = 51.00 \text{ m} \qquad x_{1b} = 7.00 \text{ m} \qquad x_{2b} = 41.00 \text{ m}$$

$$x_{1c} = 4.00 \text{ m} \qquad x_{2c} = 47.00 \text{ m} \qquad x_{1d} = 4.00 \text{ m} \qquad x_{2d} = 48.00 \text{ m}$$

The a priori cofactor matrix will be selected at different values in order to observe how closely the final estimates approach the correct answers obtained from the classical solution. The results of these different possibilities are summarized in Table 12-3.

TABLE 12-3

$\mathbf{Q}_{xx} =$ (m)	$10\mathbf{I}$	$10^2\mathbf{I}$	$10^3\mathbf{I}$	CLASSICAL $\infty\mathbf{I}$
$\hat{x}_{1a}$	6.9972	6.9970	6.9969	6.9969
$\hat{x}_{2a}$	49.7221	49.7152	49.7145	49.7144
$\hat{x}_{1b}$	6.9839	6.9834	6.9833	6.9833
$\hat{x}_{2b}$	41.9581	41.9675	41.9674	41.9674
$\hat{x}_{1c}$	2.9955	2.9955	2.9955	2.9955
$\hat{x}_{2c}$	46.1555	46.1511	46.1507	46.1506
$\hat{x}_{1d}$	4.0318	4.0814	4.0313	4.0313
$\hat{x}_{2d}$	48.6448	48.6441	48.6440	48.6440

Similar to the situation encountered in Example 12.6, with $\mathbf{Q}_{xx} = 10^3\mathbf{I}$, we get the same answers (to within 0.1 mm in the object space) as those obtained classically in Example 9.6.

The a posteriori matrix $\mathbf{Q}_{\Delta\Delta}$ was also computed for several cases and the value obtained for the case of $\mathbf{Q}_{xx} = 10^3\mathbf{I}$ was practically the same as $\mathbf{Q}_{\Delta\Delta(f)}$ computed in Example 9.6.

Example 12.9. Continue in this example the problem of the preceding example by introducing the circle constraints and solve the problem by the unified approach. These constraint equations enforce the fact that each of the given four points lies on the circumference of a circle. The parameters of the circle, taken to be the two coordinates of its center and its radius, are not known but will be introduced as added parameters. In the unified solution we simply take them as observations and select the following values for their a priori estimates:

$$x_{10} = 7.00 \text{ m} \qquad x_{20} = 46.00 \text{ m} \qquad R = 4.00 \text{ m}$$

In the a priori $\mathbf{Q}_{xx}$ we shall also include the added parameters. Table 12-4 summarizes several cases of values of $\mathbf{Q}_{xx}$, and $\mathbf{Q}_{cc}$ is taken to be $0.0001\mathbf{I}$. This means that deviation between a point and the circumference of the constraint circle is allowed to an a priori variance (or cofactor, since $\sigma_0^2 = 1$) of 0.0001 m^2, which is an adequate and practical value. The results in Table 12-4 show that a value of $\mathbf{Q}_{xx} = 10^3\mathbf{I}$ gives answers within 0.1 mm from the classical answers.

Again as before, the cofactor matrix $\mathbf{Q}_{\Delta\Delta}$ was computed for each of the above cases with the conclusions corresponding to those given in the preceding three examples.

TABLE 12-4

$\mathbf{Q}_{xx} =$	$10\mathbf{I}$	$10^2\mathbf{I}$	$10^3\mathbf{I}$	$10^5\mathbf{I}$	CLASSICAL $\infty\mathbf{I}$
$\hat{x}_{1a}$	6.99464	6.99424	6.99420	6.99419	6.99419
$\hat{x}_{2a}$	49.78021	49.77425	49.77365	49.77358	49.77358
$\hat{x}_{1b}$	6.98327	6.98270	6.98264	6.98263	6.98263
$\hat{x}_{2b}$	41.98046	41.97989	41.97982	41.97982	41.97981
$\hat{x}_{1c}$	2.99173	2.99157	2.99155	2.99155	2.99155
$\hat{x}_{2c}$	46.17063	46.16641	46.16598	46.16593	46.16593
$\hat{x}_{1d}$	4.04346	4.04338	4.04337	4.04337	4.04337
$\hat{x}_{2d}$	48.55662	48.55233	48.55189	48.55184	48.55184
$\hat{x}_{10}$	6.88226	6.87954	6.87926	6.87923	6.87923
$\hat{x}_{20}$	45.88049	45.87723	45.87690	45.87686	45.87686
$\hat{R}$	3.90134	3.89871	3.89844	3.89841	3.89841

12.8. SUMMARY OF EQUATIONS AND CONCLUDING REMARKS

12.8.1. Summary of Equations Given n observations $\boldsymbol{l}$ with a cofactor matrix $\mathbf{Q}$, u parameters with a priori estimates $\mathbf{x}$ and cofactor matrix $\mathbf{Q}_{xx}$, t observational elements in a vector $\boldsymbol{l}_c$ associated with constraints and a cofactor matrix $\mathbf{Q}_{cc}$, and the following nonlinear equations:

$$F(\boldsymbol{l}, \mathbf{x}) = \mathbf{0} \qquad (c \text{ equations}) \tag{12.17}$$

$$F_c(\boldsymbol{l}_c, \mathbf{x}) = \mathbf{0} \qquad (s \text{ equations}) \tag{12.20}$$

With $\boldsymbol{l}^0$, $\mathbf{x}^0$, and $\boldsymbol{l}_c^0$ as approximation vectors, the linearized form of these equations is

$$\mathbf{A}\mathbf{v} + \mathbf{B}\boldsymbol{\Delta} = \mathbf{f}^0 \tag{12.23a}$$

$$\mathbf{A}_c\,\mathbf{v}_c + \mathbf{C}\boldsymbol{\Delta} = \mathbf{f}_c^0 \tag{12.23b}$$

$\mathbf{A}$ is $c \times n$, rank $(\mathbf{A}) = c$

$\mathbf{A}_c$ is $s \times t$, rank $(\mathbf{A}_c) = s$

$\mathbf{B}$ is $c \times u$, rank $(\mathbf{B}) = u$

$\mathbf{C}$ is $s \times u$, rank $(\mathbf{C}) = s$

$$\mathbf{f}^0 = -[F(\boldsymbol{l}^0, \mathbf{x}^0) - \mathbf{A}(\boldsymbol{l}^0 - \boldsymbol{l})] \qquad \text{is } c \times 1 \tag{12.24a}$$

$$\mathbf{f}_c^0 = -[F_c(l_c^0, \mathbf{x}^0) - \mathbf{A}_c(l_c^0 - l_c)] \quad \text{is } s \times 1 \tag{12.24b}$$

$$\mathbf{f}_x = \mathbf{x}^0 - \mathbf{x} \tag{12.24c}$$

$$\mathbf{W}_e = \mathbf{Q}_e^{-1} = (\mathbf{AQA}^t)^{-1} \tag{12.26a}$$

$$\mathbf{W}_{ec} = (\mathbf{A}_c\, \mathbf{Q}_{cc}\, \mathbf{A}_c^t)^{-1} \tag{12.26b}$$

$$\mathbf{N} = \mathbf{B}^t\mathbf{W}_e\,\mathbf{B} \tag{12.28a}$$

$$\mathbf{t} = \mathbf{B}^t\mathbf{W}_e\,\mathbf{f}^0 \tag{12.28b}$$

$$\mathbf{N}_c = \mathbf{C}^t\mathbf{W}_{ec}\,\mathbf{C} \tag{12.28c}$$

$$\mathbf{t}_c = \mathbf{C}^t\mathbf{W}_{ec}\,\mathbf{f}_c^0 \tag{12.28d}$$

$$\mathbf{\Delta} = (\mathbf{N} + \mathbf{N}_c + \mathbf{W}_{xx})^{-1}(\mathbf{t} + \mathbf{t}_c - \mathbf{W}_{xx}\,\mathbf{f}_x) \tag{12.29b}$$

$$\mathbf{v} = \mathbf{QA}^t\mathbf{W}_e(\mathbf{f}^0 - \mathbf{B\Delta}) \tag{12.29c}$$

$$\mathbf{v}_c = \mathbf{Q}_{cc}\,\mathbf{A}_c^t\,\mathbf{W}_{ec}(\mathbf{f}_c^0 - \mathbf{C\Delta}) \tag{12.30a}$$

$$\mathbf{v}_x = \mathbf{f}_x + \mathbf{\Delta} \tag{12.30b}$$

$$\mathbf{Q}_{\Delta\Delta} = (\mathbf{N} + \mathbf{N}_c + \mathbf{W}_{xx})^{-1} \tag{12.33a}$$

$$\begin{bmatrix} \mathbf{Q}_{vv} & \mathbf{Q}_{vv_c} & \mathbf{Q}_{vv_x} \\ & \mathbf{Q}_{v_cv_c} & \mathbf{Q}_{v_cv_x} \\ \text{symmetric} & & \mathbf{Q}_{v_xv_x} \end{bmatrix} = \text{refer directly to equation (12.38)}$$

$$\begin{bmatrix} \mathbf{Q}_{\hat{l}\hat{l}} & \mathbf{Q}_{\hat{l}\hat{l}_c} & \mathbf{Q}_{\hat{l}\hat{x}} \\ & \mathbf{Q}_{\hat{l}_c\hat{l}_c} & \mathbf{Q}_{\hat{l}_c\hat{x}} \\ \text{symmetric} & & \mathbf{Q}_{\hat{x}\hat{x}} \end{bmatrix} = \begin{bmatrix} (\mathbf{Q} - \mathbf{Q}_{vv}) & -\mathbf{Q}_{v_cv_c} & -\mathbf{Q}_{vv_x} \\ & (\mathbf{Q}_{cc} - \mathbf{Q}_{v_cv_c}) & -\mathbf{Q}_{v_cv_x} \\ \text{symmetric} & & (\mathbf{Q}_{xx} - \mathbf{Q}_{v_xv_x}) \end{bmatrix}$$

$$\phi = \frac{(\mathbf{v}^t\mathbf{W}\mathbf{v} + \mathbf{v}_c^t\,\mathbf{W}_{cc}\,\mathbf{v}_c + \mathbf{v}_x^t\,\mathbf{W}_{xx}\,\mathbf{v}_x)}{r}$$

r = redundancy or degrees of freedom (12.41)

12.8.2. Concluding Remarks The unified approach given in this chapter is the culmination of the developments recently undertaken in least squares adjustment. Allowing each variable to be treated as an observation and varying the a priori weight to account for practical considerations can result in quite a flexible method which is also in a standardized form. This alleviates having to make special decisions for different problems. In fact the procedure applies equally to extensive adjustment problems with varying inputs as well as relatively limited problems. While working with the unified approach, care must be exercised when setting up the equations and, more importantly, when determining the relative magnitudes of the a priori weights and values of the variables. Although the values and weights of the actual observations are usually well specified, those for the conventionally termed parameters must often be estimated. Here, the chosen weights, in relation to the weights of the actual observations, must be consistent with the selected a priori values. After the adjustment, care must also be taken in interpreting the a posteriori estimate of the reference variance, particularly in regard to the number of degrees of freedom.

13

Sequential Data Processing

13.1. INTRODUCTION

We shall introduce our discussion of sequential data processing by referring to the problem of the two-parameter transformation as treated in Example 8.7. Under the assumption that both the **A** and **Q** matrices are block diagonal (by point), equations (8.14) and (8.15) show that both the coefficient matrix **N** and the constant vector **t** are formed by summing the contributions of each of the three given points. This summation process can obviously be extended to any number of points m,

$$\mathbf{N} = \sum_{i=1}^{m} (\mathbf{B}^t \mathbf{W}_e \mathbf{B})_i \tag{13.1a}$$

$$\mathbf{t} = \sum_{i=1}^{m} (\mathbf{B}^t \mathbf{W}_e \mathbf{f})_i \tag{13.1b}$$

It should be emphasized that equations (13.1a) and (13.1b) are possible *only when the* **A** *and* **Q** *matrices are block diagonal.* This means that the conditions arising for the ith point contain observations that belong *only to* *those conditions* (thus the block diagonality of $\mathbf{A}$) and that are *uncorrelated with all other observations* in the model (hence the block diagonality of $\mathbf{Q}$).

Stopping the summation at an intermediate step and denoting the partially formed (or incomplete) normal equations by

$$\mathbf{N}_{(i-1)} = \sum_{j=1}^{i-1} (\mathbf{B}^t\mathbf{W}_e\,\mathbf{B})_j \tag{13.2a}$$

$$\mathbf{t}_{(i-1)} = \sum_{j=1}^{i-1} (\mathbf{B}^t\mathbf{W}_e\,\mathbf{f})_j \tag{13.2b}$$

it then follows from the pair of equations in (13.1) that the succeeding step would be

$$\mathbf{N}_{(i)} = \mathbf{N}_{(i-1)} + (\mathbf{B}^t\mathbf{W}_e\,\mathbf{B})_i \tag{13.3a}$$

$$\mathbf{t}_{(i)} = \mathbf{t}_{(i-1)} + (\mathbf{B}^t\mathbf{W}_e\,\mathbf{f})_i \tag{13.3b}$$

An interesting possibility now arises if a solution is assumed possible from both pairs of equations (13.2) and (13.3), respectively, that is,

$$\mathbf{\Delta}_{(i-1)} = \mathbf{N}_{(i-1)}^{-1}\,\mathbf{t}_{(i-1)} \tag{13.4a}$$

$$\mathbf{\Delta}_{(i)} = \mathbf{N}_{(i)}^{-1}\,\mathbf{t}_{(i)} \tag{13.4b}$$

assuming that both inverses $\mathbf{N}_{(i-1)}^{-1}$ and $\mathbf{N}_{(i)}^{-1}$ exist. The possibility is to find the value of $\mathbf{\Delta}_{(i)}$, not from $\mathbf{N}_{(i)}$ and $\mathbf{t}_{(i)}$, but from a previous solution vector $\mathbf{\Delta}_{(i-1)}$. Such a scheme would lead to obtaining a least squares solution in a sequential or recursive manner. The following are examples of why and under what circumstances we would use such a scheme.

The first case is when an older adjustment has been performed and new information becomes available. One possibility would be to combine the new information with all the old information and perform a complete readjustment. This can be rather uneconomical, particularly if the new information is relatively limited compared to the old information. Instead, it would be better to perform a sort of an "add-on" adjustment if the *results* of the preceding larger adjustment can be used in combination with the *new observational data.* This is possible with the sequential techniques to be developed here.

Another possible use of sequential methods is in the operation of designing the actual observational experiment before the adjustment is performed. Given present-day modern computers with their large capacities and procedures of data simulation, the technique of sequential data reduction can be used to great advantage to assist in an optimum design. For example, in planning a trilateration net with a fixed set of points, various line lengths can be inserted or removed (or both) and numerous observational plans can all be tested in a computer program. The one configuration that would most economically meet selected accuracy requirements would be chosen for field execution. Any computer simulation expense would be more than offset by savings accrued from selecting an optimum scheme.

In the above example of design of observational experiments it was implied that the sequential technique to be developed would be capable of handling both problems of adding information as well as deleting information. This is true, as the algorithm can be used with only minor sign changes to remove the effect of any designated condition equation(s) from the adjustment. Such a possibility offers another rather useful practical application, the automatic editing of data during the reduction process. Large and involved computer programs used in photogrammetric and geodetic net adjustments often incorporate techniques of data editing. The ability to eliminate the effect of undesirable information, or to reinstate the effect of that which has proven to be useful, sequentially and at the same time of the adjustment, can lead to computational savings.

Yet another possibility of using sequential techniques is in the field of activity of on-line systems of data acquisition. Rapid advances in computer technology are leading more and more to acquisition systems that use electronic computers interfaced with regular electro-optical-mechanical equipment of measurement. One such example is a dedicated computer attached to a multiple-plate comparator for the acquisition of photogrammetric plate measurements. Here the sequential nature of the measurement process and the availability of a computer that allows for real-time or near real-time computational capability leave no doubt as to the desirability of developing sequential data processing techniques.

13.2. MATHEMATICAL DERIVATION

In the example discussed in the introduction, the parameter vector $\mathbf{\Delta}$ was the same during all the steps of adding more condition equations. This represents one class of problems in which the size of the partially reduced normal equations remains unaltered during both addition and deletion of condition equations. In a second class, the change in the number of condition equations causes a corresponding change in the size of the normal equations; that is, the number of parameters increases with the addition of conditions and decreases when conditions are deleted. We shall treat each case separately.

13.2.1. Fixed Number of Parameters There are several procedures of deriving a sequential algorithm depending in part on the type of information assumed known in the preceding step. For example, given $\mathbf{N}_{(i-1)}^{-1}$ and $\mathbf{t}_{(i-1)}$, we would like to get $\mathbf{N}_{(i)}^{-1}$ and $\mathbf{t}_{(i)}$. Note that we assume that we know the inverse $\mathbf{N}_{(i-1)}^{-1}$ itself. This is a most useful case since the inversion of such a matrix during each step can then be avoided. However, one inverse must be assumed possible at the beginning of the recursive scheme. (This assumption will be eliminated later when the unified approach is considered.) In order to get $\mathbf{N}_{(i)}^{-1}$ from $\mathbf{N}_{(i-1)}^{-1}$, we simply apply the relations (A68) and (A69) (Appendix A) to equation (13.3a), or

$$\underset{u,u}{\mathbf{N}_{(i)}^{-1}} = \underset{u,u}{\mathbf{N}_{(i-1)}^{-1}}\left[\underset{u,u}{\mathbf{I}} \mp \underset{u,r_i}{\mathbf{B}_i^t}\left(\underset{r_ir_i}{\mathbf{Q}_{ei}} \pm \underset{r_i,u}{\mathbf{B}_i}\ \underset{u,u}{\mathbf{N}_{(i-1)}^{-1}}\ \underset{u,r_i}{\mathbf{B}_i^t}\right)^{-1}\underset{r_i,u}{\mathbf{B}_i}\ \underset{u,u}{\mathbf{N}_{(i-1)}^{-1}}\right] \tag{13.5a}$$

where u is the total number of parameters and r_i is the number of conditions involved in the ith step. Included in equation (13.5a) are cases of both addition and deletion of conditions: the *upper signs refer to condition addition* and the lower signs are for condition deletion. Although $\mathbf{N}_{(i)}^{-1}$ is computed from equation (13.5a), the updated constant term vector can be directly written from equation (13.3b), or

$$\mathbf{t}_{(i)} = \mathbf{t}_{(i-1)} \pm \mathbf{B}_i^t\,\mathbf{W}_{ei}\,\mathbf{f}_i \tag{13.5b}$$

where the upper and lower signs in equation (13.5b) refer to condition addition and deletion, respectively. Once $\mathbf{N}_{(i)}^{-1}$ and $\mathbf{t}_{(i)}$ are evaluated, $\mathbf{\Delta}_i$ may be obtained, if desired, from equation (13.4b).

The second case to be considered is concerned with $\mathbf{N}_{(i-1)}^{-1}$ and $\mathbf{\Delta}_{i-1}$. Obviously, $\mathbf{N}_{(i)}^{-1}$ can be evaluated from equation (13.5a) for both situations of adding as well as deleting conditions. For developing $\mathbf{\Delta}_i$ from $\mathbf{\Delta}_{i-1}$ we shall take the case of adding conditions in the following derivation as an example.

From equation (13.4b) we write

$$\mathbf{\Delta}_i = \mathbf{N}_{(i)}^{-1}\,\mathbf{t}_{(i)}$$

and using equations (13.5a) (with the upper signs) and (13.5b) leads to

$$\mathbf{\Delta}_i = [\mathbf{N}_{(i-1)}^{-1} - \mathbf{N}_{(i-1)}^{-1}\,\mathbf{B}_i^t(\mathbf{Q}_{ei} + \mathbf{B}_i\,\mathbf{N}_{(i-1)}^{-1}\,\mathbf{B}_i^t)^{-1}\mathbf{B}_i\,\mathbf{N}_{(i-1)}^{-1}] \times (\mathbf{t}_{(i-1)} + \mathbf{B}_i^t\,\mathbf{W}_{ei}\,\mathbf{f}_i)$$

For simplicity we temporarily drop the subscript e from $\mathbf{Q}_e$ and reinstate it at the end of the derivation. Introducing the auxiliary,

$$\mathbf{J} = \mathbf{B}_i\,\mathbf{N}_{i-1}^{-1}\,\mathbf{B}_i^t \tag{13.6a}$$

then

$$\begin{aligned}\mathbf{\Delta}_i = {} & \mathbf{N}_{i-1}^{-1}\,\mathbf{t}_{i-1} \\ & - \mathbf{N}_{i-1}^{-1}\,\mathbf{B}_i^t(\mathbf{Q}_i + \mathbf{J})^{-1}\mathbf{B}_i\,\mathbf{N}_{i-1}^{-1}\,\mathbf{t}_{i-1} \\ & + \mathbf{N}_{i-1}^{-1}\,\mathbf{B}_i^t\,\mathbf{W}_i\,\mathbf{f}_i \\ & - \mathbf{N}_{i-1}^{-1}\,\mathbf{B}_i^t(\mathbf{Q}_i + \mathbf{J})^{-1}(\mathbf{B}_i\,\mathbf{N}_{i-1}^{-1}\,\mathbf{B}_i^t)\mathbf{W}_i\,\mathbf{f}_i\end{aligned} \tag{13.6b}$$

The last term in equation (13.6b) is

$$\mathbf{K} = -\mathbf{N}_{i-1}^{-1}\,\mathbf{B}_i^t(\mathbf{Q}_i + \mathbf{J})^{-1}\mathbf{J}\mathbf{W}_i\,\mathbf{f}_i$$

which when using equation (A70) (Appendix A) gives

$$\mathbf{K} = -\mathbf{N}_{i-1}^{-1}\,\mathbf{B}_i^t\,\mathbf{W}_i(\mathbf{W}_i + \mathbf{J}^{-1})^{-1}\mathbf{W}_i\,\mathbf{f}_i$$

and applying equation (A69), realizing that $\mathbf{U} = \mathbf{V} = \mathbf{I}$ in the present case, leads to

$$\begin{aligned}\mathbf{K} &= -\mathbf{N}_{i-1}^{-1}\,\mathbf{B}_i^t\,\mathbf{W}_i[\mathbf{Q}_i - \mathbf{Q}_i(\mathbf{J} + \mathbf{Q}_i)^{-1}\mathbf{Q}_i]\mathbf{W}_i\,\mathbf{f}_i \\ &= -\mathbf{N}_{i-1}^{-1}\,\mathbf{B}_i^t\,\mathbf{W}_i\,\mathbf{f}_i + \mathbf{N}_{i-1}^{-1}\,\mathbf{B}_i^t(\mathbf{Q}_i + \mathbf{J})^{-1}\mathbf{f}_i\end{aligned}$$

which when used in equation (13.6b) results in

$$\boldsymbol{\Delta}_i = \boldsymbol{\Delta}_{i-1} - \mathbf{N}_{i-1}^{-1}\,\mathbf{B}_i^t(\mathbf{Q}_i + \mathbf{J})^{-1}\mathbf{B}_i\,\boldsymbol{\Delta}_{i-1} + \mathbf{N}_{i-1}^{-1}\,\mathbf{B}_i^t(\mathbf{Q}_i + \mathbf{J})^{-1}\mathbf{f}_i$$

Finally,

$$\boldsymbol{\Delta}_i = \boldsymbol{\Delta}_{i-1} + \mathbf{N}_{i-1}^{-1}\,\mathbf{B}_i^t(\mathbf{Q}_i + \mathbf{B}_i\,\mathbf{N}_{i-1}^{-1}\,\mathbf{B}_i^t)^{-1}(\mathbf{f}_i - \mathbf{B}_i\,\boldsymbol{\Delta}_{i-1}) \tag{13.6c}$$

Restoring the subscript e and including the case of condition deletion yields (the reader is advised to work out the case of deletion for himself)

$$\boldsymbol{\Delta}_i = \boldsymbol{\Delta}_{i-1} \pm \mathbf{N}_{(i-1)}^{-1}\,\mathbf{B}_i^t(\mathbf{Q}_{ei} \pm \mathbf{B}_i\,\mathbf{N}_{(i-1)}^{-1}\,\mathbf{B}_i^t)^{-1}(\mathbf{f}_i - \mathbf{B}_i\,\boldsymbol{\Delta}_{i-1}) \tag{13.7}$$

Thus with $\mathbf{N}_{(i-1)}^{-1}$ and $\boldsymbol{\Delta}_{i-1}$ known, the new matrices $\mathbf{N}_{(i)}^{-1}$ and $\boldsymbol{\Delta}_i$ can be computed from the recursive formulas (13.5a) and (13.7).

All the preceding derivations were accomplished using only operations of matrix algebra. It should be recognized that the same relation must also be obtainable from applying the basic least squares criterion. As a demonstration we show the case of getting $\mathbf{N}_{(i)}^{-1}$ from $\mathbf{N}_{(i-1)}^{-1}$ for the addition of conditions and leave the other relations for the reader to derive as an exercise. Let

$$\mathbf{A}_{i-1}\,\mathbf{v}_{i-1} + \mathbf{B}_{i-1}\,\boldsymbol{\Delta} = \mathbf{f}_{i-1}$$

represent the total conditions up to and including the $(i-1)$ step; and

$$\mathbf{A}_i\,\mathbf{v}_i + \mathbf{B}_i\,\boldsymbol{\Delta} = \mathbf{f}_i$$

denote the conditions to be added at the ith step. These two equations may be combined into the form

$$\begin{bmatrix}\mathbf{A}_{i-1} & \mathbf{0}\\ \mathbf{0} & \mathbf{A}_i\end{bmatrix}\begin{bmatrix}\mathbf{v}_{i-1}\\ \mathbf{v}_i\end{bmatrix} + \begin{bmatrix}\mathbf{B}_{i-1}\\ \mathbf{B}_i\end{bmatrix}\boldsymbol{\Delta} = \begin{bmatrix}\mathbf{f}_{i-1}\\ \mathbf{f}_i\end{bmatrix} \tag{13.8}$$

With the total cofactor matrix (keeping the original assumption of no correlation)

$$\mathbf{Q} = \begin{bmatrix}\mathbf{Q}_{(i-1)} & \mathbf{0}\\ \mathbf{0} & \mathbf{Q}_{(i)}\end{bmatrix}$$

then

$$\mathbf{Q}_e = \begin{bmatrix} (\mathbf{AQA}^t)_{(i-1)} & \mathbf{0} \\ \mathbf{0} & (\mathbf{AQA}^t)_{(i)} \end{bmatrix} = \begin{bmatrix} \mathbf{Q}_{e(i-1)} & \mathbf{0} \\ \mathbf{0} & \mathbf{Q}_{e(i)} \end{bmatrix}$$

The scalar to be minimized is in this case

$$\begin{aligned} \phi = (\mathbf{v}^t\mathbf{W}\mathbf{v})_{i-1} + (\mathbf{v}^t\mathbf{W}\mathbf{v})_i - 2\mathbf{k}_{i-1}^t(\mathbf{A}_{i-1}\,\mathbf{v}_{i-1} + \mathbf{B}_{i-1}\,\boldsymbol{\Delta} - \mathbf{f}_{i-1}) \\ - 2\mathbf{k}_i^t(\mathbf{A}_i\,\mathbf{v}_i + \mathbf{B}_i\,\boldsymbol{\Delta} - \mathbf{f}_i) \end{aligned}$$

which must be differentiated with respect to $\mathbf{v}_{i-1}$, $\mathbf{v}_i$, and $\boldsymbol{\Delta}$, and the resulting differentials equated to zero. These three equations combined with the pair in equation (13.8) make up five equations in five unknown vectors, $\mathbf{v}_{i-1}$, $\mathbf{v}_i$, $\mathbf{k}_{i-1}$, $\mathbf{k}_i$, and $\boldsymbol{\Delta}$. After eliminating $\mathbf{v}_{i-1}$, $\mathbf{v}_i$, and $\mathbf{k}_{i-1}$, the partially reduced normal equations take the form

$$\begin{bmatrix} -\mathbf{N}_{(i-1)} & \mathbf{B}_i^t \\ \mathbf{B}_i & \mathbf{W}_{e(i)} \end{bmatrix} \begin{bmatrix} \boldsymbol{\Delta} \\ \mathbf{k}_i \end{bmatrix} = \begin{bmatrix} -\mathbf{t}_{(i-1)} \\ \mathbf{f}_i \end{bmatrix} \tag{13.9}$$

in which $\mathbf{N}_{(i-1)}$ and $\mathbf{t}_{(i-1)}$ are equivalent to those given by equations (13.2a) and (13.2b). The solution of equation (13.9) by partitioning can be carried out as follows:

$$\begin{bmatrix} -\mathbf{N}_{(i-1)} & \mathbf{B}_i^t \\ \mathbf{B}_i & \mathbf{W}_{e(i)} \end{bmatrix}^{-1} = \begin{bmatrix} \mathbf{P} & \mathbf{R}^t \\ \mathbf{R} & \mathbf{S} \end{bmatrix}$$

and according to equation (A37a) (Appendix A)

$$-\mathbf{P} = (\mathbf{N}_{(i-1)} + \mathbf{B}_i^t\,\mathbf{W}_{e(i)}\,\mathbf{B}_i)^{-1}$$

which is equal to $\mathbf{N}_{(i)}^{-1}$ from equation (13.3a). Multiplying the coefficient matrix in equation (13.9) by its inverse and equating the result to the identity matrix gives

$$\begin{aligned} -\mathbf{P} &= \mathbf{N}_{(i-1)}^{-1}(\mathbf{I} - \mathbf{B}_i^t\,\mathbf{R}) \\ -\mathbf{B}_i &= \mathbf{N}_{(i-1)}^{-1}(\mathbf{I} - \mathbf{B}_i^t\,\mathbf{R}) + \mathbf{Q}_{e(i)}\,\mathbf{R} = \mathbf{0} \\ \mathbf{R} &= (\mathbf{Q}_{e(i)} + \mathbf{B}_i\,\mathbf{N}_{(i-1)}^{-1}\,\mathbf{B}_{i-1}^t)^{-1}\mathbf{B}_i\,\mathbf{N}_{(i-1)}^{-1} \end{aligned}$$

then

$$\mathbf{N}_{(i)}^{-1} = \mathbf{N}_{(i-1)}^{-1}[\mathbf{I} - \mathbf{B}_i^t(\mathbf{Q}_{e(i)} + \mathbf{B}_i\,\mathbf{N}_{(i-1)}^{-1}\,\mathbf{B}_i^t)^{-1}\mathbf{B}_i\,\mathbf{N}_{(i-1)}^{-1}]$$

which is identical to equation (13.5a) with the upper signs for addition of conditions.

The recursive formulas that have been derived exhibit an apparent advantage in avoiding the inversion of the normal equations coefficient matrix. On the other hand these sequential relationships, for example, equation (13.5a), involve an apparently large number of matrix multiplications and an inversion of a matrix of order r_i. As regards the inversion, r_i is, in general, much smaller than u, since we would usually add or eliminate a few conditions at a time. Of course, if only one equation is involved, then the inversion degenerates into computing a reciprocal of a scalar. Concerning the number of matrix multiplications, the question becomes a matter of computational efficiency of the sequential scheme compared to the conventional "batch" least squares. As an example, 90 linear equations in 60 unknowns, of the form

$$\underset{90,1}{\mathbf{v}} + \underset{90,60}{\mathbf{B}}\ \underset{60,1}{\mathbf{\Delta}} = \underset{90,1}{\mathbf{f}}$$

were generated. As a start, a 60 × 60 normal equations coefficient matrix from all 90 condition equations was formed. The inverse of this matrix, which is used as a standard, consumed approximately 8 seconds on the computer used (UNIVAC 1108). Next, several cases of deleting a progressively larger number of condition equations (up to the limit of 30 equations) were computed using both methods of conventional batch least squares and by the sequential scheme developed here. To emphasize the difference between these two methods we can consider as an example the case of deleting 5 equations from the given 90 equations. In the batch process we would use the remaining 85 equations, form a 60 × 60 normal equations matrix and invert. In this case, the time used for these operations is a little less than 7 seconds. By contrast, in the sequential procedure the standard inverse (that is, that computed from all 90 condition equations) is used together with the 5 equations to be deleted to compute the inverse that would correspond to the remaining 85 equations. This sequential operation, which leads to the same inverse, consumed less than 1 second of computer time. The same process of taking off equations was performed several times by both methods; each time a larger number of equations was deleted, and the computing time was noted. The results are presented in Figure 13.1 in a form of two graphs, one for the batch solution and the other for the sequential solution.

Examining the graph for the batch solution first, we can see that the difference between the time required for getting an inverse from 90 equations and that required for getting it from 60 equations is rather small, being only slightly over 2 seconds. This indicates that the batch method is not strongly dependent either on the total number of condition equations or the number of equations to be deleted, but rather on the size of the reduced normal equation coefficient matrix under the assumption that all the condition equations are already available; that is, the time for their formulation is not considered here. By contrast, the sequential method is more strongly dependent on the number of condition equations, particularly those to be deleted. The time increases relatively rapidly as the number of such equations increases, until it reaches a cutoff point where the two graphs intersect and where the use of sequential solution offers no advantage over the conventional batch method. Therefore it would be advisable to limit the number of

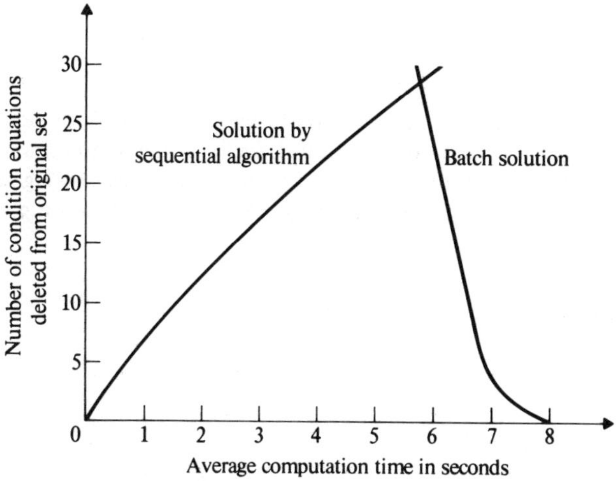

Figure 13.1. Example on Computational Time for Batch and Sequential Solutions (A 60-Parameter Problem)

condition equations by which the problem is changed in relation to the number of normal equations.

It should be emphasized that these results are only a demonstration of the possibilities and hold only for the data used (60 normal equations, inversion by Gauss elimination algorithm, UNIVAC 1108 computer, and so on). Different data may produce different results, but it is felt that the general characteristics of Figure 13.1 will remain. It is suspected further that if the condition equations were to be formed within the adjustment process, the sequential solution would offer additional computational time savings.

Example 13.1. Consider the following set of four condition equations:

$$v_1 - 2x_1 + x_2 = 1$$

$$v_2 - x_1 - 3x_2 = -9$$

$$v_3 + x_1 + 2x_2 = 8$$

$$\text{-----------------------}$$

$$v_4 - 4x_1 - x_2 = -7$$

or in matrix form

$$\begin{bmatrix} v_1 \\ v_2 \\ v_3 \\ \hdashline v_4 \end{bmatrix} + \begin{bmatrix} -2 & +1 \\ -1 & -3 \\ +1 & +2 \\ \hdashline -4 & -1 \end{bmatrix} \begin{bmatrix} x_1 \\ x_2 \end{bmatrix} = \begin{bmatrix} 1 \\ -9 \\ 8 \\ \hdashline -7 \end{bmatrix}$$

which corresponds to

$$\underset{4,1}{\mathbf{v}} + \underset{4,2}{\mathbf{B}}\ \underset{2,1}{\Delta} = \underset{4,1}{\mathbf{f}}$$

Assume for simplicity that $\mathbf{Q} = \mathbf{I}_4$. With this data we seek to demonstrate the concepts developed in the preceding section for both the addition and deletion of conditions. First consider *the case of condition addition*, by solving with all four conditions, then with three plus one added sequentially, and then comparing the final results.

For all four conditions,

$$(\mathbf{B}^t\mathbf{W}\mathbf{B})\boldsymbol{\Delta} = (\mathbf{B}^t\mathbf{W}\mathbf{f}) \qquad \text{or} \qquad \mathbf{N}\boldsymbol{\Delta} = \mathbf{t}$$

or

$$\begin{bmatrix} 22 & 7 \\ 7 & 15 \end{bmatrix} \begin{bmatrix} x_1 \\ x_2 \end{bmatrix} = \begin{bmatrix} 43 \\ 51 \end{bmatrix}$$

With

$$\mathbf{N}^{-1} = \frac{1}{281}\begin{bmatrix} 15 & -7 \\ -7 & 22 \end{bmatrix} \qquad \text{and} \qquad \mathbf{t} = \begin{bmatrix} 43 \\ 51 \end{bmatrix}$$

then

$$\begin{bmatrix} x_1 \\ x_2 \end{bmatrix} = \frac{1}{281}\begin{bmatrix} 288 \\ 821 \end{bmatrix}$$

For the first three equations,

$$\mathbf{B}_0 = \begin{bmatrix} -2 & 1 \\ -1 & -3 \\ 1 & +2 \end{bmatrix}, \qquad \mathbf{f}_0 = \begin{bmatrix} 1 \\ -9 \\ 8 \end{bmatrix}, \qquad \mathbf{N}_0 = \begin{bmatrix} 6 & 3 \\ 3 & 14 \end{bmatrix}, \qquad \mathbf{t}_0 = \begin{bmatrix} 15 \\ 44 \end{bmatrix}$$

$$\mathbf{N}_0^{-1} = \frac{1}{75}\begin{bmatrix} 14 & -3 \\ -3 & 6 \end{bmatrix}$$

thus

$$\boldsymbol{\Delta}_0 = \frac{1}{75}\begin{bmatrix} 78 \\ 219 \end{bmatrix}$$

In order to effect the addition of the fourth condition equation, its matrices are

$$\mathbf{B}_1 = [-4 \quad -1] \qquad \mathbf{f}_1 = [-7]$$

with the subscript 1 referring to the first (and only, in this case) sequential addition. First, use equation (13.5a) (with upper signs) to compute $\mathbf{N}_1^{-1}$, as follows. Because of having $\mathbf{Q} = \mathbf{I}$, the recursive equation becomes

$$\mathbf{N}_1^{-1} = \mathbf{N}_0^{-1}[\mathbf{I} - \mathbf{B}_1^t(1 + \mathbf{B}_1\,\mathbf{N}_0^{-1}\mathbf{B}_1^t)^{-1}\mathbf{B}_1\,\mathbf{N}_0^{-1}]$$

noting that the parenthetical expression to be inverted is only a scalar equal to

$$1 + \frac{1}{75}[-4 \quad -1]\begin{bmatrix} 14 & -3 \\ -3 & 6 \end{bmatrix}\begin{bmatrix} -4 \\ -1 \end{bmatrix} = \frac{281}{75}$$

Performing the arithmetic operations, yields $\mathbf{N}_1^{-1}$ which is identical to $\mathbf{N}^{-1}$ obtained directly from the four equations. As for the constant term vector, apply equation (13.5b)

$$\mathbf{t}_1 = \mathbf{t}_0 + \mathbf{B}_1^t \mathbf{W}\mathbf{f}_1 = [43 \quad 51]^t$$

which is also identical to $\mathbf{t}$ computed from the four condition equations. Hence the estimates for x_1, x_2 will be identical from this sequential scheme as from the batch solution.

The next step is to apply equation (13.7) (with upper signs), which in this example becomes

$$\boldsymbol{\Delta}_1 = \boldsymbol{\Delta}_0 + \mathbf{N}_0^{-1}\mathbf{B}_1^t(1 + \mathbf{B}_1 \mathbf{N}_0^{-1}\mathbf{B}_1^t)^{-1}(\mathbf{f}_1 - \mathbf{B}_1 \boldsymbol{\Delta}_0)$$

First,

$$(\mathbf{f}_1 - \mathbf{B}_1 \boldsymbol{\Delta}_0) = \frac{2}{25}$$

and

$$\mathbf{N}_0^{-1}\mathbf{B}_1^t(\mathbf{I} + \mathbf{B}_1\mathbf{N}_0^{-1}\mathbf{B}_1^t)^{-1}(\mathbf{f}_1 - \mathbf{B}_1 \boldsymbol{\Delta}_0) = \frac{1}{75 \times 281}[-318 \quad 36]^t$$

then

$$\boldsymbol{\Delta}_1 = \frac{1}{281}[288 \quad 821]^t$$

which is identical to the solution obtained before. Consequently, both schemes of sequential solution for addition of condition equations are valid. Now consider the same data for the *case of condition deletion.* The solution for all four conditions is modified by eliminating the contribution of the fourth equation, and the results obtained are compared to those directly available from solving the first three equations together. Thus

$$\mathbf{N}_0^{-1} = \frac{1}{281}\begin{bmatrix} 15 & -7 \\ -7 & 22 \end{bmatrix} \qquad \mathbf{t}_0 = \begin{bmatrix} 43 \\ 51 \end{bmatrix} \qquad \boldsymbol{\Delta}_0 = \frac{1}{281}\begin{bmatrix} 288 \\ 821 \end{bmatrix}$$

and

$$\mathbf{B}_1 = [-4 \quad -1] \qquad \boldsymbol{\Delta}_1 = [-7] \qquad \mathbf{Q}_1 = 1$$

First apply equation (13.5a) with the lower signs to compute $\mathbf{N}_1^{-1}$ from $\mathbf{N}_0^{-1}$.

$$\mathbf{N}_1^{-1} = \frac{1}{75}\begin{bmatrix} 14 & -3 \\ -3 & 6 \end{bmatrix}$$

then from equation (13.5b) compute

$$\mathbf{t}_1 = [15 \quad 44]^t$$

Both $\mathbf{N}_1^{-1}$ and $\mathbf{t}_1$ are identical to $\mathbf{N}_0^{-1}$ and $\mathbf{t}_0$ for the case of addition for the batch solution of the first three conditions. Finally, we can also demonstrate the applicability of equation (13.7) with lower signs for the case of condition deletion, which is left as an exercise.

13.2.2. Variable Number of Parameters In many problems, such as aerial triangulation in photogrammetry and net design with variable number of control points in surveying, the increase or decrease in the number of condition equations may change the number of parameters in the adjustment accordingly. For example, if the collinearity type of condition equations is used in a photogrammetric triangulation adjustment, each time a pass point is added three new parameters corresponding to its coordinates will also be added. In the following development first the basic foundation is given then each of the two situations of condition addition and deletion will be treated separately.

Consider the two sets of conditions

$$\mathbf{A}_{i-1}\,\mathbf{v}_{i-1} + \mathbf{B}_{i-1}\,\boldsymbol{\Delta} = \mathbf{f}_{i-1} \tag{13.10a}$$

$$\mathbf{A}_i\,\mathbf{v}_i + \mathbf{B}_i\,\boldsymbol{\Delta} + \mathbf{b}_i\,\boldsymbol{\delta} = \mathbf{f}_i \tag{13.10b}$$

in which $\boldsymbol{\delta}$ is a subvector of parameters that appears in the set of conditions to be added or subtracted sequentially. This pair of equations may be combined into

$$\begin{bmatrix} \mathbf{A}_{i-1} & \mathbf{0} \\ \mathbf{0} & \mathbf{A}_i \end{bmatrix} \begin{bmatrix} \mathbf{v}_{i-1} \\ \mathbf{v}_i \end{bmatrix} + \begin{bmatrix} \mathbf{B}_{i-1} & \mathbf{0} \\ \mathbf{B}_i & \mathbf{b}_i \end{bmatrix} \begin{bmatrix} \boldsymbol{\Delta} \\ \boldsymbol{\delta} \end{bmatrix} = \begin{bmatrix} \mathbf{f}_{i-1} \\ \mathbf{f}_i \end{bmatrix} \tag{13.11}$$

for which the reduced normal equations may be written directly as a coefficient matrix $\mathbf{N}_i$ and constant vector $\mathbf{t}_i$

$$\mathbf{N}_i = \begin{bmatrix} (\mathbf{B}^t\mathbf{W}_e\,\mathbf{B})_{i-1} + (\mathbf{B}^t\mathbf{W}_e\,\mathbf{B})_i & (\mathbf{B}^t\mathbf{W}_e\,\mathbf{b})_i \\ (\mathbf{b}^t\mathbf{W}_e\,\mathbf{B})_i & (\mathbf{b}^t\mathbf{W}_e\,\mathbf{b})_i \end{bmatrix}$$

$$= \begin{bmatrix} \mathbf{N}_{i-1} + \delta\mathbf{N}_i & \bar{\mathbf{n}}_i \\ \bar{\mathbf{n}}_i^t & \mathbf{n}_i \end{bmatrix} \tag{13.12a}$$

$$\mathbf{t}_i = \begin{bmatrix} (\mathbf{B}^t\mathbf{W}_e\,\mathbf{f})_{i-1} + (\mathbf{B}^t\mathbf{W}_e\,\mathbf{f})_i \\ (\mathbf{b}^t\mathbf{W}_e\,\mathbf{f})_i \end{bmatrix} = \begin{bmatrix} \mathbf{t}_{i-1} + \delta\mathbf{t}_i \\ \bar{\mathbf{t}}_i \end{bmatrix} = \begin{bmatrix} \mathbf{t}_i' \\ \bar{\mathbf{t}}_i \end{bmatrix} \tag{13.12b}$$

The inverse of $\mathbf{N}_i$ may be evaluated by partitioning as

$$\mathbf{N}_i^{-1} = \begin{bmatrix} \bar{\mathbf{N}}_i & \bar{\mathbf{n}}_i \\ \bar{\mathbf{n}}_i^t & \mathbf{n}_i \end{bmatrix}^{-1} = \begin{bmatrix} \mathbf{F}_i & \mathbf{G}_i \\ \mathbf{G}_i^t & \mathbf{H}_i \end{bmatrix} \tag{13.13}$$

with [see equations (A37), Appendix A]

$$\mathbf{F}_i = (\bar{\mathbf{N}}_i - \bar{\mathbf{n}}_i\,\mathbf{n}_i^{-1}\bar{\mathbf{n}}_i^t)^{-1} \tag{13.14a}$$

$$\mathbf{G}_i = -\mathbf{F}_i\,\bar{\mathbf{n}}_i\,\mathbf{n}_i^{-1} \tag{13.14b}$$

$$\mathbf{H}_i = \mathbf{n}_i^{-1} - \mathbf{n}_i^{-1}\bar{\mathbf{n}}_i^t\,\mathbf{G}_i \tag{13.14c}$$

If $\mathbf{G}_i$ and $\mathbf{F}_i$ are replaced by their expressions, $\mathbf{H}_i$ would become

$$\mathbf{H}_i = (\mathbf{n}_i - \bar{\mathbf{n}}_i^t\,\bar{\mathbf{N}}_i^{-1}\bar{\mathbf{n}}_i)^{-1} \tag{13.14d}$$

which is an alternate form. Equation (A69) may be applied to equation (13.14a) to give

$$\mathbf{F}_i = \bar{\mathbf{N}}_i^{-1} + \bar{\mathbf{N}}_i^{-1}\bar{\mathbf{n}}_i(\mathbf{n}_i - \bar{\mathbf{n}}_i^t \bar{\mathbf{N}}_i^{-1}\bar{\mathbf{n}}_i)^{-1}\bar{\mathbf{n}}_i^t \bar{\mathbf{N}}_i^{-1} \tag{13.15a}$$

which in view of equation (13.14d) becomes

$$\mathbf{F}_i = \bar{\mathbf{N}}_i^{-1} + \bar{\mathbf{N}}_i^{-1}\bar{\mathbf{n}}_i \mathbf{H}_i \bar{\mathbf{n}}_i^t \bar{\mathbf{N}}_i^{-1} \tag{13.15b}$$

and since from equations (13.12a) and (13.13)

$$\bar{\mathbf{N}}_i = \mathbf{N}_{i-1} + \delta\mathbf{N}_i = \mathbf{N}_{i-1} + \mathbf{B}_i^t \mathbf{W}_{ei} \mathbf{B}_i \tag{13.16}$$

the application of equation (A69) to equation (13.16) leads to

$$\bar{\mathbf{N}}_i^{-1} = \mathbf{N}_{i-1}^{-1} - \mathbf{N}_{i-1}^{-1} \mathbf{B}_i^t(\mathbf{Q}_{ei} + \mathbf{B}_i \mathbf{N}_{i-1}^{-1} \mathbf{B}_i^t)^{-1}\mathbf{B}_i \mathbf{N}_{i-1}^{-1} \tag{13.17}$$

The foregoing development establishes all the fundamental relationships necessary to handle this case of variable size of parameters. Because the case of addition increases the order of the normal equations coefficient matrix, whereas the case of deletion effects the opposite, each case must be handled separately.

ADDING CONDITION EQUATIONS At the $(i-1)$ step of the sequential processing we would have $\mathbf{N}_{i-1}^{-1}$ and $\mathbf{t}_{i-1}$, and we would want to add the new conditions given by equation (13.10b). These conditions include $\boldsymbol{\delta}_i$ which increases the size of the parameters from $\boldsymbol{\Delta}_{i-1}$ which is $u_{i-1} \times 1$ to

$$\boldsymbol{\Delta}_i = \begin{bmatrix} \boldsymbol{\Delta}_{i-1} \\ \boldsymbol{\delta}_i \end{bmatrix}$$

which is $u_i \times 1$, where u_i is $u_{i-1} + w_i$ and w_i is the number of parameters in $\boldsymbol{\delta}_i$. It is important to note that the new observations in equation (13.10b) are assumed to be uncorrelated with all the original observations. This allows for constructing the total cofactor matrix as a block diagonal matrix. After the new conditions of equation (13.10b) are added, we would have $\mathbf{N}_i^{-1}$ and $\mathbf{t}_i$, which are obtained as follows.

First, equation (13.17) is used to compute $\bar{\mathbf{N}}_i^{-1}$ from $\mathbf{N}_{i-1}^{-1}$ and the matrices of the added conditions of equation (13.10b), noting that $\mathbf{Q}_{ei} = \mathbf{A}_i \mathbf{Q}_i \mathbf{A}_i^t$. Second, according to equation (13.12a) the auxiliary matrices $\bar{\mathbf{n}}_i$ and $\mathbf{n}_i$ are evaluated from

$$\bar{\mathbf{n}}_i = \mathbf{B}_i^t \mathbf{W}_{ei} \mathbf{b}_i \tag{13.18a}$$

$$\mathbf{n}_i = \mathbf{b}_i^t \mathbf{W}_{ei} \mathbf{b}_i \tag{13.18b}$$

Next, from $\bar{\mathbf{N}}_i^{-1}$ and these auxiliary matrices, $\mathbf{F}_i$, $\mathbf{G}_i$, and $\mathbf{H}_i$ are computed from equations (13.15), (13.14b), and (13.14c), respectively. These three matrices are then collected together to give $\mathbf{N}_i^{-1}$ according to equation

(13.13). As regards to the constant term vector, by reference to equation (13.12b) we compute

$$\mathbf{t}'_i = \mathbf{t}_{i-1} + \mathbf{B}_i \mathbf{W}_{ei} \mathbf{f}_i \tag{13.19a}$$

$$\bar{\mathbf{t}}_i = \mathbf{b}_i^t \mathbf{W}_{ei} \mathbf{f}_i \tag{13.19b}$$

and collect them to form $\mathbf{t}_i$ as shown in equation (13.12b). This completes the ith step of the sequential solution.

Deleting Condition Equations Here, in order to use the same basic relations developed at the beginning of this section, we shall for the moment go from the ith step to the $(i-1)$ step, and then at the end change the subscript variable to give the more conventional relations. Thus at the ith step we would have $\mathbf{N}_i^{-1}$ and $\mathbf{t}_i$ and would wish to eliminate the effect of the conditions of equation (13.10b) to get $\mathbf{N}_{i-1}^{-1}$ and $\mathbf{t}_{i-1}$. Since this elimination will reduce the parameters by $\boldsymbol{\delta}$, then both $\mathbf{N}_{i-1}^{-1}$ and $\mathbf{t}_{i-1}$ will be correspondingly of smaller dimensions than $\mathbf{N}_i^{-1}$ and $\mathbf{t}_i$, respectively. Consequently, from the partitioning of $\mathbf{N}_i^{-1}$ in equation (13.13) only the submatrix $\mathbf{F}_i$ would be of interest, and the remaining matrices are simply discarded. Inverting equation (13.14a) yields

$$\mathbf{F}_i^{-1} = \bar{\mathbf{N}}_i - \bar{\mathbf{n}}_i \mathbf{n}_i^{-1} \bar{\mathbf{n}}_i^t$$

or

$$\bar{\mathbf{N}}_i = \mathbf{F}^{-1} + \bar{\mathbf{n}}_i \mathbf{n}_i^{-1} \bar{\mathbf{n}}_i^t \tag{13.20a}$$

and from equation (A69) the inverse of $\bar{\mathbf{N}}_i$ would be

$$\bar{\mathbf{N}}_i^{-1} = \mathbf{F}_i - \mathbf{F}_i \bar{\mathbf{n}}_i(\mathbf{n}_i + \bar{\mathbf{n}}_i^t \mathbf{F}_i \bar{\mathbf{n}}_i)^{-1} \bar{\mathbf{n}}_i^t \mathbf{F}_i \tag{13.20b}$$

Next, from equation (13.16)

$$\mathbf{N}_{i-1} = \bar{\mathbf{N}}_i - \mathbf{B}_i^t \mathbf{W}_{ei} \mathbf{B}_i \tag{13.21a}$$

which again may be inverted according to equation (A69) to give

$$\mathbf{N}_{i-1}^{-1} = \bar{\mathbf{N}}_i^{-1} + \bar{\mathbf{N}}_i^{-1} \mathbf{B}_i^t (\mathbf{Q}_{ei} - \mathbf{B}_i \bar{\mathbf{N}}_i^{-1} \mathbf{B}_i^t)^{-1} \mathbf{B}_i \bar{\mathbf{N}}_i^{-1} \tag{13.21b}$$

In a similar manner only the subvector $\mathbf{t}'_i$ of $\mathbf{t}_i$ is pertinent here, from which $\mathbf{t}_{i-1}$ may be directly obtained from equation (13.19a), or

$$\mathbf{t}_{i-1} = \mathbf{t}'_i - \mathbf{B}_i \mathbf{W}_{ei} \mathbf{f}_i \tag{13.21c}$$

This completes showing how $\mathbf{N}_{i-1}^{-1}$ and $\mathbf{t}_{i-1}$ may be obtained from $\mathbf{N}_i$ and $\mathbf{t}_i$. However, in a sequential procedure we are conventionally accustomed to ascending subscripts instead of this case of decreasing subscripts. Consequently, we recapitulate the above development while changing the subscript to correspond to starting with $\mathbf{N}_{j-1}^{-1}$ and $\mathbf{t}_{j-1}$ and seeking $\mathbf{N}_j^{-1}$ and $\mathbf{t}_j$ due to eliminating the conditions,

$$\mathbf{A}_j \mathbf{v}_j + \mathbf{B}_j \boldsymbol{\Delta} + \mathbf{b}_j \boldsymbol{\delta} = \mathbf{f}_j \tag{13.22}$$

The sequence is summarized as follows:

1. Partition $\mathbf{N}_{j-1}^{-1}$ and take the upper left matrix $\mathbf{F}_{j-1}$ of order corresponding to the reduced number of parameters.
2. Compute the auxiliary matrices $\bar{\mathbf{n}}_j$ and $\mathbf{n}_j$ using the matrices of equation (13.22) in equations (13.18a) and (13.18b), and noting that $\mathbf{W}_{ej} = (\mathbf{A}_j \mathbf{Q}_j \mathbf{A}_j^t)^{-1}$.
3. Evaluate $\bar{\mathbf{N}}_{j-1}^{-1}$ from [see equation (13.20b)],

$$\bar{\mathbf{N}}_{j-1}^{-1} = \mathbf{F}_{j-1} - \mathbf{F}_{j-1}\,\bar{\mathbf{n}}_j(\mathbf{n}_j + \bar{\mathbf{n}}_j^t\,\mathbf{F}_{j-1}\,\bar{\mathbf{n}}_j)^{-1}\bar{\mathbf{n}}_j^t\,\mathbf{F}_{j-1} \tag{13.23}$$

4. Compute $\mathbf{N}_j^{-1}$ from [see equation (13.21b)],

$$\mathbf{N}_j^{-1} = \bar{\mathbf{N}}_{j-1}^{-1} + \bar{\mathbf{N}}_{j-1}^{-1}\mathbf{B}_j^t(\mathbf{Q}_{ej} - \mathbf{B}_j\bar{\mathbf{N}}_{j-1}^{-1}\mathbf{B}_j^t)^{-1}\mathbf{B}_j\bar{\mathbf{N}}_{j-1}^{-1} \tag{13.24}$$

5. Compute $\mathbf{t}_j$ from [see equation (13.21c)],

$$\mathbf{t}_j = \mathbf{t}_{j-1}' - \mathbf{B}_j^t\,\mathbf{W}_{ej}\,\mathbf{f}_j \tag{13.25}$$

   in which $\mathbf{t}_{j-1}'$ is the upper subvector of $\mathbf{t}_{j-1}$ with as many elements as the number of remaining parameters.

   This concludes the steps of sequentially eliminating conditions that reduce the number of parameters.

Example 13.2. To demonstrate the applicability of the derivations in this section, take a numerical example and consider both *cases of adding and deleting information.* Assuming for simplicity that **A** and **Q** are identity matrices for all conditions and observations, let the total system of condition equations be denoted by [see equation (13.11)]

$$\begin{bmatrix}\dot{\mathbf{v}}\\ \ddot{\mathbf{v}}\end{bmatrix} + \begin{bmatrix}\dot{\mathbf{B}} & \mathbf{0}\\ \ddot{\mathbf{B}} & \mathbf{b}\end{bmatrix}\begin{bmatrix}\boldsymbol{\Delta}\\ \boldsymbol{\delta}\end{bmatrix} = \begin{bmatrix}\dot{\mathbf{f}}\\ \ddot{\mathbf{f}}\end{bmatrix} \qquad \text{or} \qquad \mathbf{v} + \mathbf{B}\boldsymbol{\Delta}' = \mathbf{f}$$

where the total coefficient matrix is

$$\mathbf{B} = \left[\begin{array}{cc:c} 0.44 & 0.76 & 0\\ 1.97 & 3.18 & 0\\ 8.35 & 3.52 & 0\\ \hdashline 5.40 & 7.24 & 2.70\\ 9.99 & 2.69 & 2.69\\ 5.39 & 8.08 & 3.47\\ 1.55 & 5.03 & 6.59 \end{array}\right]$$

For simplicity we shall not consider the constant vector, as its modification is relatively straightforward. For batch processing, with $\mathbf{Q} = \mathbf{I}_7$, we get

$$\underset{3,3}{\mathbf{N}^{-1}} = (\mathbf{B}^t\mathbf{B})^{-1} = \begin{bmatrix} 0.01036 & -0.01066 & 0.00301\\ -0.01066 & 0.02681 & -0.02303\\ 0.00301 & -0.02303 & 0.04025 \end{bmatrix}$$

For sequential processing, we originally have three condition equations with the coefficient matrix

$$\underset{3,2}{\dot{\mathbf{B}}} = \begin{bmatrix} 0.44 & 0.76 \\ 1.97 & 3.18 \\ 8.35 & 3.52 \end{bmatrix}$$

from which the inverse of the normal equations coefficient matrix is

$$\dot{\mathbf{N}}^{-1} = (\dot{\mathbf{B}}^t\dot{\mathbf{B}})^{-1} = \begin{bmatrix} 0.05658 & 0.08823 \\ 0.08823 & 0.18091 \end{bmatrix}$$

The new set of equations to be added sequentially (as one group) has the following coefficient matrices:

$$\ddot{\mathbf{B}} = \begin{bmatrix} 5.40 & 7.24 \\ 9.99 & 2.69 \\ 5.39 & 8.08 \\ 1.55 & 5.03 \end{bmatrix} \qquad \mathbf{b} = \begin{bmatrix} 2.7 \\ 2.69 \\ 3.47 \\ 6.59 \end{bmatrix}$$

In order to construct the total augmented matrix $\mathbf{N}^{-1}$ [see equation (13.13)], first compute $\bar{\mathbf{N}}^{-1}$ from equation (13.17) using $\dot{\mathbf{N}}^{-1}$ (representing $\mathbf{N}_{i-1}^{-1}$) and $\ddot{\mathbf{B}}$, thus

$$\bar{\mathbf{N}}^{-1} = \begin{bmatrix} 0.01013 & -0.00894 \\ -0.00894 & 0.01363 \end{bmatrix}$$

Next, from equations (13.18a) and (13.18b) compute the following auxiliary matrices

$$\bar{\mathbf{n}} = \underset{2,4}{\ddot{\mathbf{B}}}\ \underset{4,1}{\mathbf{b}} = \begin{bmatrix} 70.37090 \\ 88.10440 \end{bmatrix} \qquad \mathbf{n} = \underset{1,4}{\mathbf{b}^t}\ \underset{4,1}{\mathbf{b}} = [69.99510]$$

and use them and $\bar{\mathbf{N}}^{-1}$ above to compute $\mathbf{F}$ from equation (13.15)

$$\mathbf{F} = \begin{bmatrix} 0.01036 & -0.01066 \\ -0.01066 & 0.92681 \end{bmatrix}$$

which is the first submatrix of the required inverse. We finish up by using equations (13.14b) and (13.14c) to compute the remaining submatrices

$$\mathbf{G} = -\mathbf{F}\bar{\mathbf{n}}\mathbf{n}^{-1} = \begin{bmatrix} 0.00301 \\ -0.02303 \end{bmatrix}$$

$$\mathbf{H} = \mathbf{n}^{-1} - \mathbf{n}^{-1}\bar{\mathbf{n}}^t\mathbf{G} = [0.04025]$$

Thus the total updated inverse is

$$\mathbf{N}^{-1} = \begin{bmatrix} \mathbf{F} & \mathbf{G} \\ \mathbf{G}^t & \mathbf{H} \end{bmatrix} = \begin{bmatrix} 0.01036 & -0.01066 & 0.00301 \\ -0.01066 & 0.02681 & -0.02303 \\ 0.00301 & -0.02303 & 0.04025 \end{bmatrix}$$

which is identical to $\mathbf{N}^{-1}$ computed directly from all seven condition equations. The same example may be reworked to demonstrate *the case of deleting information.* Having $\underset{3,3}{\mathbf{N}^{-1}}$ which corresponds to all seven equations involving three parameters, we wish to obtain the inverse $\underset{2,2}{\dot{\mathbf{N}}^{-1}}$ for only the top three condition equations and two parameters. Consequently, from $\mathbf{N}^{-1}$ only the top left 2×2 block diagonal submatrix, or $\mathbf{F}$ is relevant. Referring to equation (13.23) first compute $\bar{\mathbf{N}}^{-1}$, using $\bar{\mathbf{n}}$ and $\mathbf{n}$ already evaluated

$$\mathbf{F}\bar{\mathbf{n}} = \begin{bmatrix} -0.21015 \\ 1.61193 \end{bmatrix}, \qquad \bar{\mathbf{n}}^t\mathbf{F}\bar{\mathbf{n}} = 127.22924, \qquad \mathbf{n}\bar{\mathbf{n}}^t\mathbf{F}\bar{\mathbf{n}} = 197.22434$$

$$\mathbf{F}\bar{\mathbf{n}}(\mathbf{n} + \bar{\mathbf{n}}^t\mathbf{F}\bar{\mathbf{n}})^{-1}\bar{\mathbf{n}}^t\mathbf{F} = \begin{bmatrix} 0.000224 & -0.001718 \\ -0.001718 & 0.013174 \end{bmatrix}$$

and

$$\bar{\mathbf{N}}^{-1} = \begin{bmatrix} 0.01013 & -0.00894 \\ -0.00894 & 0.01363 \end{bmatrix}$$

which is the same as computed before. Finally, using equation (13.24), compute

$$\dot{\mathbf{N}}^{-1} = \begin{bmatrix} 0.05658 & 0.08823 \\ 0.08823 & 0.18091 \end{bmatrix}$$

which is identical to $\dot{\mathbf{N}}^{-1}$ computed directly from the first three condition equations.

13.3. SEQUENTIAL PROCESSING AND THE UNIFIED APPROACH

The concept of sequential data reduction may be extended to the case of having a priori estimates of functionally independent parameters and their weights. The question of parameter constraints will be dealt with later. Cases of both fixed number of parameters and of variable number of parameters are addressed separately.

13.3.1. Fixed Number of Parameters In the unified approach with functionally independent parameters, we have the observations $\boldsymbol{l}$ and their cofactor matrix $\mathbf{Q}$, and a priori estimates $\mathbf{x}$ for the parameters and an a priori cofactor matrix $\mathbf{Q}_{xx}$. With an approximation vector $\mathbf{x}^0$, a constant vector $\mathbf{f}_x$ was defined by equation (12.24c) in the preceding chapter as

$$\mathbf{f}_x = \mathbf{x}^0 - \mathbf{x} \tag{12.24c}$$

The effect of these additional data is that $\mathbf{W}_{xx}(= \mathbf{Q}_{xx}^{-1})$ is added to the coefficient matrix, and the vector $(\mathbf{W}_{xx}\mathbf{f}_x)$ is subtracted from the constant term vector of the reduced normal equations. Thus if these matrices are redefined as

$$\mathbf{N} = \mathbf{B}^t\mathbf{W}_e\mathbf{B} + \mathbf{W}_{xx} \tag{13.26a}$$

$$\mathbf{t} = \mathbf{B}^t\mathbf{W}_e\mathbf{f} - \mathbf{W}_{xx}\mathbf{f}_x \tag{13.26b}$$

then all the relations in Section 13.2.1 apply directly without any change. For example, equation (13.5a) is used to update the inverse of the coefficient matrix, equation (13.5b) for updating the constant term vector, and equation (13.7) for directly updating the value of the parameter vector. In the earlier case when no a priori information was available, the sequential algorithm was possible only if a set of conditions, which is *more than or equal to the minimum necessary for a unique solution*, was used to evaluate the initial values $\mathbf{N}_0^{-1}$, $\mathbf{t}_0^{-1}$ or $\mathbf{\Delta}_0$. If such was not the case, neither $\mathbf{N}_0^{-1}$ nor $\mathbf{\Delta}_0$ could be computed and the sequential process could not be started.

By contrast, in the presence of a priori information this restriction need not be imposed. In fact, we can begin the sequential procedure *with zero conditions*. In this case

$$\mathbf{N}_0^{-1} = \mathbf{Q}_{xx}, \qquad \mathbf{t}_0 = -\mathbf{W}_{xx}\,\mathbf{f}_x$$

with $\mathbf{x}^0 = \mathbf{0}$ the first value of the parameters, $\mathbf{\Delta}_0$ would simply be equal to $\mathbf{x}$, their a priori estimates. The following rather simple example demonstrates this useful concept.

Example 13.3. Consider a much simplified problem of seeking the elevation of a point C (Figure 13.2) by measuring two differences in elevation from two known points A and B, whose elevations are 5.00 m and 4.00 m, respectively. The observations are $l_1 = 1.74$ m and $l_2 = 2.76$ m with a covariance matrix $\mathbf{\Sigma} = 10^{-4}\mathbf{I}_2$ m^2. Assume further that an a priori value for the elevation of point C (that is, the unknown parameter) is $x = 6.70$ m with a variance of 0.01 m^2. Compute the elevation x of point C.

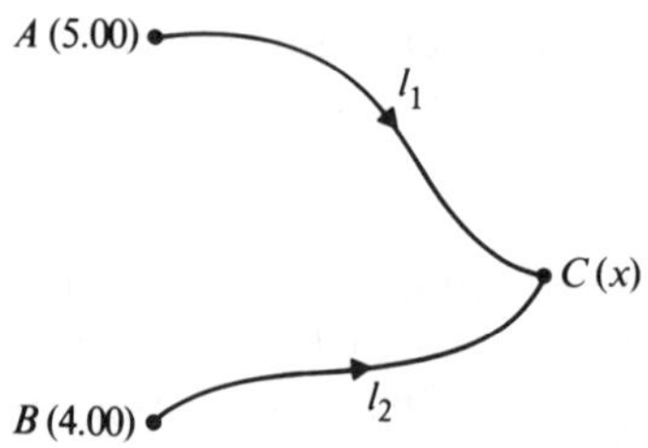

Figure 13.2

Solution: Several procedures are considered.

1. The simplest and most direct way is to begin by writing two condition equations, one for each observation, without regard to the given a priori information about point C. (Note that here $r = 1$, $u = 1$; thus $c = 2 = n$). Hence

$$5 + l_1 - x = 0 \qquad 4 + l_2 - x = 0$$

or

$$\mathbf{v} + \begin{bmatrix} -1 \\ 1 \end{bmatrix} x = \begin{bmatrix} -l_1 & -5 \\ -l_2 & -4 \end{bmatrix} = \begin{bmatrix} -6.74 \\ -6.76 \end{bmatrix}$$

Let $\sigma_0^2 = 10^{-4}$; thus $\mathbf{Q} = \mathbf{I}$ and $q_x = 100$ or $W_{xx} = 0.01$. Thus the normal equation is $2x = 13.50$ or $x = 6.75$ m.

2. Next, consider the given a priori information and write the normal equation

$$(\mathbf{N} + \mathbf{W}_{xx})\boldsymbol{\Delta} = (\mathbf{t} - \mathbf{W}_{xx}\,\mathbf{f}_x)$$

where, according to equation (13.24) and with $x^0 = 0$, $\mathbf{f}_x = -6.7$. Thus

$$(2 + 0.01)x = 13.5 - (0.01)(-6.7) = 13.567$$

and $x \cong 6.75$ m which is the same as in (1) above to two decimal places.

3. In another approach we may perform the solution *in two steps*: the first, using only the first condition equation, with the a priori information, and in the second step computing the contribution of the second condition equation. The first condition equation is

$$v_1 + [-1]x = -6.74$$

which with

$$W_0 = 1, \qquad W_{xx} = 0.01, \qquad \text{and} \qquad f_x = -6.7$$

yields

$$\mathbf{N}_0 = (\mathbf{B}_0^t\,\mathbf{W}_0\,\mathbf{B}_0 + \mathbf{W}_{xx}) = (-1)(1)(-1) + (0.01) = 1.01$$

$$\mathbf{t}_0 = (\mathbf{B}_0^t\,\mathbf{W}_0\,\mathbf{f}_0 - \mathbf{W}_{xx}\,\mathbf{f}_x) = (-1)(1)(-6.74) - (0.01)(-6.7) = 6.807$$

$$\boldsymbol{\Delta}_0 = \mathbf{N}_0^{-1}\mathbf{t}_0 = (1.01)^{-1}(6.807) = 6.7396$$

The second condition equation is

$$v_2 + [-1]x = -6.76$$

with $W_1 = 1$, for which the second term of equation (13.7) becomes

$$\boldsymbol{\delta\Delta} = \mathbf{N}_0^{-1}\mathbf{B}_1^t(\mathbf{A}_1\mathbf{Q}_1\mathbf{A}_1^t + \mathbf{B}_1\mathbf{N}_0^{-1}\mathbf{B}_1^t)^{-1}(\mathbf{f}_1 - \mathbf{B}_1\boldsymbol{\Delta}^0) = 0.01015$$

Hence

$$x = \boldsymbol{\Delta}_0 + \boldsymbol{\delta\Delta} = 6.74975 \cong 6.75 \text{ m}$$

which checks with the above answers to two decimal places.

4. The solution in (3) above was based on using a minimum set of conditions, which is one condition in this simple example, to obtain $\boldsymbol{\Delta}_0$. Here we shall begin with no conditions, then add one condition equation at a time. Therefore for $\boldsymbol{\Delta}_0$

$$\mathbf{B}_0 = \mathbf{0}\mathbf{f}_0 = \mathbf{0}$$

$$\mathbf{N}_0 = (\mathbf{0} + \mathbf{W}_{xx}) = 0.01$$

$$\mathbf{t}_0 = (\mathbf{0} - \mathbf{W}_{xx}\,\mathbf{f}_x) = 0.067$$

$$\boldsymbol{\Delta}_0 = \mathbf{N}_0^{-1}\mathbf{t}_0 = (0.01)^{-1}(0.067) = 6.7 \text{ m}$$

(which was said to be so in the text). Now, with $\mathbf{B}_1 = -1$ and $\mathbf{f}_1 = -6.74$, equation (13.7) yields $\delta\Delta_1 = 0.0396$.

$$\boldsymbol{\Delta}_1 = \boldsymbol{\Delta}_0 + \boldsymbol{\delta\Delta}_1 = 6.7396 \text{ m}$$

[the same as $\boldsymbol{\Delta}_0$ in the two-step solution of (3)].

In order to compute the contribution of the second condition equation, evaluate first $\mathbf{N}_1^{-1}$ from $\mathbf{N}_0^{-1}$, using equation (13.5a) with upper signs,

$$\mathbf{N}_1^{-1} = \mathbf{N}_0^{-1} - \mathbf{N}_0^{-1}\mathbf{B}_1^t(\mathbf{A}_1\mathbf{Q}\mathbf{A}_1^t + \mathbf{B}_1\mathbf{N}_0^{-1}\mathbf{B}_1^t)^{-1}\mathbf{B}_1\mathbf{N}_0^{-1} = (1.01)^{-1}$$

The second condition equation has $\mathbf{B}_2 = -1$ and $\mathbf{f}_2 = -6.76$, for which equation (13.7) gives $\delta\Delta_2 = 0.01015$. Hence

$$x = \mathbf{\Delta}_1 + \boldsymbol{\delta}\mathbf{\Delta}_2 = 6.74975 \cong 6.75 \text{ m}$$

For the sake of completeness, four different techniques were given, all leading to essentially the same solution. A casual consideration for a choice would almost certainly pick up the first technique as it is simple, short, and most direct. This is undeniably true for this nearly trivial example. Further consideration, however, is necessary when large size adjustment problems are involved. Furthermore, when a priori information is available and is realistic, we must use it, which would exclude the first technique.

If the observational data are in fact generated in a sequential manner, it is advisable, and even necessary in certain cases, to perform the adjustment sequentially. In this case several options are available. We may wait until enough observations are accumulated for a minimum unique solution, or we may simply begin with the a priori estimates that are given. The choice may not be simple since the overriding factor will be computational efficiency. Therefore for a given problem and its corresponding computational configuration, both hardware and software, we must analyze the different steps involved to determine which procedure would be more appropriate. As regards the redundant data, the choice of the number of equations to be added (or deleted) at a time would also influence the time of computation, as may be implied from Figure 13.1. Therefore, in a similar manner we ought to look into this aspect as well when designing the sequential scheme to be used for a particular adjustment application in order to arrive at an optimum configuration.

13.3.2. Variable Number of Parameters In the case of a variable number of parameters the condition equations are of the form given in equations (13.10a) and (13.10b). The a priori information would be $\mathbf{x}$ and $\mathbf{W}_{xx}$ for $\mathbf{\Delta}$, and in addition $\mathbf{x}_\delta$ and $\mathbf{W}_{\delta\delta}$ for the new parameters $\boldsymbol{\delta}$. Consequently, besides redefining $\mathbf{N}$ and $\mathbf{t}$ as given in equations (13.26a) and (13.26b), we also need to redefine $\mathbf{n}$ and $\mathbf{t}$ appearing in equations (13.12a) and (13.12b). If $\mathbf{x}_\delta^0$ denotes an approximation vector, then

$$\mathbf{f}_\delta = \mathbf{x}_\delta^0 - \mathbf{x}_\delta \tag{13.27a}$$

and

$$\mathbf{n}_i = \mathbf{b}_i^t \mathbf{W}_{ei} \mathbf{b}_i + \mathbf{W}_{\delta\delta} \tag{13.27b}$$

$$\bar{\mathbf{t}}_i = \mathbf{b}_i^t \mathbf{W}_{ei} \mathbf{f}_i - \mathbf{W}_{\delta\delta} \mathbf{f}_\delta \tag{13.27c}$$

With all the new definitions, every relation developed in Section 13.2.2 can be used directly and without alteration for the situation when a priori parameter information is available.

Example 13.4. Suppose that we have the following six linear condition equations:

$$\begin{aligned}
v_1 + x_1 &= 1.1 \\
v_2 + 2x_1 &= 2.0 \\
v_3 + x_1 + x_2 &= 2.1 \\
v_4 + 2x_1 + x_2 &= 3.0 \\
v_5 + 2x_1 + 2x_2 + x_3 &= 5.0 \\
v_6 + x_1 + x_2 + 2x_3 &= 4.1
\end{aligned}$$

with

$$\underset{6,\,6}{\mathbf{Q}} = \mathbf{I}_6 \qquad \text{and} \qquad \underset{3,\,3}{\mathbf{Q}_{xx}} = 100\mathbf{I}_3$$

and the a priori estimates of the three parameters are all equal to 1.

Required: For the parameters (1) compute the estimates in one step using the a priori information, and (2) compute the estimates sequentially adding one equation at a time.

Solution

1. The matrix **B** and vector **f** are

$$\mathbf{B} = \begin{bmatrix} 1 & 2 & 1 & 2 & 2 & 1 \\ 0 & 0 & 1 & 1 & 2 & 1 \\ 0 & 0 & 0 & 0 & 1 & 2 \end{bmatrix}^t \qquad \mathbf{f} = [1.1 \quad 2 \quad 2.1 \quad 3 \quad 5 \quad 4.1]^t$$

$$\mathbf{N} = \mathbf{B}^t\mathbf{B} = \begin{bmatrix} 15 & 8 & 4 \\ 8 & 7 & 4 \\ 4 & 4 & 5 \end{bmatrix} \qquad \mathbf{t} = \mathbf{B}^t\mathbf{f} = \begin{bmatrix} 27.3 \\ 19.2 \\ 13.2 \end{bmatrix}$$

Because of the a priori estimates of the parameters and the fact that the conditions are linear, $\mathbf{x}^0 = \mathbf{0}$, from equation (13.24c)

$$\mathbf{f}_x = -\mathbf{x} = [-1 \quad -1 \quad -1]^t$$

The corresponding normal equations would therefore be

$$(\mathbf{B}^t\mathbf{B} + \mathbf{W}_{xx})\mathbf{x} = (\mathbf{B}^t\mathbf{f} - \mathbf{W}_{xx}\,\mathbf{f}_x)$$

$$\begin{bmatrix} 15.01 & 8 & 4 \\ 8 & 7.01 & 4 \\ 4 & 4 & 5.01 \end{bmatrix} \begin{bmatrix} x_1 \\ x_2 \\ x_3 \end{bmatrix} = \begin{bmatrix} 27.31 \\ 19.21 \\ 13.21 \end{bmatrix}$$

$$[x_1 \quad x_2 \quad x_3] = [1.015537 \quad 0.990998 \quad 1.034703].$$

2. We begin here by having $\mathbf{W}_{xx} = 0.01$ for the normal equations coefficient matrix and $-\mathbf{W}_{xx}\mathbf{f}_x = 0.01$ for the constant vector, then we add the first equation for which

$$\mathbf{B}_1 = 1, \qquad \mathbf{f}_1 = 1.1, \qquad \mathbf{Q}_1 = 1,$$

therefore

$$\mathbf{N}_1^{-1} = (1.01)^{-1} \qquad \text{and} \qquad \mathbf{t}_1 = 1.11$$

Next,

$$\mathbf{B}_2 = 2, \qquad \mathbf{f}_2 = 2, \qquad \mathbf{Q}_2 = 1, \qquad \mathbf{N}_2^{-1} = (5.01)^{-1}, \qquad \text{and} \qquad \mathbf{t}_2 = 5.11$$

At this point the parameter number will increase from one to two. Thus with a priori $\mathbf{W}_{\delta_2\delta_2} = 0.01$ the matrix $\mathbf{N}_2^{-1}$ should now be expanded to

$$\bar{\mathbf{N}}_2^{-1} = \begin{bmatrix} (5.01)^{-1} & 0 \\ 0 & 100 \end{bmatrix} = \begin{bmatrix} 0.1996 & 0 \\ 0 & 100 \end{bmatrix}$$

$$\mathbf{B}_3 = [1 \quad 1] \qquad \mathbf{f}_3 = 2.1 \qquad \mathbf{Q}_3 = 1 \qquad \mathbf{f}_{\delta_3} = -1 \qquad \mathbf{W}_{\delta_3\delta_3} = 0.01$$

$$\mathbf{N}_3^{-1} = \bar{\mathbf{N}}_2^{-1}[\mathbf{I}_2 - \mathbf{B}_3^t(\mathbf{I} + \mathbf{B}_3\,\bar{\mathbf{N}}_2^{-1}\mathbf{B}_3^t)^{-1}\mathbf{B}_3\,\bar{\mathbf{N}}_2^{-1}] = \begin{bmatrix} 0.199206 & -0.197225 \\ -0.197225 & 1.19 \end{bmatrix}$$

The constant term vector also needs to be expanded to

$$\bar{\mathbf{t}}_2 = \begin{bmatrix} 5.11 \\ -\mathbf{W}_{\delta_3\delta_3}\,\mathbf{f}_{\delta_3} \end{bmatrix} = \begin{bmatrix} 5.11 \\ 0.01 \end{bmatrix} \qquad \mathbf{t}_3 = \bar{\mathbf{t}}_2 + \mathbf{B}_3^t\,\mathbf{f}_3 = \begin{bmatrix} 7.21 \\ 2.11 \end{bmatrix}$$

$$\mathbf{B}_4 = [2 \quad 1] \qquad \mathbf{f}_4 = 3.0 \qquad \mathbf{Q}_3 = 1$$

$$\mathbf{N}_4^{-1} = \begin{bmatrix} 0.18079 & -0.270045 \\ -0.270045 & 0.902047 \end{bmatrix} \qquad \text{and} \qquad \mathbf{t}_4 = \begin{bmatrix} 13.21 \\ 5.11 \end{bmatrix}$$

$$\mathbf{B}_5 = [2 \quad 2 \quad 1] \qquad \mathbf{f}_5 = 5.0 \qquad \mathbf{Q}_5 = 1 \qquad \mathbf{f}_{\delta_5} = -1$$

$$\bar{\mathbf{N}}_4^{-1} = \begin{bmatrix} 0.18079 & -0.270045 & 0 \\ & 0.902047 & 0 \\ \text{symmetric} & & 100 \end{bmatrix} \qquad \text{and} \qquad \bar{\mathbf{t}}_4 = \begin{bmatrix} 13.21 \\ 5.11 \\ 0.01 \end{bmatrix}$$

$$\mathbf{N}_5^{-1} = \begin{bmatrix} 0.180446 & -0.267606 & 0.172593 \\ & 0.884764 & -1.222095 \\ \text{symmetric} & & 3.068322 \end{bmatrix} \qquad \mathbf{t}_5 = \begin{bmatrix} 23.21 \\ 15.11 \\ 5.01 \end{bmatrix}$$

$$\mathbf{B}_6 = [1 \quad 1 \quad 2] \qquad \mathbf{f}_6 = 4.1 \qquad \mathbf{Q}_6 = 1$$

$$\mathbf{N}_6^{-1} = \begin{bmatrix} 0.173515 & -0.218526 & 0.035937 \\ & 0.537242 & -0.254463 \\ \text{symmetric} & & 0.374073 \end{bmatrix} \qquad \mathbf{t}_6 = \begin{bmatrix} 27.31 \\ 19.21 \\ 13.21 \end{bmatrix}$$

and

$$[x_1 \quad x_2 \quad x_3] = [1.015538 \quad 0.991017 \quad 1.034710]$$

which agree well with the values obtained previously, the differences being due to round-off errors.

13.3.3. An Alternative Direct Scheme The solution in the second part of Example 7.4 suggests a general scheme that may be applied equally well to cases of both a fixed and variable number of parameters, and for addition as well as deletion of condition equations. The only requisite for this scheme is that *all the parameters involved must have a priori estimates and corresponding a priori weight matrices.* Because of the allowance for varying the number of parameters, the case of adding conditions differs somewhat from that of deleting them.

THE DIRECT SCHEME FOR ADDING CONDITIONS The number of parameters is assumed to be variable since the case of fixed parameters is a special case (see also Section 13.3.1 and Example 13.3). Thus when a group of parameters are to be included in the adjustment an a priori estimate of their values and their weight matrix, $\mathbf{W}_{xx}$, must be given. If two subvectors of parameters $\mathbf{\Delta}_i$ and $\mathbf{\Delta}_j$ enter the adjustment sequentially ($j = i + 1$), the assumption will be made that only their respective a priori weight matrices $\mathbf{W}_{x_ix_i}$ and $\mathbf{W}_{x_jx_j}$ are given. This implies that they are not correlated, which is a totally practical assumption. At the end of the $(i - 1)$ cycle we have $\mathbf{N}_{i-1}^{-1}$ with dimensions $(u_{i-1} \times u_{i-1})$ and $\mathbf{t}_{i-1}$ for a computation of $\mathbf{\Delta}_{i-1}$ of size u_{i-1}. For the ith cycle, we add the conditions

$$\underset{s_i, t_i}{\mathbf{A}_i}\ \underset{t_i, 1}{\mathbf{v}_i} + \underset{s_i, u_i}{\mathbf{B}_i}\ \underset{u_i, 1}{\mathbf{\Delta}_i} = \underset{s_i, 1}{\mathbf{f}_i} \tag{13.28}$$

where $\mathbf{\Delta}_i$ is increased over $\mathbf{\Delta}_{i-1}$ by $\mathbf{w}_i = u_i - u_{i-1}$. Note that s_i must be larger than w_i in order that equation (13.28) may contribute to the adjustment. The added subvector of parameters $\boldsymbol{\delta}_i$ would have an a priori estimate. Furthermore, we must also have the a priori weight matrix $\mathbf{W}_{\delta_i\delta_i}$ which is $(w_i \times w_i)$. Thus before computing $\mathbf{N}_i^{-1}$ from $\mathbf{N}_{i-1}^{-1}$, the latter must be expanded from a size of $u_{i-1} \times u_{i-1}$ to the larger size of $u_i \times u_i$, as follows:

$$\underset{u_i, u_i}{\bar{\mathbf{N}}_{i-1}^{-1}} = \begin{bmatrix} \mathbf{N}_{i-1}^{-1} & 0 \\ \mathbf{0} & \mathbf{W}_{\delta_i\delta_i}^{-1} \end{bmatrix} \tag{13.29}$$

Then

$$\mathbf{N}_i^{-1} = \bar{\mathbf{N}}_{i-1}^{-1}[\mathbf{I} - \mathbf{B}_i^t(\mathbf{A}_i\,\mathbf{Q}_i\,\mathbf{A}_i^t + \mathbf{B}_i\,\bar{\mathbf{N}}_{i-1}^{-1}\mathbf{B}_i^t)^{-1}\mathbf{B}_i\,\bar{\mathbf{N}}_{i-1}^{-1}] \tag{13.30}$$

realizing, of course, that $\mathbf{Q}_i$ is the cofactor matrix of the new set of observations which is assumed by necessity to be uncorrelated with the preceding set already used. If the a priori estimate corresponding to $\boldsymbol{\delta}_i$ is $\mathbf{x}_{\delta_i}$ and $\mathbf{x}_{\delta_i}^0$ is an approximate vector, then $\mathbf{f}_{\delta_i}$ is computed from equation (13.27a). The constant term vector $\mathbf{t}_{i-1}$ must also be expanded from a $u_{i-1} \times 1$ to a $u_i \times 1$ vector, or

$$\bar{\mathbf{t}}_{i-1} = \begin{bmatrix} \mathbf{t}_{i-1} \\ -\mathbf{W}_{\delta_i\delta_i}\ \mathbf{f}_{\delta_i} \end{bmatrix} \tag{13.31}$$

and then

$$\mathbf{t}_i = \bar{\mathbf{t}}_{i-1} + \mathbf{B}_i^t(\mathbf{A}_i \mathbf{Q}_i \mathbf{A}_i^t)^{-1}\mathbf{f}_i$$

which completes the computation of $\mathbf{N}_i$ and $\mathbf{t}_i$ necessary for evaluating $\mathbf{\Delta}_i$, the answer at the ith step.

THE DIRECT SCHEME FOR DELETING CONDITIONS At the ith step of editing, for example, we may delete the effect of the condition equations (13.28) from the solution available as given by $\mathbf{N}_{i-1}^{-1}$ and $\mathbf{t}_{i-1}$, both of size u_{i-1}. For the sake of generality, we assume that u_{i-1} is larger than u_i, that is, the number of parameters decreases, by say w_i, when the effect of equation (13.28) is removed from the solution. Obviously, then, $u_i = u_{i-1} - w_i$. The matrix $\mathbf{N}_{i-1}^{-1}$ is first partitioned such that

$$\underset{u_{i-1},\, u_{i-1}}{\mathbf{N}_{i-1}^{-1}} = \begin{bmatrix} \underset{u_i,\, u_i}{\bar{\mathbf{N}}_{i-1}} & \underset{u_i,\, w_i}{\bar{\mathbf{n}}_{i-1}} \\ \underset{w_i,\, u_i}{\bar{\mathbf{n}}_{i-1}^t} & \underset{w_i,\, w_i}{\mathbf{n}_{i-1}} \end{bmatrix} \tag{13.32}$$

then the new inverse is obtained from submatrix $\bar{\mathbf{N}}_{i-1}$

$$\mathbf{N}_i^{-1} = \bar{\mathbf{N}}_{i-1}[\mathbf{I} + \mathbf{B}_i^t(\mathbf{A}_i \mathbf{Q}_i \mathbf{A}_i^t - \mathbf{B}_i \bar{\mathbf{N}}_{i-1}\mathbf{B}_i^t)^{-1}\mathbf{B}_i \bar{\mathbf{N}}_{i-1}] \tag{13.33}$$

in which $\mathbf{Q}_i$ is the cofactor matrix of the observations associated with $\mathbf{v}_i$. The remaining submatrices $\bar{\mathbf{n}}_{i-1}$ and $\mathbf{n}_{i-1}$ are not used, as they become irrelevant with regard to the new set of parameters $\mathbf{\Delta}_i$. It should also be pointed out that if no change in parameters is warranted in the ith step (because the conditions to be deleted do not reduce the number of parameters), $\bar{\mathbf{N}}_{i-1}$ will be the total matrix $\mathbf{N}_{i-1}^{-1}$ which must then be used directly in equation (13.33) to compute $\mathbf{N}_i^{-1}$. In a similar fashion $\mathbf{t}_{i-1}$ is partitioned

$$\mathbf{t}_{i-1} = \begin{bmatrix} \bar{\mathbf{t}}_{i-1} \\ \mathbf{t}'_{i-1} \end{bmatrix} \tag{13.34}$$

and the new constant term vector computed from

$$\mathbf{t}_i = \bar{\mathbf{t}}_{i-1} - \mathbf{B}_i^t(\mathbf{A}_i \mathbf{Q}_i \mathbf{A}_i^t)^{-1}\mathbf{f}_i \tag{13.35}$$

The same remark about the case of no change in the number of parameters also applies here, thus $\bar{\mathbf{t}}_{i-1}$ would be equal to $\mathbf{t}_{i-1}$.

Example 13.5. Given the following four linear condition equations, $\mathbf{Q} = \mathbf{I}_4$, $\mathbf{Q}_{xx} = 100\mathbf{I}_2$, and the a priori estimates are both equal to 1.

$$\begin{aligned} v_1 + x_1 \qquad\quad &= 1.1 \\ v_2 + 2x_1 \qquad\quad &= 2.0 \\ v_3 + x_1 + x_2 &= 2.1 \\ v_4 + 2x_1 + x_2 &= 3.0 \end{aligned}$$

Compute first the estimates of x_1 and x_2 from all four conditions; then removing sequentially the last two equations verify that the value of x_1 thus obtained is the same as that obtained from solving the first two conditions directly by least squares.

Solution

$$\mathbf{B}_0 = \begin{bmatrix} 1 & 2 & 1 & 2 \\ 0 & 0 & 1 & 1 \end{bmatrix}^t \qquad \mathbf{f}_0 = [1.1 \quad 2.0 \quad 2.1 \quad 3.0]^t$$

$$\mathbf{N}_0 = \mathbf{B}_0^t \mathbf{B}_0 + \mathbf{Q}_{xx}^{-1} = \begin{bmatrix} 10.01 & 3. \\ 3. & 2.01 \end{bmatrix}$$

$$\mathbf{N}_0^{-1} = \frac{1}{11.1201} \begin{bmatrix} 2.01 & -3 \\ -3 & 10.01 \end{bmatrix} \qquad \mathbf{f}_x = \begin{bmatrix} -1 \\ -1 \end{bmatrix}$$

$$\mathbf{t}_0 = \mathbf{B}_0^t \mathbf{f}_0 - \mathbf{W}_{xx} \mathbf{f}_x = \begin{bmatrix} 13.21 \\ 5.11 \end{bmatrix} \qquad \begin{bmatrix} x_1 \\ x_2 \end{bmatrix} = \begin{bmatrix} 1.009 \\ 1.036 \end{bmatrix}$$

Now to remove the effect of the fourth equation

$$\mathbf{A}_1 = 1 \qquad \mathbf{B}_1 = [2 \quad 1] \qquad \mathbf{f}_1 = 3.0 \qquad \mathbf{Q}_1 = 1$$

Because the number of parameters remains unaltered in this step, then $\bar{\mathbf{N}}_0^{-1} = \mathbf{N}_0^{-1}$ and we apply equation (13.33) to compute $\mathbf{N}_1^{-1}$

$$\mathbf{N}_1^{-1} = \begin{bmatrix} 0.19920711 & -0.19723476 \\ \text{symmetric} & 1.18538095 \end{bmatrix} \qquad \mathbf{t}_1 = \begin{bmatrix} 7.21 \\ 2.11 \end{bmatrix}$$

To remove the effect of the third equation

$$\mathbf{A}_2 = 1 \qquad \mathbf{B}_2 = [1 \quad 1] \qquad \mathbf{f}_2 = 2.1 \qquad \mathbf{Q}_2 = 1$$

Since the number of parameters still remains unchanged we need not modify the solution from the above step but simply take $\bar{\mathbf{N}}_1 = \mathbf{N}_1^{-1}$ and $\bar{\mathbf{t}}_1 = \mathbf{t}_1$. Equation (13.33) thus yields

$$\mathbf{N}_2^{-1} = \begin{bmatrix} 0.19960079 & 0.00000027 \\ \text{symmetric} & 100.00001894 \end{bmatrix}, \qquad \mathbf{t}_2 = \begin{bmatrix} 5.11 \\ 0.01 \end{bmatrix} \qquad \text{and}$$

$$\begin{bmatrix} x_1 \\ x_2 \end{bmatrix}_2 = \begin{bmatrix} 1.019960 \\ 1.000002 \end{bmatrix}$$

Returning to the original set of conditions, the first *two equations* yield

$$\mathbf{B} = \begin{bmatrix} 1 \\ 2 \end{bmatrix}, \qquad \mathbf{f} = \begin{bmatrix} 1.1 \\ 2.0 \end{bmatrix}, \qquad \mathbf{W} = \mathbf{I}_2, \qquad \mathbf{W}_{x_1x_1} = 0.01, \qquad \text{and} \qquad \mathbf{f}_{x_1} = -1$$

$$\mathbf{N} = \mathbf{B}^t\mathbf{B} + \mathbf{W}_{x_1x_1} = 5.01 \qquad \mathbf{N}^{-1} = 0.19960079$$

$$\mathbf{t} = \mathbf{B}^t\mathbf{f} - \mathbf{W}_{x_1x_1}\mathbf{f}_{x_1} = 5.11 \qquad x_1 = 1.019961$$

Note that $\mathbf{N}^{-1}$ is exactly the same value as the (1, 1) element of $\mathbf{N}_2^{-1}$ and $\mathbf{x}_1$ differs only in the sixth decimal place. This shows that aside from rounding off error the direct scheme is both rigorous and general. Its generality is demonstrated by the fact that $\mathbf{x}_2$ computed after deleting the third and fourth equations is nothing but its a priori estimate of 1.00, except for round-off error.

To demonstrate the reduction in the number of parameters, we extend the problem one more step by subtracting the effect of the second equation for which $\mathbf{B}_3 = 2$, $\mathbf{f}_3 = 2.0$, and $\mathbf{Q}_3 = 1$. Here only a submatrix of appropriate dimensions (in this case a scalar) from $\mathbf{N}_2^{-1}$ is needed. Thus $\bar{\mathbf{N}}_2 = 0.19960079$ and equation (13.33) leads to $\mathbf{N}_3^{-1} = 0.99009885$. Similarly, a subvector of $\mathbf{t}_2$, $\bar{\mathbf{t}}_2 = 5.11$, is relevant, hence $\mathbf{t}_3 = 1.11$. From the first equation directly

$$\mathbf{N} = \mathbf{B}'\mathbf{B} + \mathbf{W}_{xx} = 1.01 \qquad \mathbf{N}^{-1} = (1.01)^{-1} = 0.99009900$$

$$\mathbf{t} = \mathbf{B}'\mathbf{B} - \mathbf{W}_{xx}\,\mathbf{f}_x = 1.11$$

which agree well with those computed by the sequential scheme.

13.4. SEQUENTIAL REDUCTION, THE UNIFIED APPROACH, AND PARAMETER CONSTRAINTS

The presence of parameter constraints poses no particular difficulties as long as the unified approach is applied. If the classical methods were used, there would be several special cases depending on whether the constraints were available at the beginning or arose with the new conditions. Since in the unified approach no distinction is made between conditions and constraints, the present case can be considered as essentially the same as that given in Section 13.3. Therefore no special derivation is necessary here and a numerical example is given instead to illustrate the points of treatment that are applicable to the present case.

Example 13.6. Given are two conditions and one "constraint,"

$$\mathbf{v} + \begin{bmatrix} 1 & 1 \\ 1 & -1 \end{bmatrix} \begin{bmatrix} x_1 \\ x_2 \end{bmatrix} = \begin{bmatrix} 2.1 \\ 0.1 \end{bmatrix}$$

of the form

$$\mathbf{v} + \mathbf{B}\boldsymbol{\Delta} = \mathbf{f}$$

and

$$\mathbf{v}_c + [1 \quad 0] \begin{bmatrix} x_1 \\ x_2 \end{bmatrix} = 1.0$$

of the form

$$\mathbf{v}_c + \mathbf{C}\boldsymbol{\Delta} = \mathbf{g}$$

The a priori weight matrix associated with the first two equations is to be taken as $\mathbf{W} = \mathbf{I}_2$, and the weight associated with the third equation is $\mathbf{w}_c = 100$; $\mathbf{w}_c$ must be large relative to $\mathbf{W}$ in order to effect a practical enforcement of the third equation as a parameter constraint. The a priori estimate of the parameters is

$$\mathbf{x} = \begin{bmatrix} 1.0 \\ 1.0 \end{bmatrix}$$

with the a priori weight matrix $\mathbf{W}_{xx} = 0.01\mathbf{I}_2$.

With the above data a solution is assumed to be found at the start. Then that solution is to be modified due to the addition of another equation,

$$\mathbf{v}_3 + [-1 \quad -1]\begin{bmatrix} x_1 \\ x_2 \end{bmatrix} = -2.2$$

with $\mathbf{w}_3 = 1.0$.

Solution: For the first part of the data we compute the normal equations as

$$\mathbf{N}_1 = \mathbf{B}^t\mathbf{W}\mathbf{B} + \mathbf{C}^t\mathbf{W}_c\,\mathbf{C} + \mathbf{W}_{xx} = \begin{bmatrix} 102.01 & 0 \\ 0 & 2.01 \end{bmatrix}$$

$$\mathbf{N}_1^{-1} = \begin{bmatrix} (102.01)^{-1} & 0 \\ 0 & (2.01)^{-1} \end{bmatrix}, \qquad \mathbf{f}_x = \begin{bmatrix} -1 \\ -1 \end{bmatrix}$$

$$\mathbf{t}_1 = \mathbf{B}^t\mathbf{W}\mathbf{f} + \mathbf{C}^t\mathbf{W}_c\,\mathbf{B} - \mathbf{W}_{xx}\,\mathbf{f}_x = \begin{bmatrix} 102.21 \\ 2.01 \end{bmatrix}$$

The answer for the first part is therefore

$$\boldsymbol{\Delta}_1 = \begin{bmatrix} x_1 \\ x_2 \end{bmatrix} = \mathbf{N}_1^{-1}\mathbf{t}_1 = \begin{bmatrix} 1.002 \\ 1.000 \end{bmatrix}$$

For the second part

$$\mathbf{N}_2^{-1} = \frac{1}{309.06}\begin{bmatrix} 3.01 & -1 \\ -1 & 103.01 \end{bmatrix}, \qquad \mathbf{t}_2 = \begin{bmatrix} 104.41 \\ 4.21 \end{bmatrix}$$

thus

$$\boldsymbol{\Delta}_2 = \begin{bmatrix} x_1 \\ x_2 \end{bmatrix} = \begin{bmatrix} 1.003 \\ 1.065 \end{bmatrix}$$

In Example 13.6 the number of parameters remained fixed through the sequential treatment. As an illustration of how to apply the unified approach sequentially with constraints in the case of variable number of parameters, we shall take another numerical example.

Example 13.7. The original data include two conditions and one constraint:

$$\mathbf{v} + \begin{bmatrix} 1 \\ 2 \end{bmatrix} x_1 = \begin{bmatrix} 1.1 \\ 2.0 \end{bmatrix}$$

of the form

$$\mathbf{v} + \mathbf{B}\boldsymbol{\Delta} = \mathbf{f}$$

and

$$\mathbf{v}_c + x_1 = 1$$

of the form

$$\mathbf{v}_c + \mathbf{C}\mathbf{\Delta} = \mathbf{g}$$

for which

$$\mathbf{W} = \mathbf{I}_2, \qquad w_c = 100, \qquad x_1 = 1.0, \qquad \text{and} \qquad w_{xx} = 0.01$$

After obtaining the solution for this data the following conditions are to be added:

$$\mathbf{v}_1 + \begin{bmatrix} 1 \\ 2 \end{bmatrix} x_1 + \begin{bmatrix} 1 \\ 1 \end{bmatrix} x_2 = \begin{bmatrix} 2.1 \\ 3.0 \end{bmatrix}$$

of the form

$$\mathbf{v}_1 + \mathbf{B}_1 \mathbf{\Delta} + \mathbf{b}\boldsymbol{\delta} = \mathbf{f}_1$$

with

$$\mathbf{W} = \mathbf{I}_2, \qquad x_2 = 1.0, \qquad \text{and} \qquad w = 0.01$$

Solution: For the first part,

$$\mathbf{f}_{x_1} = -x_1 = -1.0, \qquad \text{and} \qquad \mathbf{f}_\delta = 1.0$$

thus

$$\mathbf{N} = \mathbf{B}'\mathbf{W}\mathbf{B} + \mathbf{C}'\mathbf{W}_c\,\mathbf{C} + \mathbf{W}_{xx} = 105.01$$

$$\mathbf{t} = \mathbf{B}'\mathbf{W}\mathbf{f} + \mathbf{C}'\mathbf{W}_c\,\mathbf{g} - \mathbf{W}_{xx}\,\mathbf{f}_{x_1} = 105.11$$

$$\mathbf{\Delta} = 1.0010$$

With $\mathbf{A}_1 = \mathbf{I}$, $(\mathbf{AQA}') = 1$, thus

$$\mathbf{n}_1^{-1} = (\mathbf{b}'(\mathbf{AQA}')_1^{-1}\mathbf{b} + \mathbf{W}_{\delta\delta})^{-1} = (2.01)^{-1}$$

$$\bar{\mathbf{n}}_1 = \mathbf{B}_1'(\mathbf{AQA}')_1^{-1}\mathbf{b} = 3, \qquad \mathbf{B}_1'(\mathbf{AQA}')_1^{-1}\mathbf{f}_1 = 8.1$$

$$\bar{\mathbf{t}} = \mathbf{b}'(\mathbf{AQA}')_1^{-1}\mathbf{f}_1 - \mathbf{W}_{\delta\delta}\,\mathbf{f}_\delta = 5.11$$

Now equation (13.17) yields $\bar{\mathbf{N}}^{-1} = (110.01)^{-1}$.

To compute $\mathbf{F}$ use equation (13.15),

$$(\mathbf{n}_1 - \bar{\mathbf{n}}_1'\bar{\mathbf{N}}_1^{-1}\bar{\mathbf{n}}_1)^{-1} = \frac{110.01}{212.1201}$$

$$\bar{\mathbf{n}}_1(\mathbf{n}_1 - \bar{\mathbf{n}}_1'\bar{\mathbf{N}}_1^{-1}\bar{\mathbf{n}}_1)^{-1}\bar{\mathbf{n}}_1'\mathbf{n}_1^{-1} = \frac{9}{212.1201}$$

$$\mathbf{F}_1 = (110.01)^{-1}\frac{221.1201}{212.1201}$$

From equation (13.14b)

$$\mathbf{G}_1 = -\mathbf{F}_1\bar{\mathbf{n}}_1\mathbf{n}_1^{-1} = -\frac{3}{313.1201}$$

$$\mathbf{H}_1 = \mathbf{n}_1^{-1} - \mathbf{n}_1^{-1}\bar{\mathbf{n}}_1^t\mathbf{G}_1 = (2.01)^{-1}\frac{221.1201}{212.1201}$$

$$\mathbf{N}_1^{-1} = \begin{bmatrix} \mathbf{F} & \mathbf{G} \\ \mathbf{G}^t & \mathbf{H} \end{bmatrix} = \frac{1}{212.1201}\begin{bmatrix} 2.01 & -3 \\ -3 & 110.01 \end{bmatrix}$$

Finally, according to equation (13.12b),

$$\mathbf{t} = \begin{bmatrix} \mathbf{t} + \mathbf{B}_1^t(\mathbf{AQA}^t)_1^{-1}\mathbf{f}_1 \\ \mathbf{t} \end{bmatrix} = \begin{bmatrix} 113.21 \\ 5.11 \end{bmatrix}, \qquad \boldsymbol{\Delta} = \begin{bmatrix} 1.0005 \\ 1.0490 \end{bmatrix}$$

In both Examples 13.6 and Example 13.7 the constraints were given at the start. The cases in which constraints arise later pose no new points for consideration since both types of conditions are the same. Consequently, no particular treatment of such cases is necessary. Finally, cases of information deletion are sufficiently straightforward and need no further elaboration.

13.5. PRECISION ESTIMATION

The question of a posteriori cofactor matrices is not much different from the regular practice since at the end of a sequential treatment we end up with the same information as would be obtained from the direct batch least squares. For fixed number of parameters, $\mathbf{Q}_{\Delta\Delta}$ will simply be equal to the new inverse of the normal equations matrix after either addition or deletion of information. Thus in this case equation (13.5a) may be used for evaluating $\mathbf{Q}_{\Delta\Delta}$. In case we are working with a variable number of parameters, equation (13.13) may be used for the addition of information and equation (13.21b) for deleting information. All we need to remember is to use the *updated* inverse of the normal equations coefficient matrix as the required cofactor matrix $\mathbf{Q}_{\Delta\Delta}$.

With regard to Section 13.3 and the use of the unified approach, the same equations referred to above also apply. The only difference is that a priori $\mathbf{W}_{xx}$ is included here in $\mathbf{N}^{-1}$.

Precision estimation is not much different for the situation when parameter constraints are present as given in Section 13.4. Since the unified approach is applied, no modification to the above statements is needed.

13.6. SEQUENTIAL PROCESSING AND NONLINEAR EQUATIONS

All derivations in this chapter concerned linear functions. For nonlinear problems, in addition to linearization by series expansion, further discussion is needed when considering sequential or recursive procedures. As an illustration we shall consider the case of addition of information when the number of parameters is fixed. Let a set of nonlinear conditions be

$$F(l, \mathbf{x}) = \mathbf{0} \tag{13.36}$$

and its linearized form

$$\underset{c,n}{\mathbf{A}}\ \underset{n,1}{\mathbf{v}} + \underset{c,u}{\mathbf{B}}\ \underset{u,1}{\mathbf{\Delta}} = \underset{c,1}{\mathbf{f}^0} \tag{13.37}$$

where **A**, **B**, and $\mathbf{f}^0$ are evaluated at approximate vector $\mathbf{x}_0^0$ for the parameters (disregarding the aspect of iterating on the observations, for the sake of simplicity). From equation (13.37) a system of u normal equations may be formed and solved for the first vector of parameter corrections $\mathbf{\Delta}_1$. Updated approximations to be used in the second iteration would be

$$\mathbf{x}_1 = \mathbf{x}_0^0 + \mathbf{\Delta}_1$$

The process may be repeated many (m) times leading to the eventual answer

$$\hat{\mathbf{x}} = \mathbf{x}_0^0 + \sum_{k=1}^{m} \mathbf{\Delta}_k \tag{13.38}$$

Let an additional set of conditions be

$$\mathbf{F}_a(l, \mathbf{x}) = 0 \tag{13.39}$$

The combination of equations (13.36) and (13.39) gives a larger set of condition equations from which the final estimate $\mathbf{x}_f$ of the u parameters will obviously be different from that given by equation (13.38). The most direct process (which may not necessarily be the most efficient) is to linearize equation (13.39) at an approximation vector $\mathbf{x}_f^0$ equal to $\hat{\mathbf{x}}$ obtained from equation (13.38) to give

$$\underset{p,n}{\mathbf{A}_a}\ \underset{n,1}{\mathbf{v}} + \underset{p,u}{\mathbf{B}_a}\ \underset{u,1}{\mathbf{\Delta}_t} = \underset{p,1}{\mathbf{f}^0} \tag{13.40}$$

Next, equation (13.37) is re-evaluated at the same value $\mathbf{x}_f^0$ and combined with equation (13.40) to form a total set of $(c + p)$ condition equations. The corresponding normal equations are formed and solved to obtain a correction, $\mathbf{\Delta}_t$, and the solution is repeated, if necessary, until convergence is achieved. It should be noted that in this process the normal equation matrix of size u will have to be inverted every time a correction vector is computed.

Another possibility occurs when the equations in (13.37) are stored and kept until all the additional equations in (13.39) become available. In this case, equation (13.39) is linearized at the approximation vector $\mathbf{x}_0^0$ used for equation (13.36) and the set of linear equations is combined with equation (13.37). Normal equations are formed and solved and relinearizations carried out until the solution converges to a preset tolerance.

A number of possibilities arise when sequential procedures are used for nonlinear problems. Some of these possibilities entail exact solutions whereas others involve certain approximations. To begin with, the sequential algorithm is exact when used inside each linearization. For example, suppose that $\mathbf{\Delta}_{t1}$ is obtained from the first cycle of the direct solution for equations (13.36) and (13.39) together (both linearized at the same approximation vector $\mathbf{x}_0^0$). If, then, the inverse of the normal equation matrix and constant term vector arising from the set of equations (13.36) (linearized at $\mathbf{x}_0^0$) are modified by the sequential algorithm to include equation (13.39) (also linearized at $\mathbf{x}_0^0$), the product of the resulting inverse and constant term vector will give $\mathbf{\Delta}_{t1}$ exactly. For the nonlinear case the set of equations in (13.37) is solved by the linearization process as shown in equation (13.38). The new set of equations in (13.40) is then added by the sequential algorithm after being linearized with the best estimates of the parameters from the original set of conditions, that is, $\hat{\mathbf{x}}$ from equation (13.38). In this case the resulting estimated parameter $\mathbf{\Delta}_t$ will not necessarily be precisely the same $\mathbf{\Delta}_{t1}$ estimated directly from linear least squares solution for all condition equations considered together. Obviously, the amount of difference between $\mathbf{\Delta}_t$ from the sequential solution and $\mathbf{\Delta}_{t_1}$ from the batch solution will depend, among other things, on the quality of the first vector of approximations, $\mathbf{x}_0^0$.

To obtain an appreciation of this method of sequential reduction in the case of nonlinear models, the solution of different cases of a relative orientation problem of a pair of aerial photographs is given in Table 13-1. First, six points were used in a regular batch least squares and the solution iterated until convergence. The answers—that is, the values of the five relative orientation parameters—from this case are given in the first line. Adding three more points, the second line gives the results from a batch solution in which all nine points are used directly, and the fifth line gives the corresponding

TABLE 13-1 Comparison of Batch and Sequential Solution of the Nonlinear Model of Relative Orientation

TYPE OF SOLUTION		ESTIMATED PARAMETERS				
BATCH POINTS	SEQUENTIAL POINTS	Y(mm)	Z(mm)	ω	ϕ	κ
6	0	0.179	151.766	1′ 42.851″	36′ 0.070″	6′ 17.920″
9	0	0.168	151.770	1′ 56.564″	35′ 59.249″	6′ 19.278″
15	0	0.158	151.768	2′ 07.057″	36′ 33.944″	6′ 12.202″
40	0	0.152	151.773	2′ 12.462″	36′ 38.931″	6′ 12.551″
6	3	0.168	151.770	1′ 56.573″	35′ 59.238″	6′ 19.277″
6	9	0.158	151.768	2′ 07.097″	36′ 33.939″	6′ 12.207″
6	34	0.152	151.773	2′ 12.462″	36′ 38.931″	6′ 12.551″

results when the three points are added sequentially. In the third and sixth lines the results of a total of 15 points are given, and those in the fourth and seventh lines are for 40 points. The reader will note that when extensive redundancy exists, such as the case of 40 points, the two solutions are identical. For other cases there is some difference between the two methods of solution reflecting the already mentioned fact that the sequential algorithm is not exact for nonlinear cases. However, such differences, at least for the example given, are so small they can be neglected. Of course, it must be emphasized that this is only one example that shows tendencies, and other examples from different applications would demonstrate the concept.

13.7. SUMMARY OF EQUATIONS AND CONCLUDING REMARKS

13.7.1. Summary of Equations For a case of fixed number of parameters, given $\mathbf{N}_{(i-1)}^{-1}$, $\mathbf{t}_{(i-1)}$, and a new set of conditions $\mathbf{A}_i \mathbf{v}_i + \mathbf{B}_i \mathbf{\Delta} = \mathbf{f}_i$ with $\mathbf{Q}_i$, then

$$\mathbf{Q}_{ei} = (\mathbf{AQA}^t)_i \qquad \mathbf{W}_{ei} = \mathbf{Q}_{ei}^{-1}$$

$$\mathbf{N}_i^{-1} = \mathbf{N}_{(i-1)}^{-1}[\mathbf{I} \mp \mathbf{B}_i^t(\mathbf{Q}_{ei} \pm \mathbf{B}_i \mathbf{N}_{(i-1)}^{-1} \mathbf{B}_i^t)^{-1}\mathbf{B}_i \mathbf{N}_{(i-1)}^{-1}] \tag{13.5a}$$

$$\mathbf{t}_i = \mathbf{t}_{(i-1)} \pm \mathbf{B}_i^t \mathbf{W}_{ei} \mathbf{f}_i \tag{13.5b}$$

with upper signs for addition of conditions whereas the lower signs are for subtraction of conditions. Alternatively, given $\mathbf{N}_{(i-1)}^{-1}$ and $\mathbf{\Delta}_{i-1}$, the effect of the new conditions would be

$$\mathbf{\Delta}_i = \mathbf{\Delta}_{i-1} \pm \mathbf{N}_{(i-1)}^{-1} \mathbf{B}_i^t(\mathbf{Q}_{ei} \pm \mathbf{B}_i \mathbf{N}_{(i-1)}^{-1} \mathbf{B}_i^t)^{-1}(\mathbf{f}_i - \mathbf{B}_i \mathbf{\Delta}_{i-1}) \tag{13.7}$$

again with upper signs for condition addition and vice versa.

For a case of variable number of parameters, given $\mathbf{N}_{i-1}^{-1}$, $\mathbf{t}_{i-1}$, and the new conditions $\mathbf{A}_i \mathbf{v}_i + \mathbf{B}_i \mathbf{\Delta} + \mathbf{b}_i \mathbf{\delta} = \mathbf{f}_i$ with $\mathbf{Q}_i$, and where $\mathbf{\delta}$ is an added set of parameters, then for condition addition,

$$\bar{\mathbf{N}}_i^{-1} = \mathbf{N}_{i-1}^{-1}[\mathbf{I} - \mathbf{B}_i^t(\mathbf{Q}_{ei} + \mathbf{B}_i \mathbf{N}_{i-1}^{-1} \mathbf{B}_i^t)^{-1}\mathbf{B}_i \mathbf{N}_{i-1}^{-1}] \tag{13.17}$$

$$\bar{\mathbf{n}}_i = (\mathbf{B}^t\mathbf{W}_e \mathbf{b})_i \tag{13.18a}$$

$$\mathbf{n}_i = (\mathbf{b}^t\mathbf{W}_e \mathbf{b})_i \tag{13.18b}$$

$$\mathbf{F}_i = \bar{\mathbf{N}}_i^{-1}[\mathbf{I} + \bar{\mathbf{n}}_i(n_i - \bar{\mathbf{n}}_i^t \bar{\mathbf{N}}_i^{-1}\bar{\mathbf{n}}_i)^{-1}\bar{\mathbf{n}}_i^t \bar{\mathbf{N}}_i^{-1}] \tag{13.15a}$$

$$\mathbf{G}_i = -\mathbf{F}_i \bar{\mathbf{n}}_i \mathbf{n}_i^{-1} \tag{13.14b}$$

$$\mathbf{H}_i = \mathbf{n}_i^{-1} - \mathbf{n}_i^{-1}\bar{\mathbf{n}}_i^t \mathbf{G}_i = (\mathbf{n}_i - \bar{\mathbf{n}}_i^t \bar{\mathbf{N}}_i^{-1}\bar{\mathbf{n}}_i)^{-1} \qquad (13.14c),\ (13.14d)$$

$$\mathbf{N}_i^{-1} = \begin{bmatrix} \mathbf{F}_i & \mathbf{G}_i \\ \mathbf{G}_i^t & \mathbf{H}_i \end{bmatrix} \tag{13.13}$$

$$\mathbf{t}'_i = \mathbf{t}_{i-1} + \mathbf{B}_i \mathbf{W}_{ei} \mathbf{f}_i \tag{13.19a}$$

$$\bar{\mathbf{t}}_i = \mathbf{b}_i^t \mathbf{W}_{ei} \mathbf{f}_i \tag{13.19b}$$

$$\mathbf{t}_i = \begin{bmatrix} \mathbf{t}'_i \\ \bar{\mathbf{t}}_i \end{bmatrix} \tag{13.12b}$$

$$\begin{bmatrix} \boldsymbol{\Delta}_i \\ \boldsymbol{\delta} \end{bmatrix} = \mathbf{N}_i^{-1}\mathbf{t}_i$$

For deletion of the conditions $\mathbf{A}_j \mathbf{v}_j + \mathbf{B}_j \boldsymbol{\Delta} + \mathbf{b}_j \boldsymbol{\delta} = \mathbf{f}_j$ from the solution given by $\mathbf{N}_{j-1}^{-1}$ and $\mathbf{t}_{j-1}$, then partition $\mathbf{N}_{j-1}^{-1}$ and $\mathbf{t}_{j-1}$,

$$\mathbf{N}_{j-1}^{-1} = \begin{bmatrix} \mathbf{F}_{j-1} & \mathbf{G}_{j-1} \\ \mathbf{G}_{j-1}^t & \mathbf{H}_{j-1} \end{bmatrix}, \qquad \mathbf{t}_{j-1} = \begin{bmatrix} \mathbf{t}'_{j-1} \\ \bar{\mathbf{t}}_{j-1} \end{bmatrix}$$

where $\mathbf{H}$ and $\mathbf{t}$ are of order equal to that of $\boldsymbol{\delta}$.

Compute $\bar{\mathbf{n}}_j$ and $\mathbf{n}_j$ from equations (13.18a) and (13.18b) above,

$$\bar{\mathbf{N}}_{j-1}^{-1} = \mathbf{F}_{j-1}[\mathbf{I} - \bar{\mathbf{n}}_j(\mathbf{n}_j + \bar{\mathbf{n}}_j^t \mathbf{F}_{j-1} \bar{\mathbf{n}}_j)^{-1} \bar{\mathbf{n}}_j^t \mathbf{F}_{j-1}] \tag{13.23}$$

$$\mathbf{N}_j^{-1} = \bar{\mathbf{N}}_{j-1}^{-1}[\mathbf{I} + \mathbf{B}_j^t(\mathbf{Q}_{ej} - \mathbf{B}_j \bar{\mathbf{N}}_{j-1}^{-1} \mathbf{B}_j^t)^{-1} \mathbf{B}_j \bar{\mathbf{N}}_{j-1}^{-1}] \tag{13.24}$$

$$\mathbf{t}_j = \mathbf{t}'_{j-1} - \mathbf{B}_j^t \mathbf{W}_{ej} \mathbf{f}_j \tag{13.25}$$

When unified approach is used, apply all formulas above but redefining

$$\mathbf{N} = \mathbf{B}^t \mathbf{W}_e \mathbf{B} + \mathbf{W}_{xx} \qquad \mathbf{t} = \mathbf{B}^t \mathbf{W}_e \mathbf{f} - \mathbf{W}_{xx} \mathbf{f}_x \qquad (13.26a), (13.26b)$$

$$\mathbf{n} = \mathbf{b}^t \mathbf{W}_e \mathbf{b} + \mathbf{W}_{\delta\delta} \qquad \bar{\mathbf{t}} = \mathbf{b}^t \mathbf{W}_e \mathbf{f} - \mathbf{W}_{\delta\delta} \mathbf{f}_\delta \qquad (13.27b), (13.27c)$$

where $\mathbf{f}_x = \mathbf{x}^0 - \mathbf{x}$ and $\mathbf{f}_\delta = \mathbf{x}_\delta^0 - \mathbf{x}_\delta$; $\mathbf{x}^0$ and $\mathbf{x}_\delta^0$ are vectors of parameter approximation; and $\mathbf{x}$, $\mathbf{x}_\delta$ and $\mathbf{W}_{xx}$, $\mathbf{W}_{\delta\delta}$ are a priori estimates of the parameters and their weight matrices.

13.7.2. Concluding Remarks We introduced in this chapter several modern concepts that may be regarded as an extension of regularly practiced least squares adjustment procedures. Although these concepts may appear to be somewhat theoretical or academic, they may in fact take a place of prominence in adjustment methodology because of current rapid advances in on-line modes of data acquisition and the ever-increasing role played by electronic computers.

Because the sequential methods are relatively recent, we have by no means exhausted the subject. Actually, this chapter is mostly an introductory exposition. One point that must be emphasized is that the entire concept of derivation as given was built on the lack of correlation between the new set of observations and those preceding it. This assumption allowed for the block diagonality of the total cofactor matrices and made possible the derivation of the algorithms as given. Obviously, further development is needed to deal with the effect on these algorithms of having correlated observations that need to be sequentially processed.

Another area that requires further elaboration concerns the nonlinearity of the model, and how it can best be handled by sequential procedures. Numerical factors and computational efficiency, further considerations of cases of variable number of parameters, and the unified approach combined with model nonlinearity all need more development. Therefore the material given here is intended to present the reader with enough basic concepts to prepare him for further study of more advanced work.

14

An Introduction to Least Squares Interpolation, Filtering, and Collocation

14.1. INTRODUCTION

The method of least squares has been used extensively in the major portions of this book for solving *adjustment* problems involving random variables and specified mathematical models. In such problems the selected mathematical model is treated as superior to the given sample, and the least squares criterion is used to estimate another set of random variables that is consistent with the model. Therefore *adjustment* procedures may be considered as one type of statistical estimation in cases in which a mathematical model is given a priori.

There are other estimation procedures that may also apply the least squares criterion and which rely on the basic concepts of stochastic processes. We may speak of "least squares prediction" when "time" is the variable involved, and therefore we estimate (or predict) what would happen in the future on the basis of past occurrences (history). On the other hand many of the applications in photogrammetry, geodesy, and surveying often involve location instead of time, and hence either least squares interpolation or extrapolation is used. The task of interpolation is to estimate (interpolate) at locations other than those for which (observed) data are given. Because this situation is more often in practice, the term "interpolation" will be selected for use hereafter. At data points, measuring errors may also occur. If the estimation procedure "filters" these errors when estimating at new points, we speak of "least squares filtering."

When interpolating at new locations from data at given points, nonlinear relationships could, in general, be assumed. However, for practical reasons nonlinear interpolation is not often applied and instead "linear interpolation" is more commonly used. For this reason consideration will be given only to the linear case in the developments given in this chapter.

Least squares adjustment has been applied to problems in photogrammetry, geodesy, and surveying for a long time. On the other hand least squares interpolation and filtering have found their way into these fields only recently. In addition to applying interpolation to geodetic problems a general least squares technique combining classical adjustment with interpolation and filtering was devised. This technique has been given the name "least squares collocation."

Instead of commencing with the relatively advanced method of least squares collocation, least squares interpolation is introduced first. And in order to facilitate its introduction several basic concepts and definitions are concisely presented, without proofs.

14.1.1. Random or Stochastic Functions The reader will recall that a *random variable* is defined as a variable that assumes various values depending on the result of an experiment (or observation, or trial) when such an experiment is repeated many times under similar conditions. The random variable has therefore a *probability distribution* associated with it, and it is considered specified when its *distribution function* is known. When the outcome of the experiment is a set of quantities (or numbers), we may consider them collectively in a vector as a single *multidimensional random vector*. In this case such random variables are specified by a multidimensional distribution function. Random variables, whether one dimensional or multidimensional, were sufficient for the treatment of different techniques of least squares adjustment. In order to explain least squares interpolation and filtering, however, the concept of random functions is required. A *random function* (also called "stochastic" function) is a function that is specified by the outcome of an experiment, and which may have different realizations as the experiment is repeated. It is composed of a set of random variables. An example of a random function is the diameter of the cross section of a thin wire (see Yaglom, 1962, in the Bibliography). Due to many factors in manufacturing a wire, its cross section is never a constant but is instead a *random function* along the wire. Another example concerns the position of image points (reseau marks) on an aerial photograph.

A *realization* of a random function (or a *sample function*) is one given set of values for the function. A given segment of a thin wire is one realization since the diameters of various cross sections along this specified segment are known values in the sense that they can be measured. Another segment (of the same length) would be another realization which may, in general, have different values for the diameters of the wire cross sections. Similarly, a given photograph is a realization of a random function where the position of each specific point (reseau mark) is determined. The random variation is therefore from one realization to another.

The general term "random function" is used to designate either a *random sequence* or a *random process*. In the case of random sequences, the variable(s) in the function (for example, the position of a reseau point on an aerial photograph) assume only discrete values (when properly scaled, they become integer values). Random processes (often termed "stochastic" processes) on the other hand consider all values of such variable(s) (for example, the position of a wire cross section). Random functions are specified by the moments, particularly the first two, of their probability distributions. The first moment is the mean whereas the second moment would represent the covariance if the mean is zero. If $s(u)$ represents a random function, then the mean $\mu_s(u_1)$ at a particular value u_1 is defined as

$$\mu_s(u_1) = \lim_{m\to\infty} \frac{1}{m} \sum_{k=1}^{m} s_k(u_1) \tag{14.1}$$

Similarly, the joint moment (which is called *autocorrelation function*) $C_s(u_1, u_1 + \Delta)$ at two different values u_1 and $(u_1 + \Delta)$ is given by

$$C_s(u_1, u_1 + \Delta) = \lim_{m\to\infty} \frac{1}{m} \sum_{k=1}^{m} s_k(u_1)s_k(u_1 + \Delta) \tag{14.2}$$

For practical applications, μ_s and C_s are often sufficient for specifying the random function in question (in equations (14.1) and (14.2) the summation is over realizations).

14.1.2. Stationarity, Nonstationarity, and Ergodicity If the conditions of the experiment, on whose outcome the values of the random function depend, remain unaltered (within one experiment and from one experiment to the next), then the function is called "stationary"; otherwise it is called "nonstationary." Consequently, if the random function $s(x)$ is nonstationary, then both μ_s and C_s of equations (14.1) and (14.2) would vary as u_1 varies. On the other hand for the special case of a stationary function, μ_s and C_s do not change in value as u_1 varies. It follows that for stationary functions the mean is constant (often taken as zero) and the autocorrelation function depends only on the increment Δ. An example of stationary function is when the manufacturing environment of the thin wire does not change with time. In this case the mean wire diameter is constant and the autocorrelation function depends only on the distance between points along the wire. Any two different segments of the wire would have the same autocorrelation function.

For the purposes of the current presentation, consideration will be given mostly to stationary random functions. It is therefore convenient to take the mean as zero and speak of the autocorrelation function as the *autocovariance function.* If $\mu_s \neq 0$, a new random function $s'(u) = s(u) - \mu_s$, which has a zero mean, would replace $s(u)$.

The mean and autocorrelation would require a large number of realizations of the function s_k in order to be estimated from the data. This is often not practical and an alternative is desirable. In many cases in practice it is possible to describe the characteristics of a *stationary* function from *only one realization.* Here the mean and autocorrelation are given by

$$\mu_s(k) = \lim_{U \to \infty} \frac{1}{U} \int_0^U s_k(u)\, du \tag{14.3}$$

$$C_s(\Delta, k) = \lim_{U \to \infty} \frac{1}{U} \int_0^U s_k(u) s_k(u + \Delta)\, du \tag{14.4}$$

If the mean and autocorrelation as given by equations (14.3) and (14.4) turn out to be the same no matter which realization k is used, the stationary random function is called *ergodic.* The property of ergodicity applies only to those functions that are stationary.

A characteristic that is useful in practice pertains to *normal stationary* random functions. When the function mean is *zero,* the ergodic property will be satisfied when

$$\lim_{\Delta \to \infty} C_s(\Delta) = 0 \tag{14.5}$$

In other words, when the function is stationary and has a normal probability function† (which occurs frequently in practice) with zero mean and a covariance function that converges to zero as the interval Δ increases indefinitely, then the function is ergodic.

If the example of aerial photography is recalled, when ergodicity applies we may use data at different (reseau) points of the same photograph (one realization) to derive properties pertaining to the multitude of aerial photographs. In photogrammetry this is frequently carried out in tests on photographic quality without having to obtain an excessive number of photographic frames. In such tests ergodicity is assumed.

14.1.3. Multidimensional Random (Stochastic) Functions—Random Fields

So far, consideration has been given to single random functions such as

† A random function $s(x)$ has a normal distribution if the multidimensional joint distribution function $f[s(u_1), s(u_2), \ldots, s(u_n)]$ is normal for any arbitrary set of $u_1, u_2, \ldots, u_n$.

$s(u)$ with only one variable u which may or may not be time. Random vectors have been introduced before, however, and in a like manner a *random vector function* (or vector process) may be introduced, again of a single independent variable u. For example,

$$\mathbf{s}(u) = [s_1(u) \quad s_2(u) \cdots s_n(u)]^t \tag{14.6}$$

is an n-dimensional vector process that is a function of the one variable u. In this case we would have a mean n-vector $\boldsymbol{\mu}_s$, but the concept of correlation function must be expanded. Here, in addition to the autocorrelation (autocovariance) functions for each $s_i(u)$, there will be crosscorrelation (crosscovariance) functions for each pair $s_i(u)$ and $s_j(u)$, $i \neq j$. Consequently, the covariance function $C(u)$ is replaced by a covariance matrix $\mathbf{C}(u)$. This matrix is *symmetric* for *stationary* random vectors (of real functions). Each element $s_i(u)$ is a stationary random function, and each pair, $s_i(u)$ and $s_j(u)$, $i \neq j$, is only *stationarily correlated*, that is, the crosscovariance function depends only on the difference $(u_i - u_j)$.

Although the random vector function consists of several random functions all of which are in terms of only one variable (u), a *random field* consists of only *one* random function *but* in terms of more than one independent variable. For example, $s(\mathbf{u})$ is a random field that is a function of a vector of independent variables $\mathbf{u}$. Wind speed is an example of a four-dimensional random field, where $\mathbf{u}$ includes four independent variables—that is, time and the three dimensions of Euclidean space. For fields, the concept of stationarity is replaced by the concepts of "homogeneity" and "isotropy". A field is called "homogeneous" if its mean is constant and its autocovariances depend only on the vector separation (both length and direction) between different points of the field, not on their absolute location. If, in addition, autocovariances are invariant under rotation—that is, they depend only on the length of the vector separation—then the field is not only homogeneous, it is also *isotropic*.

A last kind of multidimensional random function combines both of the types that have been presented so far. A *random vector field* consists of several random functions (or processes), each of which is a function of a vector of variables $\mathbf{u}$, or

$$\mathbf{s}(\mathbf{u}) = [s_1(\mathbf{u}) \quad s_2(\mathbf{u}) \cdots s_n(\mathbf{u})]^t \tag{14.7}$$

The properties of such a relatively complex field would be expressed by averages, autocorrelations, and crosscorrelations in a manner similar to those presented previously.

14.1.4. Linear Interpolation and Filtering In many branches of applied science and engineering, problems often occur in which an estimate of a random quantity is required from certain available data. The random quantity sought could, in general, be written as a nonlinear function of the variables representing the data, but the solution in such a case would be difficult analytically. Instead, linear functions are often used to simplify the problem, hence the term "linear interpolation and filtering." For an explanation of interpolation and filtering, we define three random functions $l(u)$, $s(u)$, and $r(u)$ such that

$$l(u) = s(u) + r(u) \tag{14.8}$$

The observable function is $l(u)$, whereas $r(u)$ represents the measuring error or "noise," thus $s(u)$ is the so-called "signal" or response that is to be estimated. Interpolation and filtering, therefore, are the problems of finding an estimate $\hat{s}(u_0)$ of the random function $s(u_0)$, at $u = u_0$, when a discrete set of function values $l(u_1)$, $l(u_2)$, ..., $l(u_n)$ from a given realization $l(u)$ are available. The estimate is sought such that $\hat{s}(u_0)$ is a linear combination of $l(u_i)$, the function values, or

$$\hat{s}_0 = \hat{s}(u_0) = \mathbf{a}^t \boldsymbol{l} \tag{14.9}$$

where

$$\mathbf{a}^t = [a_1 \quad a_2 \quad \cdots \quad a_n] \tag{14.10a}$$

is a vector of coefficients, and

$$\boldsymbol{l} = [l(u_1) \quad l(u_2) \cdots l(u_n)]^t \tag{14.10b}$$

is the vector of given data values. Special cases of this general formulation include

1. If $r(u) = 0$, then $l(u) = s(u)$ implies that either no measuring errors occurred or if they did, they are of sufficiently small magnitude to be neglected. This is the problem of interpolation, where $s(u_0)$ is also $l(u_0)$.
2. If the noise $r(u) \neq 0$ and we are interested in estimating $s(u_0)$ at $u = u_0$ from the data values $l(u_i)$, then this is the filtering problem.

If the independent variable u is time, then the estimation process is called "prediction." In this case the data represent the "past" of the random function, and signal values at "future" times are to be estimated. In all cases the type of estimation considered is linear inasmuch as equation (14.9) represents the linear operator used for estimation.

Before dwelling on the details of interpolation and filtering, the concept of correlation (covariance) functions should first be introduced. Dealing with stationary random functions, it is convenient to assume zero means, or

$$\mu_s = E(s) = 0 \tag{14.11a}$$

$$\mu_r = E(r) = 0 \tag{14.11b}$$

$$\mu_l = E(l) = 0 \tag{14.11c}$$

Furthermore, stationarity means that autocorrelation C is only a function of the separation Δ. The functional expression relating C and Δ is called the "correlation" or "covariance" function. There are in general three correlation functions—$C_l(\Delta)$, $C_s(\Delta)$, and $C_r(\Delta)$—which, under the assumption of zero means and that r and s are stationarily uncorrelated, are related by

$$C_l(\Delta) = C_s(\Delta) + C_r(\Delta) \tag{14.12a}$$

and

$$C_{ls}(\Delta) = C_s(\Delta) \tag{14.12b}$$

Figure 14.1(a) depicts graphically the relation of equation (14.12a) where the shaded area represents the correlation function of the noise.

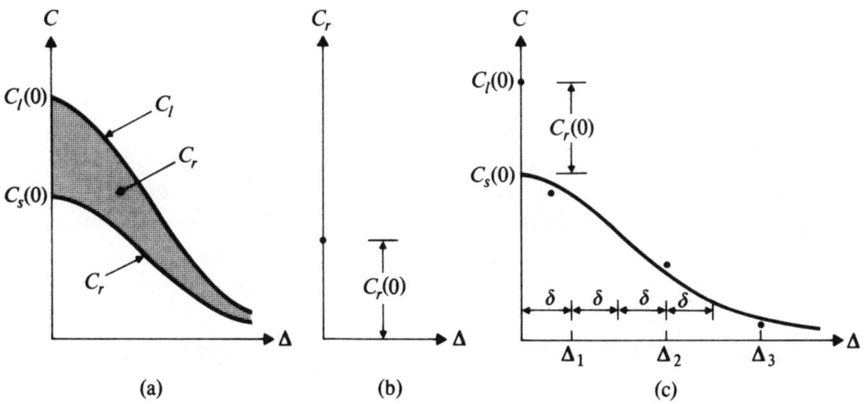

Figure 14.1

In practice, however, the noise $r(u)$ is more realistically assumed to be uncorrelated and therefore $C_r(\Delta)$ would be a single point (representing the variance of the noise) as depicted in Figure 14.1(b). Consequently, the three covariance functions in this case would be as shown in Figure 14.1(c), and for $\Delta \to 0$ the relation between variances is

$$C_l(0) = C_s(0) + C_r(0) \tag{14.12c}$$

Covariance functions are either theoretically known, or they may be derived from the experimental data as will be explained in Section 14.5. Once the functions are established, the covariance $C(\Delta)$ may be evaluated for any two points separated by a distance Δ.

14.2. LINEAR INTERPOLATION OF STATIONARY FUNCTIONS

As given under the special case (1) of the preceding section the data pertain to $s(u)$, and hence equation (14.9) may be used replacing $\mathbf{l}$ by $\mathbf{s}$. Let the deviation between the estimate $\hat{s}$ and s be designated by

$$v = s(u_0) - \hat{s}(u_0) = s_0 - \hat{s}_0 = s_0 - \mathbf{a}^t\mathbf{s}(u)$$

or

$$v = [1 \quad -\mathbf{a}^t]\begin{bmatrix} s_0 \\ \mathbf{s}(u) \end{bmatrix} \tag{14.13}$$

for which the variance is (applying the technique of variance propagation, see Chapter 4, Part I)

$$\sigma_v^2 = \mathbf{J}\boldsymbol{\Sigma}\mathbf{J}^t \tag{14.14}$$

with the Jacobian, $\mathbf{J} = [1 \quad -\mathbf{a}^t]$ and the covariance matrix of the random variables involved given by

$$\boldsymbol{\Sigma} = \begin{bmatrix} \sigma_{s0}^2 & \boldsymbol{\sigma}_{s0\,s} \\ \boldsymbol{\sigma}_{ss0} & \boldsymbol{\Sigma}_{ss} \end{bmatrix} \tag{14.15}$$

This is an $(n + 1)$ square symmetric matrix of autocovariances and crosscovariances, with σ_{s0}^2 the variance of s_0 (a scalar) which is equal to $C_s(0)$ [see Figure 14.1(a)]; $\boldsymbol{\sigma}_{ss0} = \boldsymbol{\sigma}_{s0\,s}^t$ the $n \times 1$ vector of crosscovariances between s_0 and each element $s(u_i)$ of $\mathbf{s}$ (each value is obtained from the covariance function $C_s(\Delta)$ for each distance $\Delta_i = u_0\,u_i$); and $\boldsymbol{\Sigma}_{ss}$ the $n \times n$ symmetric autocovariance matrix of $\mathbf{s}$. Its diagonal elements are all equal to $C_s(0)$, and each off-diagonal element ij is obtained from the covariance function using $\Delta_{ij} = u_i\,u_j$ as arguments.

Using equation (14.13) equation (14.12) may be expanded to

$$\sigma_v^2 = \sigma_{s0}^2 - 2\mathbf{a}^t\boldsymbol{\sigma}_{ss0} + \mathbf{a}^t\boldsymbol{\Sigma}_{ss}\,\mathbf{a} \tag{14.16}$$

which represents the mean square interpolation error. For a minimum interpolation error, we select a vector $\mathbf{a}$ such that

$$\frac{\partial \sigma_v^2}{\partial \mathbf{a}} = -2\boldsymbol{\sigma}_{ss0}^t + 2\mathbf{a}^t\boldsymbol{\Sigma}_{ss} = 0$$

or, after transposing and rearranging

$$\mathbf{a} = \boldsymbol{\Sigma}_{ss}^{-1}\boldsymbol{\sigma}_{ss0} \tag{14.17}$$

This leads to the estimate of the interpolated value [see equation (14.9)]

$$\hat{s}_0 = \boldsymbol{\sigma}_{s0\,s}\boldsymbol{\Sigma}_{ss}^{-1}\mathbf{s} \tag{14.18}$$

The estimate of the mean square error of estimation is

$$\sigma_{\hat{s}0}^2 = \sigma_v^2 = \sigma_{s0}^2 - 2\mathbf{a}^t\boldsymbol{\sigma}_{ss0} + \mathbf{a}^t\boldsymbol{\Sigma}_{ss}\,\mathbf{a}$$

or in view of equation (14.17) it becomes

$$\sigma_{\hat{s}0}^2 = \sigma_{s0}^2 - \mathbf{a}^t\boldsymbol{\sigma}_{ss0} = C_s(0) - \mathbf{a}^t\boldsymbol{\sigma}_{ss0} = C_s(0) - \boldsymbol{\sigma}_{ss0}^t\boldsymbol{\Sigma}_{ss}^{-1}\boldsymbol{\sigma}_{ss0} \tag{14.19a}$$

and alternatively

$$\sigma_{\hat{s}0}^2 = \sigma_{s0}^2 - \mathbf{a}^t\boldsymbol{\Sigma}_{ss}\,\mathbf{a} = C_s(0) - \mathbf{a}^t\boldsymbol{\Sigma}_{ss}\,\mathbf{a} \tag{14.19b}$$

14.3. LINEAR FILTERING OF STATIONARY FUNCTIONS

Linear filtering is actually a more interesting problem since it is of more practical usefulness than interpolation. In a manner similar to the derivation for interpolation, the estimate $\hat{s}_0$ for filtering may be obtained as follows:

$$v = s_0 - \mathbf{a}^t \boldsymbol{l} = [1 \quad -\mathbf{a}^t] \begin{bmatrix} s_0 \\ \boldsymbol{l} \end{bmatrix}$$

$$\sigma_v^2 = [1 \quad -\mathbf{a}^t] \Sigma \begin{bmatrix} 1 \\ -\mathbf{a} \end{bmatrix}$$

Here, the covariance matrix Σ is constructed as

$$\Sigma = \begin{bmatrix} \sigma_{so}^2 & \boldsymbol{\sigma}_{sol} \\ \boldsymbol{\sigma}_{lso} & \Sigma_{ll} \end{bmatrix}$$

The variance σ_{so}^2, in Σ, is equal to $C_s(0)$. Next, the vector $\boldsymbol{\sigma}_{sol}$ contains crosscovariances $C_{sl}(\Delta)$ which are equal to $C_s(\Delta)$ according to equation (14.12b). Finally, Σ_{ll} is the covariance matrix for the given data. Its diagonal elements are all equal to $C_l(0)$. Its off-diagonal elements are the same as those for Σ_{ss} because $C_l(\Delta) = C_s(\Delta)$ for $\Delta \neq 0$ [see Figure 14.1(c)].

Expanding σ_v^2 and differentiating to get a minimum, we get an equation similar to equation (14.17)

$$\mathbf{a} = \Sigma_{ll}^{-1} \boldsymbol{\sigma}_{lso} \tag{14.20}$$

Finally, as in the case of interpolation

$$\hat{s}_0 = \boldsymbol{\sigma}_{sol} \Sigma_{ll}^{-1} \boldsymbol{l} \tag{14.21}$$

$$\sigma_{\hat{s}0}^2 = \sigma_{so}^2 - \mathbf{a}^t \boldsymbol{\sigma}_{lso}$$

or

$$\sigma_{\hat{s}0}^2 = C_s(0) - \mathbf{a}^t \boldsymbol{\sigma}_{lso} = C_s(0) - \boldsymbol{\sigma}_{lso}^t \Sigma_{ll}^{-1} \boldsymbol{\sigma}_{lso} \tag{14.22a}$$

and

$$\sigma_{\hat{s}0}^2 = C_s(0) - \mathbf{a}^t \Sigma_{ll} \, \mathbf{a} \tag{14.22b}$$

This concludes discussion of filtering.

In the treatments of both interpolation and filtering, attention was limited to estimating one random function only. In practice, two- and three-dimensional random vectors may occur. Therefore both interpolation and filtering will be extended in the next section for multidimensional cases.

14.4. EXTENSION TO MULTIDIMENSIONAL CASES

Suppose that at each point more than one random variable are defined. For practical reasons only two- and three-dimensional random vectors will be considered. We will denote the components of these vectors by $x(u)$, $y(u)$, and $z(u)$. Thus for the two-dimensional case,

$$\begin{bmatrix} \hat{x}_0 \\ \hat{y}_0 \end{bmatrix} = \begin{bmatrix} \underset{1,n}{\boldsymbol{\sigma}_{x0x}} & \underset{1,n}{\boldsymbol{\sigma}_{x0y}} \\ \underset{1,n}{\boldsymbol{\sigma}_{y0x}} & \underset{1,n}{\boldsymbol{\sigma}_{y0y}} \end{bmatrix} \begin{bmatrix} \underset{n,n}{\boldsymbol{\Sigma}_{xx}} & \underset{n,n}{\boldsymbol{\Sigma}_{xy}} \\ \underset{n,n}{\boldsymbol{\Sigma}_{yx}} & \underset{n,n}{\boldsymbol{\Sigma}_{yy}} \end{bmatrix}^{-1} \begin{bmatrix} \mathbf{x}(u) \\ \mathbf{y}(u) \end{bmatrix} \tag{14.23}$$

and for the three-dimensional case

$$\begin{bmatrix} \hat{x}_0 \\ \hat{y}_0 \\ \hat{z}_0 \end{bmatrix} = \begin{bmatrix} \boldsymbol{\sigma}_{x0x} & \boldsymbol{\sigma}_{x0y} & \boldsymbol{\sigma}_{x0z} \\ \boldsymbol{\sigma}_{y0x} & \boldsymbol{\sigma}_{y0y} & \boldsymbol{\sigma}_{y0z} \\ \boldsymbol{\sigma}_{z0x} & \boldsymbol{\sigma}_{z0y} & \boldsymbol{\sigma}_{z0z} \end{bmatrix} \begin{bmatrix} \boldsymbol{\Sigma}_{xx} & \boldsymbol{\Sigma}_{xy} & \boldsymbol{\Sigma}_{xz} \\ \boldsymbol{\Sigma}_{yx} & \boldsymbol{\Sigma}_{yy} & \boldsymbol{\Sigma}_{yz} \\ \boldsymbol{\Sigma}_{zx} & \boldsymbol{\Sigma}_{zy} & \boldsymbol{\Sigma}_{zz} \end{bmatrix}^{-1} \begin{bmatrix} \mathbf{x}(u) \\ \mathbf{y}(u) \\ \mathbf{z}(u) \end{bmatrix} \tag{14.24}$$

$\hat{x}_0$, $\hat{y}_0$, $\hat{z}_0$ are the estimates of the three components of the signal at $u = u_0$.

$\boldsymbol{\sigma}_{x0x}$, $\boldsymbol{\sigma}_{x0y}$, $\boldsymbol{\sigma}_{x0z}$ are each a $1 \times n$ row vector of covariances between $x(u_0)$ and the data values $x(u_i)$, $y(u_i)$, and $z(u_i)$, respectively.

$\boldsymbol{\sigma}_{y0x}$, $\boldsymbol{\sigma}_{y0y}$, ..., $\boldsymbol{\sigma}_{z0z}$ are vectors similarly defined.

$\boldsymbol{\Sigma}_{xx}$, $\boldsymbol{\Sigma}_{yy}$, $\boldsymbol{\Sigma}_{zz}$ are each an $n \times n$ square symmetric matrix of autocovariances of the given data.

$\boldsymbol{\Sigma}_{xy}$, $\boldsymbol{\Sigma}_{xz}$, $\boldsymbol{\Sigma}_{yz}$ are each an $n \times n$ square matrix of crosscovariances of the given data.

Equations (14.23) and (14.24) may be written more concisely as

$$\underset{m,1}{\mathbf{s}_0} = \underset{m,mn}{\boldsymbol{\Sigma}_{s0s}} \underset{mn,mn}{\boldsymbol{\Sigma}_{ss}^{-1}} \underset{mn,1}{\mathbf{s}} \qquad \text{(interpolation)} \tag{14.25}$$

$$\underset{m,1}{\mathbf{s}_0} = \underset{m,mn}{\boldsymbol{\Sigma}_{s0l}} \underset{mn,mn}{\boldsymbol{\Sigma}_{ll}^{-1}} \underset{mn,1}{\boldsymbol{l}} \qquad \text{(filtering)} \tag{14.26}$$

and the corresponding covariance matrices

$$\begin{aligned} \underset{m,m}{\boldsymbol{\Sigma}_{\hat{s}0\hat{s}0}} &= \underset{m,m}{\boldsymbol{\Sigma}_{s0s0}} - \underset{m,mn}{\mathbf{a}^t} \underset{mn,m}{\boldsymbol{\Sigma}_{ss0}} \\ &= \underset{m,m}{\boldsymbol{\Sigma}_{s0s0}} - \underset{m,mn}{\boldsymbol{\Sigma}_{s0s}} \underset{mn,mn}{\boldsymbol{\Sigma}_{ss}^{-1}} \underset{mn,m}{\boldsymbol{\Sigma}_{ss0}} \end{aligned} \tag{14.27}$$

$$\boldsymbol{\Sigma}_{\hat{s}0\hat{s}0} = \boldsymbol{\Sigma}_{s0s0} - \mathbf{a}^t\boldsymbol{\Sigma}_{ls0} = \boldsymbol{\Sigma}_{s0s0} - \boldsymbol{\Sigma}_{s0l}\boldsymbol{\Sigma}_{ll}^{-1}\boldsymbol{\Sigma}_{ls0} \tag{14.28}$$

where $m = 2$ for the two-dimensional case, and $m = 3$ for the three-dimensional problem. The covariance matrix $\boldsymbol{\Sigma}_{\hat{s}0\hat{s}0}$ is the computed matrix for the estimated signals $x(u_0)$, $y(u_0)$, and so on. The matrix $\boldsymbol{\Sigma}_{s0s0}$ is a covariance matrix whose elements are the zero values $C_s(0)$ of the auto- and crosscovariance functions, or, for three dimensions,

$$\Sigma_{s0s0} = \begin{bmatrix} C_{xx}(0) & C_{xy}(0) & C_{xz}(0) \\ C_{yx}(0) & C_{yy}(0) & C_{yz}(0) \\ C_{zx}(0) & C_{zy}(0) & C_{zz}(0) \end{bmatrix} \tag{14.29}$$

In case the multidimensionality pertain to the independent variable, that is, in case of *random fields*, the relations of equations (14.18), (14.19) and equations (14.21), (14.22) for interpolation and filtering, respectively, would directly apply if the fields are assumed to be homogeneous and isotropic. Similarly, for homogeneous isotropic random vector fields, all equations in this section can be directly used. The increment Δ would be the scalar distance between the points in the field.

14.5. TREND SURFACES, COVARIANCE FUNCTIONS, AND EXPERIMENTAL RESULTS

Stationary random functions are characterized by a constant mean and the fact that the correlation function $C(\Delta)$ approaches the mean as $\Delta \to \infty$. For simplicity, the presented formulation of interpolation and filtering assumed zero mean; thus $C(\Delta) \to 0$ as $\Delta \to \infty$. In order to ascertain that these properties hold for given observational data, the *trend* must be removed first. A general definition of "trend" is that it is that component of a random phenomenon which has a period larger than the recorded data sample.

14.5.1. Trend Surfaces The trend is often represented by linear or low-order polynomial terms. The given data must therefore be preprocessed to remove the trend by transforming the data to another surface called "reference" or "trend" surface as shown in Figure 14.2. In this figure the signal surface (to be interpolated) would be referred to the trend surface instead of to the original datum. Consequently, the given values l_i at the data or reference points, would be measured with respect to the trend surface. Both the signal surface and data points vary randomly about the trend surface such that their means are zero, one of the requirements for removing the trend.

Trend removal is an important operation. If the trend is not properly removed, distortions which cause erroneous results will occur. The type of function selected to represent the trend surface is essentially determined from the nature of the given data. In fact, caution should be exercised not to remove trend arbitrarily when there is no reason to believe that the data

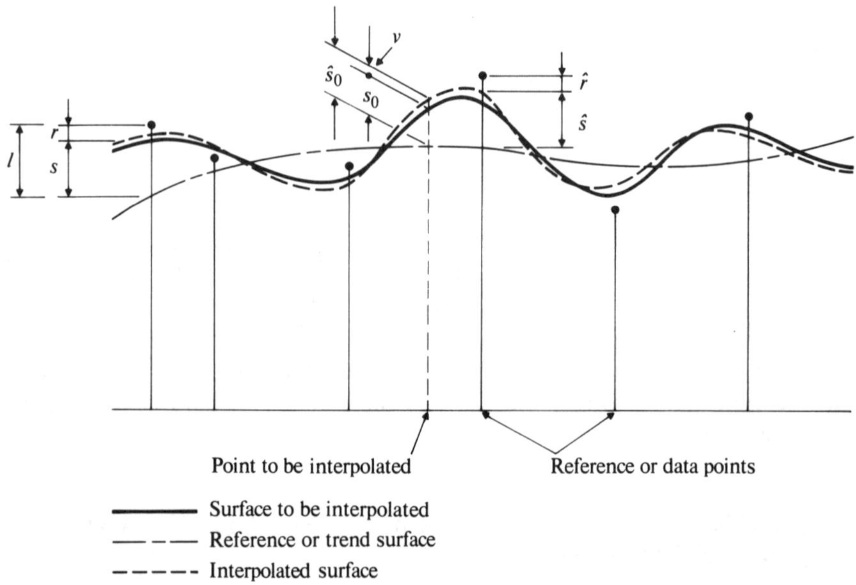

Figure 14.2

include a trend. On the other hand once it is recognized that a particular deformation exists in the data, it should be accounted for in the trend removal. For example, residual film deformation often causes affine distortion of a triangulated photogrammetric block. Consequently, we would choose an affine transformation to remove the trend when we are interested in external block adjustment using the filtering method presented here.

Having selected an appropriate expression for the trend surface, the transformation of the data is usually accomplished using the method of least squares adjustment. This is advisable because the number of reference (data) points is usually so large as to yield considerable redundancy. Therefore a least squares surface (curve) fitting would be the most suitable procedure. Once the fitting is accomplished, the data values are then referenced to the trend surface and become ready for the computation of the covariance function. Trend computation may be combined with interpolation or filtering in one operation. This operation is the so-called collocation, which will be given in detail in Section 14.7.

14.5.2. Covariance Functions Although in some special instances it is possible to have a theoretically determined covariance function, it is often necessary to derive such a function from the given data. Since consideration is limited here to stationary random functions (or fields), the covariance function would usually take the general form shown in Figure 14.1. Addressing first the one-dimensional case, we may compute the variances and covariances from the given data l_i as follows:

$$\text{Variance:} \qquad C_l(0) = \frac{1}{n} \sum_{k=1}^{n} l_k^2 \tag{14.30}$$

$$\text{Covariance:} \qquad C_l(\Delta_k) = \frac{1}{n_k} \sum_{i<j} l_i l_j \tag{14.31}$$

Equation (14.30) shows that the variance is computed by simply summing up the squares of all given data values l_k and dividing by the total number of reference points, n. If there is no noise, or measuring errors, in the data, the computed $C_l(0)$ will be identical to $C_s(0)$, the variance of the signal to be interpolated. Otherwise these two variances will be different by the amount $C_r(0)$, the error variance, as shown in Figure 14.1(c).

With respect to the covariance, reference is made to equation (14.12a) which shows that the covariance computed from the data $C_l(\Delta)$ is composed of two covariances $C_s(\Delta)$ and $C_r(\Delta)$, when s and r are uncorrelated. In practice the measuring errors are usually uncorrelated, thus $C_r(\Delta) = 0$ for $\Delta \neq 0$, and $C_l(\Delta) = C_s(\Delta)$ for all $\Delta > 0$, as shown in Figure 14.1(c). In the figure, Δ is the distance between the two points whose correlation is considered.

To compute the values $C_l(\Delta)$ of the covariance function from the data, Δ is divided into equal intervals. For example, as shown in Figure 14.1(c). the first is $\Delta_1 = \delta$, the second is $\Delta_2 = 3\delta$, and so on, with the interval being 2δ. Therefore the mean of all possible n_k products $l_i l_j$, at points whose separation Δ_{ij} is within an interval $(\Delta_k \pm \delta)$, is computed and regarded as the covariance $C_l(\Delta_k)$ at Δ_k according to equation (14.31). The increment δ is rather arbitrary and its value can usually be chosen on the basis of the knowledge of the given data. Once the number of covariance points is thus evaluated from the data, thc corresponding covariance function may be derived. In order to determine an appropriate expression for the function, it is advisable to plot these covariance points as shown in Figure 14.1(c). When studying the disposition of the plotted points, and from knowledge of the conditions of the problem, we can usually elect the suitable function to fit the covariance points. Of the functions commonly used for autocovariance we may use a constant, sinusoidal, Gaussian, exponential, exponential cosine, and exponential sine and cosine. The function depicted in Figure 14.1(c) is the Gaussian with the equation

$$C(\Delta) = C(0) \exp(-k^2\Delta^2) \tag{14.32}$$

which has been used frequently in applications to photogrammetric problems.

Usually the number of covariance points exceeds the number of coefficients in the covariance function, which are to be determined numerically. Therefore a least squares curve fitting is applied. For example, in the case of the Gaussian curve of Figure 14.1(c), both $C(0)$ and k would be determined using all covariance points but not the point $C_l(0)$ along the covariance axis if filtering is to be effected. It is worthwhile to mention that linearization of equation (14.32) for least squares adjustment requires good approximations, otherwise difficulties with convergence may be encountered. When interpolation is to be carried out without filtering, we would use $C_l(0)$ and have the covariance function curve pass through it. This is accomplished by using a constraint equation.

Once $C(0)$ and k (as an example) are determined, then the covariance for any value of Δ is computed using equation (14.32). Consequently, we can construct the autocovariance matrices Σ for both interpolation and filtering. In case of interpolation only, the diagonal elements of Σ_{ss} [see equation (14.18)] would all be equal to $C_s(0)$—which is also equal to $C_l(0)$ in this case. Next, each off diagonal element in Σ_{ss} is obtained by evaluating Δ for each corresponding pair of points, and using it to compute the corresponding covariance using equation (14.32). For instance, considering points 3 and 5, we compute Δ_{35} and use it as an argument in equation (14.32) to get the element σ_{35}. The same procedure is repeated for all other off diagonal elements of Σ_{ss}.

In case of filtering, the diagonal elements of Σ_{ll} [see equation (14.21)] are equal to $C_l(0)$ and not $C_s(0)$. This is the major difference that leads to filtering. Other than that, the remaining off diagonal elements of Σ_{ll} would be obtained in the same manner as for Σ_{ss}.

To have an appreciation of the importance of the covariance function a simulated problem of one-dimensional filtering of a profile was set up. Figure 14.3 shows the empirical covariance functions used, limiting consideration to Gaussian and exponential functions only. Figure 14.4 shows several interpolated profiles corresponding to different covariance functions with different ratios between $C_s(0)$ and $C_l(0)$. Points to note here follow.

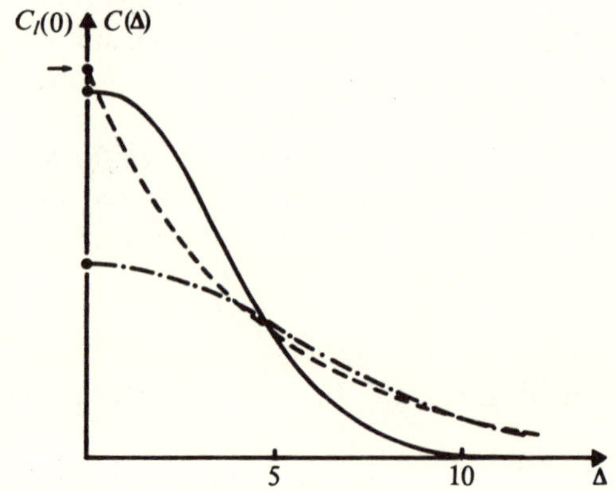

Figure 14.3. Experimentation with Different Covariance Functions

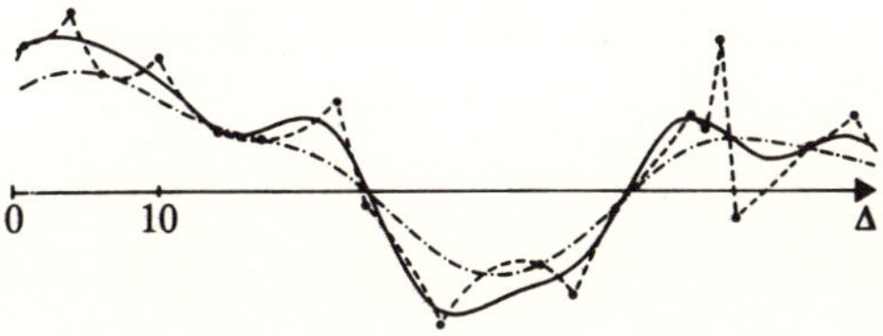

Figure 14.4. Interpolated Profiles Corresponding to the Different Covariance Functions in Figure 14.3

1. The Gaussian function produces estimated profiles with continuous first derivative, whereas the exponential function leads to profiles that exhibit discontinuous slope changes (that is, breaks or sharp peaks) at the locations of the reference points. This is one reason for using exponential covariance functions for interpolation problems in the digital terrain model, for example.
2. When $C_s(0) = C_l(0)$, the estimated profiles pass through the reference points (that is, no filtering). In practice, however, it is always better to filter even with $C_s(0) = 0.99C_l(0)$ in order to avoid potential numerical problems.
3. As the value of $C_s(0)$ gets smaller relative to $C_l(0)$ the profiles in general get farther from the reference points (that is, more filtering).

14.5.3. Multidimensional Cases So far discussion has been limited to covariance functions for one-dimensional random functions. Extension to random fields, that is, when the argument of the one random function is multidimensional, is not problematic if the field is homogeneous and isotropic. In such a case the increment Δ becomes the distance between the points. If the field is not isotropic, then we must consider both distance and azimuth in constructing the covariance function—a situation that is not quite so common in practice. When considering random vector functions (or isotropic random vector fields), we must evaluate both autocovariance as well as crosscovariance functions. For each element of the random vector, an autocovariance function may be evaluated in a manner similar to the development given in the preceding section. In addition, for each pair of elements, a crosscovariance function also needs to be evaluated. If all such crosscovariances, when evaluated from the data, turned out to be small enough not to be significant, then an n-dimensional random vector would break down to n one-dimensional functions. Consequently, both interpolation and filtering of such random vectors can be performed as n separate operations each of one-dimensional random function. This turns out to be the case in many applications in practice.

When crosscorrelation is not insignificant, the evaluation of a crosscovariance function is done in a similar manner as the autocovariance function. As an example, consider the case of a two-dimensional random vector with components $x(u)$ and $y(u)$. The autocovariance functions $C_{xx}(\Delta)$ and $C_{yy}(\Delta)$ can be derived from the data using equation (14.31), or

$$C_{xx}(\Delta_k) = \frac{1}{n_k} \sum_{i<j} x_i x_j \tag{14.33}$$

$$C_{yy}(\Delta_k) = \frac{1}{n_k} \sum_{i<j} y_i y_j \tag{14.34}$$

The crosscovariance functions $C_{xy}(\Delta)$ and $C_{yx}(\Delta)$ may be computed from the data using

$$C_{yx}(\Delta_k) = \frac{1}{n_k} \sum_{i<j} x_i y_i \tag{14.35}$$

$$C_{yx}(\Delta_k) = \frac{1}{n_k} \sum_{i<j} y_i x_j \tag{14.36}$$

Once the autocovariance and crosscovariance functions are evaluated, the elements of the covariance matrices $\boldsymbol{\Sigma}_{xx}$, $\boldsymbol{\Sigma}_{xy}$, and so on, may be computed from the functions in a straightforward manner.

14.5.4. Experimentation and Evaluation The filtering method presented in Section 14.3 can be practically evaluated when the possibility of filtering at each of the given n reference points is considered. Hence, for the ith point, the estimated value according to equation (14.21) is given by

$$\hat{s}_i = \boldsymbol{\sigma}_{il}\boldsymbol{\Sigma}_{ll}^{-1}\boldsymbol{l} \tag{14.37}$$

Referring to the discussion in Section 14.5.2 on covariance functions, the vector $\boldsymbol{\sigma}_{li}$ is the same as the ith column of $\boldsymbol{\Sigma}_{ll}$ except for the diagonal element σ_{ii}. For completeness we give both,

$$\boldsymbol{\sigma}_{li} = \begin{bmatrix} \sigma_{i1} \\ \sigma_{i2} \\ \vdots \\ \sigma_{ii} = C_s(0) \\ \vdots \\ \sigma_{in} \end{bmatrix} \quad \text{and} \quad \boldsymbol{\Sigma}_{ll} = \begin{bmatrix} C_l(0) & \sigma_{12} & \cdots & \sigma_{1n} \\ & C_l(0) & \cdots & \sigma_{2n} \\ & & \cdots & \\ \text{symmetric} & & & C_l(0) \end{bmatrix} \tag{14.38}$$

The vector $\boldsymbol{\sigma}_{li}$ may be considered as a column of a square symmetric matrix $\bar{\boldsymbol{\Sigma}}_{ll}$,

$$\bar{\boldsymbol{\Sigma}}_{ll} = \begin{bmatrix} C_s(0) & \sigma_{12} & \cdots & \sigma_{1n} \\ & C_s(0) & \cdots & \sigma_{2n} \\ & & \cdots & \\ \text{symmetric} & & & C_s(0) \end{bmatrix} \tag{14.39}$$

and the total vector of estimated signals at the n reference points becomes

$$\hat{\mathbf{s}} = \bar{\Sigma}_{ll}\Sigma_{ll}^{-1}\boldsymbol{l} \tag{14.40}$$

Recalling that $s + r = l$ [see equation (14.8)], an estimate of the error vector $\mathbf{r}$ would be

$$\hat{\mathbf{r}} = \boldsymbol{l} - \hat{\mathbf{s}} = (\mathbf{I} - \bar{\Sigma}_{ll}\Sigma_{ll}^{-1})\boldsymbol{l} \tag{14.41}$$

$$\hat{\mathbf{r}} = (\Sigma_{ll} - \bar{\Sigma}_{ll})\Sigma_{ll}^{-1}\boldsymbol{l}$$

The matrices Σ_{ll} and $\bar{\Sigma}_{ll}$ differ only in the diagonal elements [see equations (14.38) and (14.39)], and from equation (14.12c) and Figure 14.1(c), equation (14.41) reduces to

$$\hat{\mathbf{r}} = C_r(0)\Sigma_{ll}^{-1}\boldsymbol{l} \tag{14.42}$$

This is the vector of the components filtered out at the reference points. If the assumption of zero mean is adhered to, an a posteriori estimate of $C_r(0)$ may be computed from

$$\hat{C}_r(0) = \frac{1}{n}\hat{\mathbf{r}}^t\hat{\mathbf{r}} \tag{14.43}$$

Since $C_r(0)$ is often obtained a priori from the data (when establishing the covariance function) whereas $\hat{C}_r(0)$ is an estimate after filtering, their relative magnitudes may be considered as an indication of how well the filtering was performed. As an example with a one-dimensional problem, the a priori variance was $C_r(0) = 0.0026$ and the estimate was $\hat{C}_r(0) = 0.0019$. In terms of standard deviation, 86% of r was filtered out. This type of evaluation has the same practical usefulness as the reference variance (σ_0^2) in least squares adjustment. Just as σ_0^2 may be used to judge both the model and the given data, the test on $C_r(0)$ helps evaluate the overall adequacy of the method, particularly the amount of filtering.

14.6. EXAMPLES OF FILTERING APPLICATIONS

14.6.1. Compensation for Aerial Film Deformation The mathematical model in computational photogrammetry relies on the collinearity condition that the camera position, the image point, and the object point all lie on a straight line. This is, in fact, a simplification of the actual physical shape of the light ray from the object point to the image point, which is usually a curve. Instead of formulating much too complex an expression for that curved path, the collinear form is introduced together with shifting the image position in the plane of the photograph to make its application realistic. In addition to the image shift to compensate for the curvature of the projection ray, there is also an image shift to account for the deformation of the emulsion and film base between the moment of exposure and the time of image coordinate measurement.

To account for image deformation, reference points with calibrated positions, such as fiducial marks or reseau points (or both) are used. Although procedures using polynomials and other functions have been applied for some time, only recently has linear filtering been used. Figure 14.5 shows the

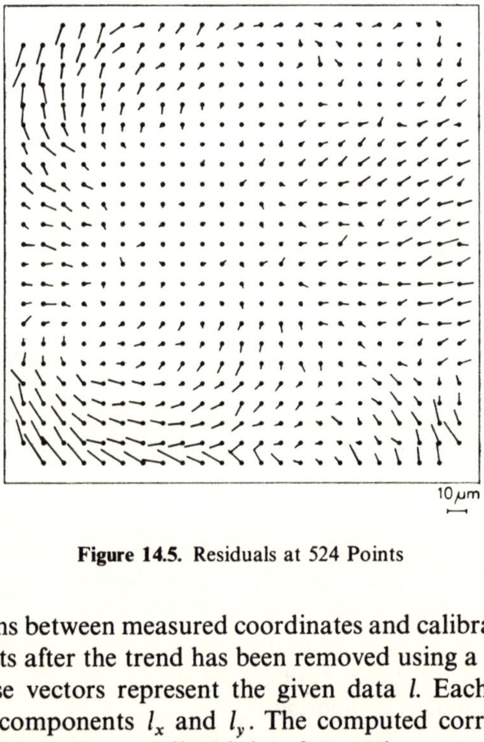

Figure 14.5. Residuals at 524 Points

field of deviations between measured coordinates and calibrated positions at 524 reseau points after the trend has been removed using a similarity transformation. These vectors represent the given data l. Each l_i can be considered in two components l_x and l_y. The computed correlation between these two components was small and therefore each component was treated independently as a one-dimensional filtering problem. From the given data, Gaussian covariance functions [see equation (14.32)] were evaluated for both the x and y components. Figure 14.6 shows the covariance function for the y component, as an example, for which $C_l(0) = 18.4\ \mu m^2$, $C_s(0) = 12.2\ \mu m^2$, and $k = 0.0173\ mm^{-2}$. In fact, the experiment involved three photographs, the results of which are summarized in Table 14.1, with

$$\sigma_s = \sqrt{C_s(0)} \qquad \text{and} \qquad \sigma_r = \sqrt{C_r(0)} = \sqrt{C_l(0) - C_s(0)}$$

Continuing with the results on one photograph, Figure 14.8 shows the computed $\hat{s}$ component at all 524 reseau points when 144 reference points are used. The vectors depicted in the figure are drawn as the resultants of the separately computed $\hat{s}_x$ and $\hat{s}_y$. Comparison of Figure 14.7 to Figure 14.5

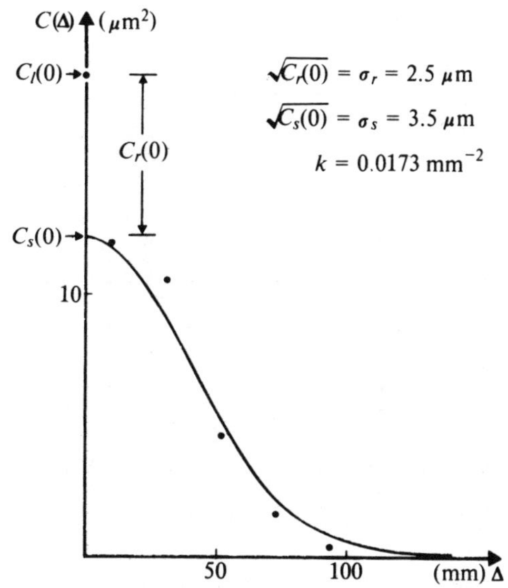

Figure 14.6. Covariance Function for Y Component of Film Deformation

TABLE 14.1 Uncorrelated Random Component of Film Deformation—Standard Deviation σ_r. Correlated Random Component for Film Deformation—Standard Deviation σ_s. Constant k of the Covariance Function

PLATE	x COMPONENT			y COMPONENT		
	σ_r (μm)	σ_s (μm)	k (mm^{-2})	σ_r (μm)	σ_s (μm)	k (mm^{-2})
302	1.7	2.8	0.017	2.8	3.9	0.017
358	1.8	3.3	0.014	2.5	3.5	0.017
412	2.1	3.9	0.015	3.0	4.9	0.021

shows that the original field is approximated quite well. The filtered uncorrelated random components, $\hat{r}$, are plotted in Figure 14.8 and show how effective filtering is for this type of problem.

14.6.2. Radial Lens Distortion Similar to film deformation, lens distortion is another factor for which measured image coordinates are treated before utilizing them in an analytical photogrammetry application. Limiting consideration to radial distortion, Figure 14.9 shows the distortion curves along

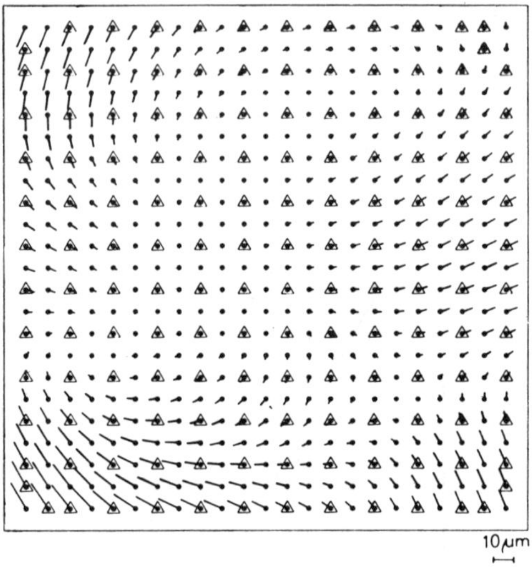

Figure 14.7. Computed $\hat{s}$ Component of Film Deformation Using 144 Reference Points

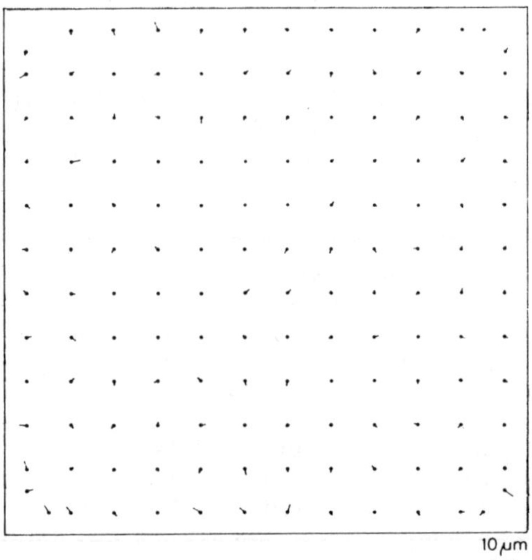

Figure 14.8. Filtered $\hat{r}$ Component of Film Deformation Using 144 Reference Points

the four radii of the format. It is not uncommon in practice simply to take the mean of all four values (see Figure 14.9) and apply a uniform radial correction to each point depending on its radial distance from the center of the format. This is obviously inadequate and better results can be obtained by a more effective method that operates directly with the given distortions along the different radii.

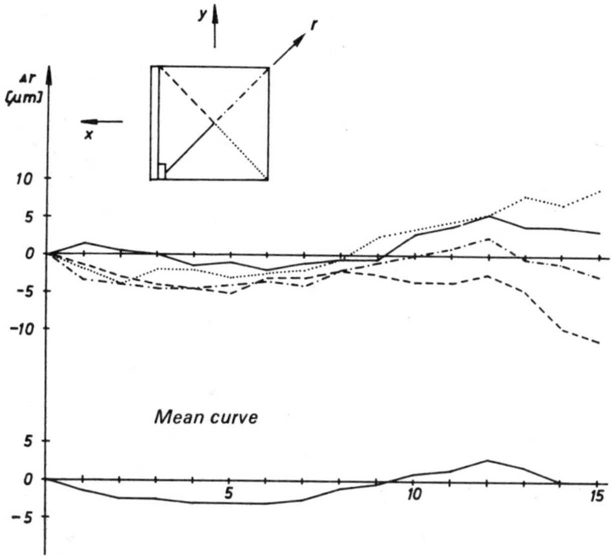

Figure 14.9. Radial Lens Distortion Curves

The first step is to seek an appropriate trend surface that was found to be the polynomial (see Vlcek, 1969, in the Bibliography),

$$d_r = b_1 r + b_2 r^3 + b_3 r^5 + a_1 r^2 \cos \phi + a_2 r^2 \sin \phi \tag{14.44}$$

where d_r is the radial distortion, r is the radial distance, and ϕ is the direction angle.

After removing the trend using this polynomial, the remaining field of residuals is shown, for 20-mm points, in Figure 14.10. As usual an appropriate covariance function, in this case a Gaussian, was found for both the x and y directions as shown in Figure 14.11. Using least squares filtering the resulting correlated $\hat{s}$ components are added to those from the polynomial and the combined effects are plotted for the 2-cm grid of points in Figure 14.12. Finally, the uncorrelated random $\hat{r}$ components filtered out during estimation are shown in Figure 14.13.

14.6.3. External Photogrammetric Block Adjustment After block triangulation, particularly by relatively simplified methods such as Anblock, residual vectors remain at the check control points (or at used control points if they were not assigned infinite weights during triangulation). Residuals at different points invariably include correlated components (s components) and usually uncorrelated random errors (r components). In order to account for these correlated residuals and to compute corresponding quantities at

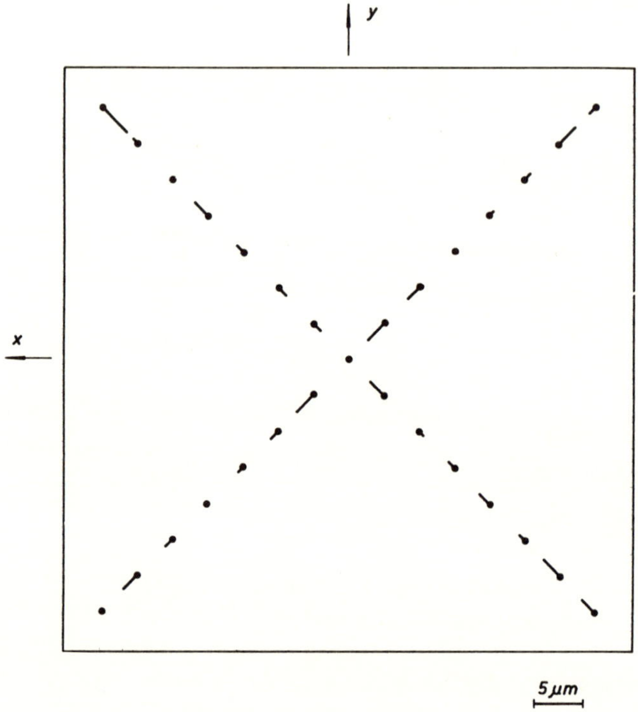

Figure 14.10. Lens Distortion Residuals after Trend Removal

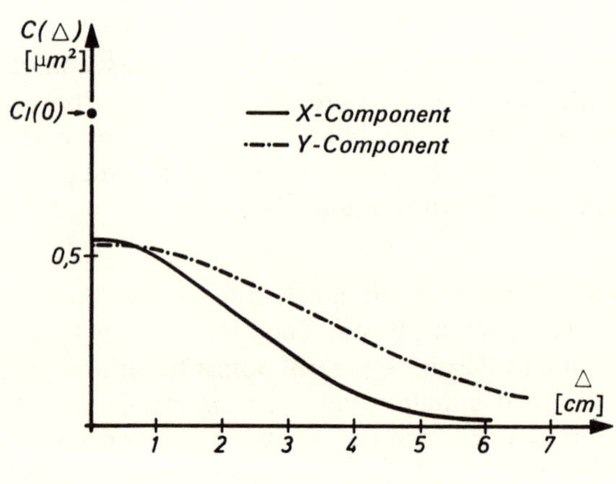

Figure 14.11. Gaussian Covariance Functions for Lens Distortion

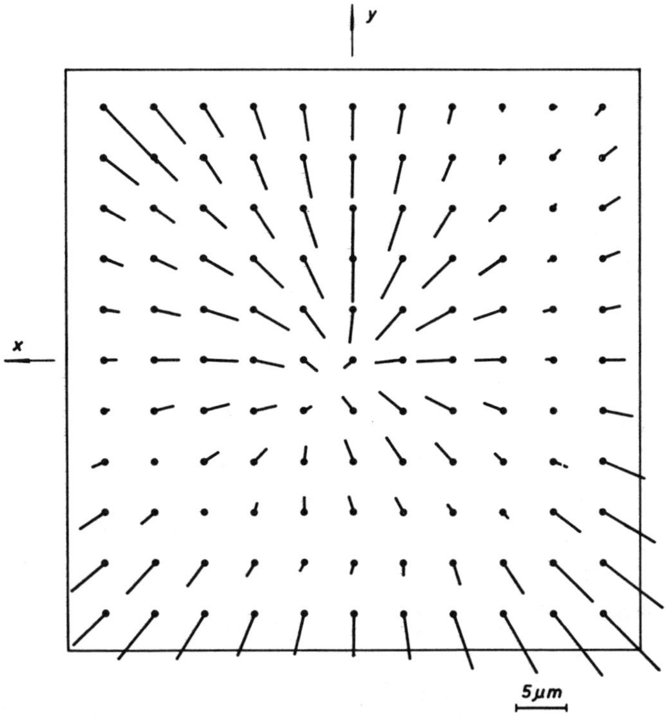

Figure 14.12. Combined Trend and $\hat{s}$ Component (Total Signal) for Lens Distortion

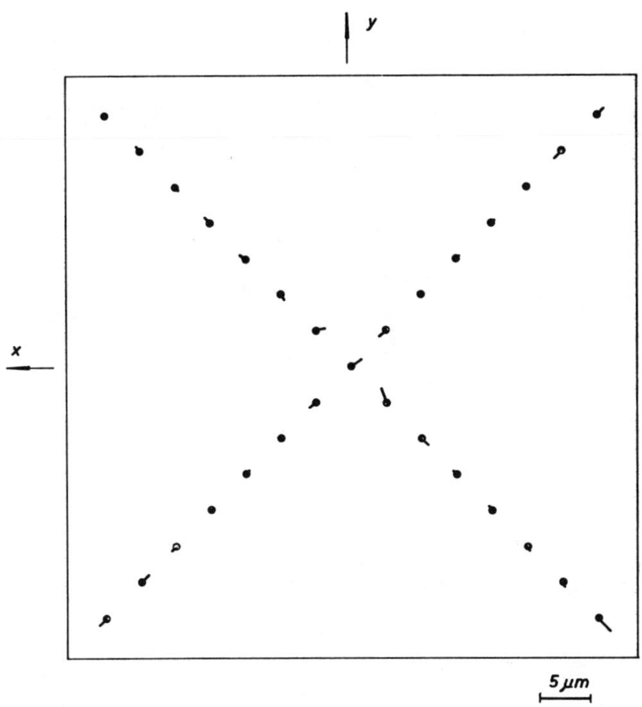

Figure 14.13. Filtered Uncorrelated $\hat{r}$ Component of Lens Distortion

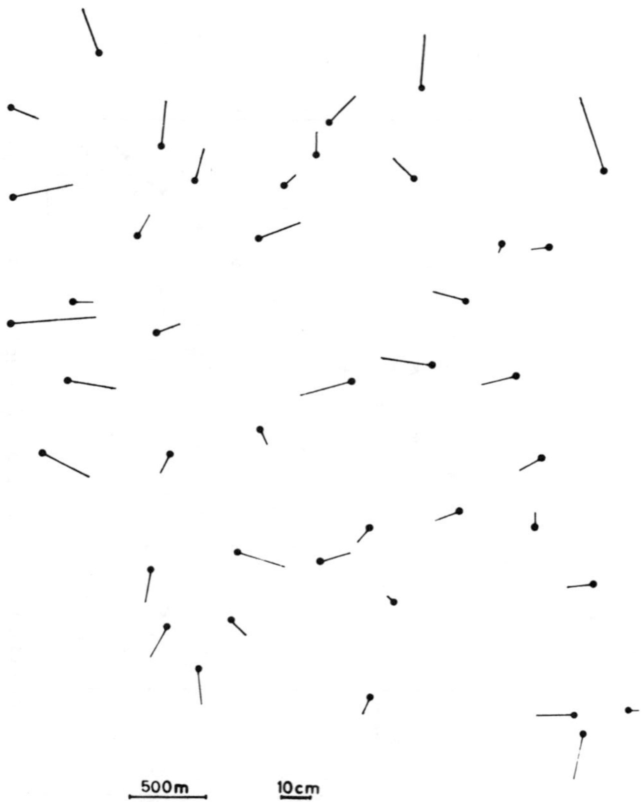

Figure 14.14. External Block Adjustment (Residuals at $n = 42$ Horizontal Check Control Points)

Figure 14.15. Estimated $\hat{s}$ Components at Control Points ($\cdot$) and Pass Points ($\circ$)

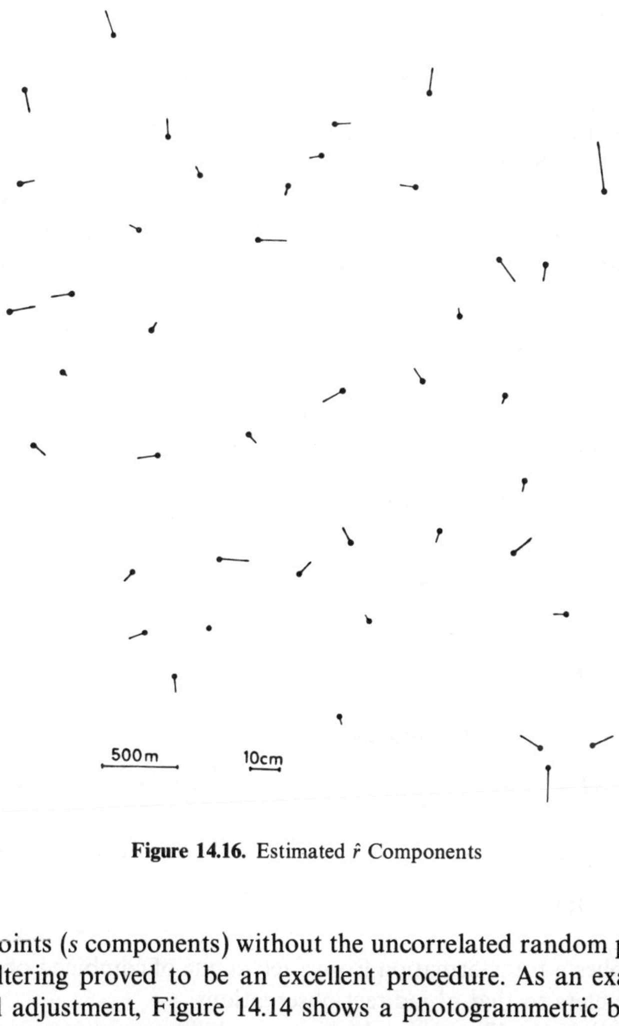

Figure 14.16. Estimated $\hat{r}$ Components

the pass points (s components) without the uncorrelated random part, least squares filtering proved to be an excellent procedure. As an example for horizontal adjustment, Figure 14.14 shows a photogrammetric block with the residuals after triangulation at 42 horizontal check control (reference) points. Using a Gaussian covariance function the values $C_s(0) = 0.72$, $C_l(0) = 1.00$, and $k = 0.00086\ \text{m}^2$ were estimated from the data. Both X and Y components were estimated independently and separately, but then combined to plot total horizontal vectors.

Figure 14.15 shows the estimated $\hat{s}$ for both the check control (reference) points (designated by solid circles) and the interpolated pass points (hollow circles ∘). Figure 14.16 shows the remaining $\hat{r} = l - \hat{s}$ component of the residuals at the reference points. It can be ascertained by comparing Figures 14.15 and 14.16 how the $\hat{s}$ and $\hat{r}$ components vary in appearance and behavior. Although the $\hat{s}$ components are quite correlated, the r components obviously are not.

14.7. LEAST SQUARES COLLOCATION

It has been mentioned earlier in this chapter that least squares collocation combines the well-established techniques of adjustment with those of interpolation and filtering. The presentation by Moritz (1972, see Bibliography) considers the technique of adjustment of indirect observation and interpolation. Although we shall rely mainly on this work, we shall extend it slightly to include the more general least squares technique of adjustment of observations and functionally independent parameters, for which the conditions are

$$\underset{c,\,n}{\mathbf{A}}(\boldsymbol{l}+\mathbf{v})+\underset{c,\,u}{\mathbf{B}}\;\underset{u,\,1}{\boldsymbol{\Delta}}=\underset{c,\,1}{\mathbf{d}} \tag{14.45}$$

with $\boldsymbol{\Sigma}$ as the covariance matrix of the original observations. A new set of equivalent observations $\boldsymbol{l}_e$ may be introduced so that

$$\boldsymbol{l}_e+\mathbf{v}_e=(-\mathbf{A}\boldsymbol{l}+\mathbf{d})+(\mathbf{A}\mathbf{v})$$

Thus

$$\underset{c,\,1}{\boldsymbol{l}_e}=-\mathbf{A}\boldsymbol{l}+\mathbf{d} \tag{14.46a}$$

$$\underset{c,\,1}{\mathbf{v}_e}=\mathbf{A}\mathbf{v} \tag{14.46b}$$

and from propagation techniques

$$\underset{c,\,c}{\boldsymbol{\Sigma}_e}=\mathbf{A}\boldsymbol{\Sigma}\mathbf{A}^t \tag{14.46c}$$

is the covariance matrix of $\boldsymbol{l}_e$. Equation (14.45) may now be rewritten as

$$\mathbf{v}_e=-\mathbf{B}\boldsymbol{\Delta}+\boldsymbol{l}_e \tag{14.47}$$

which implies a transformation from a technique of combined observations and parameters to that of indirect observations.

In equation (14.47) the residuals $\mathbf{v}_e$ are decomposed into a correlated component (signal) $\mathbf{s}$ and noise $\mathbf{r}$, or $\mathbf{v}_e=\mathbf{s}+\mathbf{r}$, hence

$$\mathbf{r}+\mathbf{s}=\boldsymbol{l}_e-\mathbf{B}\boldsymbol{\Delta} \tag{14.48}$$

Comparing equations (14.48) to (14.8) shows that subtracting $\mathbf{B}\boldsymbol{\Delta}$ from $\boldsymbol{l}_e$ amounts to trend computation. Thus collocation means combining estimating the trend simultaneously with interpolation and filtering.

Since both $\mathbf{r}$ and $\mathbf{s}$ are random vectors whose expectations are zero vectors, or

$$\boldsymbol{\mu}_r=E(\mathbf{r})=\mathbf{0} \tag{14.49}$$

$$\boldsymbol{\mu}_s=E(\mathbf{s})=\mathbf{0} \tag{14.50}$$

then $\boldsymbol{\mu}_{v_e} = \mathbf{0}$, and $\mathbf{B\Delta}$ represents the mean vector of $\boldsymbol{l}_e$ or $E(\boldsymbol{l}_e)$. Consequently,

$$\underset{c,\,c}{\boldsymbol{\Sigma}_{v_e}} = E(\boldsymbol{l}_e - \mathbf{B\Delta})(\boldsymbol{l}_e - \mathbf{B\Delta})^t$$

$$= E(\boldsymbol{l}_e - E(\boldsymbol{l}_e))(\boldsymbol{l}_e - E(\boldsymbol{l}_e))^t$$

which is by definition the covariance matrix of $\boldsymbol{l}_e$, or

$$\boldsymbol{\Sigma}_{v_e v_e} = \boldsymbol{\Sigma}_e \tag{14.51}$$

The vector of observations is assumed to be known at n points (the reference or data points). In addition to estimating $\boldsymbol{\Delta}$, we may also wish to interpolate and filter at another set of m points for which the vector $\mathbf{s}_0$ will be used. In other words $\mathbf{s}_0$ is the vector of estimated signals at m points that may be different from the data points. Combining the two random vectors

$$\underset{m,\,1}{\mathbf{s}_0} \quad \text{and} \quad \underset{c,\,1}{\mathbf{v}_e}$$

into one, let

$$\underset{p,\,1}{\dot{\mathbf{v}}} = \begin{bmatrix} \mathbf{s}_0 \\ \mathbf{v}_e \end{bmatrix} \qquad \text{with } p = m + c \tag{14.52}$$

and so equation (14.47) becomes

$$[\mathbf{0} \quad \mathbf{I}] \begin{bmatrix} \mathbf{s}_0 \\ \mathbf{v}_e \end{bmatrix} + \mathbf{B\Delta} = \boldsymbol{l}_e = \mathbf{f} \tag{14.53}$$

or

$$\dot{\mathbf{A}}\dot{\mathbf{v}} + \mathbf{B\Delta} = \mathbf{f} \tag{14.54}$$

Taking the a priori reference variance equal to one, the covariance and cofactor matrices associated with equation (14.54) may be constructed as

$$\boldsymbol{\Sigma} = \begin{bmatrix} \boldsymbol{\Sigma}_{s_0 s_0} & \boldsymbol{\Sigma}_{s_0 e} \\ \boldsymbol{\Sigma}^t_{s_0 e} & \boldsymbol{\Sigma}_e \end{bmatrix} = \mathbf{Q} = \begin{bmatrix} \mathbf{Q}_{s_0 s_0} & \mathbf{Q}_{s_0 e} \\ \mathbf{Q}^t_{s_0 e} & \mathbf{Q}_e \end{bmatrix} \tag{14.55}$$

According to equation (6.20), the least squares estimate of $\boldsymbol{\Delta}$ from equation (14.54) is

$$\boldsymbol{\Delta} = [\mathbf{B}^t(\dot{\mathbf{A}}\mathbf{Q}\dot{\mathbf{A}}^t)^{-1}\mathbf{B}]^{-1}[\mathbf{B}^t(\dot{\mathbf{A}}\mathbf{Q}\dot{\mathbf{A}}^t)^{-1}\mathbf{f}] \tag{14.56}$$

From equations (14.53), (14.54), and (14.55)

$$(\dot{\mathbf{A}}\mathbf{Q}\dot{\mathbf{A}}^t) = [\mathbf{0} \quad \mathbf{I}] \begin{bmatrix} \mathbf{Q}_{s_0 s_0} & \mathbf{Q}_{s_0 e} \\ \mathbf{Q}^t_{s_0 e} & \mathbf{Q}_e \end{bmatrix} \begin{bmatrix} \mathbf{0} \\ \mathbf{I} \end{bmatrix} = \mathbf{Q}_e$$

hence equation (14.56) becomes

$$\boldsymbol{\Delta} = (\mathbf{B}^t\mathbf{W}_e\,\mathbf{B})^{-1}(\mathbf{B}^t\mathbf{W}_e\,\mathbf{f}) \tag{14.57}$$

Equation (14.57) is identical to the results in equations (6.21), (6.22), and (6.24), which shows that the extension to least squares collocation still yields the proper adjustment estimate for the parameters. It can also be shown that the a posteriori cofactor matrix $\mathbf{Q}_{\Delta\Delta}$ is in fact $(\mathbf{B}^t\mathbf{W}_e\mathbf{B})^{-1}$, which is consistent with the results in Chapter 6 (Part II).

To continue with filtering as a part of collocation, we may write

$$\dot{\mathbf{v}} = \mathbf{Q}\dot{\mathbf{A}}^t\mathbf{W}_e(\boldsymbol{l}_e - \mathbf{B}\boldsymbol{\Delta})$$

or

$$\begin{bmatrix} \hat{\mathbf{s}}_0 \\ \mathbf{v}_e \end{bmatrix} = \begin{bmatrix} \boldsymbol{\Sigma}_{s_0s_0} & \boldsymbol{\Sigma}_{s_0e} \\ \boldsymbol{\Sigma}^t_{s_0e} & \boldsymbol{\Sigma}_e \end{bmatrix} \begin{bmatrix} \mathbf{0} \\ \mathbf{I} \end{bmatrix} \boldsymbol{\Sigma}_e^{-1}(\boldsymbol{l}_e - \mathbf{B}\boldsymbol{\Delta})$$

Thus the estimate $\hat{\mathbf{s}}_0$ of the filtered signal at the m points is

$$\hat{\mathbf{s}}_0 = \boldsymbol{\Sigma}_{s_0e}\boldsymbol{\Sigma}_e^{-1}(\boldsymbol{l}_e - \mathbf{B}\boldsymbol{\Delta}) \tag{14.58}$$

If the original least squares adjustment were of the indirect observation technique, then $\boldsymbol{l}_e$ in equation (14.58) would be replaced by the observations themselves (data) $\boldsymbol{l}$, thus

$$\hat{\mathbf{s}}_0 = \boldsymbol{\Sigma}_{s_0l}\boldsymbol{\Sigma}_{ll}^{-1}(\boldsymbol{l} - \mathbf{B}\boldsymbol{\Delta}) \tag{14.59}$$

in which $\hat{\mathbf{s}}_0$ is the $m \times 1$ vector of interpolated or filtered values at the m new points; $\boldsymbol{\Sigma}_{s_0l} = \boldsymbol{\Sigma}^t_{ls_0}$ is the $m \times n$ cross covariance matrix between the signal and the given data; $\boldsymbol{\Sigma}_{ll}$ is the $n \times n$ symmetric autocovariance matrix of the observations (data); $\boldsymbol{l}$ is the vector of observations ($= \mathbf{s}$ for interpolation); and $\mathbf{B}\boldsymbol{\Delta}$ is the mean of the observations (representing the trend surface) as the product of the $m \times u$ coefficient matrix and the $u \times 1$ vector of parameters. If the trend has been removed before interpolation and filtering, $\mathbf{B}\boldsymbol{\Delta}$ would be missing and equation (14.59) becomes identical to equation (14.26).

As a further demonstration if only one value s_0 is to be estimated, the least squares collocation reduces to

$$\hat{s}_0 = \underset{1,n}{\boldsymbol{\sigma}_{s_0l}}\boldsymbol{\Sigma}_{ll}^{-1}(\boldsymbol{l} - \mathbf{B}\boldsymbol{\Delta}) \tag{14.60a}$$

in which $\boldsymbol{\sigma}_{s_0l}$ is a row vector. Furthermore, when the vector of parameters is zero due to a priori trend removal, least squares collocation reduces to interpolation and filtering, and equation (14.60a) becomes

$$\hat{s}_0 = \boldsymbol{\sigma}_{s_0l}\boldsymbol{\Sigma}_{ll}^{-1}\boldsymbol{l} \tag{14.60b}$$

which is identical to equation (14.21). Consequently, it is obvious that both least squares adjustment [see equation (14.57)] and least squares interpolation and filtering are special cases of collocation.

By way of summary then, least squares collocation makes possible the estimation of parameters, $\boldsymbol{\Delta}$, as well as estimating filtered values for the variables representing the observations (signals) not only at the data points, but also at other points. These filtered (or interpolated) values will by necessity be referred to the trend surface **BΔ**. Consequently, if we are interested in the total estimate of the signal **t** with respect to the original datum of the data, then

$$\hat{\mathbf{t}}_0 = \hat{\mathbf{s}}_0 + \mathbf{D\Delta} \tag{14.61}$$

where **D** is the appropriate coefficient matrix necessary for computing values of the function at the *m* new filtered points. Obviously, if filtering is done at the original *n* reference or data points, the matrix **D** would be identical to the original matrix **B**. The computation of the total signal is demonstrated by the example succeeding the following section.

A last remark concerns the fact that all the functions given in the preceding derivations were linear. This is in keeping with the premise that least squares is practical mainly in linear estimation. Naturally, if the original functional model is nonlinear, as it often is in practice, a linearized form must be derived and used in the developed algorithms.

14.7.1. Precision Estimation for Least Squares Collocation The a posteriori cofactor (covariance) matrix of the parameters is

$$\mathbf{Q}_{\Delta\Delta} = (\mathbf{B}^t\mathbf{W}_e\,\mathbf{B})^{-1} = [\mathbf{B}^t(\mathbf{A}\mathbf{Q}_{ll}\,\mathbf{A}^t)^{-1}\mathbf{B}]^{-1} \tag{14.62}$$

In order to evaluate the a posteriori cofactor matrix of the estimated signals $\hat{\mathbf{s}}_0$, we should recall that in general the a posteriori cofactor matrix of the estimated observations is equal to the a priori cofactor matrix minus the a posteriori cofactor matrix of the residuals. According to equation (6.30) the cofactor matrix $\mathbf{Q}_{\dot{v}\dot{v}}$ may be evaluated

$$\mathbf{Q}_{\dot{v}\dot{v}} = \mathbf{Q}\dot{\mathbf{A}}^t(\mathbf{W}_e - \mathbf{W}_e\,\mathbf{B}\mathbf{Q}_{\Delta\Delta}\,\mathbf{B}^t\mathbf{W}_e)\dot{\mathbf{A}}\mathbf{Q} \tag{14.63}$$

From the structure of $\dot{\mathbf{A}}$ and **Q**, see equations (14.53), (14.54), and (14.55), the cofactor submatrix relating to $\hat{\mathbf{s}}_0$ is

$$\mathbf{Q}_{v_sv_s} = \mathbf{Q}_{soe}\,\mathbf{W}_e\,\mathbf{Q}^t_{soe} - \mathbf{Q}_{soe}\,\mathbf{W}_e\,\mathbf{B}\mathbf{Q}_{\Delta\Delta}\,\mathbf{B}^t\mathbf{W}_e\,\mathbf{Q}^t_{soe} \tag{14.64}$$

Therefore the estimated cofactor matrix of $\hat{\mathbf{s}}_0$ is

$$\mathbf{Q}_{\hat{s}_0\hat{s}_0} = \mathbf{Q}_{soso} - \mathbf{Q}_{v_sv_s}$$

or

$$\mathbf{Q}_{\hat{s}_0\hat{s}_0} = \mathbf{Q}_{soso} - \mathbf{Q}_{soe}\,\mathbf{Q}_e^{-1}\mathbf{Q}^t_{soe} + \mathbf{Q}_{soe}\,\mathbf{Q}_e^{-1}\,\mathbf{B}\mathbf{Q}_{\Delta\Delta}\,\mathbf{B}^t\mathbf{Q}_e^{-1}\mathbf{Q}^t_{soe} \tag{14.65}$$

with the last term in equation (14.65) representing the contribution of the errors in the parameters **Δ**. In the absence of parameters, equation (14.65) in terms of covariance matrices becomes

$$\mathbf{\Sigma}_{\hat{s}_0\hat{s}_0} = \mathbf{\Sigma}_{soso} - \mathbf{\Sigma}_{soe}\mathbf{\Sigma}_e^{-1}\mathbf{\Sigma}^t_{soe} \tag{14.66}$$

which is very similar to equation (14.28) except that here $\boldsymbol{l}_e$ replaces the $\boldsymbol{l}$ used in the earlier treatment. Finally, if only one value is estimated, and the conditions are of the indirect observation type, then (using the original symbols for covariances)

$$\sigma_{\hat{s}0}^2 = C_s(0) - \boldsymbol{\sigma}_{lso}^t \boldsymbol{\Sigma}_{ll}^{-1} \boldsymbol{\sigma}_{lso} \tag{14.67}$$

which is identical to equation (14.22a). This shows that the results from least squares collocation will reduce to the results of interpolation and filtering once the parameters are not considered, which means that the trend has been removed a priori.

A crosscofactor matrix $\mathbf{Q}_{\Delta s_0}$ may also be evaluated as follows. Let the vector of observational variables associated with equation (14.54) be denoted by $\dot{\boldsymbol{l}}$. The estimated observations are therefore

$$\hat{\boldsymbol{l}} = \dot{\boldsymbol{l}} + \dot{\mathbf{v}} = \begin{bmatrix} 0 \\ \boldsymbol{l}_e \end{bmatrix} + \begin{bmatrix} \mathbf{s}_0 \\ \mathbf{v}_e \end{bmatrix} = \begin{bmatrix} \mathbf{s}_0 \\ \hat{\boldsymbol{l}}_e \end{bmatrix} \tag{14.68}$$

Referring to equation (6.33), the crossfactor matrix between the parameters Δ and $\dot{\boldsymbol{l}}$ is

$$\mathbf{Q}_{\Delta l} = -\mathbf{N}^{-1}\mathbf{B}^t\mathbf{W}_e\dot{\mathbf{A}}\mathbf{Q} = -\mathbf{Q}_{\Delta\Delta}\,\mathbf{B}^t\mathbf{W}_e[\mathbf{0}\quad \mathbf{I}]\begin{bmatrix} \mathbf{Q}_{soso} & \mathbf{Q}_{soe} \\ \mathbf{Q}_{soe}^t & \mathbf{Q}_e \end{bmatrix}$$

or

$$[\mathbf{Q}_{\Delta\hat{s}_0}\,\mathbf{Q}_{\Delta l_e}] = -\mathbf{Q}_{\Delta\Delta}\,\mathbf{B}^t\mathbf{W}_e[\mathbf{Q}_{soe}^t\quad \mathbf{Q}_e]$$

from which

$$\mathbf{Q}_{\Delta\hat{s}_0} = -\mathbf{Q}_{\Delta\Delta}\,\mathbf{B}^t\mathbf{W}_e\,\mathbf{Q}_{soe}^t \tag{14.69}$$

or in terms of covariance matrices

$$\boldsymbol{\Sigma}_{\Delta\hat{s}_0} = -\boldsymbol{\Sigma}_{\Delta\Delta}\,\mathbf{B}^t\mathbf{W}_e\boldsymbol{\Sigma}_{soe} \tag{14.70}$$

Finally, an a posteriori cofactor matrix is needed for the final estimate $\hat{\mathbf{t}}$ of the total signal given by equation (14.61). Applying the propagation principle

$$\mathbf{Q}_{tt} = \mathbf{Q}_{\hat{s}_0\hat{s}_0} + \mathbf{D}\mathbf{Q}_{\Delta\Delta}\,\mathbf{D}^t + \mathbf{Q}_{\hat{s}_0\Delta}\,\mathbf{D}^t + \mathbf{D}\mathbf{Q}_{\Delta\hat{s}_0}$$

which according to equations (14.65) and (14.69) becomes

$$\begin{aligned}\mathbf{Q}_{\hat{t}\hat{t}} &= \mathbf{Q}_{soso} - \mathbf{Q}_{soe}\,\mathbf{Q}_e^{-1}\mathbf{Q}_{soe}^t + \mathbf{Q}_{soe}\,\mathbf{Q}_e^{-1}\mathbf{B}\mathbf{Q}_{\Delta\Delta}\,\mathbf{B}^t\mathbf{Q}_e^{-1}\mathbf{Q}_{soe}^t \\ &\quad + \mathbf{D}\mathbf{Q}_{\Delta\Delta}\,\mathbf{D}^t - \mathbf{Q}_{soe}\mathbf{Q}_e^{-1}\mathbf{B}\mathbf{Q}_{\Delta\Delta}\,\mathbf{D}^t - \mathbf{D}\mathbf{Q}_{\Delta\Delta}\,\mathbf{B}^t\mathbf{Q}_e^{-1}\mathbf{Q}_{soe}^t \\ &= \mathbf{Q}_{soso} - \mathbf{Q}_{soe}\,\mathbf{Q}_e^{-1}\mathbf{Q}_{soe}^t + (\mathbf{Q}_{soe}\,\mathbf{Q}_e^{-1}\mathbf{B} - \mathbf{D})\mathbf{Q}_{\Delta\Delta} \\ &\quad \times (\mathbf{B}^t\mathbf{Q}_e^{-1}\mathbf{Q}_{soe}^t - \mathbf{D}^t)\end{aligned}$$

or

$$\mathbf{Q}_{\hat{i}\hat{i}} = \mathbf{Q}_{soso} - \mathbf{Q}_{soe}\,\mathbf{Q}_e^{-1}\mathbf{Q}_{soe}^t + (\mathbf{Q}_{soe}\,\mathbf{Q}_e^{-1}\mathbf{B} - \mathbf{D})\mathbf{Q}_{\Delta\Delta} \times (\mathbf{Q}_{se}\,\mathbf{Q}_e^{-1}\mathbf{B} - \mathbf{D})^t \tag{14.71}$$

in which $\mathbf{Q}_{\Delta\Delta}$ is as given by equation (14.62). In terms of covariance matrices, equation (14.71) becomes

$$\boldsymbol{\Sigma}_{\hat{i}\hat{i}} = \boldsymbol{\Sigma}_{soso} - \boldsymbol{\Sigma}_{soe}\boldsymbol{\Sigma}_e^{-1}\boldsymbol{\Sigma}_{soe}^t + (\boldsymbol{\Sigma}_{soe}\boldsymbol{\Sigma}_e^{-1}\mathbf{B} - \mathbf{D})\boldsymbol{\Sigma}_{\Delta\Delta} \times (\boldsymbol{\Sigma}_{soe}\boldsymbol{\Sigma}_e^{-1}\mathbf{B} - \mathbf{D})^t \tag{14.72}$$

14.7.2. Example on Least Squares Collocation It was pointed out previously that the removal of the trend in the data is an important prerequisite to the application of interpolation and filtering. The operation of trend removal has usually been performed prior to the process of interpolation or filtering. Least squares collocation as presented here allows the determination of the trend surface, simultaneous with filtering. The trend surface therefore becomes the functional part of the model.

This example considers a one-dimensional filtering problem with the trend surface limited to a straight line. It is taken from the report by Moritz (1972, see Bibliography) in which Dr. K. P. Schwarz is acknowledged as having performed the computations. Figure 14.17 is a schematic of the problem.

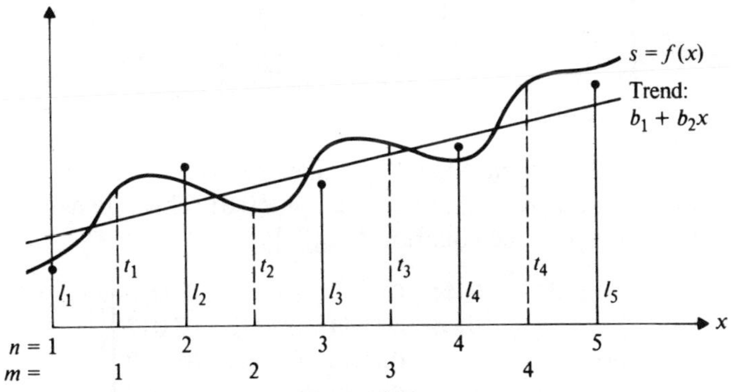

Figure 14.17

The data, or observations, are given at $n = 5$ points in Table 14.2. Assuming stationarity and ergodicity, a covariance function may be written in terms of distances between points, d. The expression

$$C_l(d) = C_l(0)e^{-a^2d^2} \tag{14.73}$$

is used for such a covariance function with the constants

$$C_l(0) = 0.1260 \qquad a = 0.6$$

TABLE 14-2

x	l
0.000	0.6108
1.445	1.0863
2.890	2.9034
4.335	4.5925
5.780	6.2714

Taking the covariance function for the uncorrelated random component r to be a constant equal to 0.01, then

$$C_r(0) = 0.01 \qquad C_r(d) = 0 \qquad \text{for all } d \neq 0$$

Consequently, the a priori covariance matrices are

$$\Sigma_{ll} = \begin{bmatrix} 0.1360 & 0.0594 & 0.0062 & 0.0002 & 0.0000 \\ & 0.1360 & 0.0594 & 0.0062 & 0.0002 \\ & & 0.1360 & 0.0594 & 0.0062 \\ & & & 0.1360 & 0.0594 \\ \text{symmetric} & & & & 0.1360 \end{bmatrix}$$

$$\Sigma_{ss} = \begin{bmatrix} 0.1260 & 0.0594 & 0.0062 & 0.0002 \\ & 0.1260 & 0.0594 & 0.0062 \\ & & 0.1260 & 0.0594 \\ \text{symmetric} & & & 0.1260 \end{bmatrix}$$

The reader should note that the diagonal elements of Σ_{ll} are equal and each is the sum of one diagonal element of Σ_{ss} plus 0.01. This satisfies the fact that $C_l(0) = C_s(0) + C_r(0)$ [see equation (14.12c)],

$$\Sigma_{sol} = \begin{bmatrix} 0.1044 & 0.1044 & 0.0232 & 0.0012 & 0.0000 \\ 0.0232 & 0.1044 & 0.1044 & 0.0232 & 0.0012 \\ 0.0012 & 0.0232 & 0.1044 & 0.1044 & 0.0232 \\ 0.0000 & 0.0012 & 0.0232 & 0.1044 & 0.1044 \end{bmatrix}$$

Before presenting the results, the appropriate equations are first given. The equation for the trend surface is

$$y = b_1 + b_2 x$$

Thus with $\mathbf{\Delta} = [b_1 \quad b_2]^t$, the functional model is expressed by the equation [see equation (14.48)] for each data point

$$s_i + r_i = l_i - [1 \quad x_i]\mathbf{\Delta} = l_i - \mathbf{B}_i\mathbf{\Delta}$$

with $\mathbf{B}_i = [1 \quad x_i]$. The total coefficient matrix $\mathbf{B}_{5,2}$ is

$$\mathbf{B} = \begin{bmatrix} 1 & 1 & 1 & 1 & 1 \\ x_1 & x_2 & x_3 & x_4 & x_5 \end{bmatrix}^t$$

and the estimate of the parameters is [see equation (14.57)]

$$\Delta = (\mathbf{B}^t\Sigma_{ll}^{-1}\mathbf{B})(\mathbf{B}^t\Sigma_{ll}^{-1}\boldsymbol{l}) = \begin{bmatrix} b_1 \\ b_2 \end{bmatrix} = \begin{bmatrix} 0.3252 \\ 0.9891 \end{bmatrix}$$

Note that since the model is that of indirect observations, $\mathbf{W}_e$ becomes Σ_{ll}^{-1}. Next, the filtered values are obtained at the $m = 4$ points shown in Figure 14.17, from [see equation (14.59)]

$$\hat{\mathbf{s}}_0 = \Sigma_{s_0l}\Sigma_{ll}^{-1}(\boldsymbol{l} - \mathbf{B}\Delta) = [-0.221 \quad -0.5590 \quad -0.1052 \quad 0.1082]^t$$

The points at which filtering is performed are taken as halfway between data points. Their coordinates are needed to compute the values of the total signal at them. For example $\hat{t}_1$ at the point $x_1 = 0.7225$ is computed as [see equation (14.61)]

$$\hat{t}_1 = 0.2221 + [1 \quad 0.7225]\Delta = 0.8177$$

Collecting all values for computing the total signal we get the data in Table 14-3.

TABLE 14-3

x	s_0	t
0.7225	−0.2221	0.8177
2.1675	−0.5590	1.9101
3.6125	−0.1052	3.7932
5.0575	0.1082	5.4359

Finally, the a posteriori covariance matrices are

$$\Sigma_{\Delta\Delta} = (\mathbf{B}^t\Sigma_{ll}^{-1}\mathbf{B})^{-1} = \begin{bmatrix} 0.1136 & -0.0234 \\ -0.0234 & 0.0081 \end{bmatrix}$$

$$\Sigma_{\hat{s}_0\hat{s}_0} = \begin{bmatrix} 0.1034 & 0.0532 & 0.0343 & 0.0025 \\ & 0.0564 & 0.0369 & 0.0343 \\ & & 0.0564 & 0.0532 \\ \text{symmetric} & & & 0.1034 \end{bmatrix}$$

$$\Sigma_{\hat{t}\hat{t}} = \begin{bmatrix} 0.0131 & -0.0024 & 0.0015 & -0.0009 \\ & 0.0121 & -0.0019 & 0.0015 \\ & & 0.0121 & -0.0024 \\ \text{symmetric} & & & 0.0131 \end{bmatrix}$$

The value of **D** used to compute $\mathbf{\Sigma}_{\hat{t}\hat{t}}$ is

$$\mathbf{D}^t = \begin{bmatrix} 1.0000 & 1.0000 & 1.0000 & 1.0000 \\ 0.7225 & 2.1675 & 3.6125 & 5.0575 \end{bmatrix}$$

$$\mathbf{\Sigma}_{\Delta s_0} = \begin{bmatrix} 0.1007 & 0.0578 & 0.0298 & -0.0080 \\ -0.0188 & -0.0048 & 0.0048 & 0.0188 \end{bmatrix}$$

Problems for Part III

1. Using the unified approach rework Problem 16 of Part II taking the a priori coordinates of point B as $x_1 = 500.00$ m and $x_2 = 866.00$ m with a priori covariance matrix $\mathbf{\Sigma}_{xx} = 25\mathbf{I}_2(\mathrm{m}^2)$.

2. Using the unified approach rework Problem 10 of Part II taking the a priori elevations as follows:

Station	1	2	3	4
Elevation (m)	542.0	529.5	487.2	495.7

Assume that the cofactor for each of these a priori elevations is 20 times the largest cofactor for the given observed differences in elevation.

3. Rework Problem 20, Part II, using a priori value for x as 60 degrees with a priori weight $w_{xx} = 10$ times the weight of the observations.

4. Refer to Problem 39, Part II. Let $x_1 = 10$, $x_2 = 4$, $y_2 = 2$ be a priori values for the parameters, with a priori $\mathbf{W}_{xx} = 100\mathbf{I}_3$. Using the unified approach compute the estimates of x_1, x_2, y_2 and their a posteriori cofactor matrix.

5. For the traverse in Problem 36, Part II, the coordinates of the control points B and E are assumed to have a priori covariance matrix $\mathbf{\Sigma} = 4 \times 10^{-6}\mathbf{I}_4$ m². In order

to apply the unified approach totally to the problem, a priori values for the coordinates of points C and D are needed. Compute these using a minimum subset of the given observations, and take the estimate of their a priori covariance matrix as $0.04\mathbf{I}_4$ m^2.

6. The figure shows the geometry of lines and points in a plane. The observations are $l_1 = 7.07$ m; $l_2 = 7.08$ m; $l_3 = 10.00$ m; $l_4 = 10.02$ m; with $\sigma_1 = \sigma_2 = \sigma_3 = \sigma_4 = 0.01$ m; $l_5 = 30°$; $\sigma_5 = 0.01°$. Points A and B are control points with *perfectly known*

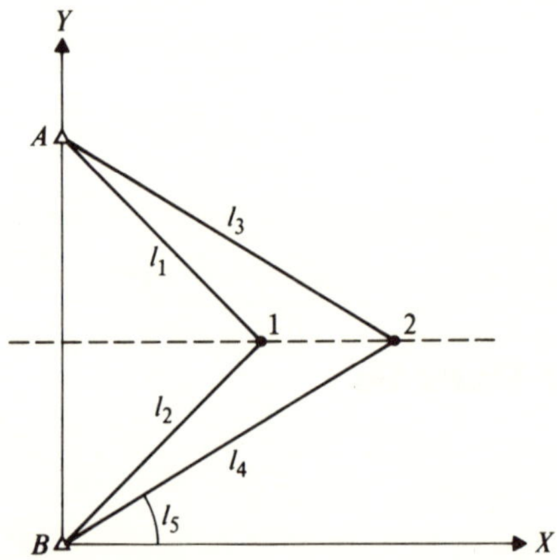

coordinates $X_A = 0$, $Y_A = 10$ m, $X_B = Y_B = 0$. For *all the parts to follow* we are interested in the least squares estimates of the coordinates of points 1 and 2.

a. Write the elements of the *functional* and *stochastic* models. (Give the equations, but do not linearize if nonlinear. Instead write symbolically the linearized form and give the dimensions of the matrices).

b. If, in addition to the information above, points 1 and 2 are on a line parallel to the X axis, give the elements of the *functional* model (again no linearization—only matrices' dimensions).

c. If, instead of assuming the coordinates of points A and B to be perfectly known, the given coordinates (that is, $X_A = X_B = Y_B = 0$ and $Y_B = 10.0$ m) *each* has a standard deviation of 0.001 m, give the elements of the *functional* and *stochastic* models. (No linearization—only matrices' dimensions).

d. In addition to the information in (b) and (c) a priori coordinates for points 1 and 2 are $X_1 = Y_1 = 5.0$ m, $X_2 = 8.7$ m, $Y_2 = 5.0$ m, *each* with a standard deviation of 0.2 m. Taking the approximations for all the observations equal to their a priori values, and the approximations for the parameters to be $X_1^0 = Y_1^0 = 5.0$ m, $X_2^0 = 8.0$ m, $Y_2^0 = 5.0$ m, give the *linearized* form of the functional model, as well as the stochastic model. Show how you would construct the reduced normal equations, giving the numerical values of the contributions from the a priori information on the parameters X_1, Y_1, X_2, Y_2.

7. For Problem 30, Part II, use enough points to determine uniquely the parameters, then add the contribution of the remaining points sequentially.

8. For Problem 7 above use the parameter values from a unique solution as a priori estimates and assume a suitable a priori weight matrix for these estimates. Add

the contribution of all five points sequentially using two points, then two points, then one point.

9. For Problem 38, Part II, find the least squares solution for the first two points, then add the contribution of the third point sequentially.

10. Rework Problem 31, Part II, beginning with a minimum solution and adding the remaining data sequentially.

11. Using the a priori data in Problem 3 above, add *sequentially* the conditions arising from the given observational data to get the least squares estimate of the angle x.

12. The figure shows a schematic of a simulated level net involving a bench mark (elevation 100.000 m) and five points 1, 2, 3, 4, 5. There is a total of 13 observed

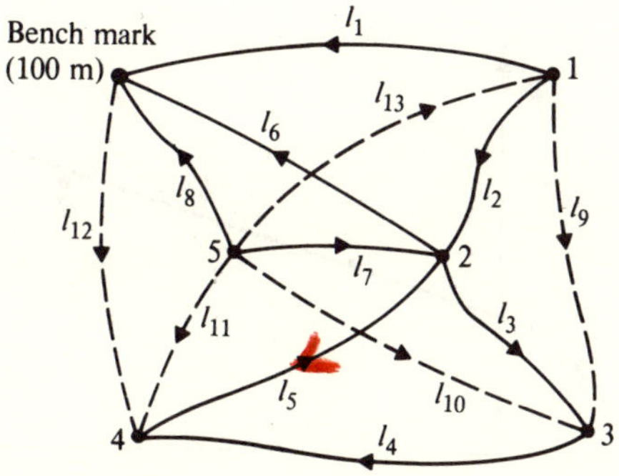

differences in elevation, the values of which as well as corresponding lengths of the lines are given in the following table:

LINE	l_1	l_2	l_3	l_4	l_5	l_6
Observations (m)	5.013	2.985	6.987	10.014	17.013	1.982
Length (km)	1.4	1.03	1.2	1.71	1.42	1.27

LINE	l_7	l_8	l_9	l_{10}	l_{11}	l_{12}	l_{13}
Observations (m)	7.982	10.014	10.014	15.016	25.017	14.982	4.981
Length (km)	0.6	0.95	1.82	1.66	1.14	2.0	1.42

The variance of each observation in square metres is taken proportional to the corresponding line length in kilometres. Use the first eight observations, shown in solid lines in the figure, to perform a batch least squares for estimating the elevations of the five points. Then sequentially add one observation at a time, l_9 through l_{13} (dotted lines in the figure), and compute new estimates for the five elevations as well as the reference variance after each observation. Use an F test to determine at which observation, if any, is the change in the estimate of the reference variance, $\hat{\sigma}_0^2$, significant at $\alpha = 0.05$.

13. With respect to the figure the coordinates are as follows:

POINT	X	Y
A	1	1
1	2	1.2
2	3	1.3

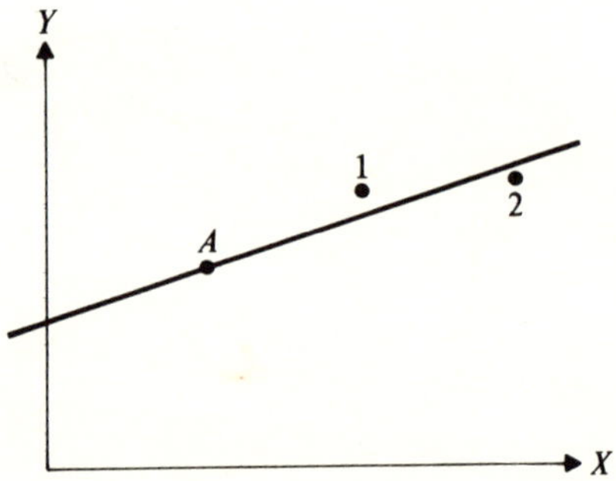

A straight line is to pass through point A *exactly*. Points 1 and 2, whose Y coordinates are taken as the observations, are to be used to determine the straight line. Get a solution from point 1 first, then sequentially add the contribution of the data from point 2. Assume $\mathbf{Q} = \mathbf{I}$.

APPENDIX

A

Review of Matrix Theory

A1. DEFINITIONS

A *matrix* is an array of numbers of some algebraic system, which in general is written in the geometric form of a rectangle of specified number of rows and columns. Thus

$$\underset{m,\,n}{\mathbf{A}} = \begin{bmatrix} a_{11} & a_{12} & \cdots & a_{1n} \\ a_{21} & a_{22} & \cdots & a_{2n} \\ \vdots & & & \vdots \\ a_{m1} & a_{m2} & \cdots & a_{mn} \end{bmatrix}$$

is a rectangular matrix of order $m \times n$. The first subscript always refers to the number of the row and the second refers to the column number. The capital boldface letter refers to the whole matrix, whereas the lowercase letter refers to one element only. It follows then that a_{ij} lies at the intersection of the ith row and the jth column. Another way of expressing the matrix **A** is by typical element such as

$$\mathbf{A} = \{a_{ij}\} i = 1, 2, \ldots, m; \qquad j = 1, 2, \ldots, n$$

If the matrix is of the order 1×1, it is called a "scalar."

A2. TYPES OF MATRICES

A2.1. Square Matrix A square matrix is the matrix with an equal number of rows and columns. In this case $\mathbf{A}_{m,m}$ is a square matrix of order j. The principal (or main, or leading) diagonal is that composed of the elements a_{ij} for $i = j$. Two special cases of the square matrix are the *symmetric* and *skew-symmetric* ones. These will be introduced later.

A2.2. Row Matrix or Row Vector A row matrix or row vector is the matrix composed of only one row. It is designated by a lowercase boldface roman letter, for example,

$$\mathbf{a} = [a_1 \quad a_2 \quad \cdots \quad a_n] \qquad \text{and} \qquad \underset{1,3}{\mathbf{a}} = [1 \quad 2 \quad 4]$$

A2.3. Column Matrix, or Column Vector A column matrix or column vector is composed of only one column, and is denoted by

$$\mathbf{a} = \begin{bmatrix} a_1 \\ a_2 \\ \vdots \\ a_m \end{bmatrix} \qquad \text{for example, } \underset{2,1}{\mathbf{b}} = \begin{bmatrix} -1 \\ 3 \end{bmatrix}$$

A2.4. Diagonal Matrix A diagonal matrix is the square matrix with all elements off the main diagonal being zero:

$$\mathbf{D} = \begin{bmatrix} d_{11} & 0 & \cdots & 0 \\ 0 & d_{22} & \cdots & 0 \\ \vdots & \vdots & & \vdots \\ 0 & 0 & & d_{mm} \end{bmatrix}$$

where $d_{ij} = 0$ for all $i \neq j$, and $d_{ij} \neq 0$ for some or all $i = j$.

Example

$$\mathbf{G} = \begin{bmatrix} 1 & 0 & 0 \\ 0 & 0 & 0 \\ 0 & 0 & -3 \end{bmatrix}$$

We will also denote a diagonal matrix by the form

$$\mathbf{B} = \text{diag.}\ \{b_1, b_2, b_3, \ldots, b_n\}.$$

A2.5. Scalar Matrix A scalar matrix is a diagonal matrix whose elements are *all* equal to the same scalar:

$$\mathbf{A} = \begin{bmatrix} a & & & \\ & a & & \\ & & \ddots & \\ & & & a \end{bmatrix} \qquad \begin{array}{ll} a_{ij} = 0 & \text{for } all\ i \neq j \\ a_{ij} = a & \text{for } all\ i = j \end{array}$$

Thus

$$\mathbf{H} = \begin{bmatrix} 2 & 0 & 0 \\ 0 & 2 & 0 \\ 0 & 0 & 2 \end{bmatrix}$$

is a scalar matrix.

A2.6. Unit or Identity Matrix A unit or identity matrix is a diagonal matrix whose elements are all equal to 1. It is always referred to by

$$\mathbf{I} = \begin{bmatrix} 1 & 0 & \cdots & 0 \\ 0 & 1 & \cdots & 0 \\ \vdots & \vdots & & \\ 0 & 0 & \cdots & 1 \end{bmatrix} \qquad \begin{array}{ll} a_{ij} = 0 & \text{for } all\ i \neq j \\ a_{ij} = 1 & \text{for } all\ i = j \end{array}$$

A2.7. Null or Zero Matrix A null or zero matrix is a matrix whose elements are all zero. It is denoted by a boldface zero, **0**.

A2.8. Triangular Matrix A triangular matrix is a square matrix whose elements above (or below), but not including, the main diagonal are all zero:

$$\mathbf{A} = \begin{bmatrix} a_{11} & a_{12} & \cdots & a_{1m} \\ 0 & a_{22} & \cdots & a_{2m} \\ \vdots & & & \\ 0 & 0 & \cdots & a_{mm} \end{bmatrix} = \text{upper triangular} \quad a_{ij} = 0,\ i > j$$

Example

$$\mathbf{A} = \begin{bmatrix} -1 & 3 & 4 \\ 0 & 1 & 0 \\ 0 & 0 & 7 \end{bmatrix}$$

$$\mathbf{A} = \begin{bmatrix} a_{11} & 0 & \cdots & 0 \\ a_{21} & a_{22} & \cdots & 0 \\ \vdots & & & \\ a_{m1} & a_{m2} & \cdots & a_{mm} \end{bmatrix} = \text{lower triangular} \quad a_{ij} = 0,\ i < j$$

$$\mathbf{B} = \begin{bmatrix} 8 & 0 \\ 2 & -1 \end{bmatrix}$$

A3. MATRIX OPERATIONS

A3.1. Equality Two matrices **A** and **B** of the *same order* are *equal* if $a_{ij} = b_{ij}$ for all i and j. Matrices of different orders cannot be equated. The laws of equality that hold are

$$\mathbf{A} = \mathbf{A} \qquad \text{for all } \mathbf{A} \qquad \text{(reflexive law)} \tag{A1}$$

$$\text{If} \quad \mathbf{A} = \mathbf{B}, \quad \text{then} \quad \mathbf{B} = \mathbf{A} \qquad \text{for all } \mathbf{A} \text{ and } \mathbf{B} \qquad \text{(symmetric law)} \tag{A2}$$

$$\text{If} \quad \mathbf{A} = \mathbf{B}, \mathbf{B} = \mathbf{C} \quad \text{then} \quad \mathbf{A} = \mathbf{C} \qquad \text{for all } \mathbf{A}, \mathbf{B}, \mathbf{C} \tag{A3}$$

A3.2. Sums The *sum* of two matrices **A** and **B**, of the *same order*, is a matrix **C** of that order whose elements are $c_{ij} = a_{ij} + b_{ij}$ for all i and j. Matrices of different orders cannot be added. The laws of addition that hold are

$$\mathbf{A} + \mathbf{B} = \mathbf{B} + \mathbf{A} \qquad \text{(commutative law)} \tag{A4}$$

$$\mathbf{A} + (\mathbf{B} + \mathbf{C}) = (\mathbf{A} + \mathbf{B}) + \mathbf{C} = \mathbf{A} + \mathbf{B} + \mathbf{C} \qquad \text{(associative law)} \tag{A5}$$

With the null matrix **0**, we have

$$\mathbf{A} + \mathbf{0} = \mathbf{0} + \mathbf{A} = \mathbf{A} \tag{A6}$$

$$\mathbf{A} + (-\mathbf{A}) = \mathbf{0} \tag{A7}$$

where $(-\mathbf{A})$ is the matrix composed of $(-a_{ij})$ as elements.

A3.3. Scalar Multiplication Scalar multiplication of a matrix **A** by a scalar α is another matrix **B** whose elements are $b_{ij} = \alpha a_{ij}$, for all i and j. (We will in general denote a scalar by a lowercase Greek letter.) Thus

$$\mathbf{B} = \alpha\mathbf{A} \qquad \text{and} \qquad \{b_{ij}\} = \{\alpha a_{ij}\}$$

Example. For

$$\mathbf{A} = \begin{bmatrix} 1 & 2 \\ 3 & -4 \end{bmatrix}$$

$$\mathbf{B} = 2\mathbf{A} = \begin{bmatrix} 2 & 4 \\ 6 & -8 \end{bmatrix}$$

The following laws hold true:

$$\alpha(\mathbf{A} + \mathbf{B}) = \alpha\mathbf{A} + \alpha\mathbf{B} \tag{A8}$$

$$(\alpha + \beta)\mathbf{A} = \alpha\mathbf{A} + \beta\mathbf{A} \tag{A9}$$

$$\alpha(\mathbf{AB}) = (\alpha\mathbf{A})\mathbf{B} = \mathbf{A}(\alpha\mathbf{B}) \tag{A10}$$

$$\alpha(\beta\mathbf{A}) = (\alpha\beta)\mathbf{A} \tag{A11}$$

A3.4. Matrix Multiplication The product of two matrices **A** and **B**, *in that order*, whose dimensions are $m \times k$ and $k \times n$, respectively, is another matrix **C** of order $m \times n$ and elements,

$$c_{ij} = \sum_{r=1}^{r=k} a_{ir} b_{rj} \qquad \text{for all } i \text{ and } j \tag{A12}$$

or

$$c_{ij} = a_{i1} b_{1j} + a_{i2} b_{2j} + \cdots + a_{ik} b_{kj}$$

This process is shown schematically in the following relation:

$$\begin{bmatrix} a_{11} & a_{12} & \cdots & a_{1k} \\ \vdots & \vdots & & \vdots \\ a_{i1} & a_{i2} & \cdots & a_{ik} \\ \vdots & \vdots & & \vdots \\ a_{m1} & a_{m2} & \cdots & a_{mk} \end{bmatrix} \begin{bmatrix} b_{11} & \cdots & b_{1j} & \cdots & b_{1n} \\ b_{21} & & & & b_{2n} \\ & \cdots & b_{2j} & \cdots & \\ \vdots & & \vdots & & \\ b_{k1} & \cdots & b_{kj} & \cdots & b_{kn} \end{bmatrix} = \begin{bmatrix} c_{11} & \cdots & c_{1j} & \cdots & c_{1n} \\ \vdots & & \vdots & & \vdots \\ c_{i1} & \cdots & \boxed{c_{ij} = \sum_{r=1}^{r=k} a_{ir} b_{rj}} & \cdots & c_i \\ & & \vdots & & \vdots \\ c_{m1} & \cdots & c_{mj} & \cdots & c_{mn} \end{bmatrix}$$

It is important to note that for matrix multiplication to be defined the *number of columns of the first matrix must be equal to the number of rows of the second matrix.*

Example

$$\underset{3,\,2}{\mathbf{A}} = \begin{bmatrix} 1 & 2 \\ 3 & 0 \\ 1 & 1 \end{bmatrix} \qquad \text{and} \qquad \underset{2,\,2}{\mathbf{B}} = \begin{bmatrix} 5 & 1 \\ 2 & 3 \end{bmatrix}$$

$$\underset{3,\,2}{\mathbf{C}} = \underset{3,\,2}{\mathbf{A}}\ \underset{2,\,2}{\mathbf{B}} = \begin{bmatrix} 1 & 2 \\ 3 & 0 \\ 1 & 1 \end{bmatrix} \begin{bmatrix} 5 & 1 \\ 2 & 3 \end{bmatrix}$$

$$= \begin{bmatrix} [1(5) + 2(2)] & [1(1) + 2(3)] \\ [3(5) + 0(2)] & [3(1) + 0(3)] \\ [1(5) + 1(2)] & [1(1) + 1(3)] \end{bmatrix} = \begin{bmatrix} 9 & 7 \\ 15 & 3 \\ 7 & 4 \end{bmatrix}$$

The following relationships regarding matrix multiplication hold:

$$\mathbf{AI} = \mathbf{IA} = \mathbf{A} \qquad \text{with } \mathbf{I} = \text{identity matrix} \tag{A13}$$

$$\mathbf{A(BC)} = \mathbf{(AB)C} = \mathbf{ABC} \qquad \text{(associative law)} \tag{A14}$$

$$\mathbf{A(B + C)} = \mathbf{AB} + \mathbf{AC} \qquad \text{(distributive law)} \tag{A15}$$

$$\mathbf{(A + B)C} = \mathbf{AC} + \mathbf{BC} \qquad \text{(distributive law)} \tag{A16}$$

In all these relations the sequence of the matrices is strictly preserved. Note that in general the commutative law does not hold even if multiplication is defined in both orders, i.e.,

$$\mathbf{AB} \neq \mathbf{BA} \qquad \text{in general} \tag{A17}$$

Example

$$\mathbf{A} = \begin{bmatrix} 1 & 2 \\ 3 & 4 \end{bmatrix} \quad \text{and} \quad \mathbf{B} = \begin{bmatrix} 1 & 3 \\ 2 & 4 \end{bmatrix}$$

$$\mathbf{AB} = \begin{bmatrix} 5 & 11 \\ 11 & 25 \end{bmatrix} \quad \text{and} \quad \mathbf{BA} = \begin{bmatrix} 10 & 14 \\ 14 & 20 \end{bmatrix}$$

Of course if orders are different (that is, component matrices are not square) the equality is not even defined, for example,

$$\underset{m,\,k}{\mathbf{A}}\ \underset{k,\,m}{\mathbf{B}} = \underset{m,\,m}{\mathbf{C}}$$

and

$$\underset{k,\,m}{\mathbf{B}}\ \underset{m,\,k}{\mathbf{A}} = \underset{k,\,k}{\mathbf{D}}$$

It is important to note that the product of two matrices can be the null matrix without either matrix being the zero matrix.

Examples

1. $$\begin{bmatrix} 1 & 1 \\ 0 & 0 \end{bmatrix} \begin{bmatrix} 1 & 2 \\ -1 & -2 \end{bmatrix} = \begin{bmatrix} 0 & 0 \\ 0 & 0 \end{bmatrix}$$

2. $$\begin{bmatrix} 1 & 2 & 3 \\ 1 & -1 & 0 \end{bmatrix} \begin{bmatrix} 2 & -3 \\ 2 & -3 \\ -2 & 3 \end{bmatrix} = \begin{bmatrix} 0 & 0 \\ 0 & 0 \end{bmatrix}$$

3. $$[1 \quad 1 \quad -1] \begin{bmatrix} 1 & 2 & 1 \\ 0 & -2 & 1 \\ 1 & 0 & 2 \end{bmatrix} = [0 \quad 0 \quad 0]$$

It is useful to note some properties of diagonal matrices as regards multiplication. If **A** is a square matrix and **D** is a diagonal matrix, then

1. **DA** causes each *row* $\mathbf{A}_i$ of **A** to be multiplied by the corresponding element d_{ii} of **D**.
2. **AD** causes each *column* $\mathbf{A}_j$ of **A** to be multiplied by the corresponding element d_{jj} of **D**.

Example. For

$$\mathbf{A} = \begin{bmatrix} a_{11} & a_{12} \\ a_{21} & a_{22} \end{bmatrix} \quad \text{and} \quad \mathbf{D} = \begin{bmatrix} \alpha & 0 \\ 0 & \beta \end{bmatrix}$$

$$\mathbf{DA} = \begin{bmatrix} \alpha a_{11} & \alpha a_{12} \\ \beta a_{21} & \beta a_{22} \end{bmatrix} \quad \text{and} \quad \mathbf{AD} = \begin{bmatrix} \alpha a_{11} & \beta a_{12} \\ \alpha a_{21} & \beta a_{22} \end{bmatrix}$$

The concept of the *fractional power of a diagonal matrix* applies to diagonal matrices with nonnegative elements. Thus if $d_{ii} \geq 0$ and given a scalar $\gamma > 0$, then the matrix $\mathbf{D}^\gamma$ is

$$\underset{n,\, n}{\mathbf{D}^\gamma} = \begin{bmatrix} d_{11}^\gamma & 0 & \cdots & 0 \\ 0 & d_{22}^\gamma & \cdots & 0 \\ 0 & 0 & \cdots & d_{nn}^\gamma \end{bmatrix}$$

Obviously, if $\alpha > 0$ and $\beta > 0$, then

$$\mathbf{D}^\alpha \cdot \mathbf{D}^\beta = D^{(\alpha+\beta)}$$

and in particular

$$\mathbf{D}^{1/2}\mathbf{D}^{1/2} = \mathbf{D}$$

A3.5. Matrix Transpose The transpose of a matrix $\mathbf{A}$ of order $m \times n$ is an $n \times m$ matrix formed from $\mathbf{A}$ by interchanging rows and columns such that row i of $\mathbf{A}$ becomes column i of the transposed matrix. We denote the transpose by $\mathbf{A}^t$. If $\mathbf{B} = \mathbf{A}^t$, it follows that $b_{ij} = a_{ji}$ for all i and j.

Examples

$$\mathbf{A} = \begin{bmatrix} 1 & 3 \\ -6 & 5 \end{bmatrix} \quad \text{and} \quad \mathbf{A}^t = \begin{bmatrix} 1 & -6 \\ 3 & 5 \end{bmatrix}$$

$$\mathbf{B} = \begin{bmatrix} -2 & 7 \\ 0 & 8 \\ 1 & 4 \end{bmatrix} \quad \text{and} \quad \mathbf{B}^t = \begin{bmatrix} -2 & 0 & 1 \\ 7 & 8 & 4 \end{bmatrix}$$

The following relationships hold true

$$(\mathbf{A} + \mathbf{B})^t = \mathbf{A}^t + \mathbf{B}^t \tag{A18}$$

$$(\mathbf{AB})^t = \mathbf{B}^t\mathbf{A}^t \tag{A19}$$

$$(\alpha\mathbf{A})^t = \alpha\mathbf{A}^t \tag{A20}$$

$$(\mathbf{A}^t)^t = \mathbf{A} \tag{A21}$$

If $\mathbf{D}$ is a diagonal matrix, then $\mathbf{D}^t = \mathbf{D}$. Similarly, for the scalar matrix $\mathbf{H}$ and the identity matrix $\mathbf{I}$ we have $\mathbf{H}^t = \mathbf{H}$ and $\mathbf{I}^t = \mathbf{I}$.

If $\mathbf{x}$ is a column vector, then $\mathbf{x}^t\mathbf{x}$ is a nonnegative scalar that is equal to the sum of squares of the vector's components. On the other hand, $\mathbf{x}\mathbf{x}^t$ is a *square symmetric* matrix of the same order as the vector $\mathbf{x}$.

A3.6. Symmetry and Skew Symmetry A matrix is symmetric if $\mathbf{A}^t = \mathbf{A}$ and it is skew symmetric if $\mathbf{A}^t = -\mathbf{A}$. These matrices must always be square. Thus

$$\begin{bmatrix} 1 & 2 & 3 \\ 2 & 4 & 5 \\ 3 & 5 & 6 \end{bmatrix}$$

is symmetric and

$$\begin{bmatrix} 0 & 1 & 2 \\ -1 & 0 & -3 \\ -2 & 3 & 0 \end{bmatrix}$$

is skew symmetric.

There are at most $(n^2 + n)/2$ distinct elements in a symmetric matrix of order n, although a skew-symmetric matrix of the same order would have at most $(n^2 - n)/2$ non-zero elements. The only matrix that is both symmetric and skew symmetric is the null matrix $\mathbf{0}$. Every square matrix can be uniquely decomposed into the sum of a symmetric and a skew symmetric matrix:

$$\mathbf{A} = \mathbf{A} + \tfrac{1}{2}\mathbf{A}^t - \tfrac{1}{2}\mathbf{A}^t = \tfrac{1}{2}(\mathbf{A} + \mathbf{A}^t) + \tfrac{1}{2}(\mathbf{A} - \mathbf{A}^t) \tag{A22}$$

The matrix

$$\mathbf{A}_s = \tfrac{1}{2}(\mathbf{A} + \mathbf{A}^t)$$

is symmetric because

$$\mathbf{A}_s^t = \tfrac{1}{2}(\mathbf{A} + \mathbf{A}^t)^t = \tfrac{1}{2}(\mathbf{A} + \mathbf{A}^t) = \mathbf{A}_s$$

and the matrix

$$\mathbf{A}_a = \tfrac{1}{2}(\mathbf{A} - \mathbf{A}^t)$$

is skew symmetric since

$$\mathbf{A}_a^t = \tfrac{1}{2}(\mathbf{A} - \mathbf{A}^t)^t = -\tfrac{1}{2}(\mathbf{A} - \mathbf{A}^t) = -\mathbf{A}_a$$

Thus we have expressed the matrix $\mathbf{A}$ as the sum of the symmetric matrix $\mathbf{A}_s$ and the skew symmetric matrix $\mathbf{A}_a$.

Example

$$\mathbf{A} = \begin{bmatrix} 1 & 2 \\ 0 & 3 \end{bmatrix} \quad \text{and} \quad \mathbf{A}^t = \begin{bmatrix} 1 & 0 \\ 2 & 3 \end{bmatrix}$$

$$\mathbf{B} = \tfrac{1}{2}(\mathbf{A} + \mathbf{A}^t) = \begin{bmatrix} 1 & 1 \\ 1 & 3 \end{bmatrix} \quad \text{symmetric}$$

$$\mathbf{C} = \tfrac{1}{2}(\mathbf{A} - \mathbf{A}^t) = \begin{bmatrix} 0 & 1 \\ -1 & 0 \end{bmatrix} \quad \text{skew symmetric}$$

and

$$\mathbf{B} + \mathbf{C} = \begin{bmatrix} 1 & 2 \\ 0 & 3 \end{bmatrix} = \mathbf{A}$$

For *any matrix* **A** and any *symmetric matrix* **B**, the matrices $\mathbf{AA}^t$, $\mathbf{A}^t\mathbf{A}$, $\mathbf{ABA}^t$, and $\mathbf{A}^t\mathbf{BA}$ are all symmetric. The same applies for skew symmetry if **B** were skew symmetric. Note that the sum and difference of symmetric (or skew symmetric) matrices are also symmetric (or skew symmetric). On the other hand the product of symmetric matrices is symmetric only if the matrices commute.

A3.7. Partitioning of Matrices We often wish to study a subset of elements from a given matrix **A**. The matrix formed by that subset of elements is called a *submatrix* of **A**. For example, the matrix

$$\begin{bmatrix} a_{11} & a_{13} & a_{15} \\ a_{31} & a_{33} & a_{35} \end{bmatrix}$$

is a 2×3 submatrix obtained by deleting the second row and second and fourth columns of the 3×5 matrix

$$\mathbf{A} = \begin{bmatrix} a_{11} & a_{12} & a_{13} & a_{14} & a_{15} \\ a_{21} & a_{22} & a_{23} & a_{24} & a_{25} \\ a_{31} & a_{32} & a_{33} & a_{34} & a_{35} \end{bmatrix}$$

The scheme called *partitioning* allows for the writing of a matrix not in terms of its original elements but in terms of submatrices. The submatrices will be denoted by boldface uppercase (or lowercase in the event that they are vectors) letters with appropriate subscripts. Thus the matrix

$$\mathbf{A} = \left[\begin{array}{ccc:ccc} a_{11} & a_{12} & a_{13} & a_{14} & \cdots & a_{1n} \\ a_{21} & a_{22} & a_{23} & a_{24} & \cdots & a_{2n} \\ \hdashline a_{31} & a_{32} & a_{33} & a_{34} & \cdots & a_{3n} \\ \vdots & & & & & \\ a_{m1} & a_{m2} & a_{m3} & a_{m4} & \cdots & a_{mn} \end{array}\right]$$

may be partitioned several ways. Considering the vertical dotted line only, we write

$$\mathbf{A} = [\mathbf{A}_1 \quad \mathbf{A}_2]$$

where $\mathbf{A}_1$ is an $m \times 3$ submatrix and $\mathbf{A}_2$ an $m \times (n-3)$ submatrix. Similarly, considering the horizontal dotted line only,

$$\mathbf{A} = \begin{bmatrix} \mathbf{A}_1 \\ \mathbf{A}_2 \end{bmatrix}$$

where in this case $\mathbf{A}_1$ is a $2 \times n$ submatrix and $\mathbf{A}_2$ an $(m-2) \times n$ submatrix. Considering both horizontal and vertical partition lines

$$\mathbf{A} = \begin{bmatrix} \mathbf{A}_{11} & \mathbf{A}_{12} \\ \mathbf{A}_{21} & \mathbf{A}_{22} \end{bmatrix}$$

where $\mathbf{A}_{11}$ is 2×3, $\mathbf{A}_{12}$ is $2 \times (n-3)$, $\mathbf{A}_{21}$ is $(m-2) \times 3$, and $\mathbf{A}_{22}$ is $(m-2) \times (n-3)$.

All matrix operations given in the preceding sections can be performed on the submatrices as if they are elements, provided necessary precautions are exercised regarding dimensions. Thus, for example,

$$\underset{n,\,m}{\mathbf{B}} = \mathbf{A}^t = \begin{bmatrix} \underset{3,\,2}{\mathbf{A}_{11}^t} & \underset{3,\,(m-2)}{\mathbf{A}_{21}^t} \\ \underset{(n-3),\,2}{\mathbf{A}_{12}^t} & \underset{(n-3),\,(m-2)}{\mathbf{A}_{22}^t} \end{bmatrix} = \begin{bmatrix} \mathbf{B}_{11} & \mathbf{B}_{12} \\ \mathbf{B}_{21} & \mathbf{B}_{22} \end{bmatrix}$$

Example. Let

$$\underset{3,\,4}{\mathbf{A}} = \left[\begin{array}{cc:cc} 1 & 2 & 3 & 4 \\ 4 & 3 & 2 & 1 \\ \hdashline 2 & 1 & 3 & 4 \end{array}\right] \quad \text{and} \quad \underset{4,\,2}{\mathbf{B}} = \left[\begin{array}{cc} 1 & 2 \\ 3 & 4 \\ \hdashline 2 & 1 \\ 1 & 3 \end{array}\right]$$

then

$$\underset{3,\,4}{\mathbf{A}}\ \underset{4,\,2}{\mathbf{B}} = \underset{3,\,2}{\mathbf{C}}$$

$$\mathbf{A} = \begin{bmatrix} \mathbf{A}_{11} & \mathbf{A}_{12} \\ \mathbf{A}_{21} & \mathbf{A}_{22} \end{bmatrix} \quad \text{and} \quad \mathbf{B} = \begin{bmatrix} \mathbf{B}_{11} \\ \mathbf{B}_{21} \end{bmatrix}$$

then

$$\mathbf{AB} = \begin{bmatrix} (\mathbf{A}_{11}\mathbf{B}_{11} + \mathbf{A}_{12}\,\mathbf{B}_{21}) \\ (\mathbf{A}_{21}\mathbf{B}_{11} + \mathbf{A}_{22}\,\mathbf{B}_{21}) \end{bmatrix} = \begin{bmatrix} \mathbf{C}_{11} \\ \mathbf{C}_{21} \end{bmatrix}$$

where

$$\underset{2,\,2}{\mathbf{A}_{11}} = \begin{bmatrix} 1 & 2 \\ 4 & 3 \end{bmatrix}, \quad \underset{2,\,2}{\mathbf{A}_{12}} = \begin{bmatrix} 3 & 4 \\ 2 & 1 \end{bmatrix}, \quad \underset{1,\,2}{\mathbf{A}_{21}} = [2 \quad 1], \quad \underset{1,\,2}{\mathbf{A}_{22}} = [3 \quad 4]$$

$$\underset{2,\,2}{\mathbf{B}_{11}} = \begin{bmatrix} 1 & 2 \\ 3 & 4 \end{bmatrix} \quad \underset{2,\,2}{\mathbf{B}_{21}} = \begin{bmatrix} 2 & 1 \\ 1 & 3 \end{bmatrix}$$

Thus

$$\mathbf{A}_{11}\mathbf{B}_{11} = \begin{bmatrix} 7 & 10 \\ 13 & 20 \end{bmatrix}, \quad \mathbf{A}_{12}\,\mathbf{B}_{21} = \begin{bmatrix} 10 & 15 \\ 5 & 5 \end{bmatrix},$$

$$\mathbf{A}_{21}\mathbf{B}_{11} = [5 \quad 8], \quad \mathbf{A}_{22}\,\mathbf{B}_{21} = [10 \quad 15]$$

leading to

$$\mathbf{AB} = \begin{bmatrix} \begin{bmatrix} 7 & 10 \\ 13 & 20 \end{bmatrix} + \begin{bmatrix} 10 & 15 \\ 5 & 5 \end{bmatrix} \\ [5 \quad 8] + [10 \quad 15] \end{bmatrix} = \begin{bmatrix} \begin{bmatrix} 17 & 25 \\ 18 & 25 \end{bmatrix} \\ [15 \quad 23] \end{bmatrix} = \begin{bmatrix} 17 & 25 \\ 18 & 25 \\ 15 & 23 \end{bmatrix}$$

which is the same answer as that obtained directly from

$$\mathbf{AB} = \begin{bmatrix} 1 & 2 & 3 & 4 \\ 4 & 3 & 2 & 1 \\ 2 & 1 & 3 & 4 \end{bmatrix} \begin{bmatrix} 1 & 2 \\ 3 & 4 \\ 2 & 1 \\ 1 & 3 \end{bmatrix} = \begin{bmatrix} 17 & 25 \\ 18 & 25 \\ 15 & 23 \end{bmatrix}$$

A3.8. Matrix Inverse† Unlike scalars, *division* of matrices is *not defined.* In fact, we may have $\mathbf{AB} = \mathbf{AC}$ without having $\mathbf{B} = \mathbf{C}$. This implies that the operation of "dividing" by $\mathbf{A}$, even if $\mathbf{A} \neq \mathbf{0}$, is not possible. As an example let

$$\mathbf{A} = \begin{bmatrix} 1 & 0 \\ 4 & 0 \end{bmatrix}, \quad \mathbf{B} = \begin{bmatrix} 1 & -1 \\ 2 & 2 \end{bmatrix}, \quad \mathbf{C} = \begin{bmatrix} 1 & -1 \\ 1 & 1 \end{bmatrix}$$

where obviously $\mathbf{B} \neq \mathbf{C}$, but computing we get

$$\mathbf{AB} = \begin{bmatrix} 1 & -1 \\ 4 & -4 \end{bmatrix} = \mathbf{AC}$$

In place of division, we introduce the concept of *inverse*, $\mathbf{A}^{-1}$, of a matrix $\mathbf{A}$ (similar to the reciprocal α^{-1} of the scalar α).

The *inverse* of a *square* matrix $\mathbf{A}$, *if it exists*, is the unique matrix $\mathbf{A}^{-1}$ with the property

$$\mathbf{AA}^{-1} = \mathbf{A}^{-1}\mathbf{A} = \mathbf{I} \tag{A23}$$

The following rules hold:

$$(\mathbf{AB})^{-1} = \mathbf{B}^{-1}\mathbf{A}^{-1} \tag{A24}$$

$$(\mathbf{A}^{-1})^{-1} = \mathbf{A} \tag{A25}$$

$$(\mathbf{A}^t)^{-1} = (\mathbf{A}^{-1})^t \tag{A26}$$

$$(\alpha\mathbf{A})^{-1} = \frac{1}{\alpha}\mathbf{A}^{-1} \tag{A27}$$

A square matrix that has an inverse is called *nonsingular*; one that does not have an inverse is called *singular*.

† Consideration in this Appendix is limited to Cayley matrix algebra where an inverse is only defined for a square matrix. Although a square matrix may not have an inverse in this algebra, it has a generalized inverse. However, generalized matrix algebra is not utilized in this book.

We showed in Section A3.4 that $\mathbf{AB} = \mathbf{0}$ did not necessarily imply that $\mathbf{A} = \mathbf{0}$ or $\mathbf{B} = \mathbf{0}$. If, however, either $\mathbf{A}$ or $\mathbf{B}$ is nonsingular then the other matrix must be a null matrix. Hence, the product of two nonsingular matrices cannot be a null matrix.

To appreciate a method of inverting matrices we need to introduce the concept of determinants of square matrices. We associate with each square matrix $\mathbf{A}$ a uniquely defined scalar that is called the *determinant* of $\mathbf{A}$ and is denoted by either "det $\mathbf{A}$" or $|\mathbf{A}|$. The determinant of order n (for an $n \times n$ matrix) will be defined in terms of determinants of order $(n - 1)$. In order to apply this procedure we need to define the determinant of the least-order matrix, namely, that of a 1×1 matrix. For a matrix consisting of a single element,

$$\underset{1,\,1}{\mathbf{A}} = [a_{11}]$$

the determinant is defined as

$$|\mathbf{A}| = \det \mathbf{A} = a_{11}$$

An $(n - 1) \times (n - 1)$ determinant obtained from the $n \times n$ matrix is called a *minor*. The minor m_{ij} of an element a_{ij} of a square matrix $\mathbf{A}$ is the determinant of the square submatrix formed from $\mathbf{A}$ by deleting the ith row and jth column.

$$\mathbf{A} = \begin{bmatrix} a_{11} & a_{12} & a_{13} \\ a_{21} & a_{22} & a_{23} \\ a_{31} & a_{32} & a_{33} \end{bmatrix} \qquad m_{12} \text{ (for } a_{12}) = \begin{vmatrix} a_{21} & a_{23} \\ a_{31} & a_{33} \end{vmatrix}$$

The *cofactor* c_{ij} of an element a_{ij} in a matrix $\mathbf{A}$ is given by

$$c_{ij} = (-1)^{i+j} m_{ij}$$

where m_{ij} is the minor of the element a_{ij}. We may now define a determinant as

$$|\mathbf{A}| = \det \mathbf{A} = \sum_{j=1}^{n} a_{1j} c_{1j}$$

or that $|\mathbf{A}|$ is the sum of the products of the elements of the first row times their respective cofactors. For the matrix

$$\mathbf{A} = \begin{bmatrix} a_{11} & a_{12} \\ a_{21} & a_{22} \end{bmatrix}$$

the cofactors for the elements a_{11} and a_{12} are

$$c_{11} = |a_{22}| = a_{22} \qquad \text{and} \qquad c_{12} = -|a_{21}| = -a_{21}$$

Thus

$$|\mathbf{A}| = a_{11} a_{22} - a_{12} a_{21}$$

Similarly, for

$$\mathbf{A} = \begin{bmatrix} a_{11} & a_{12} & a_{13} \\ a_{21} & a_{22} & a_{23} \\ a_{31} & a_{32} & a_{33} \end{bmatrix}$$

$$\begin{aligned} \mathbf{A} &= a_{11} \cdot \begin{vmatrix} a_{22} & a_{23} \\ a_{32} & a_{33} \end{vmatrix} - a_{12} \cdot \begin{vmatrix} a_{21} & a_{23} \\ a_{31} & a_{33} \end{vmatrix} + a_{13} \cdot \begin{vmatrix} a_{21} & a_{22} \\ a_{31} & a_{32} \end{vmatrix} \\ &= a_{11}(a_{22}a_{33} - a_{23}a_{32}) - a_{12}(a_{21}a_{33} - a_{23}a_{31}) \\ &\quad + a_{13}(a_{21}a_{32} - a_{22}a_{31}) \end{aligned}$$

There are several properties of determinants that make their evaluation relatively less complex.

1. The determinant of a triangular matrix is the product of its diagonal elements (same for a diagonal matrix).
2. $|\mathbf{A}_1 \cdot \mathbf{A}_2 \cdot \cdots \cdot \mathbf{A}_k| = |\mathbf{A}_1| \cdot |\mathbf{A}_2| \cdot \cdots \cdot |\mathbf{A}_k|$
3. The determinant of a matrix with a zero vector is zero.
4. The determinant of a matrix with two rows or two columns that are identical is zero.
5. The determinant of a matrix with one row (or column) a multiple of another row (or column) is zero.
6. $|\mathbf{A}| = |\mathbf{A}^t|$
7. Interchanging two rows or two columns changes the sign of the determinant.
8. If a row or a column is multiplied by a constant, the value of the determinant is multiplied by that constant.
9. The value of a determinant does not change if a multiple of a row (or column) is added to another row (or column).
10. If a matrix **A** can be partitioned to form a triangular matrix with square matrices $\mathbf{A}_{11}, \mathbf{A}_{22}, \ldots, \mathbf{A}_{kk}$ on main diagonal, then

$$|\mathbf{A}| = |\mathbf{A}_{11}| \cdot |\mathbf{A}_{22}| \cdot |\mathbf{A}_{33}| \cdot \cdots \cdot |\mathbf{A}_{kk}|$$

Having introduced the concept and some properties of determinants, we may now give some of the more direct methods of evaluating an inverse of a matrix.

1. The *cofactor matrix* **C** of a matrix **A** is the square matrix of the same order as **A** and in which each element a_{ij} is replaced by its cofactor c_{ij}.

$$\mathbf{A} = \begin{bmatrix} a_{11} & a_{12} & \cdots & a_{1m} \\ a_{21} & a_{22} & \cdots & a_{2m} \\ \vdots & \vdots & & \vdots \\ a_{m1} & a_{m2} & \cdots & a_{mm} \end{bmatrix}$$

and the cofactor matrix

$$\mathbf{C} = \begin{bmatrix} c_{11} & c_{12} & \cdots & c_{1m} \\ c_{21} & c_{22} & \cdots & c_{2m} \\ \vdots & & & \vdots \\ c_{m1} & c_{m2} & \cdots & c_{mm} \end{bmatrix}$$

2. The *adjoint* matrix of a matrix **A** is the transpose of its cofactor matrix, that is, $\mathbf{C}^t$, and is denoted by adj **A**.

$$\text{adj } \mathbf{A} = \mathbf{C}^t = \begin{bmatrix} c_{11} & c_{21} & \cdots & c_{m1} \\ c_{12} & c_{22} & \cdots & c_{m2} \\ \vdots & \vdots & & \vdots \\ c_{1m} & c_{2m} & \cdots & c_{mm} \end{bmatrix}$$

It can be shown that

$$\mathbf{A}(\text{adj } \mathbf{A}) = (\text{adj } \mathbf{A})\mathbf{A} = |\mathbf{A}| \cdot \mathbf{I}$$

Example. Let

$$\mathbf{A} = \begin{bmatrix} 2 & 1 & 0 \\ 1 & 3 & 2 \\ 1 & 2 & 1 \end{bmatrix}$$

$$|\mathbf{A}| = 2(-1) - 1(-1) + 0(-1) = -1$$

The elements of the cofactor matrix are

$$c_{11} = +(-1) \qquad c_{12} = -(-1) \qquad c_{13} = +(-1)$$

$$c_{21} = -(1) \qquad c_{22} = +(2) \qquad c_{23} = -(3)$$

$$c_{31} = +(2) \qquad c_{32} = -(4) \qquad c_{33} = +(5)$$

Thus

$$\mathbf{C} = \begin{bmatrix} -1 & 1 & -1 \\ -1 & 2 & -3 \\ 2 & -4 & 5 \end{bmatrix}$$

and

$$\text{adj } \mathbf{A} = \mathbf{C}^t = \begin{bmatrix} -1 & -1 & 2 \\ 1 & 2 & -4 \\ -1 & -3 & 5 \end{bmatrix}$$

$$\mathbf{A}(\text{adj } \mathbf{A}) = \begin{bmatrix} 2 & 1 & 0 \\ 1 & 3 & 2 \\ 1 & 2 & 1 \end{bmatrix} \begin{bmatrix} -1 & -1 & 2 \\ 1 & 2 & -4 \\ -1 & -3 & 5 \end{bmatrix} = \begin{bmatrix} -1 & 0 & 0 \\ 0 & -1 & 0 \\ 0 & 0 & -1 \end{bmatrix}$$

$$= (-1)\mathbf{I} = |\mathbf{A}| \cdot \mathbf{I}$$

The reader can ascertain that (adj **A**) **A** is also equal $|\mathbf{A}| \cdot \mathbf{I}$. From the preceding relation between **A**, adj **A**, and $|\mathbf{A}|$, and the basic definition of a matrix inverse given in equation (A23), it follows readily that the inverse of a square matrix **A** is computed from

$$\mathbf{A}^{-1} = \frac{\text{adj } \mathbf{A}}{|\mathbf{A}|} \tag{A28}$$

It is apparent from equation (A28) that $|\mathbf{A}|$ *must not vanish* in order that $\mathbf{A}^{-1}$ may exist.

Example. Given the simultaneous pair of linear equations

$$2X_1 + 3X_2 = 8$$

$$3X_1 - X_2 = 1$$

find a unique solution.

First the two equations are written in matrix form as

$$\begin{bmatrix} 2 & 3 \\ 3 & -1 \end{bmatrix} \begin{bmatrix} X_1 \\ X_2 \end{bmatrix} = \begin{bmatrix} 8 \\ 1 \end{bmatrix} \quad \text{or} \quad \mathbf{Ax} = \mathbf{b}$$

Evaluating the cofactors,

$$c_{11} = (-1)^2(-1) = -1$$

$$c_{12} = (-1)^3(3) = -3$$

$$c_{21} = (-1)^3(3) = -3$$

$$c_{22} = (-1)^4(2) = 2$$

and

$$|\mathbf{A}| = \begin{bmatrix} 2 & 3 \\ 3 & -1 \end{bmatrix} = -2 - 9 = -11$$

$$\mathbf{C} = \begin{bmatrix} -1 & -3 \\ -3 & 2 \end{bmatrix} \qquad \text{adj } \mathbf{A} = \mathbf{C}^t = \begin{bmatrix} -1 & -3 \\ -3 & 2 \end{bmatrix}$$

$$\mathbf{A}^{-1} = \frac{\text{adj } \mathbf{A}}{|\mathbf{A}|} = -\tfrac{1}{11}\begin{bmatrix} -1 & -3 \\ -3 & 2 \end{bmatrix} = \begin{bmatrix} \frac{1}{11} & \frac{3}{11} \\ \frac{3}{11} & -\frac{2}{11} \end{bmatrix}$$

From $\mathbf{Ax} = \mathbf{b}$ we may write

$$\mathbf{A}^{-1}\mathbf{Ax} = \mathbf{x} = \mathbf{A}^{-1}\mathbf{b} = \begin{bmatrix} \frac{1}{11} & \frac{3}{11} \\ \frac{3}{11} & -\frac{2}{11} \end{bmatrix} \begin{bmatrix} 8 \\ 1 \end{bmatrix} = \begin{bmatrix} 1 \\ 2 \end{bmatrix}$$

Hence $X_1 = 1$ and $X_2 = 2$.

A very important matrix which is encountered frequently in practice is the *orthogonal* matrix. A square matrix **M** is orthogonal if

$$\mathbf{M}^t\mathbf{M} = \mathbf{M}\mathbf{M}^t = \mathbf{I} \tag{A29}$$

which implies that

$$\mathbf{M}^{-1} = \mathbf{M}^t \tag{A30}$$

The columns of an orthogonal matrix are mutually orthogonal vectors of unit length. Also,

$$|\mathbf{M}| = \pm 1 \tag{A31}$$

When $|\mathbf{M}| = +1$, then $\mathbf{M}$ is called proper orthogonal, otherwise it is termed "improper."

The product of two orthogonal matrices is also an orthogonal matrix.

MATRIX INVERSE BY PARTITIONING Let $\mathbf{A}$ be an $n \times n$ square nonsingular matrix whose inverse is to be evaluated. We partition $\mathbf{A}$ in the form,

$$\begin{array}{cc} & \begin{array}{cc} s & m \end{array} \\ \mathbf{A} = & \begin{bmatrix} \mathbf{A}_{11} & \mathbf{A}_{12} \\ \mathbf{A}_{21} & \mathbf{A}_{22} \end{bmatrix} \begin{array}{c} s \\ m \end{array} \end{array}$$

where $\mathbf{A}_{11}$ is $s \times s$, $\mathbf{A}_{12}$ is $s \times m$, $\mathbf{A}_{21}$ is $m \times s$, $\mathbf{A}_{22}$ is $m \times m$, and $m + s = n$. The inverse $\mathbf{A}^{-1}$ exists and we shall denote it, in the correspondingly partitioned form, by

$$\mathbf{A}^{-1} = \mathbf{B} = \begin{bmatrix} \mathbf{B}_{11} & \mathbf{B}_{12} \\ \mathbf{B}_{21} & \mathbf{B}_{22} \end{bmatrix}$$

From the basic definition of an inverse we have $\mathbf{A}\mathbf{A}^{-1} = \mathbf{A}\mathbf{B} = \mathbf{I}$, or in the partitioned form

$$\begin{bmatrix} \mathbf{A}_{11} & \mathbf{A}_{12} \\ \mathbf{A}_{21} & \mathbf{A}_{22} \end{bmatrix} \begin{bmatrix} \mathbf{B}_{11} & \mathbf{B}_{12} \\ \mathbf{B}_{21} & \mathbf{B}_{22} \end{bmatrix} = \begin{bmatrix} \mathbf{I}_s & \mathbf{0} \\ \mathbf{0} & \mathbf{I}_m \end{bmatrix}$$

which leads to

$$\mathbf{A}_{11}\mathbf{B}_{11} + \mathbf{A}_{12}\mathbf{B}_{21} = \mathbf{I}_s \tag{A32a}$$

$$\mathbf{A}_{11}\mathbf{B}_{12} + \mathbf{A}_{12}\mathbf{B}_{22} = \mathbf{0} \tag{A32b}$$

$$\mathbf{A}_{21}\mathbf{B}_{11} + \mathbf{A}_{22}\mathbf{B}_{21} = \mathbf{0} \tag{A32c}$$

$$\mathbf{A}_{21}\mathbf{B}_{12} + \mathbf{A}_{22}\mathbf{B}_{22} = \mathbf{I}_m \tag{A32d}$$

From equation (A32c) we have

$$\mathbf{B}_{21} = -\mathbf{A}_{22}^{-1}\mathbf{A}_{21}\mathbf{B}_{11} \tag{A33}$$

Substituting into equation (A32a), we obtain the value of $\mathbf{B}_{11}$ as

$$\mathbf{B}_{11} = [\mathbf{A}_{11} - \mathbf{A}_{12}\mathbf{A}_{22}^{-1}\mathbf{A}_{21}]^{-1} \tag{A34}$$

From equation (A32d) we have

$$\mathbf{B}_{22} = \mathbf{A}_{22}^{-1} - \mathbf{A}_{22}^{-1}\mathbf{A}_{21}\,\mathbf{B}_{12} \tag{A35}$$

which, with equation (A32b) and taking equation (A34) into account, gives

$$\mathbf{B}_{12} = -\mathbf{B}_{11}\,\mathbf{A}_{12}\,\mathbf{A}_{22}^{-1} \tag{A36}$$

Thus all the submatrices have been evaluated and their formulas may be collected as

$$\mathbf{B}_{11} = [\mathbf{A}_{11} - \mathbf{A}_{12}\,\mathbf{A}_{22}^{-1}\mathbf{A}_{21}]^{-1} \tag{A37a}$$

$$\mathbf{B}_{12} = -\mathbf{B}_{11}\,\mathbf{A}_{12}\,\mathbf{A}_{22}^{-1} \tag{A37b}$$

$$\mathbf{B}_{21} = -\mathbf{A}_{22}^{-1}\mathbf{A}_{21}\,\mathbf{B}_{11} \tag{A37c}$$

$$\mathbf{B}_{22} = \mathbf{A}_{22}^{-1} - \mathbf{A}_{22}^{-1}\mathbf{A}_{21}\,\mathbf{B}_{12} \tag{A37d}$$

If $\mathbf{A}$ is originally a symmetric matrix, then $\mathbf{A}_{21} = \mathbf{A}_{12}^{t}$ and correspondingly $\mathbf{B}_{21} = \mathbf{B}_{12}^{t}$.

In inverting by partitioning we end up computing directly the inverse of a matrix $\mathbf{A}_{22}$ of a smaller order ($m < n$) than the original matrix. In fact, if m is taken equal to 1, then $\mathbf{A}_{22}^{-1}$ is simply the reciprocal of a scalar. Inversion by partitioning can, of course, be performed in more than one step.

A4. RANK OF A MATRIX

The rank of a matrix is the order of the largest nonzero determinant that can be formed from the elements of the matrix by appropriate deletion of rows or columns (or both). Thus a matrix is said to be of *rank m* if and only if it has *at least one nonsingular submatrix of order m*, but has no nonsingular submatrix of order more than m. A nonsingular matrix of order n has a rank n. A matrix with zero rank has elements that must all be zero.

A4.1 A rectangular matrix has a rank that is less than or equal to the smallest dimension. Therefore if $\mathbf{A}$ is an $n \times k$ matrix and $k < n$, then rank $(\mathbf{A}) \leq k$.

A4.2 The rank of the product of a number of matrices does not exceed the *least* rank of individual multiplicands, or

$$\text{rank}\,(\mathbf{A}_1\,\mathbf{A}_2\,\mathbf{A}_3\,\cdots\,\mathbf{A}_k) \leq \min\,[\text{rank}\,(\mathbf{A}_1), \text{rank}\,(\mathbf{A}_2), \ldots, \text{rank}\,(\mathbf{A}_k)] \tag{A38}$$

A4.3 Multiplication of a matrix by a nonsingular matrix does not change its rank.

A4.4 If $\mathbf{A}$ and $\mathbf{B}$ are $m \times k$ and $k \times n$, respectively, and each is of rank k, then $\mathbf{AB}$ is of rank k.

Example. Given

$$\mathbf{A} = \begin{bmatrix} 1 & 0 & 0 \\ 0 & 1 & 0 \\ 0 & 0 & 0 \end{bmatrix}, \qquad \mathbf{B} = \begin{bmatrix} 0 & 0 & 0 \\ 0 & 1 & 0 \\ 0 & 0 & 1 \end{bmatrix}, \qquad r(\mathbf{A}) = 2, \qquad r(\mathbf{B}) = 2$$

then

$$\mathbf{C} = \mathbf{A}^2 = \begin{bmatrix} 1 & 0 & 0 \\ 0 & 1 & 0 \\ 0 & 0 & 0 \end{bmatrix} \begin{bmatrix} 1 & 0 & 0 \\ 0 & 1 & 0 \\ 0 & 0 & 0 \end{bmatrix} = \begin{bmatrix} 1 & 0 & 0 \\ 0 & 1 & 0 \\ 0 & 0 & 0 \end{bmatrix} \qquad r(\mathbf{C}) = 2$$

and

$$\mathbf{D} = \mathbf{AB} = \begin{bmatrix} 1 & 0 & 0 \\ 0 & 1 & 0 \\ 0 & 0 & 0 \end{bmatrix} \begin{bmatrix} 0 & 0 & 0 \\ 0 & 1 & 0 \\ 0 & 0 & 1 \end{bmatrix} = \begin{bmatrix} 0 & 0 & 0 \\ 0 & 1 & 0 \\ 0 & 0 & 0 \end{bmatrix}$$

$$r(\mathbf{D}) = 1 < 2$$

An interesting property for the product of two matrices is that if

$$r_c = \text{rank}\,(\underset{m,\,s}{\mathbf{C}}), \qquad r_d = \text{rank}\,(\underset{s,\,n}{\mathbf{D}}), \qquad \text{and} \qquad \mathbf{CD} = \mathbf{0}$$

then

$$r_c + r_d \leq s \tag{A39}$$

This relation can be verified by the three examples given in Section A3.4. As an illustration, let us consider the third example, where $\mathbf{CD} = \mathbf{0}$, or

$$\underset{1,\,3}{[1 \quad 1 \quad -1]} \underset{3,\,3}{\begin{bmatrix} 1 & 2 & 1 \\ 0 & -2 & 1 \\ 1 & 0 & 2 \end{bmatrix}} = [0 \quad 0 \quad 0]$$

The rank of the first matrix is $r_c = 1$, and for the second matrix, $r_d = 2$, leading to $r_c + r_d = 3 = s$.

A4.5 An important fact is that $\mathbf{AA}'$ and $\mathbf{A}'\mathbf{A}$ have the same rank as $\mathbf{A}$. It follows that if $\mathbf{A}$ is of order $n \times k$ and $n > k$ and is of (full) rank k, then $\mathbf{A}'\mathbf{A}$ is nonsingular, whereas $\mathbf{AA}'$ is singular with rank k. Furthermore, $\mathbf{A}'\mathbf{BA}$ is nonsingular if $\mathbf{B}$ is also nonsingular.

A4.6 The rank of the sum of two matrices cannot exceed the sum of their ranks.

The determination of the *rank* of a matrix by direct application of the definition is not practical, particularly when we are dealing with other than simple small matrices. Instead, we obtain a much faster answer if we realize that *primary row or column operations* do not change either the order or the rank of the matrix. These operations are as follows:

1. The interchange of any two rows (or columns).
2. The multiplication of all the elements of any row (or column) by the same nonzero constant.
3. The addition to any row (column) of an arbitrary multiple of any other row (column).

Example. Find the rank of the matrix

$$\begin{bmatrix} 1 & 2 & -1 & 3 \\ 2 & 4 & -4 & 7 \\ -1 & -2 & -1 & -2 \end{bmatrix}$$

Subtracting twice the first row from the second and adding the first row to the third yields

$$\begin{bmatrix} 1 & 2 & -1 & 3 \\ 0 & 0 & -2 & 1 \\ 0 & 0 & -2 & 1 \end{bmatrix}$$

Subtracting the new second row from the new third row gives

$$\begin{bmatrix} 1 & 2 & -1 & 3 \\ 0 & 0 & -2 & 1 \\ 0 & 0 & 0 & 0 \end{bmatrix}$$

Primary row operations may be stopped at this point since

$$\begin{vmatrix} 2 & -1 \\ 0 & -2 \end{vmatrix} \neq 0$$

leading to rank = 2. However, the primary operations may be continued until

$$\begin{bmatrix} 1 & 0 & 0 & 0 \\ 0 & 1 & 0 & 0 \\ 0 & 0 & 0 & 0 \end{bmatrix}$$

leading directly to the rank being 2.

Two matrices are said to be *equivalent* if one can be transformed to the other by elementary operations. Hence they must be of equal order and rank.

The primary row (or column) operations are also useful in several methods of solving sets of *linear equations*. We shall briefly give here two of the more common procedures.

THE GAUSS METHOD In the Gauss method, elementary row operations are used to reduce the matrix to an upper triangular form or canonical form. This is classically called the *forward solution*, which is followed by a *back-*

ward solution to determine the values of the unknowns. Let us consider as an example the two equations solved by the method of inverse in Section A3.8.

$$\mathbf{Ax} = \mathbf{b}$$

with

$$\mathbf{A} = \begin{bmatrix} 2 & 3 \\ 3 & -1 \end{bmatrix} \quad \text{and} \quad \mathbf{b} = \begin{bmatrix} 8 \\ 1 \end{bmatrix}$$

First we set the matrices

$$\mathbf{A} \,\vdots\, \mathbf{b} \,\vdots\, \mathbf{I}$$

side by side

$$\left[\begin{array}{cc:c:cc} 2 & 3 & 8 & 1 & 0 \\ 3 & -1 & 1 & 0 & 1 \end{array}\right]$$

Augmentation by **I** is useful for the following method. Divide the first row by 2 and multiply by -1 ($R'_1 = -\frac{1}{2}R_1$); multiply R'_1 by 3 and add to negative of second row ($R'_2 = -R_2 + 3R'_1$)

$$\left[\begin{array}{cc:c:cc} 1 & \frac{3}{2} & 4 & \frac{1}{2} & 0 \\ 0 & \frac{11}{2} & 11 & \frac{3}{2} & -1 \end{array}\right]$$

At this point we can readily write from the second row that

$$\tfrac{11}{2}x_2 = 11 \qquad \text{or} \qquad x_2 = 2$$

and substituting backward

$$x_1 = 4 - \tfrac{3}{2}x_2 = 1$$

We could also continue until the upper triangular matrix (in **A**) has ones along the diagonal by

$$R''_2 = \tfrac{2}{11}R'_2$$

$$\left[\begin{array}{cc:c:cc} 1 & \frac{3}{2} & 4 & \frac{1}{2} & 0 \\ 0 & 1 & 2 & \frac{3}{11} & -\frac{2}{11} \end{array}\right]$$

From which $x_2 = 2$ directly and x_1 computed as above.

THE GAUSS-JORDAN METHOD In the Gauss-Jordan method, the elementary row operations are continued until **A** becomes a unit matrix, **b** thus becoming the answer vector for **x**.

$$R''_2 = \tfrac{2}{11}R'_2$$

$$R''_1 = R'_1 - \tfrac{3}{2}R''_2$$

$$\left[\begin{array}{cc:c:cc} 1 & 0 & 1 & \frac{1}{11} & \frac{3}{11} \\ 0 & 1 & 2 & \frac{3}{11} & \frac{1}{11} \end{array}\right]$$

giving $x_1 = 1$ and $x_2 = 2$ directly. It is also quite interesting to note that although **A** reduced to **I**, the augmented identity matrix **I** became the inverse $\mathbf{A}^{-1}$. This is useful when, in addition to solving the equations, the inverse is also needed.

A5. TRACE OF A MATRIX

If **A** is an $n \times n$ *square* matrix, then the sum of its diagonal elements is a scalar called the *trace* of **A** and is denoted by tr (**A**). (Some textbooks on matrices use the German word *Spur* instead of the English translation "trace.")

$$\operatorname{tr}(\mathbf{A}) = a_{11} + a_{22} + \cdots + a_{nn}$$

The definition leads directly to the fact that $\operatorname{tr}(\mathbf{A}') = \operatorname{tr}(\mathbf{A})$. If **A** and **B** are both of order m, then

$$\operatorname{tr}(\mathbf{A} + \mathbf{B}) = \operatorname{tr}(\mathbf{A}) + \operatorname{tr}(\mathbf{B}) \tag{A40}$$

Also, if **E** is of order $p \times q$ and **F** of order $q \times p$, then

$$\operatorname{tr}(\mathbf{EF}) = \operatorname{tr}(\mathbf{FE}) \tag{A41}$$

Finally, if **F** is nonsingular, that is, if

$$|\mathbf{F}| \neq 0 \qquad \text{and} \qquad \mathbf{B} = \mathbf{FAF}^{-1}$$

then

$$\operatorname{tr}(\mathbf{B}) = \operatorname{tr}(\mathbf{A}) \tag{A42}$$

A6. THE EIGENVALUE PROBLEM

For a square matrix **A** of order n, we seek a nonzero vector **x** and a scalar λ such that

$$\mathbf{Ax} = \lambda\mathbf{x} \tag{A43}$$

This is called the "eigenvalue problem." A solution λ_0 and $\mathbf{x}_0$ to this problem is called an *eigenvalue* (proper value, characteristic value) and the corresponding *eigenvector* (proper vector, characteristic vector) of the matrix **A**. An eigenvector, if one exists, can be determined only to a scalar multiplication, for if λ_0, $\mathbf{x}_0$ satisfy equation (A43), then λ_0, $\alpha\mathbf{x}_0$, where α is an arbitrary scalar, will also do.

A6.1. The Characteristic Equation

Equation (A43) can be rewritten as

$$(\mathbf{A} - \lambda\mathbf{I})\mathbf{x} = \mathbf{0} \tag{A44}$$

which represents a set of homogenous linear equations. For a nontrivial solution to this set the following must be satisfied,

$$|\mathbf{A} - \lambda\mathbf{I}| = 0 \tag{A45}$$

Equation (A45) represents a real polynomial equation of degree n:

$$b_n(-\lambda)^n + b_{n-1}(-\lambda)^{n-1} + \cdots + b_0 = 0 \tag{A46}$$

where

$$\begin{aligned}
b_n &= 1 \\
b_{n-1} &= a_{11} + a_{22} + \cdots + a_{nn} = \sum_{i=1}^{n} a_{ii} = \text{tr}\ (\mathbf{A}) = \text{trace of } \mathbf{A} \\
&\ \vdots \\
b_{n-r} &= \text{sum of all principal minors of order } r \text{ of } \mathbf{A} \\
&\ \vdots \\
b_0 &= |\mathbf{A}| = \text{determinant of } \mathbf{A}
\end{aligned} \tag{A47}$$

Equation (A46) is called the *characteristic equation* of **A**, or the *eigenvalue equation*. The matrix $(\mathbf{A} - \lambda\mathbf{I})$ is called the *characteristic matrix*. There are n roots for equation (A46), counting multiplicity. These are the n eigenvalues of **A**, $\lambda_1, \lambda_2, \ldots, \lambda_n$. For an eigenvalue λ_i, we solve the set of (homogeneous) linear equations $(\mathbf{A} - \lambda_i\mathbf{I})\mathbf{x} = \mathbf{0}$ to determine the components of the corresponding eigenvector $\mathbf{x}_i$. In general, λ_i and $\mathbf{x}_i$ are either real or complex numbers and vectors, respectively.

If the matrix **A** is *symmetric*, then

1. The eigenvalues are real.
2. The eigenvectors are all mutually orthogonal, that is,

$$\mathbf{x}_i^t\,\mathbf{x}_j = \mathbf{x}_j^t\,\mathbf{x}_i = 0$$

Theorem: For every symmetric matrix **A** there exists a *rotation* (*orthogonal*) *matrix* **R** such that $(\mathbf{R}^t\mathbf{A}\mathbf{R})$ is a *diagonal* matrix. The *columns* of **R** are the *normalized* eigenvectors of **A** and the elements of the diagonal matrix are the corresponding eigenvalues of **A** (**R** is not unique). We demonstrate the validity of this fact for the case in which all eigenvalues λ_i are distinct. Here the corresponding eigenvectors

$$\mathbf{x}_i = [x_{i1} \quad x_{i2} \quad \cdots \quad x_{in}]$$

are also distinct.

Since, as we mentioned earlier, these vectors are orthogonal, if they are *normalized* so that $\mathbf{x}_i^t\mathbf{x}_i = 1$, then the matrix **R** whose columns are these normalized vectors, is an *orthogonal* matrix. Thus

$$\mathbf{R} = [\mathbf{x}_1 \quad \mathbf{x}_2 \quad \cdots \quad \mathbf{x}_2] = \begin{bmatrix} x_{11} & x_{21} & \cdots & x_{n1} \\ x_{12} & x_{22} & \cdots & x_{n2} \\ \vdots & & & \vdots \\ x_{1n} & x_{2n} & \cdots & x_{nn} \end{bmatrix} \tag{A48}$$

has the property

$$\mathbf{R}^t = \mathbf{R}^{-1}$$

and

$$\mathbf{AR} = [\mathbf{Ax}_1 \quad \mathbf{Ax}_2 \quad \cdots \quad \mathbf{Ax}_n] = [\lambda_1 \mathbf{x}_1 \quad \lambda_2 \mathbf{x}_2 \quad \cdots \quad \lambda_n \mathbf{x}_n]$$

$$= \begin{bmatrix} \lambda_1 x_{11} & \cdots & \lambda_n x_{n1} \\ \vdots & & \vdots \\ \lambda_1 x_{1n} & \cdots & \lambda_n x_{nn} \end{bmatrix} = \mathbf{R} \begin{bmatrix} \lambda_1 & 0 & \cdots & 0 \\ 0 & \lambda_2 & & \vdots \\ \vdots & & \ddots & \vdots \\ 0 & & \cdots & \lambda_n \end{bmatrix} \tag{A49}$$

Premultiplying equation (A49) by $\mathbf{R}^{-1} = \mathbf{R}^t$,

$$\mathbf{R}^{-1}\mathbf{AR} = \mathbf{R}^t\mathbf{AR} = \begin{bmatrix} \lambda_1 & 0 & \cdots & 0 \\ 0 & \lambda_2 & \cdots & 0 \\ 0 & 0 & \cdots & \lambda_n \end{bmatrix} = \mathbf{D} = \text{diagonal matrix} \tag{A50}$$

Example. Find the eigenvalues and eigenvectors for the matrix,

$$\mathbf{B} = \begin{bmatrix} 1 & -1 & 1 \\ 0 & 2 & 0 \\ 1 & -1 & 1 \end{bmatrix}$$

From equation (A47), we have

$$b_3 = 1, \qquad b_2 = 1 + 2 + 1 = 4$$

$$b_1 = \begin{vmatrix} 1 & -1 \\ 0 & 2 \end{vmatrix} + \begin{vmatrix} 2 & 0 \\ -1 & 1 \end{vmatrix} + \begin{vmatrix} 1 & 1 \\ 1 & 1 \end{vmatrix} = 2 + 2 + 0 = 4$$

$$b_0 = 2 - 2 = 0$$

Thus the characteristic polynomial is

$$-\lambda^3 + 4\lambda^2 - 4\lambda = 0 \qquad \text{or} \qquad \lambda(\lambda^2 - 4\lambda + 4) = 0$$

Hence

$$\lambda_1 = 0, \qquad \lambda_2 = 2, \qquad \lambda_3 = 2$$

are the three eigenvalues. To compute the eigenvectors, we have for $\lambda_1 = 0$

$$\begin{bmatrix} 1 & -1 & 1 \\ 0 & 2 & 0 \\ 1 & -1 & 1 \end{bmatrix} \begin{bmatrix} x_1 \\ x_2 \\ x_3 \end{bmatrix} = \begin{bmatrix} 0 \\ 0 \\ 0 \end{bmatrix}$$

from which

$$x_2 = 0, \qquad x_1 = -x_3$$

hence $[x_1, \quad 0, \quad -x_1]$ is an eigenvector.

For λ_2 and $\lambda_3 = 2$,

$$\begin{bmatrix} -1 & -1 & 1 \\ 0 & -1 & 0 \\ 1 & -1 & -1 \end{bmatrix} \begin{bmatrix} x_1 \\ x_2 \\ x_3 \end{bmatrix} = \begin{bmatrix} 0 \\ 0 \\ 0 \end{bmatrix}$$

or

$$x_2 = 0, \qquad x_1 = x_3$$

and $[x_1, \ 0, \ x_1]$ is an eigenvector.

Example. Find the eigenvalues and eigenvectors of the symmetric matrix

$$\mathbf{A} = \begin{bmatrix} 1 & 2 \\ 2 & 1 \end{bmatrix}$$

Show that $\mathbf{A}$ is orthogonally similar to a diagonal matrix whose elements are the eigenvalues of $\mathbf{A}$, with the transforming orthogonal matrix composed of the normalized eigenvectors of $\mathbf{A}$ as columns.

The characteristic polynomial is

$$\begin{vmatrix} (1-\lambda) & 2 \\ 2 & (1-\lambda) \end{vmatrix} = \lambda^2 - 2\lambda - 3 = 0$$

from which

$$\lambda_1 = -1 \qquad \text{and} \qquad \lambda_2 = 3$$

are the eigenvalues.

For $\lambda_1 = -1$, we write

$$\begin{bmatrix} 1 & 2 \\ 2 & 1 \end{bmatrix} \begin{bmatrix} x_1 \\ x_2 \end{bmatrix} = \begin{bmatrix} -x_1 \\ -x_2 \end{bmatrix}$$

or $\mathbf{x}_1 = (1, -1)$ is an eigenvector.

For $\lambda_2 = 3$, we have

$$\begin{bmatrix} 1 & 2 \\ 2 & 1 \end{bmatrix} \begin{bmatrix} x_1 \\ x_2 \end{bmatrix} = \begin{bmatrix} 3x_1 \\ 3x_2 \end{bmatrix}$$

or $\mathbf{x}_2 = (1, 1)$ is an eigenvector.

These two vectors are orthogonal since

$$\mathbf{x}_1^t \, \mathbf{x}_2 = 1 - 1 = 0$$

The normalized vector $\mathbf{x}_1$ is

$$\mathbf{y}_1 = \left(\frac{1}{\sqrt{2}}, -\frac{1}{\sqrt{2}}\right)$$

and for $\mathbf{x}_2$

$$\mathbf{y}_2 = \left(\frac{1}{\sqrt{2}}, \frac{1}{\sqrt{2}}\right).$$

Then the orthogonal matrix is

$$\mathbf{R} = \frac{1}{\sqrt{2}}\begin{bmatrix} 1 & 1 \\ -1 & 1 \end{bmatrix}$$

for which

$$\mathbf{R}^t\mathbf{A}\mathbf{R} = \tfrac{1}{2}\begin{bmatrix} 1 & -1 \\ 1 & 1 \end{bmatrix}\begin{bmatrix} 1 & 2 \\ 2 & 1 \end{bmatrix}\begin{bmatrix} 1 & 1 \\ -1 & 1 \end{bmatrix} = \begin{bmatrix} -1 & 0 \\ 0 & 3 \end{bmatrix} = \begin{bmatrix} \lambda_1 & 0 \\ 0 & \lambda_2 \end{bmatrix}$$

which demonstrates equation (A50).

A7. BILINEAR AND QUADRATIC FORMS

If $\mathbf{A}$ is a square matrix of order n and $\mathbf{x}$ and $\mathbf{y}$ are two arbitrary n vectors, then the scalar

$$u = \mathbf{x}^t\mathbf{A}\mathbf{y} \tag{A51}$$

is called a *bilinear form*. If, however, the matrix $\mathbf{A}$ is also *symmetric*, then

$$v = \mathbf{x}^t\mathbf{A}\mathbf{x} \tag{A52}$$

is called a *quadratic form* with the *kernal* $\mathbf{A}$.

The matrix $\mathbf{A}$ is called *positive definite* if $v > 0$ for all $\mathbf{x} \neq \mathbf{0}$, and we write $\mathbf{A} > \mathbf{0}$. If $v \geq 0$ for all $\mathbf{x}$ and there exists a nonzero vector $\mathbf{x}$ for which equality holds, we say $\mathbf{A}$ is *positive semidefinite* (or *nonnegative definite*) and write $\mathbf{A} \geq \mathbf{0}$. There are corresponding definitions for *negative definite* (or *nonpositive definite*). If there exist vectors $\mathbf{x}_1$ and $\mathbf{x}_2$ such that $\mathbf{x}_1^t\mathbf{A}\mathbf{x}_1 > 0$ and $\mathbf{x}_2^t\mathbf{A}\mathbf{x}_2 < 0$, we say $\mathbf{A}$ is *indefinite*.

For a positive definite matrix $\mathbf{A}$ it is necessary and sufficient that

$$a_{11} > 0, \qquad \begin{vmatrix} a_{11} & a_{12} \\ a_{21} & a_{22} \end{vmatrix} > 0, \qquad \ldots, \qquad |\mathbf{A}| > 0 \tag{A53}$$

Thus the matrix

$$\mathbf{B} = \begin{bmatrix} 3 & -2 & 1 \\ -2 & 3 & 1 \\ 1 & 1 & 4 \end{bmatrix}$$

is positive definite, because $3 > 0$; and

$$\begin{vmatrix} 3 & -2 \\ -2 & 3 \end{vmatrix} = 9 - 4 = 5 > 0$$

and

$$|\mathbf{B}| = 3(11) + 2(-9) + 1(-5) = 10 > 0$$

A quadratic form represents, in general, a conic section of some kind. Considering the two-dimensional case for simplicity we write

$$\mathbf{x}^t\mathbf{A}\mathbf{x} = b \qquad \text{with } \mathbf{A} \text{ symmetric}$$

or

$$a_{11}x_1^2 + 2a_{12}x_1x_2 + a_{22}x_2^2 = b$$

Figure A1 represents an equation of an ellipse when the quadratic form is positive definite, with axes oblique to the x_1x_2 axis system of coordinates. If we construct the orthogonal matrix $\mathbf{R}$ whose columns are the normalized eigenvectors of $\mathbf{A}$, then perform the transformation,

$$\mathbf{x} = \mathbf{R}\mathbf{y}$$

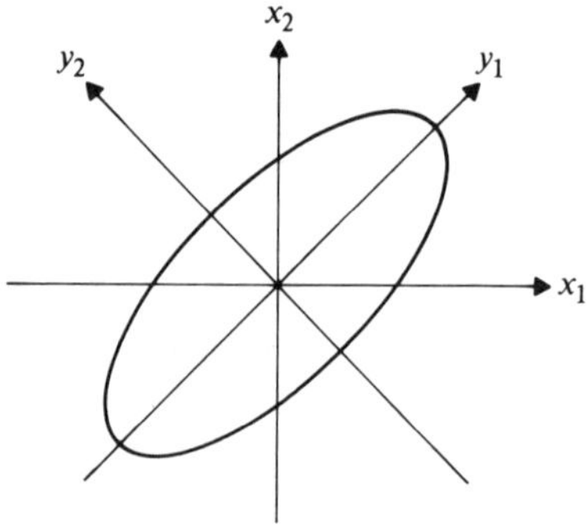

We get

$$(\mathbf{R}\mathbf{y})^t\mathbf{A}(\mathbf{R}\mathbf{y}) = \mathbf{y}^t(\mathbf{R}^t\mathbf{A}\mathbf{R})\mathbf{y} = \mathbf{y}^t\mathbf{D}\mathbf{y}$$

where

$$\mathbf{D} = \begin{bmatrix} \lambda_1 & 0 \\ 0 & \lambda_2 \end{bmatrix} \quad \text{with } \lambda_i \text{ the eigenvalues of } \mathbf{A}$$

Thus the quadratic form becomes

$$\lambda_1 y_1^2 + \lambda_2 y_2^2 = b$$

or

$$\frac{y_1^2}{(b/\lambda_1)} + \frac{y_2^2}{(b/\lambda_2)} = 1$$

which is the equation of the ellipse with respect to the $y_1\,y_2$ system as shown in Figure A1. It is seen that $\mathbf{R}$ is a rotation matrix, and that the semimajor and semiminor axes are $\sqrt{b/\lambda_1}$ and $\sqrt{b/\lambda_2}$, respectively.

A8. MATRIX DIFFERENTIATION

1. If we are given a vector $x = [x_1 \quad x_2 \quad \cdots \quad x_n]^t$ whose elements are functions of another variable u, then the derivative $d\mathbf{x}/du$ is given by

$$\underset{n,1}{\frac{d\mathbf{x}}{du}} = \begin{bmatrix} \dfrac{dx_1}{du} \\ \vdots \\ \dfrac{dx_n}{du} \end{bmatrix} \tag{A54}$$

The differential change $d\mathbf{x}$ of the vector $\underset{n,1}{\mathbf{x}}$ is defined as

$$\underset{n,1}{d\mathbf{x}} = \begin{bmatrix} dx_1 \\ dx_2 \\ \vdots \\ dx_n \end{bmatrix} \tag{A55}$$

2. If the elements of the matrix $\mathbf{A}_{m,n}$ are functions of the (scalar) variable u, then the derivative $d\mathbf{A}/du$ is given by

$$\underset{m,n}{\frac{d\mathbf{A}}{du}} = \begin{bmatrix} \dfrac{da_{11}}{du} & \cdots & \dfrac{da_{1n}}{du} \\ \vdots & & \vdots \\ \dfrac{da_{m1}}{du} & \cdots & \dfrac{da_{mn}}{du} \end{bmatrix} \tag{A56}$$

3. If a vector $\underset{m,1}{\mathbf{y}}$ represents m functions of some or all of the elements of a variable vector $\underset{n,1}{\mathbf{x}}$, then the total differential of $\mathbf{y}$ is given by

$$d\mathbf{y} = \frac{\partial \mathbf{y}}{\partial \mathbf{x}} d\mathbf{x} \tag{A57}$$

where the total differentials $d\mathbf{x}$ and $d\mathbf{y}$ follow the definition given in the first item above. By necessity the partial derivative $\partial \mathbf{y}/\partial \mathbf{x}$ is an $m \times n$ matrix, which is called the *Jacobian matrix*, and is written in detail as

$$\frac{\partial \mathbf{y}}{\partial \mathbf{x}} = \begin{bmatrix} \frac{\partial y_1}{\partial x_1} & \frac{\partial y_1}{\partial x_2} & \cdots & \frac{\partial y_1}{\partial x_n} \\ \vdots & \vdots & & \vdots \\ \frac{\partial y_m}{\partial x_1} & \frac{\partial y_m}{\partial x_2} & \cdots & \frac{\partial y_m}{\partial x_n} \end{bmatrix} \tag{A58}$$

4. For the expression

$$\underset{m,\,1}{\mathbf{v}} = \underset{m,\,n}{\mathbf{A}}\ \underset{n,\,1}{\mathbf{y}}$$

where $\mathbf{A}$ is independent of $\mathbf{y}$,

$$\underset{m,\,1}{d\mathbf{v}} = \underset{m,\,n}{\frac{\partial \mathbf{v}}{\partial \mathbf{y}}}\ \underset{n,\,1}{d\mathbf{y}} = \underset{m,\,n}{\frac{\partial(\mathbf{A}\mathbf{y})}{\partial \mathbf{y}}}\ \underset{n,\,1}{d\mathbf{y}}$$

or

$$\frac{\partial(\mathbf{A}\mathbf{y})}{\partial \mathbf{y}} = \underset{m,\,n}{\mathbf{A}} \tag{A59}$$

5. For

$$\underset{1,\,q}{\mathbf{w}} = \underset{1,\,p}{\mathbf{x}^t}\ \underset{p,\,q}{\mathbf{B}}$$

where $\underset{p,\,q}{\mathbf{B}}$ is independent of $\mathbf{x}$

$$\begin{aligned} \underset{1,\,q}{d\mathbf{w}} &= d\left(\underset{q,\,p}{\mathbf{B}^t}\ \underset{p,\,1}{\mathbf{x}}\right)^t = \left(\underset{q,\,p}{\frac{\partial(\mathbf{B}^t\mathbf{x})}{\partial \mathbf{x}}}\ \underset{p,\,1}{d\mathbf{x}}\right)^t \\ &= (\mathbf{B}^t d\mathbf{x})^t \\ &= \underset{1,\,p}{d\mathbf{x}^t}\ \underset{p,\,q}{\mathbf{B}} \end{aligned}$$

6. For the bilinear form,

$$u = \underset{1,\,n}{\mathbf{x}^t}\ \underset{n,\,n}{\mathbf{A}}\ \underset{n,\,1}{\mathbf{y}}$$

where $\mathbf{A}$ is independent of both $\mathbf{x}$ and $\mathbf{y}$,

$$du = \underset{1,\,n}{\frac{\partial u}{\partial \mathbf{y}}}\ \underset{n,\,1}{d\mathbf{y}} \qquad \text{or} \qquad \frac{\partial u}{\partial \mathbf{y}} = \mathbf{x}^t\mathbf{A} \tag{A60}$$

and

$$du = \left(\underset{1,\,n}{\frac{\partial u}{\partial \mathbf{x}}} \underset{n,\,1}{d\mathbf{x}} \right)^t \qquad \text{or} \qquad \frac{\partial u}{\partial \mathbf{x}} = \underset{1,\,n}{\mathbf{y}^t} \, \underset{n,\,n}{\mathbf{A}^t} \tag{A61}$$

7. For the quadratic form,

$$u = \mathbf{x}^t \mathbf{A} \mathbf{x}$$

where **A** is independent of **x**,

$$\underset{1,\,n}{\frac{\partial u}{\partial \mathbf{x}}} = \underset{1,\,n}{\mathbf{x}^t \mathbf{A}} + \underset{1,\,n}{\mathbf{x}^t \mathbf{A}^t} = \underset{1,\,n}{\mathbf{x}^t} \left(\underset{n,\,n}{\mathbf{A} + \mathbf{A}^t} \right) = \underset{1,\,n}{2\mathbf{x}^t \mathbf{A}} \tag{A62}$$

when $\mathbf{A}_{n,\,n}$ is symmetric.

8. The derivative of the inverse $\mathbf{A}^{-1}$, assuming that it exists, is obtained from

$$\mathbf{A}\mathbf{A}^{-1} = \mathbf{I}$$

$$\frac{d}{dx}(\mathbf{A}\mathbf{A}^{-1}) = \frac{d\mathbf{I}}{dx} = \mathbf{0}$$

$$\mathbf{A}\frac{d\mathbf{A}^{-1}}{dx} + \frac{d\mathbf{A}}{dx}\mathbf{A}^{-1} = \mathbf{0}$$

hence

$$\frac{d\mathbf{A}^{-1}}{dx} = -\mathbf{A}^{-1}\frac{d\mathbf{A}}{dx}\mathbf{A}^{-1} \tag{A63}$$

9. If the elements of a determinant $r = |\mathbf{R}|$ are functions of a variable x, then

$$\frac{dr}{dx} = \frac{d}{dx}|\mathbf{R}|$$

$$= \begin{vmatrix} \dfrac{dr_{11}}{dx} & r_{12} & \cdots & r_{1n} \\ \dfrac{dr_{21}}{dx} & r_{22} & \cdots & r_{2n} \\ \cdot & \cdot & \cdots & \cdot \\ \dfrac{dr_{n1}}{dx} & r_{n2} & \cdots & r_{nn} \end{vmatrix} + \cdots + \begin{vmatrix} r_{11} & r_{12} & \cdots & \dfrac{dr_{1n}}{dx} \\ r_{21} & r_{22} & \cdots & \dfrac{dr_{2n}}{dx} \\ \cdot & \cdot & \cdots & \cdot \\ r_{n1} & 2_{n2} & \cdots & \dfrac{dr_{nn}}{dx} \end{vmatrix} \tag{A64}$$

or

$$\frac{dr}{dx} = \begin{vmatrix} \frac{dr_{11}}{dx} & \frac{dr_{12}}{dx} & \cdots & \frac{dr_{1n}}{dx} \\ r_{21} & r_{22} & \cdots & r_{2n} \\ \cdot & \cdot & \cdots & \cdot \\ r_{n1} & r_{n2} & \cdots & r_{nn} \end{vmatrix} + \cdots + \begin{vmatrix} r_{11} & r_{12} & \cdots & r_{1n} \\ r_{21} & r_{22} & \cdots & r_{2n} \\ \cdot & \cdot & \cdots & \cdot \\ \frac{dr_{n1}}{dx} & \frac{dr_{n2}}{dx} & \cdots & \frac{dr_{nn}}{dx} \end{vmatrix} \tag{A65}$$

Note that the derivative of a determinant of order n is the sum of n determinants.

A9. SOME USEFUL MATRIX RELATIONSHIPS

1. For the bilinear and quadratic forms, we may write

$$\mathbf{x}^t\mathbf{A}\mathbf{y} = \operatorname{tr}(\mathbf{x}\mathbf{y}^t\mathbf{A}^t) \tag{A66}$$

$$\mathbf{x}^t\mathbf{A}\mathbf{x} = \operatorname{tr}(\mathbf{x}\mathbf{x}^t\mathbf{A}) \qquad (\mathbf{A}\ \text{symmetric}) \tag{A67}$$

2. Given the matrix expression

$$\underset{n,\,n}{\mathbf{X}} = \underset{n,\,n}{\mathbf{Y}} \pm \underset{n,\,p}{\mathbf{U}}\ \underset{p,\,p}{\mathbf{Z}}\ \underset{p,\,n}{\mathbf{V}} \tag{A68}$$

then

$$\mathbf{X}^{-1} = \mathbf{Y}^{-1} \mp \mathbf{Y}^{-1}\mathbf{U}(\mathbf{Z}^{-1} \pm \mathbf{V}\mathbf{Y}^{-1}\mathbf{U})^{-1}\mathbf{V}\mathbf{Y}^{-1} \tag{A69}$$

provided that all inverses in equation (A69) do exist.

3.

$$(\mathbf{A} + \mathbf{B})^{-1} = \mathbf{A}^{-1}(\mathbf{A}^{-1} + \mathbf{B}^{-1})^{-1}\mathbf{B}^{-1} \tag{A70}$$

$$= \mathbf{B}^{-1}(\mathbf{A}^{-1} + \mathbf{B}^{-1})^{-1}\mathbf{A}^{-1} \tag{A71}$$

provided that $\mathbf{A}^{-1}$ and $\mathbf{B}^{-1}$ exist.

4. In general

$$(\mathbf{A} + \mathbf{B})(\mathbf{A} - \mathbf{B}) \neq \mathbf{A}^2 - \mathbf{B}^2 \tag{A72}$$

and

$$\frac{d}{dx}\mathbf{A}^2 \neq 2\mathbf{A}\frac{d\mathbf{A}}{dx}$$

5. An idempotent matrix is a matrix that is equal to its square, or

$$\mathbf{H} = \mathbf{H}^2 \tag{A73}$$

For instance,

$$\mathbf{C} = \mathbf{I} - \mathbf{B}(\mathbf{B}^t\mathbf{B})^{-1}\mathbf{B}^t$$

is idempotent, since

$$\begin{aligned}\mathbf{C}^2 &= \mathbf{I} - 2\mathbf{B}(\mathbf{B}^t\mathbf{B})^{-1}\mathbf{B}^t + \mathbf{B}(\mathbf{B}^t\mathbf{B})^{-1}\mathbf{B}^t\mathbf{B}(\mathbf{B}^t\mathbf{B})^{-1}\mathbf{B}^t \\ &= \mathbf{I} - \mathbf{B}(\mathbf{B}^t\mathbf{B})^{-1}\mathbf{B}^t \\ &= \mathbf{C}\end{aligned}$$

6. A block diagonal matrix is a square matrix with square submatrices along the main diagonal and zero submatrices elsewhere. It may be written symbolically as

$$\mathbf{A} = \begin{bmatrix} \mathbf{A}_{11} & \mathbf{0} & \cdots & \mathbf{0} \\ \mathbf{0} & \mathbf{A}_{22} & \cdots & \mathbf{0} \\ \vdots & \vdots & & \vdots \\ \mathbf{0} & \mathbf{0} & & \mathbf{A}_{kk} \end{bmatrix}$$

as an example

$$\left[\begin{array}{c|ccc|cc} 1 & 0 & 0 & 0 & 0 & 0 \\ \hline 0 & 2 & -1 & 4 & 0 & 0 \\ 0 & 3 & 4 & 0 & 0 & 0 \\ 0 & -5 & 0 & 7 & & \\ \hline 0 & 0 & 0 & 0 & 11 & 1 \\ 0 & 0 & 0 & 0 & 4 & 0 \end{array}\right]$$

is a block diagonal matrix.

The inverse of a nonsingular block diagonal matrix is another block diagonal matrix with corresponding diagonal submatrices being inverses of the respective submatrices in the original matrix, or

$$\begin{bmatrix} \mathbf{A}_{11}^{-1} & \mathbf{0} & \cdots & \mathbf{0} \\ \mathbf{0} & \mathbf{A}_{22}^{-1} & \cdots & \mathbf{0} \\ \vdots & \vdots & & \vdots \\ \mathbf{0} & \mathbf{0} & & \mathbf{A}_{kk}^{-1} \end{bmatrix}$$

APPENDIX

B

Linearization by Series Expansion

When the original condition (and constraint) equations are nonlinear, they are usually linearized using the zero and first-order terms of Taylor series. In the following sections a number of cases are given.

B1. ONE FUNCTION OF ONE VARIABLE

For $y = f(x)$ the series expansion is

$$y = f(x^0) + \left.\frac{dy}{dx}\right|_{x_0} \Delta x + \frac{1}{2!}\left.\frac{d^2y}{dx^2}\right|_{x_0} (\Delta x)^2 + \cdots \tag{B1}$$

Any term beyond the second on the right hand side of equation (B1) is again nonlinear. Therefore, for purposes of adjustment, only the first two terms (zero and first order) are used to form a linear expression approximating the original nonlinear form. Thus

$$y \cong f(x^0) + \left.\frac{dy}{dx}\right|_{x^0} \cdot \Delta x = y^0 + j\,\Delta x \tag{B2}$$

is the linearized form to be used.

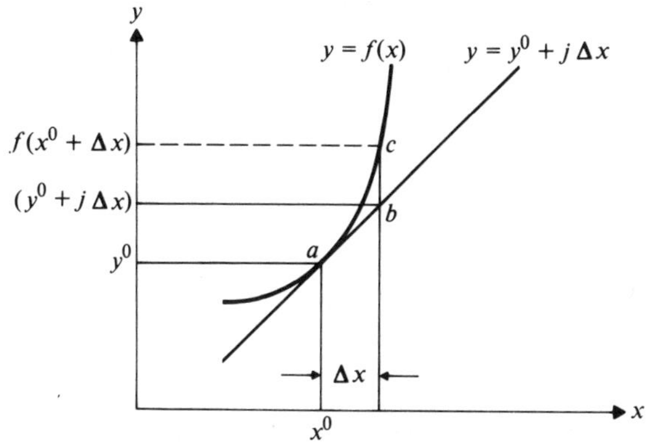

The technique of linearization is demonstrated in Figure B1. The curve represents the original nonlinear function $f(x)$, whereas the straight line represents the linearized form of equation (B2). That line is tangent to the curve at the given point a, (x^0, y^0). When Δx is given (or evaluated), the value of the function would be approximated by point b whose ordinate is $(y^0 + j \cdot \Delta x)$, and the exact value from the nonlinear function is point c, with ordinate $f(x^0 + \Delta x)$. The error arising from using the linear form is the line segment bc.

Example Given $y = x^2 + x^3$, evaluate the linearized form at $x^0 = 1$ and the value of the function and its linearized form at $x = 1.1$

$$j = \frac{dy}{dx}\bigg|_{x^0} = 2x + 3x^2\bigg|_{x^0=1} = 5$$

$$y^0 = f(x^0) = 2 \qquad \text{and} \qquad \Delta x = 0.1$$

$$y = y^0 + j\,\Delta x = 2 + 0.5 = 2.5$$

$$f(x^0 + \Delta x) = f(1.1) = 1.21 + 1.331 = 2.541$$

B2. ONE FUNCTION OF TWO VARIABLES

$$\begin{aligned} y &= f(x_1, x_2) \\ &= f(x_1^0, x_2^0) + \frac{\partial y}{\partial x_1}\bigg|_{x_1{}^0,\, x_2{}^0} \cdot \Delta x_1 + \frac{\partial y}{\partial x_2}\bigg|_{x_1{}^0,\, x_2{}^0} \cdot \Delta x_2 \\ &\quad + \frac{1}{2!}\frac{\partial^2 y}{\partial x_1^2}\bigg|_{x_1{}^0,\, x_2{}^0} \cdot (\Delta x_1)^2 + \frac{1}{2!}\frac{\partial^2 y}{\partial x_2^2}\bigg|_{x_1{}^0,\, x_2{}^0} (\Delta x_2)^2 \\ &\quad + \frac{\partial y}{\partial x_1}\bigg|_{x_1{}^0,\, x_2{}^0} \cdot \frac{\partial y}{\partial x_2}\bigg|_{x_1{}^0,\, x_2{}^0} (\Delta x_1)(\Delta x_2) + \cdots \end{aligned} \tag{B3}$$

For linearized form, equation (B3) is truncated to

$$y = y^0 + j_1 \Delta x_1 + j_2 \Delta x_2 \tag{B4}$$

where

$$y^0 = f(x_1^0, x_2^0)$$

$$j_1 = \left.\frac{\partial y}{\partial x_1}\right|_{x_1{}^0,\, x_2{}^0}$$

$$j_2 = \left.\frac{\partial y}{\partial x_2}\right|_{x_1{}^0,\, x_2{}^0}$$

Equation (B4) can be rewritten introducing vectors, or

$$y = y^0 + [j_1 \quad j_2]\begin{bmatrix}\Delta x_1 \\ \Delta x_2\end{bmatrix}$$

or

$$y = y^0 + \mathbf{J}_{yx}\, \Delta\mathbf{x} \tag{B5}$$

where

$$\mathbf{J}_{yx} = \frac{\partial y}{\partial \mathbf{x}}$$

is the Jacobian of y with respect to $\mathbf{x}$.

B3. TWO FUNCTIONS OF ONE VARIABLE

$$\begin{aligned} y_1 &= f_1(x) \cong y_1^0 + j_1 \Delta x \\ y_2 &= f_2(x) \cong y_2^0 + j_2 \Delta x \end{aligned} \tag{B6}$$

with

$$y_1 = f_1(x)^0$$

$$y_2 = f_2(x)_2^0$$

$$\mathbf{J}_{yx} = [j_1 \quad j_2] = \left[\left.\frac{dy_1}{dx}\right|_{x^0} \quad \left.\frac{dy_2}{dx}\right|_{x^0}\right]$$

B4. TWO FUNCTIONS OF TWO VARIABLES EACH

$$\begin{aligned} y_1 &= f_1(x_1, x_2) \cong y_1^0 + j_{11} \Delta x_1 + j_{12} \Delta x_2 \\ y_2 &= f_2(x_1, x_2) \cong y_2^0 + j_{21} \Delta x_1 + j_{22} \Delta x_2 \end{aligned} \tag{B7a}$$

or

$$\begin{bmatrix} y_1 \\ y_2 \end{bmatrix} \cong \begin{bmatrix} y_1^0 \\ y_2^0 \end{bmatrix} + \begin{bmatrix} j_{11} & j_{12} \\ j_{21} & j_{22} \end{bmatrix}\begin{bmatrix}\Delta x_1 \\ \Delta x_2\end{bmatrix} \tag{B7b}$$

or

$$\mathbf{y} = \mathbf{y}^0 + \mathbf{J}_{yx}\,\Delta\mathbf{x} \tag{B7c}$$

where

$$\mathbf{y}^0 = \begin{bmatrix} y_1^0 \\ y_2^0 \end{bmatrix} = \begin{bmatrix} f_1(x_1^0, x_2^0) \\ f_2(x_1^0, x_2^0) \end{bmatrix}$$

and

$$\mathbf{J}_{yx} = \frac{\partial \mathbf{y}}{\partial \mathbf{x}} = \begin{bmatrix} \dfrac{\partial y_1}{\partial x_1} & \dfrac{\partial y_1}{\partial x_2} \\[2ex] \dfrac{\partial y_2}{\partial x_1} & \dfrac{\partial y_2}{\partial x_2} \end{bmatrix}$$

evaluated at x_1^0, x_2^0.

B5. GENERAL CASE OF m FUNCTIONS OF n VARIABLES

$$\begin{aligned} y_1 &= f_1(x_1, x_2, \ldots, x_n) \\ y_2 &= f_2(x_1, x_2, \ldots, x_n) \\ &\cdots\cdots\cdots\cdots\cdots\cdots \\ y_m &= f_m(x_1, x_2, \ldots, x_n) \end{aligned} \tag{B8}$$

With the auxiliaries,

$$\mathbf{y}^0 = \begin{bmatrix} y_1^0 \\ y_2^0 \\ \vdots \\ y_m^0 \end{bmatrix} = \begin{bmatrix} f_1(x_1^0, x_2^0, \ldots, x_n^0) \\ f_2(x_2^0, x_2^0, \ldots, x_n^0) \\ \cdots\cdots\cdots\cdots\cdots\cdots \\ f_m(x_1^0, x_2^0, \ldots, x_n^0) \end{bmatrix}$$

$$\mathbf{J}_{yx} = \frac{\partial \mathbf{y}}{\partial \mathbf{x}} = \begin{bmatrix} \dfrac{\partial y_1}{\partial x_1} & \dfrac{\partial y_1}{\partial x_2} & \cdots & \dfrac{\partial y_1}{\partial x_n} \\ \cdots & \cdots & \cdots & \cdots \\ \dfrac{\partial y_m}{\partial x_1} & \dfrac{\partial y_m}{\partial x_2} & \cdots & \dfrac{\partial y_m}{\partial x_n} \end{bmatrix} \quad \text{evaluated at } \mathbf{x}^0$$

$$\Delta\mathbf{x} = \begin{bmatrix} \Delta x_1 \\ \Delta x_2 \\ \vdots \\ \Delta x_n \end{bmatrix}$$

the linearized form of equation (B8) becomes

$$\mathbf{y} \cong \mathbf{y}^0 + \mathbf{J}_{yx}\,\Delta\mathbf{x}$$

Example Given

$$y_1 = x_1^2 - x_3$$

$$y_2 = \frac{1}{x_2} + x_3^2$$

write the linearized form, in matrix notation, at the point $\mathbf{x}^0 = [1 \quad 1 \quad 1]'$

$$\mathbf{y}^0 = \begin{bmatrix} y_1^0 \\ y_2^0 \end{bmatrix} = \begin{bmatrix} 0 \\ 2 \end{bmatrix}$$

$$\mathbf{J}_{yx} = \begin{bmatrix} 2x_1^0 & 0 & -1 \\ 0 & -x^{0-2} & 2x_3^0 \end{bmatrix} = \begin{bmatrix} 2 & 0 & -1 \\ 0 & -1 & 2 \end{bmatrix}$$

Thus

$$\mathbf{y} \cong \begin{bmatrix} 0 \\ 2 \end{bmatrix} + \begin{bmatrix} 2 & 0 & -1 \\ 0 & -1 & 2 \end{bmatrix} \begin{bmatrix} \Delta x_1 \\ \Delta x_2 \\ \Delta x_3 \end{bmatrix}$$

APPENDIX

C

Ranks of a Posteriori Cofactor (Covariance) Matrices

After a least squares adjustment is performed, the residuals $\mathbf{v}$ and estimated observations $\hat{l}$ are both functionally dependent variables. Consequently, their cofactor matrices, $\mathbf{Q}_{vv}$ and $\mathbf{Q}_{\hat{l}\hat{l}}$, are singular. In the text the fact was mentioned that the rank of $\mathbf{Q}_{vv}$ is r (the redundancy) and the rank of $\mathbf{Q}_{\hat{l}\hat{l}}$ is $(n - r)$, but without proof. It is the purpose of this appendix to give the proofs of these facts for different techniques of least squares.

C1. ADJUSTMENT OF OBSERVATIONS ONLY

The conditions are:

$$\underset{r,n}{\mathbf{A}}\ \underset{n,1}{\mathbf{v}} = \underset{r,1}{\mathbf{f}}$$

with cofactor matrix $\underset{n,n}{\mathbf{Q}}$ and

$$n > r, \qquad \text{rank}\ (\mathbf{A}) = r, \qquad \text{rank}\ (\mathbf{Q}) = n$$

$$\mathbf{v} = \mathbf{QA}^t\mathbf{W}_e\,\mathbf{f}$$

$$\hat{l} = l + \mathbf{v}$$

$$\mathbf{Q}_{vv} = \mathbf{QA}^t\mathbf{W}_e\,\mathbf{AQ}$$

$$\mathbf{Q}_{\hat{l}\hat{l}} = \mathbf{Q} - \mathbf{Q}_{vv}$$

It is required to show that:

$$\text{rank } (\mathbf{Q}_{vv}) = r \qquad \text{and} \qquad \text{rank } (\mathbf{Q}_{ll}) = n - r$$

Let

$$\underset{n,\,n}{\mathbf{J}} = \mathbf{A}^t\mathbf{W}_e\,\mathbf{A}\mathbf{Q} \tag{C1}$$

which has the following properties:

1. $\mathbf{J}^2 = \mathbf{A}^t\mathbf{W}_e(\mathbf{A}\mathbf{Q}\mathbf{A}^t)\mathbf{W}_e\,\mathbf{A}\mathbf{Q} = \mathbf{J}$ (idempotent)
2. Rank $(\mathbf{J}) = r$ because

$$\text{rank } (\mathbf{J}) \leq \text{least rank of constituent matrices in equation (C1)}$$
$$\leq r$$

since

$$\mathbf{J}\mathbf{A}^t = \mathbf{A}^t \qquad \text{and} \qquad \text{rank } (\mathbf{A}^t) = r$$

then

$$r = \text{rank } (\mathbf{A}^t) \leq \min (\text{rank } (\mathbf{J}), r)$$
$$\text{rank } (\mathbf{J}) \geq r$$

Thus

$$\text{rank } (\mathbf{J}) = r$$

3. There exists an orthogonal matrix **F** that diagonalizes **J**,

$$\mathbf{F}^t\mathbf{J}\mathbf{F} = \mathbf{D}$$
$$\mathbf{D}^2 = \mathbf{F}^t\mathbf{J}\mathbf{F}\mathbf{F}^t\mathbf{J}\mathbf{F} = \mathbf{F}^t\mathbf{J}^2\mathbf{F} = \mathbf{D}$$

Thus the diagonal elements of **D** (eigenvalues) are either 0 or 1. The number of elements for which $d_i = 1$ is the same as the rank of **J**, or is r.

4. Because $\mathbf{J}\mathbf{A}^t = \mathbf{A}^t$, the row vectors of **A** (column vectors of $\mathbf{A}^t$) are the eigenvectors of **J**

Let

$$\underset{n,\,n}{\mathbf{K}} = \mathbf{I} - \mathbf{J} \tag{C2}$$

which has the properties

1. $\mathbf{K}^2 = \mathbf{I} - 2\mathbf{J} + \mathbf{J}^2 = \mathbf{I} - \mathbf{J} = \mathbf{K}$ (idempotent)
2. Rank $(\mathbf{K}) = (n - r)$, because applying the orthogonal transformation

$$\mathbf{F}^t\mathbf{K}\mathbf{F} = \mathbf{I} - \mathbf{F}^t\mathbf{J}\mathbf{F} = \mathbf{I} - \mathbf{D}$$

thus

$$\begin{aligned}\text{rank } (\mathbf{K}) &= \text{rank } (\mathbf{I} - \mathbf{D}) \\ &= \text{the number of elements for which } d_i = 0 \\ &= (n - r)\end{aligned}$$

3. If eigenvalues of **J** are λ, then for any vector **x**

$$\mathbf{Jx} = \lambda\mathbf{x}$$

and thus

$$\mathbf{Kx} = \mathbf{x} - \mathbf{Jx} = \mathbf{x} - \lambda\mathbf{x} = (\mathbf{I} - \lambda)\mathbf{x}$$

Thus the eigenvalues of **J** are eigenvalues of **K** except that those of value 1 for **J** are of value 0 for **K** and vice versa.

From the above preliminary notes the ranks of $\mathbf{Q}_{vv}$ and $\mathbf{Q}_{ll}$ may be derived.

C1.1. Rank of $\mathbf{Q}_{vv}$

$$\mathbf{Q}_{vv} = \mathbf{QJ}$$

$$\text{rank } (\mathbf{Q}_{vv}) \leq \min\,[n, r] \leq r \qquad (\text{because } r < n)$$

and

$$\mathbf{WQ}_{vv} = \mathbf{J}$$

$$\text{rank } (\mathbf{J}) = r \leq \min\,[n, \text{rank } (\mathbf{Q}_{vv})]$$

$$\text{rank } (\mathbf{Q}_{vv}) \geq r$$

Thus

$$\text{rank } (\mathbf{Q}_{vv}) = r \tag{C3}$$

which is the first required result.

C1.2. Rank of $\mathbf{Q}_{ll}$

$$\mathbf{Q}_{ll} = \mathbf{Q} - \mathbf{Q}_{vv} = \mathbf{Q} - \mathbf{QJ} = \mathbf{Q}(\mathbf{I} - \mathbf{J}) = \mathbf{Qk}$$

$$\begin{aligned}\text{rank } (\mathbf{Q}_{ll}) &\leq \min\,[n, (n - r)] \\ &\leq (n - r)\end{aligned}$$

$$\mathbf{WQ}_{ll} = \mathbf{K}$$

$$\text{rank } (\mathbf{K}) = (n - r) \leq \min\,[n, \text{rank } (\mathbf{Q}_{ll})]$$

$$\text{rank } (\mathbf{Q}_{ll}) \geq (n - r)$$

Thus

$$\text{rank}\ (\mathbf{Q}_{ll}) = (n - r) \tag{C4}$$

which is the second required result.

C2. ADJUSTMENT OF INDIRECT OBSERVATIONS

The conditions are

$$\underset{n,1}{\mathbf{v}} + \underset{n,u}{\mathbf{B}}\ \underset{u,1}{\boldsymbol{\Delta}} = \underset{n,1}{\mathbf{f}} \qquad \text{with } \underset{n,n}{\mathbf{Q}}$$

and

$$n > u, \qquad \text{rank}\ (\mathbf{B}) = u, \qquad \text{rank}\ (\mathbf{Q}) = n$$

$$\boldsymbol{\Delta} = (\mathbf{B}^t\mathbf{W}\mathbf{B})^{-1}(\mathbf{B}^t\mathbf{W}\mathbf{f})$$

$$\mathbf{v} = \mathbf{f} - \mathbf{B}\boldsymbol{\Delta}$$

$$\mathbf{Q}_{\Delta\Delta} = (\mathbf{B}^t\mathbf{W}\mathbf{B})^{-1} = \mathbf{N}^{-1}$$

$$\mathbf{Q}_{vv} = \mathbf{Q} - \mathbf{B}\mathbf{N}^{-1}\mathbf{B}^t$$

$$\mathbf{Q}_{ll} = \mathbf{B}\mathbf{N}^{-1}\mathbf{B}^t$$

$$= \mathbf{Q} - \mathbf{Q}_{vv}$$

It is required to show that

$$\text{rank}\ (\mathbf{Q}_{vv}) = (n - u) = r \qquad \text{or degrees of freedom}$$

$$\text{rank}\ (\mathbf{Q}_{ll}) = u$$

which is the minimum number of variables (observations) necessary for a unique solution.

In this case

$$\underset{n,n}{\mathbf{J}} = \mathbf{B}\mathbf{N}^{-1}\mathbf{B}^t\mathbf{W}$$

with the properties:

1. $\mathbf{J}^2 = \mathbf{J}$
2. Rank $(\mathbf{J}) = u$ because

$$\text{rank}\ (\mathbf{J}) \leq \text{min rank of constituent matrices}$$

$$< u$$

$$\mathbf{J}\mathbf{B} = \mathbf{B}$$

$$u \leq \min\ [\text{rank}\ (\mathbf{J}), u]$$

$$\text{rank}\ (\mathbf{J}) \geq u$$

or rank $(\mathbf{J})$ must equal u.

3. Orthogonal transformation

$$\mathbf{F}^t\mathbf{J}\mathbf{F} = \mathbf{D}$$

where $\mathbf{D}$ is a diagonal matrix with u of $d_i = 1$ and $(n - u)$ of $d_i = 0$, since $\mathbf{D}^2 = \mathbf{D}$.

4. $\mathbf{JB} = \mathbf{B}$, or columns of $\mathbf{B}$ are eigenvectors of $\mathbf{J}$. Next,

$$\mathbf{K} = \mathbf{I} - \mathbf{J}$$

with

$$\mathbf{K}^2 = \mathbf{K}$$

$$\text{rank}\ (\mathbf{K}) = (n - u)$$

and eigenvalues of $\mathbf{J}$ are also eigenvalues for $\mathbf{K}$ as in Section C1.

C2.1. Rank of $\mathbf{Q}_{\hat{l}\hat{l}}$

$$\mathbf{Q}_{\hat{l}\hat{l}} = \mathbf{JQ}$$

$$\text{rank}\ (\mathbf{Q}_{\hat{l}\hat{l}}) \leq u$$

$$\mathbf{Q}_{\hat{l}\hat{l}}\,\mathbf{W} = \mathbf{J}$$

$$\text{rank}\ (\mathbf{Q}_{\hat{l}\hat{l}}) \geq u$$

Thus

$$\text{rank}\ (\mathbf{Q}_{\hat{l}\hat{l}}) = u = (n - r) \tag{C5}$$

C2.2. Rank of $\mathbf{Q}_{vv}$

$$\mathbf{Q}_{vv} = \mathbf{Q} - \mathbf{Q}_{\hat{l}\hat{l}} = \mathbf{Q} - \mathbf{JQ} = \mathbf{KQ}$$

$$\text{rank}\ (\mathbf{Q}_{vv}) \leq (n - u)$$

$$\mathbf{Q}_{vv}\,\mathbf{W} = \mathbf{K}$$

$$\text{rank}\ (\mathbf{Q}_{vv}) \geq (n - u)$$

Hence

$$\text{rank}\ (\mathbf{Q}_{vv}) = (n - u) = r \tag{C6}$$

C3. ADJUSTMENT OF COMBINED OBSERVATIONS AND PARAMETERS

The conditions are

$$\underset{c,\,n}{\mathbf{A}}\ \underset{n,\,1}{\mathbf{v}} + \underset{c,\,u}{\mathbf{B}}\ \underset{u,\,1}{\boldsymbol{\Delta}} = \underset{c,\,1}{\mathbf{f}} \qquad \text{with } \underset{n,\,n}{\mathbf{Q}}$$

and

$$c < n \qquad c > u \qquad c = u + r \qquad r = \text{redundancy}$$

$$\text{rank}\ (\mathbf{A}) = c \qquad \text{rank}\ (\mathbf{B}) = u \qquad \text{rank}\ (\mathbf{Q}) = n$$

$$\Delta = \Big(\underset{u,c}{\mathbf{B}^t}\ \underset{c,c}{\mathbf{W}_e}\ \underset{c,u}{\mathbf{B}}\Big)^{-1}(\mathbf{B}^t\mathbf{W}_e\,\mathbf{f}) = \mathbf{N}^{-1}\mathbf{t}$$

$$\mathbf{v} = \mathbf{Q}\mathbf{A}^t\mathbf{W}_e(-\mathbf{B}\Delta + \mathbf{f})$$

$$\mathbf{Q}_{\Delta\Delta} = \mathbf{N}^{-1}$$

$$\underset{n,n}{\mathbf{Q}_{vv}} = \underset{n,n}{\mathbf{Q}}\ \underset{n,c}{\mathbf{A}^t}[\mathbf{W}_e - \mathbf{W}_e\,\mathbf{B}\mathbf{N}^{-1}\mathbf{B}^t\mathbf{W}_e]\mathbf{A}\mathbf{Q}$$

$$\mathbf{Q}_{ll} = \mathbf{Q} - \mathbf{Q}_{vv}$$

It is required to show that

$$\text{rank}\ (\mathbf{Q}_{vv}) = c - u = r$$

$$\text{rank}\ (\mathbf{Q}_{ll}) = n - r$$

C3.1. Rank of $\mathbf{Q}_{vv}$

$$\mathbf{Q}\mathbf{A}^t\mathbf{W}_e(\mathbf{I} - \mathbf{B}\mathbf{N}^{-1}\mathbf{B}^t\mathbf{W}_e)\mathbf{A}\mathbf{Q}$$

Let

$$\underset{c,c}{\bar{\mathbf{J}}} = \mathbf{B}\mathbf{N}^{-1}\mathbf{B}^t\mathbf{W}_e$$

$$\bar{\mathbf{J}}^2 = \mathbf{B}\mathbf{N}^{-1}(\mathbf{B}^t\mathbf{W}_e\,\mathbf{B})\mathbf{N}^{-1}\mathbf{B}^t\mathbf{W}_e = \mathbf{B}\mathbf{N}^{-1}\mathbf{B}^t\mathbf{W}_e = \bar{\mathbf{J}} \qquad \text{(idempotent)}$$

$$\text{rank}\ (\bar{\mathbf{J}}) \leq \text{min rank of constituent matrices}$$

$$\leq u$$

$$\bar{\mathbf{J}}\mathbf{B} = \mathbf{B}$$

$$\text{rank}\ (\bar{\mathbf{J}}) \geq u$$

Thus

$$\text{rank}\ (\bar{\mathbf{J}}) = u$$

Let

$$\bar{\mathbf{K}} = \mathbf{I} - \bar{\mathbf{J}}$$

$$\bar{\mathbf{K}}^2 = \bar{\mathbf{K}} \qquad \text{(idempotent)}$$

and following the scheme of orthogonal transformation

$$\bar{\mathbf{F}}^t\bar{\mathbf{K}}\bar{\mathbf{F}} = \mathbf{I} - \bar{\mathbf{D}}$$

we can get

$$\text{rank }(\bar{\mathbf{K}}) = c - u = r$$

Then

$$\mathbf{Q}_{vv} = \mathbf{Q}\mathbf{A}^t\mathbf{W}_e\,\bar{\mathbf{K}}\mathbf{A}\mathbf{Q}$$

$$\begin{aligned}\text{rank }(\mathbf{Q}_{vv}) &\le \text{min rank of constituent matrices}\\ &\le r\end{aligned}$$

Premultiplying both sides by $\mathbf{A}$, and postmultiplying by $\mathbf{A}^t$, then

$$\mathbf{A}\mathbf{Q}_{vv}\,\mathbf{A}^t = \bar{\mathbf{K}}\mathbf{Q}_e$$

or

$$\mathbf{A}\mathbf{Q}_{vv}\,\mathbf{A}^t\mathbf{W}_e = \bar{\mathbf{K}}$$

$$r = \text{rank }(\bar{\mathbf{K}}) \le \min\,(c, \text{rank }(\mathbf{Q}_{vv}))$$

$$\text{rank }(\mathbf{Q}_{vv}) \ge r$$

Hence

$$\text{rank }(\mathbf{Q}_{vv}) = r \tag{C7}$$

Next,

$$\begin{aligned}\mathbf{Q}_{ll} &= \mathbf{Q} - \mathbf{Q}_{vv} = \mathbf{Q} - \mathbf{Q}\mathbf{A}^t\mathbf{W}_e\,\bar{\mathbf{K}}\mathbf{A}\mathbf{Q}\\ &= \mathbf{Q}(\mathbf{I} - \mathbf{A}^t\mathbf{W}_e\,\bar{\mathbf{K}}\mathbf{A}\mathbf{Q})\end{aligned}$$

Let

$$\dot{\mathbf{J}} = \mathbf{A}^t\mathbf{W}_e\,\bar{\mathbf{K}}\mathbf{A}\mathbf{Q}$$

$$\dot{\mathbf{J}}^2 = \mathbf{A}^t\mathbf{W}_e\,\bar{\mathbf{K}}(\mathbf{A}\mathbf{Q}\mathbf{A}^t)\mathbf{W}_e\,\bar{\mathbf{K}}\mathbf{A}\mathbf{Q} = \mathbf{A}^t\mathbf{W}_e\,\bar{\mathbf{K}}^2\mathbf{A}\mathbf{Q} = \dot{\mathbf{J}}$$

$$\begin{aligned}\text{rank }(\dot{\mathbf{J}}) &\le \text{min rank of constituent matrices}\\ &\le r\end{aligned}$$

Premultiplying $\dot{\mathbf{J}}$ by $\mathbf{A}\mathbf{Q}$ and postmultiplying by $\mathbf{A}^t$

$$\begin{aligned}\mathbf{A}\mathbf{Q}\dot{\mathbf{J}}\mathbf{A}^t &= (\mathbf{A}\mathbf{Q}\mathbf{A}^t)\mathbf{W}_e\,\bar{\mathbf{K}}(\mathbf{A}\mathbf{Q}\mathbf{A}^t)\\ &= \bar{\mathbf{K}}\mathbf{Q}_e\end{aligned}$$

$$\mathbf{A}\mathbf{Q}\dot{\mathbf{J}}\mathbf{A}^t\mathbf{W}_e = \bar{\mathbf{K}}$$

$$r = \text{rank }(\bar{\mathbf{K}}) \le \min\,(\text{rank }(\dot{\mathbf{J}}), c)$$

$$\text{rank }(\dot{\mathbf{J}}) \ge r$$

Let

$$\dot{\mathbf{K}} = \mathbf{I} - \dot{\mathbf{J}}$$

$$\dot{\mathbf{K}}^2 = \dot{\mathbf{K}}$$

and following the scheme of orthogonal transformation

$$\mathbf{F}'\dot{\mathbf{K}}\mathbf{F} = \mathbf{I} - \mathbf{D}$$

as used previously, we conclude

$$\text{rank } (\dot{\mathbf{K}}) = (n - r)$$

Finally,

$$\mathbf{Q}_{ll} = \mathbf{Q}\dot{\mathbf{K}}$$

$$\text{rank } (\mathbf{Q}_{ll}) \leq (n - r)$$

$$\mathbf{W}\mathbf{Q}_{ll} = \dot{\mathbf{K}}$$

$$\text{rank } (\mathbf{Q}_{ll}) \geq (n - r)$$

therefore

$$\text{rank } (\mathbf{Q}_{ll}) = n - r \tag{C8}$$

APPENDIX

D

Pertinent Statistical Tables

TABLE D1 Area under the Standard Normal Density Function

$$\alpha = \int_{z_\alpha}^{\infty} f(z)\, dz = P(z > z_\alpha) = 1 - \int_{-\infty}^{z_\alpha} f(z)\, dz$$

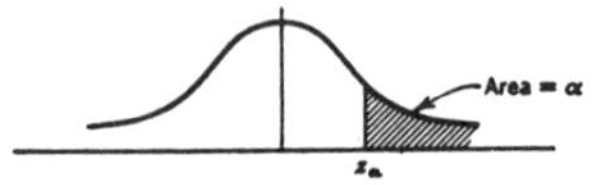

$z_\alpha \rightarrow$ ↓	0.00	0.01	0.02	0.03	0.04	0.05	0.06	0.07	0.08	0.09
0.0	0.5000	0.4960	0.4920	0.4880	0.4840	0.4801	0.4761	0.4721	0.4681	0.4641
0.1	0.4602	0.4562	0.4522	0.4483	0.4443	0.4404	0.4364	0.4325	0.4286	0.4247
0.2	0.4207	0.4168	0.4129	0.4090	0.4052	0.4013	0.3974	0.3936	0.3897	0.3859
0.3	0.3821	0.3783	0.3745	0.3707	0.3669	0.3632	0.3594	0.3557	0.3520	0.3483
0.4	0.3446	0.3409	0.3372	0.3336	0.3300	0.3264	0.3228	0.3192	0.3156	0.3121
0.5	0.3085	0.3050	0.3015	0.2981	0.2946	0.2912	0.2877	0.2843	0.2810	0.2776
0.6	0.2743	0.2709	0.2676	0.2643	0.2611	0.2578	0.2546	0.2514	0.2483	0.2451
0.7	0.2420	0.2389	0.2358	0.2327	0.2296	0.2266	0.2236	0.2206	0.2177	0.2148
0.8	0.2119	0.2090	0.2061	0.2033	0.2005	0.1977	0.1949	0.1922	0.1894	0.1867
0.9	0.1841	0.1814	0.1788	0.1762	0.1736	0.1711	0.1685	0.1660	0.1635	0.1611
1.0	0.1587	0.1562	0.1539	0.1515	0.1492	0.1469	0.1446	0.1423	0.1401	0.1379
1.1	0.1357	0.1335	0.1314	0.1292	0.1271	0.1251	0.1230	0.1210	0.1190	0.1170
1.2	0.1151	0.1131	0.1112	0.1093	0.1075	0.1056	0.1038	0.1020	0.1003	0.0985
1.3	0.0968	0.0951	0.0934	0.0918	0.0901	0.0885	0.0869	0.0853	0.0838	0.0823
1.4	0.0808	0.0793	0.0778	0.0764	0.0749	0.0735	0.0721	0.0708	0.0694	0.0681
1.5	0.0668	0.0655	0.0643	0.0630	0.0618	0.0606	0.0594	0.0582	0.0571	0.0559
1.6	0.0548	0.0537	0.0526	0.0516	0.0505	0.0495	0.0485	0.0475	0.0465	0.0455
1.7	0.0446	0.0436	0.0427	0.0418	0.0409	0.0401	0.0392	0.0384	0.0375	0.0367
1.8	0.0359	0.0351	0.0344	0.0336	0.0329	0.0322	0.0314	0.0307	0.0301	0.0294
1.9	0.0287	0.0281	0.0274	0.0268	0.0262	0.0256	0.0250	0.0244	0.0239	0.0233
2.0	0.0228	0.0222	0.0217	0.0212	0.0207	0.0202	0.0197	0.0192	0.0188	0.0183
2.1	0.0179	0.0174	0.0170	0.0166	0.0162	0.0158	0.0154	0.0150	0.0146	0.0143
2.2	0.0139	0.0136	0.0132	0.0129	0.0125	0.0122	0.0119	0.0116	0.0113	0.0110
2.3	0.0107	0.0104	0.0102	0.0099	0.0096	0.0094	0.0091	0.0089	0.0087	0.0084
2.4	0.0082	0.0080	0.0078	0.0076	0.0073	0.0071	0.0070	0.0068	0.0066	0.0064
2.5	0.0062	0.0060	0.0059	0.0057	0.0055	0.0054	0.0052	0.0051	0.0049	0.0048
2.6	0.0047	0.0045	0.0044	0.0043	0.0042	0.0040	0.0039	0.0038	0.0037	0.0036
2.7	0.0035	0.0034	0.0033	0.0032	0.0031	0.0030	0.0029	0.0028	0.0027	0.0026
2.8	0.0026	0.0025	0.0024	0.0023	0.0023	0.0022	0.0021	0.0021	0.0020	0.0019
2.9	0.0019	0.0018	0.0018	0.0017	0.0016	0.0016	0.0015	0.0015	0.0014	0.0014
3.0	0.0014	0.0013	0.0013	0.0012	0.0012	0.0011	0.0011	0.0011	0.0010	0.0010
3.1	0.0010	0.0009	0.0009	0.0009	0.0008	0.0008	0.0008	0.0008	0.0007	0.0007
3.2	0.0007	0.0007	0.0006	0.0006	0.0006	0.0006	0.0006	0.0005	0.0005	0.0005
3.3	0.0005	0.0005	0.0005	0.0004	0.0004	0.0004	0.0004	0.0004	0.0004	0.0004
3.4	0.0003	0.0003	0.0003	0.0003	0.0003	0.0003	0.0003	0.0003	0.0003	0.0002

TABLE D2 Area Under the Student t Density Function

$$t_{\alpha,m} \text{ such that } P(t_m > t_{\alpha,m}) = \alpha = \int_{t_{\alpha,m}}^{\infty} f(t)\,dt = 1 - \int_{-\infty}^{t_{\alpha,m}} f(t)\,dt$$

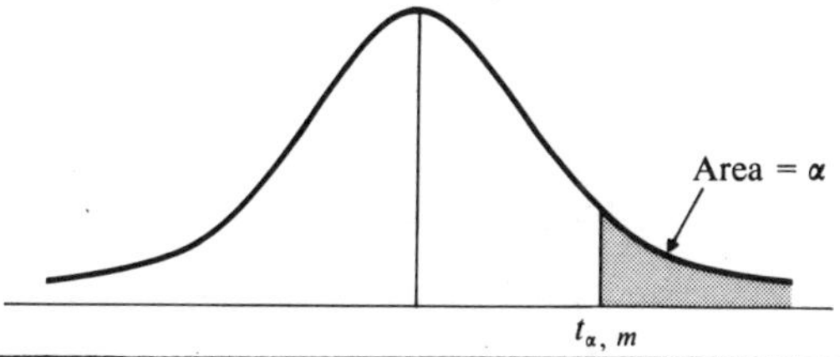

m	$\alpha = 0.25$	0.20	0.15	0.10	0.050	0.025	0.010	0.005	0.0005
1	1.001	1.376	1.963	3.078	6.314	12.706	31.821	63.657	636.619
2	0.816	1.061	1.386	1.886	2.920	4.303	6.965	9.925	31.598
3	0.765	0.978	1.250	1.638	2.353	3.182	4.541	5.841	12.941
4	0.741	0.941	1.190	1.533	2.132	2.776	3.747	4.604	8.610
5	0.727	0.920	1.156	1.476	2.015	2.571	3.365	4.032	6.859
6	0.718	0.906	1.134	1.440	1.943	2.447	3.143	3.707	5.959
7	0.711	0.896	1.119	1.415	1.895	2.365	2.998	3.499	5.405
8	0.706	0.889	1.108	1.397	1.860	2.306	2.896	3.355	5.041
9	0.703	0.883	1.100	1.383	1.833	2.262	2.821	3.250	4.781
10	0.700	0.879	1.093	1.372	1.812	2.228	2.764	3.169	4.587
11	0.697	0.876	1.088	1.363	1.796	2.201	2.718	3.106	4.437
12	0.695	0.873	1.083	1.356	1.782	2.179	2.681	3.055	4.318
13	0.694	0.870	1.079	1.350	1.771	2.160	2.650	3.012	4.221
14	0.692	0.868	1.076	1.345	1.761	2.145	2.624	2.977	4.140
15	0.691	0.866	1.074	1.341	1.753	2.131	2.602	2.947	4.073
16	0.690	0.866	1.071	1.337	1.746	2.120	2.583	2.921	4.015
17	0.689	0.863	1.069	1.333	1.740	2.110	2.567	2.898	3.965
18	0.688	0.862	1.067	1.330	1.734	2.101	2.552	2.878	3.922
19	0.688	0.861	1.066	1.328	1.729	2.093	2.539	2.861	3.883
20	0.687	0.860	1.064	1.325	1.725	2.086	2.528	2.845	3.850
21	0.686	0.859	1.063	1.323	1.721	2.080	2.518	2.831	3.819
22	0.686	0.858	1.061	1.321	1.717	2.074	2.508	2.819	3.792
23	0.685	0.858	1.060	1.319	1.714	2.069	2.500	2.807	3.762
24	0.685	0.857	1.059	1.318	1.711	2.064	2.492	2.797	3.745
25	0.684	0.856	1.058	1.316	1.708	2.060	2.485	2.787	3.725
26	0.684	0.856	1.058	1.315	1.706	2.056	2.479	2.779	3.707
27	0.684	0.855	1.057	1.314	1.703	2.052	2.473	2.771	3.690
28	0.683	0.855	1.056	1.313	1.701	2.048	2.467	2.763	3.674
29	0.683	0.854	1.055	1.311	1.699	2.045	2.462	2.756	3.659
30	0.683	0.854	1.055	1.310	1.697	2.042	2.457	2.750	3.646
40	0.681	0.851	1.050	1.303	1.684	2.021	2.423	2.704	3.551
60	0.679	0.848	1.046	1.296	1.671	2.000	2.390	2.660	3.460
120	0.677	0.844	1.042	1.289	1.658	1.980	2.358	2.617	3.380

TABLE D3 Area Under the Chi-Square Density Function

$$\chi^2_{\alpha,m} \text{ such that } P(\chi^2_m > \chi^2_{\alpha,m}) = \alpha = \int_{\chi^2_{\alpha,m}}^{\infty} f(\chi^2)\, d\chi^2 = 1 - \int_0^{\chi^2_{\alpha,m}} f(\chi^2)\, d\chi^2$$

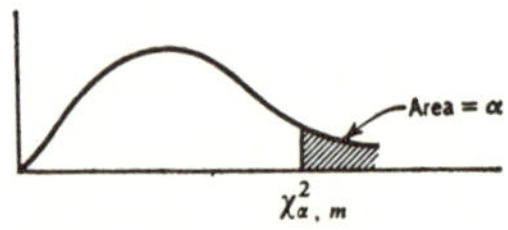

m \ α =	0.995	0.990	0.975	0.950	0.900	0.500	0.10	0.05	0.025	0.010	0.005
1	0.00	0.00	0.00	0.00	0.02	0.46	2.71	3.84	5.02	6.63	7.88
2	0.01	0.02	0.05	0.10	0.21	1.39	4.61	5.99	7.38	9.21	10.60
3	0.07	0.12	0.22	0.35	0.58	2.37	6.25	7.81	9.35	11.34	12.84
4	0.21	0.30	0.48	0.71	1.06	3.36	7.78	9.49	11.14	13.28	14.86
5	0.41	0.55	0.83	1.15	1.61	4.35	9.24	11.07	12.83	15.09	16.75
6	0.68	0.87	1.24	1.64	2.20	5.35	10.64	12.59	14.45	16.81	18.55
7	0.99	1.24	1.69	2.17	2.83	6.35	12.02	14.07	16.01	18.48	20.28
8	1.34	1.65	2.18	2.73	3.49	7.34	13.36	15.51	17.53	20.09	21.96
9	1.73	2.09	2.70	3.33	4.17	9.34	14.68	16.92	19.02	21.67	23.59
10	2.16	2.56	3.25	3.94	4.87	9.34	15.99	18.31	20.48	23.21	25.19
11	2.60	3.05	3.82	4.57	5.58	10.34	17.28	19.68	21.92	24.73	26.76
12	3.07	3.57	4.40	5.23	6.30	11.34	18.55	21.03	23.34	26.22	28.30
13	3.57	4.11	5.01	5.89	7.04	12.34	19.81	22.36	24.74	27.69	29.82
14	4.07	4.66	5.63	6.57	6.79	13.34	21.06	23.68	26.12	29.14	31.32
15	4.60	5.23	6.26	7.26	8.55	14.34	22.31	25.00	27.49	40.58	32.80
16	5.14	5.81	6.91	7'96	9.31	15.34	23.54	26.30	28.85	32.00	34.27
17	5.70	6.41	7.56	8.67	10.08	16.34	24.77	27.59	20.19	33.41	35.72
18	6.26	7.01	8.23	9.39	10.86	17.34	25.99	28.87	31.53	34.81	37.16
19	6.84	7.63	8.91	10.12	11.65	18.34	27.20	30.14	32.85	36.19	38.58
20	7.43	8.26	9.59	10.85	12.44	19.34	28.41	31.41	34.17	37.57	40.00
21	8.03	8.90	10.28	11.59	13.24	20.34	29.62	32.67	35.48	38.93	41.40
22	8.64	9.54	10.98	12.34	14.04	21.34	30.81	33.92	36.78	40.29	42.80
23	9.26	10.20	11.69	13.09	14.85	22.34	32.01	35.17	38.08	41.64	44.18
24	9.89	10.86	12.40	13.85	15.66	23.34	33.20	36.42	39.36	42.98	45.56
25	10.52	11.52	13.12	14.61	16.47	14.34	24.38	37.65	40.65	44.31	46.93
26	11.16	12.20	13.84	15.38	17.29	25.34	35.56	38.88	41.92	45.64	49.29
27	11.81	12.88	14.57	16.15	18.11	26.34	36.74	40.11	43.19	46.96	49.64
28	12.46	13.56	15.31	16.93	18.94	27.34	37.92	41.34	44.46	48.28	50.99
29	13.12	14.26	16.05	17.71	19.77	28.34	39.09	42.56	45.72	49.59	52.34
30	13.79	14.95	16.79	18.49	20.60	29.34	40.26	43.77	46.98	50.89	53.67
40	20.71	22.16	24.43	26.51	29.05	39.34	51.81	55.76	59.34	63.69	66.77
60	35.53	37.48	40.48	43.19	46.64	59.33	74.40	79.08	83.30	88.38	91.95
120	83.85	86.92	91.58	95.70	100.62	119.33	140.23	146.57	152.21	158.95	163.65

TABLE D4(a) Area Under the F Density Function

$$F_{0.05, m_1, m_2} \text{ such that } P(F_{m_1, m_2} > F_{0.05, m_1, m_2}) = 0.05$$

$$= \int_{F_{0.05, m1, m2}}^{\infty} f(F)\, dF = 1 - \int_{0}^{F_{0.05, m1, m2}} f(F)\, dF$$

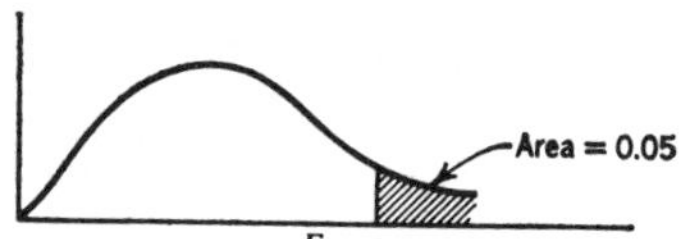

$F_{0.5, m_1, m_2}$

m_2 \ m_1	1	2	3	4	5	6	7	8	9	10	11	12	13	14	16
1	161	200	216	225	230	234	237	239	241	242	243	244	245	245	246
2	18.5	19.0	19.2	19.2	19.3	19.3	19.4	19.4	19.4	19.4	19.4	19.4	19.4	19.4	19.4
3	10.1	9.55	9.28	9.12	9.01	8.94	8.89	8.85	8.81	8.79	8.76	8.74	8.73	8.71	8.69
4	7.71	6.94	6.59	6.39	6.26	6.16	6.09	6.04	6.00	5.96	5.94	5.91	5.89	5.87	5.84
5	6.61	5.79	5.41	5.19	5.05	4.95	4.88	4.82	4.77	4.73	4.70	4.68	4.66	4.64	4.60
6	5.99	5.14	4.76	4.53	4.39	4.28	4.21	4.15	4.10	4.06	4.03	4.00	3.98	3.96	3.92
7	5.59	4.74	4.35	4.12	3.97	3.87	3.79	3.73	3.68	3.64	3.60	3.57	3.55	3.53	3.49
8	5.32	4.46	4.07	3.84	3.69	3.58	3.50	3.44	3.39	3.35	3.31	3.28	3.26	3.24	3.20
9	5.12	4.26	3.86	3.63	3.48	3.37	3.29	3.23	3.18	3.14	3.10	3.07	3.05	3.03	2.99
10	4.96	4.10	3.71	3.48	3.33	3.22	3.14	3.07	3.02	2.98	2.94	2.91	2.89	2.86	2.83
11	4.84	3.98	3.59	3.36	3.20	3.09	3.01	2.95	2.90	2.85	2.82	2.79	2.76	2.74	2.70
12	4.75	3.89	3.49	3.25	3.11	3.00	2.91	2.85	2.80	2.75	2.72	2.69	2.66	2.64	2.60
13	4.67	3.81	3.41	3.18	3.03	2.92	2.83	2.77	2.71	2.67	2.63	2.60	2.58	2.55	2.51
14	4.60	3.74	3.35	3.11	2.96	2.85	2.76	2.70	2.65	2.60	2.57	2.53	2.51	2.48	2.44
16	4.49	3.63	3.24	3.01	2.85	2.74	2.66	2.59	2.54	2.49	2.46	2.42	2.40	2.37	2.33
18	4.41	3.55	3.16	2.93	2.77	2.66	2.58	2.51	2.46	2.41	2.37	2.34	2.31	2.29	2.25
20	4.35	3.49	3.10	2.87	2.71	2.60	2.51	2.45	2.39	2.35	2.31	2.28	2.25	2.22	2.18
22	4.30	3.44	3.05	2.82	2.66	2.55	2.46	2.40	2.34	2.30	2.26	2.23	2.20	2.17	2.13
24	4.26	3.40	3.01	2.78	2.62	2.51	2.42	2.36	2.30	2.25	2.21	2.18	2.15	2.13	2.09
26	4.23	3.37	2.98	2.74	2.59	2.47	2.39	2.32	2.27	2.22	2.18	2.15	2.12	2.09	2.05
28	4.20	3.34	2.95	2.71	2.56	2.45	2.36	2.29	2.24	2.19	2.15	2.12	2.09	2.06	2.02
30	4.17	3.32	2.92	2.69	2.53	2.42	2.33	2.27	2.21	2.16	2.13	2.09	2.06	2.04	1.99
40	4.08	3.23	2.84	2.61	2.45	2.34	2.25	2.18	2.12	2.08	2.04	2.00	1.97	1.95	1.90
50	4.03	3.18	2.79	2.56	2.40	2.29	2.20	2.13	2.07	2.03	1.99	1.95	1.92	1.89	1.85
60	4.00	3.15	2.76	2.53	2.37	2.25	2.17	2.10	2.04	1.99	1.95	1.92	1.89	1.86	1.82
80	3.96	3.11	2.72	2.49	2.33	2.21	2.13	2.06	2.00	1.95	1.91	1.88	1.84	1.82	1.77
100	3.94	3.09	2.70	2.46	2.31	2.19	2.10	2.03	1.97	1.93	1.89	1.85	1.82	1.79	1.75
200	3.89	3.04	2.65	2.42	2.26	2.14	2.06	1.98	1.93	1.88	1.84	1.80	1.77	1.74	1.69
500	3.86	3.01	2.62	2.39	2.23	2.12	2.03	1.96	1.90	1.85	1.81	1.77	1.74	1.71	1.66
∞	3.84	3.00	2.60	2.37	2.21	2.10	2.01	1.94	1.88	1.83	1.79	1.75	1.72	1.69	1.64

TABLE D4(a)—(*Continued*)

$$F_{0.05,m_1,m_2} \text{ such that } P(F_{m_1,m_2} > F_{0.05,m_1,m_2}) = 0.05$$

$$= \int_{F_{0.05,m_1,m_2}}^{\infty} f(F)\,dF = 1 - \int_0^{F_{0.05,m_1,m_2}} f(F)\,dF$$

18	20	22	24	26	28	30	40	50	60	80	100	200	500	∞	m_1 / m_2
247	248	249	249	249	250	250	251	252	252	252	253	254	254	254	1
19.4	19.5	19.5	19.5	19.5	19.5	19.5	19.5	19.5	19.5	19.5	19.5	19.5	19.5	19.5	2
8.67	8.66	8.65	8.64	8.63	8.62	8.62	8.59	8.59	8.57	8.56	8.55	8.54	8.53	8.53	3
5.82	5.80	5.79	5.77	5.76	5.75	5.75	5.72	5.70	5.69	5.67	5.66	5.65	5.64	5.63	4
4.58	3.56	4.54	4.53	4.52	4.50	4.50	4.46	4.44	4.43	4.41	4.41	4.39	4.37	4.37	5
3.90	3.87	3.86	3.84	3.83	3.82	3.81	3.77	3.75	3.74	3.72	3.71	3.69	3.68	3.67	6
3.47	3.44	3.43	3.41	3.40	3.39	3.38	3.34	3.32	3.30	3.29	3.27	3.25	3.24	3.23	7
3.17	3.15	3.13	3.12	3.10	3.09	3.08	3.04	3.02	3.01	2.99	2.97	2.95	2.94	2.93	8
2.96	2.94	2.92	2.90	2.89	2.87	2.86	2.83	2.80	2.79	2.77	2.76	2.73	2.72	2.71	9
2.80	2.77	2.75	2.74	2.72	2.71	2.70	2.66	2.64	2.62	2.60	2.59	2.56	2.55	2.54	10
2.67	2.65	2.63	2.61	2.59	2.58	2.57	2.53	2.51	2.49	2.47	2.46	2.43	2.42	2.40	11
2.57	2.54	2.52	2.51	2.49	2.48	2.47	2.43	2.40	2.38	2.36	2.35	2.32	2.31	2.30	12
2.48	2.46	2.44	2.42	2.41	2.39	2.38	2.34	2.31	2.30	2.27	2.26	2.23	2.22	2.21	13
2.41	2.38	2.37	2.35	2.33	2.32	2.31	2.27	2.24	2.22	2.20	2.19	2.16	2.14	2.13	14
2.30	2.28	2.25	2.24	2.22	2.21	2.19	2.15	2.12	2.11	2.08	2.07	2.04	2.02	2.01	16
2.22	2.19	2.17	2.15	2.13	2.12	2.11	2.06	2.04	2.02	1.99	1.98	1.95	1.93	1.92	18
2.15	2.12	2.10	2.08	2.07	2.05	2.04	1.99	1.97	1.95	1.92	1.91	1.88	1.86	1.84	20
2.10	2.07	2.05	2.03	2.01	2.00	1.98	1.94	1.91	1.89	1.86	1.85	1.82	1.80	1.78	22
2.05	2.03	2.00	1.98	1.97	1.95	1.94	1.89	1.86	1.84	1.82	1.80	1.77	1.75	1.73	24
2.02	1.99	1.97	1.95	1.93	1.91	1.90	1.84	1.82	1.80	1.78	1.76	1.73	1.71	1.69	26
1.99	1.96	1.93	1.91	1.90	1.88	1.87	1.82	1.79	1.77	1.74	1.73	1.69	1.67	1.65	28
1.96	1.93	1.91	1.89	1.87	1.85	1.84	1.79	1.76	1.74	1.71	1.70	1.66	1.64	1.62	30
1.87	1.84	1.81	1.79	1.77	1.76	1.74	1.69	1.66	1.64	1.61	1.59	1.55	1.53	1.51	40
1.81	1.78	1.76	1.74	1.72	1.70	1.69	1.63	1.60	1.58	1.54	1.52	1.48	1.46	1.44	50
1.78	1.75	1.72	1.70	1.68	1.66	1.65	1.59	1.56	1.53	1.50	1.48	1.44	1.41	1.39	60
1.73	1.70	1.68	1.65	1.63	1.62	1.60	1.54	1.51	1.48	1.45	1.43	1.38	1.35	1.32	80
1.71	1.68	1.65	1.63	1.61	1.59	1.57	1.52	1.48	1.45	1.41	1.39	1.34	1.31	1.28	100
1.66	1.62	1.60	1.57	1.55	1.53	1.52	1.46	1.41	1.39	1.35	1.32	1.26	1.22	1.19	200
1.62	1.59	1.56	1.54	1.52	1.50	1.48	1.42	1.38	1.34	1.30	1.28	1.21	1.16	1.11	500
1.60	1.57	1.54	1.52	1.50	1.48	1.46	1.39	1.35	1.32	1.27	1.24	1.17	1.11	1.00	∞

TABLE D4(b)

$F_{0.025, m_1, m_2}$ such that $P(F_{m_1, m_2} > F_{0.025, m_1, m_2}) = 0.025$

$$= \int_{F_{0.025,m\ ,m}}^{\infty} f(F)\, dF = 1 - \int_{0}^{F_{0.025,m\ ,m}} f(F)\, dF$$

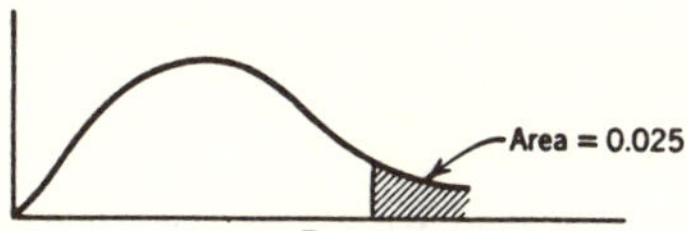

$F_{0.25, m_1, m_2}$

m_2 \ m_1	1	2	3	4	5	6	7	8	9	10	11	12	13	14	16
1	648	800	864	900	922	937	948	957	963	969	973	977	980	983	987
2	38.5	39.0	39.2	39.2	39.3	39.3	39.4	39.4	39.4	39.4	39.4	39.4	39.4	39.4	39.4
3	17.4	16.0	15.4	15.1	14.9	14.7	14.6	14.5	14.5	14.4	14.4	14.3	14.3	14.3	14.2
4	12.2	10.6	9.98	9.60	9.36	9.20	9.07	8.98	8.90	8.84	8.79	8.75	8.72	8.69	8.64
5	10.0	8.43	7.76	7.39	7.15	6.98	6.85	6.76	6.68	6.62	6.57	6.52	6.49	6.46	6.41
6	8.81	7.26	6.60	6.23	5.99	5.82	5.70	5.60	5.52	5.46	5.41	5.37	5.33	5.30	5.25
7	8.07	6.54	5.89	5.52	5.29	5.12	4.99	4.90	4.82	4.76	4.71	4.67	4.63	4.60	4.54
8	7.57	6.06	5.42	5.05	4.82	4.65	4.53	4.43	4.36	4.30	4.24	4.20	4.16	4.13	4.08
9	7.21	5.71	5.08	4.72	4.48	4.32	4.20	4.10	4.03	3.96	3.91	3.87	3.83	3.80	3.74
10	6.94	5.46	4.83	4.47	4.24	4.07	3.95	3.85	3.78	3.72	3.66	3.62	3.58	3.55	3.50
11	6.72	5.26	4.63	4.28	4.04	3.88	3.76	3.66	3.59	3.53	3.47	3.43	3.39	3.36	3.30
12	6.55	5.10	4.47	4.12	3.89	3.73	3.61	3.51	3.44	3.37	3.32	3.28	3.24	3.21	3.15
13	6.41	4.97	4.35	4.00	3.77	3.60	3.48	3.39	3.31	3.25	3.20	3.15	3.12	3.08	3.03
14	6.30	4.86	4.24	3.89	3.66	3.50	3.38	3.29	3.21	3.15	3.09	3.05	3.01	2.98	2.92
16	6.12	4.69	4.08	3.73	3.50	3.34	3.22	3.12	3.05	2.99	2.93	2.89	2.85	2.82	2.76
18	5.98	4.56	3.95	3.61	3.38	3.22	3.10	3.01	2.93	2.87	2.81	2.77	2.73	2.70	2.64
20	5.87	4.46	3.86	3.51	3.29	3.13	3.01	2.91	2.84	2.77	2.72	2.68	2.64	2.60	2.55
22	5.79	4.38	3.78	3.44	3.22	3.05	2.93	2.84	2.76	2.70	2.65	2.60	2.56	2.53	2.47
24	5.72	4.32	3.72	3.38	3.15	2.99	2.87	2.78	2.70	2.64	2.59	2.54	2.50	2.47	2.41
26	5.66	4.27	3.67	3.33	3.10	2.94	2.82	2.73	2.65	2.59	2.54	2.49	2.45	2.42	2.36
28	5.61	4.22	3.63	3.29	3.06	2.90	2.78	2.69	2.61	2.55	2.49	2.45	2.41	2.37	2.32
30	5.57	4.18	3.59	3.25	3.03	2.87	2.75	2.65	2.57	2.51	2.46	2.41	2.37	2.34	2.28
40	5.42	4.05	3.46	3.13	2.90	2.74	2.62	2.53	2.45	2.39	2.33	2.29	2.25	2.21	2.15
50	5.34	3.98	3.39	3.06	2.83	2.67	2.55	2.46	2.38	2.32	2.26	2.22	2.18	2.14	2.08
60	5.29	3.93	3.34	3.01	2.79	2.63	2.51	2.41	2.33	2.27	2.22	2.17	2.13	2.09	2.03
80	5.22	3.86	3.28	2.95	2.73	2.57	2.45	2.36	2.38	2.21	2.16	2.11	2.07	2.03	1.97
100	5.18	3.83	3.25	2.92	2.70	2.54	2.42	2.32	2.24	2.18	2.12	2.08	2.04	2.00	1.94
200	5.10	3.76	3.18	2.85	2.63	2.47	2.35	2.26	2.18	2.11	2.06	2.01	1.97	1.93	1.87
500	5.05	3.72	3.14	2.81	2.59	2.43	2.31	2.22	2.14	2.07	2.02	1.97	1.93	1.89	1.83
∞	5.02	3.69	3.12	2.79	2.57	2.41	2.29	2.19	2.11	2.05	1.99	1.94	1.90	1.87	1.80

TABLE D4(b)—(*Continued*)

$$F_{0.025,\, m_1,\, m_2} \text{ such that } P(F_{m_1,\, m_2} > F_{0.025,\, m_1,\, m_2}) = 0.025$$

$$= \int_{F_{0.025, m_1, m_2}}^{\infty} f(F)\, dF = 1 - \int_0^{F_{0.025, m_1, m_2}} f(F)\, dF$$

18	20	22	24	26	28	30	40	50	60	80	100	200	500	∞	m_1 / m_2
990	993	995	997	999	1000	1001	1006	1008	1010	1012	1013	1016	1017	1018	1
39.4	39.4	39.5	39.5	39.5	39.5	39.5	39.5	39.5	39.5	39.5	39.5	39.5	39.5	39.5	2
14.2	14.2	14.1	14.1	14.1	14.1	14.1	14.0	14.0	14.0	14.0	14.0	13.9	13.9	13.9	3
8.60	8.56	8.53	8.51	8.49	8.48	8.46	8.41	8.38	8.36	8.33	8.32	8.29	8.27	8.26	4
6.37	6.33	6.30	6.28	6.26	6.24	6.23	6.18	6.14	6.12	6.10	6.08	6.05	6.03	6.01	5
5.21	5.17	5.14	5.12	5.10	5.08	5.07	5.01	4.98	4.96	4.93	4.92	4.88	4.86	4.85	6
4.50	4.47	4.44	4.42	4.39	4.38	4.36	4.31	4.28	4.25	4.23	4.21	4.18	4.16	4.14	7
4.03	4.00	3.97	3.95	3.93	3.91	3.89	3.84	3.81	3.78	3.76	3.74	3.70	3.68	3.67	8
3.70	3.67	3.64	3.61	3.59	3.58	3.56	3.51	3.47	3.45	3.42	3.40	3.37	3.35	3.33	9
3.45	3.42	3.39	3.37	3.34	3.33	3.31	3.26	3.22	3.20	3.17	3.15	3.12	3.09	3.08	10
3.26	3.23	3.20	3.17	3.15	3.13	3.12	3.06	3.03	3.00	2.97	2.96	2.92	2.90	2.88	11
3.11	3.07	3.04	3.02	3.00	2.98	2.96	2.91	2.87	2.85	2.82	2.80	2.76	2.74	2.72	12
2.98	2.95	2.92	2.89	2.87	2.85	2.84	2.78	2.74	2.72	2.69	2.67	2.63	2.61	2.60	13
2.88	2.84	2.81	2.79	2.77	2.75	2.73	2.67	2.64	2.61	2.58	2.56	2.53	2.50	2.49	14
2.72	2.68	2.65	2.63	2.60	2.58	2.57	2.51	2.47	2.45	2.42	2.40	2.36	2.33	2.32	16
2.60	2.56	2.53	2.50	2.48	2.46	2.44	2.38	2.35	2.32	2.29	2.27	2.23	2.20	2.19	18
2.50	2.46	2.43	2.41	2.39	2.37	2.35	2.29	2.25	2.22	2.19	2.17	2.13	2.10	2.09	20
2.43	2.39	2.36	2.33	2.31	2.29	2.27	2.21	2.17	2.14	2.11	2.09	2.05	2.02	2.00	22
2.36	2.33	2.30	2.27	2.25	2.23	2.21	2.15	2.11	2.08	2.05	2.02	1.98	1.95	1.94	24
2.31	2.28	2.24	2.22	2.19	2.17	2.16	2.09	2.05	2.03	1.99	1.97	1.92	1.90	1.88	26
2.27	2.23	2.20	2.17	2.15	2.13	2.11	2.05	2.01	1.98	1.94	1.92	1.88	1.85	1.83	28
2.23	2.20	2.16	2.14	2.11	2.09	2.07	2.01	1.97	1.94	1.90	1.88	1.84	1.81	1.79	30
2.11	2.07	2.03	2.01	1.98	1.96	1.94	1.88	1.83	1.80	1.76	1.74	1.69	1.66	1.64	40
2.03	1.99	1.96	1.93	1.91	1.88	1.87	1.80	1.75	1.72	1.68	1.66	1.60	1.57	1.55	50
1.98	1.94	1.91	1.88	1.86	1.83	1.82	1.74	1.70	1.67	1.62	1.60	1.54	1.51	1.48	60
1.93	1.88	1.85	1.82	1.79	1.77	1.75	1.68	1.63	1.60	1.55	1.53	1.47	1.43	1.40	80
1.89	1.85	1.81	1.78	1.76	1.74	1.71	1.64	1.59	1.56	1.51	1.48	1.42	1.38	1.35	100
1.82	1.78	1.74	1.71	1.68	1.66	1.64	1.56	1.51	1.47	1.42	1.39	1.32	1.27	1.23	200
1.78	1.74	1.70	1.67	1.64	1.62	1.60	1.51	1.46	1.42	1.37	1.34	1.25	1.19	1.14	500
1.75	1.71	1.67	1.64	1.61	1.59	1.57	1.48	1.43	1.39	1.33	1.30	1.21	1.13	1.00	∞

TABLE D4(c)

$F_{0.01,m_1,m_2}$ such that $P(F_{m_1,m_2} > F_{0.01,m_1,m_2}) = 0.01$

$$= \int_{F_{0.01,m,m}}^{\infty} f(F)\,dF = 1 - \int_{0}^{F_{0.01,m,m}} f(F)\,dF$$

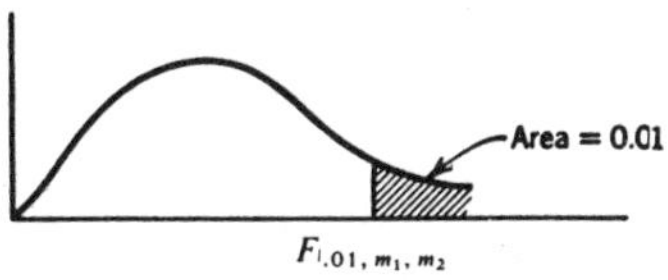

m_2 \ m_1	1	2	3	4	5	6	7	8	9	10	11	12	13	14	16
*1	405	500	540	563	576	586	593	598	602	606	608	611	613	614	617
2	98.5	99.0	99.2	99.2	99.3	99.3	99.4	99.4	99.4	99.4	99.4	99.4	99.4	99.4	99.4
3	34.1	30.8	29.5	28.7	28.2	27.9	27.7	27.5	27.3	27.2	27.1	27.1	27.0	26.9	26.8
4	21.2	18.0	16.7	16.0	15.5	15.2	15.0	14.8	14.7	14.5	14.4	14.4	14.3	14.2	14.2
5	16.3	13.3	12.1	11.4	11.0	10.7	10.5	10.3	10.2	10.1	9.96	9.89	9.82	9.77	9.68
6	13.7	10.9	9.78	9.15	8.75	8.47	8.26	8.10	7.98	7.87	7.79	7.72	7.66	7.60	7.52
7	12.2	9.55	8.45	7.85	7.46	7.19	6.99	6.84	6.72	6.62	6.54	6.47	6.41	6.36	6.27
8	11.3	8.65	7.59	7.01	6.63	6.37	6.18	6.03	5.91	5.81	5.73	5.67	5.61	5.56	5.48
9	10.6	8.02	6.99	6.42	6.06	5.80	5.61	5.47	5.35	5.26	5.18	5.11	5.05	5.00	4.92
10	10.0	7.56	6.55	5.99	5.64	5.39	5.20	5.06	4.94	4.85	4.77	4.71	4.65	4.60	4.52
11	9.65	7.21	6.22	5.67	5.32	5.07	4.89	4.74	4.63	4.54	4.46	4.40	4.34	4.29	4.21
12	9.33	6.93	5.95	5.41	5.06	4.82	4.64	4.50	4.39	4.30	4.22	4.16	4.10	4.05	3.97
13	9.07	6.70	5.74	5.21	4.86	4.62	4.44	4.30	4.19	4.10	4.02	3.96	3.91	3.86	3.78
14	8.86	6.51	5.56	5.04	4.70	4.46	4.28	4.14	4.03	3.94	3.86	3.80	3.75	3.70	3.62
16	8.53	6.23	5.29	4.77	4.44	4.20	4.03	3.89	3.78	3.69	3.62	3.55	3.50	3.45	3.37
18	8.29	6.01	5.09	4.58	4.25	4.01	3.84	3.71	3.60	3.51	3.43	3.37	3.32	3.27	3.19
20	8.10	5.85	4.94	4.43	4.10	3.87	3.70	3.56	3.46	3.37	3.29	3.23	3.18	3.13	3.05
22	7.95	5.72	4.82	4.31	3.99	3.76	3.59	3.45	3.35	3.26	3.18	3.12	3.07	3.02	2.94
24	7.82	5.61	4.72	4.22	3.90	3.67	3.50	3.36	3.26	3.17	3.09	3.03	2.98	2.93	2.85
26	7.72	5.53	4.64	4.14	3.82	3.59	3.42	3.29	3.18	3.09	3.02	2.96	2.90	2.86	2.78
28	7.64	5.45	4.57	4.07	3.75	3.53	3.36	3.23	3.12	3.03	2.96	2.90	2·84	2.79	2.72
30	7.56	5.39	4.51	4.02	3.70	3.47	3.30	3.17	3.07	2.98	2.91	2.84	2.79	2.74	7.66
40	7.31	5.18	4.31	3.83	3.51	3.29	3.12	2.99	2.89	2.80	2.73	2.66	2.61	2.56	2.48
50	7.17	5.06	4.20	3.72	3.41	3.19	3.02	2.89	2.79	2.70	2.63	2.56	2.51	2.46	2.38
60	7.08	4.98	4.13	3.65	3.34	3.12	2.95	2.82	2.72	2.63	2.56	2.50	2.44	2.39	2.31
80	6.96	4.88	4.04	3.56	3.26	3.04	2.87	2.74	2.64	2.55	2.48	2.42	2.36	2.31	2.23
100	6.90	4.82	3.98	3.51	3.21	2.99	2.82	2.69	2.59	2.50	2.43	2.37	2.31	2.26	2.19
200	6.76	4.71	3.88	3.41	3.11	2.89	2.73	2.60	2.50	2 41	2.34	2.27	2.22	2.17	2.09
500	6.69	4.65	3.82	3.36	3.05	2.84	2.68	2.55	2.44	2.36	2.28	2.22	2.17	2.12	2.04
∞	6.63	4.61	3.78	3.32	3.02	2.80	2.64	2.51	2.41	2.32	2.25	2.18	2.13	2.08	2.00

* Multiply the numbers of the first row ($m_2 = 1$) by 10.

TABLE D4(c)—(*Continued*)

$$F_{0.01,\,m_1,\,m_2} \text{ such that } P(F_{m_1,\,m_2} > F_{0.01,\,m_1,\,m_2}) = 0.01$$

$$= \int_{F_{0.01,m,m}}^{\infty} f(F)\,dF = 1 - \int_{0}^{F_{0.01,m,m}} f(F)\,dF$$

18	20	22	24	26	28	30	40	50	60	80	100	200	500	∞	m_1 / m_2
619	621	622	623	624	625	626	629	630	631	633	633	635	636	637	1
99.4	99.4	99.5	99.5	99.5	99.5	99.5	99.5	99.5	99.5	99.5	99.5	99.5	99.5	99.5	2
26.8	26.7	26.6	26.6	26.6	26.5	26.5	26.4	26.4	26.3	26.3	26.2	26.2	26.1	26.1	3
14.1	14.0	14.0	13.9	13.9	13.9	13.8	13.7	13.7	13.7	13.6	13.6	13.5	13.5	13.5	4
9.61	9.55	9.51	9.47	9.43	9.40	9.38	9.29	9.24	9.20	9.16	9.13	9.08	9.04	9.02	5
7.45	7.40	7.35	7.31	7.28	7.25	7.23	7.14	7.09	7.06	7.01	6.99	6.93	6.90	6.88	6
6.21	6.16	6.11	6.07	6.04	6.02	5.99	5.91	5.86	5.82	5.78	5.75	5.70	5.67	5.65	7
5.41	5.36	5.32	5.28	5.25	5.22	5.20	5.12	5.07	5.03	4.99	4.96	4.91	4.88	4.85	8
4.86	4.81	4.77	4.73	4.70	4.67	4.65	4.57	4.52	4.48	4.44	4.42	4.36	4.33	4.31	9
4.46	4.41	4.36	4.33	4.30	4.27	4.25	4.17	4.12	4.08	4.04	4.01	3.96	3.93	3.91	10
4.15	4.10	4.06	4.02	3.99	3.96	3.94	3.86	3.81	3.78	3.73	3.71	3.66	3.62	3.60	11
3.91	3.86	3.82	3.78	3.75	3.72	3.70	3.62	3.57	3.54	3.49	3.47	3.41	3.38	3.36	12
3.72	3.66	3.62	3.59	3.56	3.53	3.51	3.43	3.38	3.34	3.30	3.27	3.22	3.19	3.16	13
3.56	3.51	3.46	3.43	3.40	3.37	3.35	3.27	3.22	3.18	3.14	3.11	3.06	3.03	3.00	14
3.31	3.26	3.22	3.18	3.15	3.12	3.10	3.02	2.97	2.93	2.89	2.86	2.81	2.78	2.75	16
3.13	3.08	3.03	3.00	2.97	2.94	2.92	2.84	2.78	2.75	2.70	2.68	2.62	2.59	2.57	18
2.99	2.94	2.90	2.86	2.83	2.80	2.78	2.69	2.64	2.61	2.56	2.54	2.48	2.44	2.42	20
2.88	2.83	2.78	2.75	2.72	2.69	2.67	2.58	2.53	2.50	2.45	2.42	2.36	2.33	2.31	22
2.79	2.74	2.70	2.66	2.63	2.60	2.58	2.49	2.44	2.40	2.36	2.33	2.27	2.24	2.21	24
2.72	2.66	2.62	2.58	2.55	2.53	2.50	2.42	2.36	2.33	2.28	2.25	2.19	2.16	2.13	26
2.65	2.60	2.56	2.52	2.49	2.46	2.44	2.35	2.30	2.26	2.22	2.19	2.13	2.09	2.06	28
2.60	2.55	2.51	2.47	2.44	2.41	2.39	2.30	2.25	2.21	2.16	2.13	2.07	2.03	2.01	30
2.42	2.37	2.33	2.29	2.26	2.23	2.20	2.11	2.06	2.02	1.97	1.94	1.87	1.83	1.80	40
2.32	2.27	2.22	2.18	2.15	2.12	2.10	2.01	1.95	1.91	1.86	1.82	1.76	1.71	1.68	50
2.25	2.20	2.15	2.12	2.08	2.05	2.03	1.94	1.88	1.84	1.78	1.75	1.68	1.63	1.60	60
2.17	2.12	2.07	2.03	2.00	1.97	1.94	1.85	1.79	1.75	1.69	1.66	1.58	1.53	1.49	80
2.12	2.07	2.02	1.98	1.94	1.92	1.89	1.80	1.73	1.69	1.63	1.60	1.52	1.47	1.43	100
2.02	1.97	1.93	1.89	1.85	1.82	1.79	1.69	1.63	1.58	1.52	1.48	1.39	1.33	1.28	200
1.97	1.92	1.87	1.83	1.79	1.76	1.74	1.63	1.56	1.52	1.45	1.41	1.31	1.23	1.16	500
1.93	1.88	1.83	1.79	1.76	1.72	1.70	1.59	1.52	1.47	1.40	1.36	1.25	1.15	1.00	∞

* Multiply the number of the first row ($m_2 = 1$) by 10.

Bibliography

BOOKS

Anderson, V. A., and McLean, R. A. 1974. *Design of experiments—A realistic approach.* New York: Marcel Dekker.

Barry, A. 1964. *Engineering measurements.* New York: Wiley.

Beers, Y. 1953. *Introduction to the theory of error.* Reading, Mass.: Addison-Wesley.

Beightler, C. S., and Douglas, J. W. 1971. *Foundations of optimization.* Englewood Cliffs, N.J.: Prentice-Hall.

Bjerhammar, E. A. 1973. Theory of errors and generalized matrix inverses. New York: Elsevier.

Bendat, J. S., and Piersol, A. G. 1971. *Random data: Analysis and measurement procedures.* New York: Wiley.

Brownlee, K. A, 1967. *Statistical theory and methodology—In science and engineering.* New York: Wiley.

Cooper, M. A. R. 1974. *Fundamentals of survey measurements and analysis.* London: Lockwood.

Cramer, H. 1966. *Mathematical methods of statistics.* Princeton, N.J.: Princeton Univ. Press.

Deming, W. E. 1943. *Statistical adjustment of data.* New York: Wiley.

Deutsch, R. 1965. *Estimation theory.* Englewood Cliffs, N.J.: Prentice-Hall.

Draper, N. R., and Smith, J. 1967. *Applied regression analysis.* New York: Wiley.

Gelb, A. (Ed.). 1974. *Applied optimal estimation.* Cambridge, Mass.: M.I.T. Press.

Hamilton, W. C. 1967. *Statistics in physical science—Estimation, hypothesis testing, and least squares.* New York: Ronald Press. 1964.

Hazay, I. 1970. *Adjusting calculations in surveying.* Budapest: Akademiai Kiado.

Hirvonen, R. H. 1971. *Adjustment by least squares in geodesy and phytogrammetry.* New York: Ungar.

Jazwinski, A. H. 1970. *Stochastic processes and filtering theory.* New York: Academic Press.

Kendall, M. G., and Stuart, A. 1968. *The advanced theory of statistics* (3 vols.). London: Griffin.

Lawson, C. L., and Hanson, R. J. 1974. *Solving least squares problems.* Englewood Cliffs, N.J.: Prentice-Hall.

Lewis, T. O., and Odell, P. L. 1971. *Estimation in linear models.* Englewood Cliffs, N.J.: Prentice-Hall.

Liebelt, P. B. 1967. *An introduction to optimal estimation.* Cambridge, Mass.: Addison-Wesley.

Linnik, Y. V. 1961. *Method of least squares and principles of the theory of observations.* New York: Pergamon.

Ostle, B. 1963. *Statistics in research.* Ames, Iowa: Iowa State Univ. Press.

Papoulis, A. 1965. *Probability, random variables, and stochastic processes.* New York: McGraw-Hill.

Parratt, L. G. 1961. *Probability and experimental errors in science.* New York: Wiley.

Rainsford, H. F. 1957. *Survey adjustments and least squares.* London: Constable.

Rao, C. R., and Mitra, S. K. 1971. *Generalized inverse of matrices and its applications.* New York: Wiley.

Richardus, P. 1966. *Project surveying.* Amsterdam: North Holland Publishing.

Scheffe', H. 1959. *The analysis of variance.* New York: Wiley.

Shchigolev, B. M. 1965. *Mathematical analysis of observations.* New York: Elsevier.

Smart, W. M. 1958. Combination of observations. London: Cambridge Univ. Press.

Tienstra, J. M. 1966. *Theory of adjustment of normally distributed observations.* Amsterdam: Argus.

Thompson, E. H. 1969. *An introduction to the algebra of matrices with some applications.* London: Hilger.

Veress, S. A. 1974. *Adjustment by least squares.* (Monograph). Washington, D.C.: American Congress on Surveying and Mapping.

Wiener, N. 1950. *Extrapolation, interpolation, and smoothing of stationary time series.* New York: Wiley.

Yaglom, A. M. 1962. *An introduction to the theory of stationary random functions.* Englewood Cliffs, N.J.: Prentice-Hall.

ARTICLES, REPORTS, AND THESES

Ackermann, F. 1973. Experiences with applications of block adjustment for large scale surveys. *Photogram. Rec.*, Apr.

Ackerman, F., Ebner, H., and Klein, H. 1972. Combined adjustment of APR data and independent photogrammetric models. *Can. Surv.*, Sept.

Aeronautical Chart and Information Center (ACIC). 1962. *Principles of error theory and cartographic applications* (Tech. Rep. No. 96). St. Louis: ACIC, Feb. (Present name is Defence Mapping Agency Aerospace Center, DMAAC.)

——— 1963. *Error analysis by the covariance method* (Ref. Publ. No. 16). St. Louis: ACIC, Jan. (Present name is Defence Mapping Agency Aerospace Center, DMAAC.)

Aguilar, A. M. 1973. Principles of survey error analysis and adjustment. *Surv. and Mapping* (ASCE), Sept.

Allan, A. L., 1972. The error ellipse: A further note. *Surv. Rev.*, Oct.

Allman, J. S. 1971. Parametric equations in control survey adjustment. *Australian Surv.*, Mar.

——— 1974. The condition method in the adjustment of large networks. *Can. Surv.*, Dec.

Allman, J. S., and Hoar, G. J. 1973. Optimisation of geodetic networks: *Surv. Rev.*,Jan.

Allman, J. S., and Lister, R. D. 1973. Angular measurement in traverses. *Australian Surv.*, Dec.

Alpar, G. 1974. Comparison of adjustment methods of the aerial triangulation by numerical filtering techniques. Symp. of Comm. III, Int. Soc. of Photogram, Stuttgart, Sept.

Angus-Leppan, P. 1972. Adjustment of trilateration using length ratios. *Surv. Rev.*, Oct.

——— 1973. Practical application of accuracy standards in traversing. *Australian Surv.*, Mar.

Arthur, D. W. G. 1965. Interpolation of a function of many variables. *Photogram. Eng.*, Mar.

——— 1973. Interpolation of a function of many variables II. *Photogram. Eng.*, Mar.

Ashkenazi, V. 1967, 1968. Solution and error analysis of large geodetic neworks. *Surv. Rev.*, Oct. (1967) and Jan. (1968).

——— 1969. Solution and error analysis of large geodetic networks, II. *Surv. Rev.*, Jan. and Apr.

——— 1974. The adjustment and analysis of very large networks. *Can. Surv.*, Dec.

Baarda, W. 1960. *Precision, accuracy, and reliability of observations.* Report I.A.G. Symposium, Lisbon.

——— 1962. *A generalization of the concept of strength of figure.* Report SSG 1.14. Delft:

——— 1962. The connection of geodetic adjustment procedures with methods of mathematical statistics. *Bull. Geodesique*, No. 66.

——— 1967. *Statistical concepts in geodesy* (Vol. 2 No. 4). Delft: Netherlands Geodetic Commission.

——— 1968. *A testing procedure for use in geodetic networks.* (Vol. 2 No. 5), Delft: Netherlands Geodetic Commission.

——— 1971. *Criteria for precision of geodetic networks.* Wiesbaden: Int. Assoc. Surv.

Baker, J. R. 1972. Adjusting radial surveys. *Surv. and Mapping.* (ASCE) Nov.

Bell, J. F. 1968. The intersection problem—A statistical paradox? *Surv. Rev.*, Jan.

Bellamy, C., and Watterson, G. 1970. Least squares adjustment of networks, *Surv. Rev.*, Apr.

Bennett, G. G. 1964. Trilateration adjustment. *Australian Surv.*, Mar.

——— 1965. The adjustment of incomplete direction observations. *Photogram. Eng.*, Dec.

——— 1970. The least squares adjustment of a lightly damped simple harmonic motion. *Bull. Geodesique*, Mar.

Biesheuvel, H. 1972. Least squares adjustments of triangulation. *Surv. Rev.*, No. 11.

Bird, R. G. 1970. Least squares adjustment of a traverse. *Surv. Rev.*, Jan.

——— 1972. Least squares adjustment of E.D.M. traverses. *Surv. Rev.*, July.

Bjerhammar, A. 1951. Application of calculus of matrices to the method of least squares with special reference to geodetic calculations. *Trans. of Royal Inst. of Tech.* (Stockholm), No. 49.

Bomford, A. G. 1964. Geodetic adjustment of Australia. *Australian Surv.*, June and Sept.

Booker, Jr., H. 1968. Precision and accuracy in surveying. *Surv. and Mapping* (ACSM) No. 2.

Bosman, E. R. 1969. Some general remarks about the use of parametric and nonparametric statistical tests. *Photogrammetria*, Dec.

Bosman, E. R., and Kubik, K. 1971. The computation of geodetic networks with simultaneous estimation of the weights of the measured quantities. *Bull. Geodesique*, Dec.

Boyarsky, E. A. 1970. The treatment of observations having nonnormal error distribution. *Can. Surv.*, June.

Brown, D. C. 1955. *A matrix treatment of the general problem of least squares considering correlated observations.* (Report No. 937). Aberdeen, Md.: Ballistic Research Laboratories, May (pp. 22–23).

——— 1957. *A treatment of analytical photogrammetry with emphasis on ballistic camera applications* (RCA Data Reduction Technical Report No. 39, Appendix A).

——— 1969. Computational tradeoffs in a comparator. *Photogram. Eng.*, Feb.

——— 1974. Evolution, application, and potential of the bundle method of photogrammetric triangulation. *Symp. of Comm. III, Int. Soc. of Photogram.*, Stuttgart, Sept.

Brown, D. C., and Trotter, J. E. 1969. SAGA, a computer program for short arc geodetic adjustment of satellite observations. (Final Report, Contract No. F 19628-68-C-0093). Cambridge, Mass.: AF Cambridge Research Labs.

Canadian Surveyor, The. 1965. The impact of electronic computers on geodetic adjustments. *Can. Surv.*, Mar. (the entire issue).

Chetkov, B. 1955. A new method of computation in the theory of least squares. *Australian J. Appl. Sci.*, No. 6.

Chiarini, A., and Pieri, L. 1971. Statistical analysis of discrepancies in high precision leveling. *Bull. Geodesique*, Mar.

Chung-Chi, Y. 1970. On phased adjustment. *Surv. Rev.*, Apr. and July.

Conn, C. 1964. Least squares adjustment of dissimilar quantities. *Surv. Rev.*, No. 131.

Couchman, H. D. 1974. A method for evaluating cyclic errors in E.D.M. equipment, *Australian Surv.* June.

Cross, P. A. 1972. The effect of errors in weights. *Surv. Rev.*, July.

Davis, R. G. 1967. Advanced techniques for the rigorous analytical adjustment of large photogrammetric nets. *Photogrammetria*, No. 22.

Ebner, H. 1974. Analysis of covariance matrices. *Symp. of Com. III, Int. Soc. of Photogr.*, Stuttgart, Sept.

Ellipse problem—More solutions. 1972. *Surv. Rev.*, Jan.

Error ellipse: 1971. A simple example. *Surv. Rev.*, Oct.

Forester, B. C. 1975. Aerotriangulation accuracy. *Photogram. Eng.*, Apr.

Fubara, D. M. J. 1972. Three-dimensional adjustment of terrestrial geodetic networks. *Can. Surv.*, Sept.

——— 1973. Geodetic numerical and statistical analysis of data. *Bull. Geodesque*, June.

Gale, L. A. 1955. A modified-equations method for the least squares solution of condition equations. *Trans. Amer. Geophys. Un.*, No. 36.

Gazdzicki, J. 1967. On an algorithm of adjustment by means of the method of least squares. *Bull. Geodesique*, Sept.

Geisler, M., and Papo, H. 1967. Evaluation of the accuracy of precise leveling. *Surv. and Mapping* (ASCE) Oct.

Gergen, J. G. 1974. Importance of ordering systems of geodetic normal equations. *Can. Surv.*, Dec.

Grafarend, E., and Kelm, R. 1972. Point and interval estimations especially of point errors, in multidimensional least squares adjustment (Part I). *Bull. Geodesique*. June.

——— 1972. Point and interval estimation especially of point errors, in multidimensional least squares adjustment (Part II). *Bull. Geodesique*, Sept.

Grafarend, E., and Schaffrin, B. 1974. Unbiased free net adjustment. *Surv. Rev.*, Jan.

Graham, J. E., and Sunter, A. B. 1969. Error and bias in area measurement: A statistical approach, *Can. Surv.*, Dec.

Gregerson, L. F. 1967. Weighted solution of longitude equations. *Can. Surv.*, Dec.

Greggor, K. N. 1973. Computation and adjustment by geometric network. *Surv. Rev.*, Oct.

Gyer, M. S., and Kenefick, J. F. 1970. Propagation of error in blocks. *Photogram. Eng.*, Sept.

Hallert, B. 1959. Two theorems from mathematical statistics concerning the application of the theory of errors to photogrammetry and other measuring sciences. *Photogram. Rec.*, Oct.

Halmos, F. et al. 1974. Local adjustment by least squares filtering. Mar., *Bull. Geodesique*, Mar.

Heitz, S. 1971. On the astrogeodetic determination of the geoid by least squares estimation. *Bull. Geodesique*, June.

Henriksen, S. W. 1966. Application of probability limit theorems to measurements. Alexandria, Va.: Autometric/Raytheon.

——— 1967. Application of probability limit theorems to measurements. *Photogrammetria*, Jan.

Hotine, M. 1956. *Adjustment of triangulation in space*. Report to the I.U.G.G., Munich.

Hou, M. C.-Y. 1974. Linear transformation for data rejection. *Photogram. Eng.*, Sept.

Jaksic, Z. 1963. The chi square test applied to a photogrammetric sample. *Can. Surv.*, Dec.

Jones, P. B. 1970. The notion of a permissible miselose in traversing. *Australian Surv.*, Sept.

——— 1971. Standards of accuracy, and Simplified traverse design. *Australian Surv.*, Sept.

——— 1972. A comparison of the precision of traverses adjusted by bowditch rule and by least squares. *Surv. Rev.*, Apr.

Kalman, R. E. 1960. A new approach to linear filtering and prediction problems. *J. Basic Eng.*

King, C. W. B. 1967a. Traverse network adjustments Part I: Vertical angle networks. *Surv. Rev.*, Apr.

——— 1967b. Traverse network adjustments Part II: Horizontal angle networks. *Surv. Rev.*, July.

Kissam, P. 1961. Circle accuracy, its value to the engineer and the economy of its use. *Surv. and Mapping* (ASCM) No. 2.

——— 1963. The Princeton standard test for estimating the accuracy of a level. *Surv. and Mapping* (ACSM), No. 1.

Knight, N., and Steeves, P. 1974. Partial solution of the variance—Covariance matrix for geodetic networks. *Can. Surv.*, Dec.

Koch, K. R., and Pope, A. J. 1969. Least squares adjustment with zero variances. *Zeitschrift F. Vermessungswessen*, 94.

Kouba, J. 1972. Principal property and least squares adjustment design. *Can. Surv.*, June.

Krarup, T. 1967. The theory of rounding errors in the adjustment by elements of geodetic networks. *Bull. Geodesique*, June.

Kraus, K., and Mikhail, E. M. 1972. Linear least-squares interpolation. *Photogram. Eng.*, Oct.

Kubik, K. 1970. The estimation of the weights of measured quantities within the method of least squares. *Bull. Geodesique*, Mar.

Leberl, F. 1975. Photogrammetric interpolation. *Photogram. Eng.*, May.

Linkwitz, K. 1971. Zur Vermittelnden Ausgleichung mit Bedingungsgleichungen für die Unbekannten bei singulärer Teilmatrix (Parameter adjustment with constraints with a singular submatrix). *Zeitschrift für Vermessungswesen.* June (p. 246).

Maarek, A., and Konecny, G. 1973. Modulated normal distribution *Photogram. Eng.*, Sept.

Madkour, M. F. 1968. Precision of adjusted variables by least squares. *Surv. and Mapping* (ASCE), Sept.

——— 1971. Rigorous formulas for effects of weight changes. *Surv. and Mapping* (ASCE), Nov.

——— 1972. Nonlinear least squares algorithm. *Surv. and Mapping* (ASCE), Nov.

Magness, T. A., and McGuire, J. B. 1962. Comparison of least squares and minimum variance estimates of regression parameters. *Ann. Math. Stat.*, Vol. 33.

Mather, R. S. 1967. The residual error distribution. *Surv. Rev.*, Jan.

McLean, P. J. 1967. Mathematical considerations of the modified normal distribution. *Bull. Geodesique*, No. 83, Mar.

Meissl, P. 1971. *A study of covariance functions related to the earth's disturbing potential* (Report No. 151), Dept. of Geodetic Science, OSU.

——— 1972. A theoretical random error propagation model for weakly constrained angular networks. *Bull. Geodesique*, Dec.

Mikhail, E. M. 1964. Simultaneous three-dimensional transformation of higher degrees. *Photogram. Eng.*, July.

——— 1970a. Relative control for extraterrestrial work. *Photogram. Eng.*, Apr.

——— 1970b. Parameter constraints in least squares. *Photogram Eng.*, Dec.

Mikhail, E. M., and Helmering, R. J. 1973. Recursive methods in data reduction. *Photogram. Eng.*, Sept.

Miller, E. W. 1963. Errors in the three-point problem. *Surv. and Mapping.* (ACSM), No. 3.

Mittermayer, E. 1972. A generalization of the least squares method for the adjustment of free networks. *Bull. Geodesique*, June.

Mokhov, Y. V. 1967. Loss of accuracy in working out systems of normal equations by Gauss and Cholesky method. *Bull. Geodesique*, June.

Monget, J. M., and Albuisson, M. 1971. A new statistical treatment of gravity data. *Bull. Geodesique*, Dec.

Moritz, H. 1966. An extension of error theory with application to spherical harmonics. *Bull. Geodesique*, Sept.

——— 1972. *Advanced least squares methods* (Report No. 175). Dept. of Geodetic Science, OSU.

——— 1973a. Least squares collocation. *Theoretical Geodesy* (German Geodetic Comm.), No. 75, Munich.

——— 1973b. Stepwise and sequential collocation (Report No. 203). Dept. of Geodetic Science, OSU.

Murphy, B. T. 1974. The adjustment of single traverses. *Australian Surv.*, Dec.

O'Connor, D. C. 1968. *X*-, *Y*- correlation in coordinate measurement. *Photogram. Eng.*, July.

Papo, H. B. 1973. Considered parameters in a least squares adjustment process. *Bull. Geodesique*, Mar.

Parzen, E. 1961. An approach to time series analysis. *Ann. Math. Sci.*, No. 32.

Paul, C. K. 1972. Theoretical weighting of photogrammetric equations. *Surv. and Mapping* (ASCE), Nov.

Pinch, M. C., and Peterson, A. E. 1974. A method for adjusting survey networks in sections. *Can. Surv.*, Mar.

Pope, A. J. 1972. Some pitfalls to be avoided in the iterative adjustment of non-linear problems. *Proceedings of the 38th Ann. Mtg. of the Amer. Soc. of Photogr.*, Mar.

——— 1974a. Two approaches to non-linear least squares adjustments. *Can. Surv.*, Dec.

——— 1974b. *Modern trends in adjustment calculus.* Paper presented at Int. Symp. on Problems Related to the Redefinition of North American Geodetic Networks, Frederiction, New Brunswick.

Proctor, D. W. 1962. Adjustment of aerial triangulation by electronic digital computers. *Photogram. Rec.*, Apr.

Rainsford, H. F. 1968. Angles versus directions. *Surv. Rev.*, Oct.

——— 1968. Combined adjustments of angles and distances. *Surv. Rev.*, 1968.

Rapp, R. 1964. Results from least squares prediction of point and mean anomalies. *Trans. Amer. Geophys. Un.*, 30.

Remmer, O. 1971. A test of significance for the Helmert-Kubik problem of weight determination. *Bull. Geodesique*, Mar.

——— 1974. A simple geodetic imaginary experiment and its consequences for weight determination and determination of systematic errors in geodetic adjustments. *Bull. Geodesique*, June.

Roanowski, M. 1964. Suggestions on the normal law of errors. *Can. Surv.*, Mar.

Romanowski, M., and Green, E. 1967. Statistical analysis of closure errors in the primary triangulation of Central Europe. *Zeitschrift P. Vermessungswessen*, Vol. 93, No. 4.

Saito, T. 1973. The non-linear least squares of condition equations. *Bull. Geodesique*, Dec.

Sayn-Wittgenstein, L., and Aldred, A. H. 1970. Error analysis by computer simulation. *Photogram. Eng.*, Oct.

Scherz, J. P. 1974. Errors in photogrammetry. *Photogram. Eng.*, Apr.

Schmid, E. 1957. A formula for the probable error in an inaccessible base resection. *Surv. and Mapping* (ACSM), No. 2.

Schmid, H. H. 1965. Accuracy aspects of a world-wide passive satellite triangulation system. *Photogram. Eng.*, Mar.

Schrader, B. 1967. Propagation of errors in schematic chains of triangles in space with measured distances. *Bull. Geodesique*, Dec.

Schut, G. H. 1973a. Similarity transformation and least squares. *Photogram. Eng.*, June.

——— 1973b. On the choice of inverse in the adjustment of coordinates. *Bull. Geodesique*, Dec.

——— 1974. Two interpolation methods. *Photogram Eng.*, Dec.

Schwarz, K. P. 1974. Combination of spatial networks using an estimated covariance matrix. *Bull. Geodesique*, June.

Siurna, T. 1958. Simple adjustment of survey figures. *Can. Surv.*, Jan.

Smith, A. D. N. 1956. The explicit solution of the single picture resection problem with a least squares adjustment to redundant control. *Photogram. Rec.*, Oct.

Stark, E., and Mikhail, E. M. 1973. Least squares and non-linear functions. *Photogram. Eng.*, Apr.

Stearn, J. L. 1951. Iterative solutions of normal equations. *Bull. Geodesique*, Sept.

——— 1964. Test of departure from the normal distribution for theodolite errors of observations. *Can. Surv.*, Mar.

Stearn, J. L., and Richardson, H. 1962. Adjustment of conditions with parameters and error analysis. *Bull. Geodesique*, No. 64.

Stipp, D. W. 1962. Trilateration adjustment. *Surv. and Mapping* (ACSM), No. 4.

Stoch, L. L. 1969. Least squares, means, correlation. *Surv. Rev.*, Apr.

Stukkart, G. Jr. 1967. *The theory of geodetic observations.* Doctoral dissertation, The Ohio State University.

Sunter, A. B. 1961. Treatment of observations: Rejection and weight. *Can. Surv.*, May.

——— 1966. Statistical properties of least squares estimates. *Can. Surv.*, Mar.

——— 1967. Least squares adjustment of biased observations. *Can. Surv.*, Sept.

Therien, J. J. 1963. A simultaneous section adjustment for small computers. *Can. Surv.*, Dec.

Thompson, E. H. 1953–1954. A note on systematic error. *Photogrammetria*, Vol. 10.

——— 1962. The theory of the method of least squares. *Photogram. Rec.*, Apr.

Thornton-Smith, G. J. 1964. Geodetic quadrilateral adjustment. *Australian Surv.*, Mar.

Tiwari, R. S., and Rao, K. R. B. 1970. A comparative statistical study of the methods of weighting observations. *Can. Surv.*, Sept.

Uotila, U. A. 1973. Useful statistics for land surveyors. *Surv. and Mapping* (ACSM), Mar.

——— 1974. Generalized inverse as a weight matrix. *Can. Surv.*, Dec.

——— 1975. Statistical tests as guidelines in analyses of adjustment of control nets. *Surv. and Mapping* (ACSM), Mar.

Veress, S. A. 1973. Measures of accuracy for analysis and design of survey. *Surv. and Mapping* (ACSM), Dec.

Vlcek, J. 1964. An application of the principle of experimental design. *Can. Surv.*, Dec.

Vlcek, J. 1969. Systematic errors of image coordinates. *Phhotogram. Eng.*, June.

Vreeland, R. P. 1969. Adjustment of traverses. *Surv. and Mapping* (ASCE), Oct.

Wassef, A. M. 1974. On the search for reliable criteria of the accuracy of precise levelling based on statistical considerations of the discrepancies. *Bull. Geodesique*, June.

Weele, v. d. A. J. 1951–1952. Basic investigations on behalf of the application of the theory of errors in photogrammetry. *Photogrammetria*, Vol. 8.

Welch, H. J. 1957. Standardized computations for subdivision. *Surv. and Mapping* (ACSM), No. 3.

White, L. A. 1966. Simultaneous adjustment of trilaterated and triangulated figures. *Australian Surv.*, June.

——— 1973. Least squares model for differential scale factor adjustment. *Surv. Rev.*, July.

White, L. A., and Betts, N. O. 1968. Geodetic adjustment problems. *Australian Surv.*, Jan.

——— 1969. Modification of conditional adjustment solutions caused by amendment of the original observations involved. *Can. Surv.*, Dec.

Whitten, C. A. 1952. Adjustment of European triangulation. *Bull. Geodesique*, No. 18.

Wilkenson, J. H. 1963. Rounding Errors in algebraic processes. *Notes Appl. Sci.*, Vol. 32.

Williams, H. S. 1969. Rational adjustment of gyro-theodolite observations. *Bull. Geodesique*, Dec.

——— 1972. A theorem in least squares. *Photogram. Eng.*, Nov.

Wolf, H. 1950. Triangulation adjustment: General discussion and new procedure. *Bull. Geodesique*, No. 16.

Wolf, P. R. 1969. Trilaterated photo coordinates. *Photogram. Eng.*, June.

——— 1969. Horizontal position adjustment. *Surv. and Mapping* (ACSM), Dec.

Wong, K. W. 1967. Satellite triangulation accuracy. *Photogram. Eng.*, Jan.

Young, G. M. 1974. A mixed observational survey method. *Bull. Geodesique*, Dec.

Zimmerman, D. S. 1974. Least squares solution by diagonal partitioning. *Can. Surv.*, Dec.

Index